THE REPRODUCTION OF
COLOUR

When the rainbow appears in the clouds,
I will see it and remember the everlasting
covenant between Me and all living beings
on earth.   Genesis 9:16

THE REPRODUCTION OF COLOUR

IN PHOTOGRAPHY, PRINTING & TELEVISION

R. W. G. HUNT

D.Sc. · D.I.C. · A.R.C.S. · F.R.P.S. · F.R.S.A. · M.R.T.S. · F.B.K.S.T.S.

Visiting Professor of Physiological Optics, The City University, London
Formerly Assistant Director of Research, Kodak Limited, Harrow

With a Foreword by
PROFESSOR W. D. WRIGHT · D.Sc. · A.R.C.S. · D.I.C.
Formerly Head of the Applied Optics Section, Imperial College
of Science & Technology, South Kensington, London

FOUNTAIN PRESS

ENGLAND

Fountain Press
Tolworth,
England

First Published 1957
Second Impression (revised) 1961
Second Edition 1967
Third Edition 1975
Fourth Edition 1987
Second Impression 1988

ISBN 0 86343 088 0

Production by Landmark Print Consultants

Phototypeset by Wyvern Typesetting Limited, Bristol

Colour Separations by Crosfield Electronics Ltd.,
Three Cherry Trees Road,
Hemel Hempstead, Hertfordshire,
HP2 7RH, England

Printed and bound in Great Britain by
The Bath Press, Avon

Foreword

THE development of colour photography, colour television, and colour printing demands a very wide and deep understanding of the facts of colour mixture and colour perception. These methods of colour reproduction all have many interesting technical problems to solve, while physiologically the processes in the eye and the brain exercise their own subtle influences on the visual appearance of a colour reproduction. Moreover, the final assessment of a colour picture calls for aesthetic as well as scientific appraisal.

Evidently, then, a book on the reproduction of colour requires a broad outlook on the part of the author, and Dr. Hunt, with his understanding of the basic theory, his experience of commercial production, and his own contributions to fundamental research, is very well qualified to give a balanced and comprehensive account of the subject. This he has undoubtedly succeeded in doing, and in view of his original work on colour adaptation and the visual response, his comments on the subjective aspects of colour reproduction will command particular respect. However much we may regret that the requirements of a colour reproduction cannot be expressed in precise colorimetric terms, we have to recognize that engineering concepts alone are not enough.

The publication of this book will give much pleasure to Dr. Hunt's colleagues and friends, and especially to those who have had the privilege of listening to him present a paper or deliver a lecture, for the orderly presentation of his material and the clarity of his thought on such occasions deserve, and will now reach, a much wider audience. For myself, I regard it as an honour to have been invited to write this Foreword and by undertaking such an agreeable task to continue my association with Dr. Hunt, which dates from the time when he was a student at the Imperial College, and has included our co-operation in a series of courses on the Fundamentals of Colour Reproduction.

W. D. WRIGHT

Contents

PART ONE

FUNDAMENTALS

Chapter 1

SPECTRAL COLOUR REPRODUCTION

Chapter 2 ·

TRICHROMATIC COLOUR REPRODUCTION AND THE ADDITIVE PRINCIPLE

Chapter 3

ADDITIVE METHODS

Chapter 8

COLOUR STANDARDS AND CALCULATIONS

Chapter 9

THE COLORIMETRY OF SUBTRACTIVE SYSTEMS

Chapter 10

LIGHT SOURCES

Chapter 11

OBJECTIVES IN COLOUR REPRODUCTION

PART TWO

COLOUR PHOTOGRAPHY

Chapter 12

SUBTRACTIVE METHODS IN COLOUR PHOTOGRAPHY

Chapter 20

ELECTRONIC CAMERAS

Chapter 21

DISPLAY DEVICES FOR COLOUR TELEVISION

Chapter 22

THE N.T.S.C. AND SIMILAR SYSTEMS OF COLOUR TELEVISION

Chapter 23

THE USE OF COLOUR FILM IN COLOUR TELEVISION

Chapter 24

VIDEO CASSETTES

Chapter 28

PRACTICAL MASKING IN MAKING SEPARATIONS

Chapter 29

COLOUR SCANNERS

COLOUR PLATES

*For easy reference the location of the Plates
is also given at the end of the Index*

Preface to the Fourth Edition

WHEN, during the first half of the nineteenth century, a small band of indefatigable enthusiasts strove to 'fix the images of Nature', their purpose was to fix the colours as well as the tones in their pictures, and it would no doubt have seemed to them like an idle tale if some prophet had revealed that photography was to know a hundred years of black-and-white before colour began to intrude in any measure. Yet such is the case, and this in spite of a number of early milestones that held out great promise. As early as 1810, Seebeck and others knew that if a spectrum was allowed to fall on moist silver-chloride paper many of its colours were recorded, although not with any degree of permanence. In 1835 Prof. Robert Hunt published the third edition of his *Photography*, which contained a whole chapter 'on the possibility of producing photographs in their natural colours'; and he described seeing a number of 'Heliochromes' that were, he wrote, 'perfectly coloured; . . . but the colours soon faded, and it does not appear as yet that any successful mode of fixing the colours has been discovered'. By 1890, however, Prof. Gabriel Lippmann, of Paris, had not only perfected the technique of 'fixing' these colours (by the same methods as are used in black-and-white photography) but had also much improved the process in other ways, and Lippmann colour photographs of very high quality, even by present-day standards, were produced.

Seebeck, and Lippmann, however, are not the forerunners of colour photography as we know it today. That honour belongs to the British physicist James Clerk Maxwell, for it was he who, in his famous Friday Evening Discourse at the Royal Institution on May 17, 1861, demonstrated for the first time *trichromatic* colour photography. By reducing the number of variables to *three*, Maxwell laid foundations upon which practically all modern colour reproduction rests.

It was, therefore, a great honour for me when I was asked to give a Friday Evening Discourse on colour reproduction in the very same lecture theatre at the Royal Institution that Maxwell had used over a century before when delivering his famous discourse.

The object of the book can be stated quite simply. The fundamental principles of colour reproduction, whether by photography, television, or printing, are presented in the hopes that all those engaged in producing, selling, buying, or using colour pictures will be able to see the nature of the problems they encounter. It is hoped that those who want a general statement on colour reproduction will find it in the first part, and those a more detailed discussion of any one application in which they are particularly interested, in one or more of the later parts.

In this fourth edition the whole text has been revised and brought up to date, and new material has been added in many places, including a new chapter on Pictures from Computers. The colorimetric procedures have been brought up to date to be in accord with current CIE recommendations. In the interests of keeping the book to a convenient length, some descriptions of unsuccessful methods of colour reproduction have been deleted from this edition, and for this material historians should retain the third edition.

It is quite certain that this book could never have been written without the help that I have received from many colleagues and friends. I would like particularly to thank Prof. W. D. Wright for having introduced me to the fascinating subject of colour physics in so able and enjoyable a way, and for having taught me so much. I am also most grateful to all my Kodak colleagues whose friendly advice and guidance have meant so much over the years; in preparing this edition I would like to thank particularly Dr. J. Bailey, Mr. W. R. Godden, Mr. M. E. F. Howarth, Mr. L. Pate, Dr. M. R. Pointer, and Dr. A. Weissberger.

The prominence of Kodak materials and processes in the photographic sections springs naturally from the fact that the information available to me concerning other manufacturers' products was much more limited; there is no desire to minimize in any way the contributions made by the rest of the photographic industry to the development and execution of colour photography as we know it today.

I am grateful to the Physical Society for permission to reproduce Fig. 8.9, to the Optical Society of America for permission to reproduce Fig. 29.1, and to the Bell Telephone Laboratories for permission to reproduce Fig. 19.4.

In connection with the colour plates, I would like to thank Kodak Limited for kindly supplying most of the pictorial originals; I am also very grateful to Mr. Frank Judd for the originals for Plate 10; to Dr. G. C. Farnell and Mr. Frank Judd for those for Plate 16; and to the Physics Research Division of the Eastman Kodak Company for those for Plate 24. My thanks are also due to Mr. W. W. Wright of Thorn Colour Tubes Limited for supplying the originals for Plate 18; to Mr. Richard L. Sanders of the British Broadcasting Corporation for those for Plate 19; to Dr. G. Boris Townsend of the Independent Broadcasting Authority for those for Plate 21 and some of those for Plate 25; to Mr. Richard Tucker for those for Plate 22; to Mr. Nicholas Tanton of the British Broadcasting Corporation for some of those for Plate 25; to Mr. John Chapman and Mr. Paul Spence of Rediffusion Simulation Limited for those for Plate 26; to Quantel Limited for one of those for Plate 27; and to Dr. D. Clark of the University of London AV Centre for help in procuring some of those for Plates 28, 29 and 30, which were supplied by Mr. Michael Collery and Mr. Hsuen-Chung Ho of Cranston/Csuri Productions Inc., by Dr. Richard F. Voss and Dr. Benoit B. Mandelbrot of I.B.M., by Dr. James F. Blinn of the Jet Propulsion Laboratory, California, and by Mr. Ned Greene of the New York Institute of Technology. Finally, I am also greatly indebted to Crosfield Electronics Limited for kindly supplying all the corrected separations, which were made on their *Magnascan* scanners in their demonstration suite under the supervision of Mr. Graham Evelin; for providing one of the originals for Plate 27, and all of those for Plates 13, 32, 35, 36, 37, 38 and 39; and to the Chief Research Engineer of De La Rue, Mr. Peter C. Pugsley, for his encouragement and help in so many ways, and particularly with Chapter 29.

I am also grateful to Dr. Michael R. Pointer of the Kodak Limited Research Laboratories for help with the colorimetric data, and for kindly assisting with the proof reading, and to Dr. George Wakefield for suggesting many editorial improvements.

In a field that is developing as rapidly as colour reproduction, it is salutary to remember that human colour vision apparently remains remarkably constant over the centuries. For William Benson (in his *Principles of the Science of Colour*, published by Chapman & Hall in 1868) translates Aristotle, in his *Meteorologica* **3**, 2, in the following words: 'The colours of the rainbow are those that, almost alone, painters cannot make. For they compound some colours; but scarlet, green, and violet are not produced by mixture, and these are the colours of the rainbow.' Colour reproduction in the fourth century before Christ apparently suffered from the same basic limitation as it does today!

The reproduction of colour is a fascinating subject; it involves physiology, psychology, physics, chemistry, and technology; it presents complexities that seem well nigh unfathomable; it involves a wide variety of industrial enterprises; yet its climax is an event of the utmost commonplace, looking at pictures.

PART ONE
FUNDAMENTALS

CHAPTER 1

Spectral Colour Reproduction

1. Introduction – 2. The spectrum – 3. The micro-dispersion method of colour photography – 4. The Lippmann method – 5. Use of identical dyes – 6. A simplified approach

1.1 Introduction

THREE hundred years ago, a physics student at Cambridge University would have been told that

> White is that which discharges a copious light equally clear in every direction. Black is that which does not emit light at all or which does it very sparingly. Red is that which emits a light more clear than usual, but interrupted by shady interstices. Blue is that which discharges a rarefied light, as in bodies which consist of white and black particles arranged alternatively. . . . The blue colour of the sea arises from the whiteness of the salt it contains mixed with the blackness of the pure water in which the salt is dissolved (Houston, 1923).[1]

No wonder that Pope wrote:

> 'Nature and Nature's Laws lay hid in night
> God said "Let Newton be!" and all was light.'

In 1666 Newton laid the foundation-stone of colour science, when he discovered that white sunlight was composed of a mixture of all the colours of the spectrum, and this discovery is also the natural starting point to a consideration of the fundamentals of colour reproduction.

1.2 The spectrum

Suppose we are taking a colour photograph of a street in daylight. All the light falling on the street comes from the sun, either directly when the sky is clear, or after diffusion by clouds if the sky is overcast, or after scattering in the atmosphere if there is blue sky. Since sunlight is a mixture of all the colours of the spectrum,

---

[1] References will be found at the end of each Chapter, and, in the text, are identified by the author's name and the year of publication of the work referred to.

our street scene is being illuminated by such a mixture, and some of the components of this mixture will be revealed by certain natural objects. Foliage contains a dye called chlorophyll which has the property of absorbing reddish and bluish light, but transmits greenish light; hence when foliage is illuminated by daylight it suppresses the reddish and bluish components of the light so that only the greenish components are seen by the eye, and we say that the foliage looks green. Similarly, if the street contains a greengrocer's shop and tomatoes are displayed, the tomatoes look red, because they absorb most of the violet, blue, green, and yellow components of the daylight, and reflect mainly the reddish components. It is thus clear that both the quality of the illuminant and the nature of the objects contribute towards the colour seen. If we return to the street after dark, and find that it is lit by sodium lamps, we shall find that the leaves and the tomatoes now look black because the illuminant contains only yellow light and this is absorbed by the foliage and tomatoes; there being no green light for the foliage to reflect, and no red light for the tomatoes to reflect, these colours cannot be seen.

However, the sodium lamp is very exceptional as far as its colour is concerned, and most sources of light are similar to the sun in that they usually emit a mixture of all the colours of the spectrum. This is true of gas-lamps, electric filament lamps, carbon-arcs, flash bulbs, and most fluorescent lamps. This being so, the extent to which an object reflects the different colours of the spectrum provides a very useful measure of its colour properties.

So far we have only spoken loosely of reddish, or bluish, or yellow light without defining exactly to which part of the spectrum it belongs. Since all light has wave-like properties, and light in different parts of the spectrum corresponds to waves of different length, it is convenient to define each spectral colour by the wavelength of its light. The wavelengths are all extremely short and convenient units of measurement are: the micron or micro-metre (μm)which is a millionth of a metre, the milli-micron (mμ) which is one thousandth of a micron or, which is the same thing, the nano-metre (nm) which is one thousand-millionth (10^{-9}) of a metre, and the Ångström (Å) which is one ten-thousandth of a micron. In the rest of this book we shall mostly use the nano-metre. The main spectral colours occupy approximately the following wavelength bands: violet 450 nm and less; blue 450 to 480 nm; blue-green 480 to 510 nm; green 510 to 550 nm; yellow-green 550 to 570 nm; yellow 570 to 590 nm; orange 590 to 630 nm; red 630 nm and greater. These regions are shown in Fig. 1.1(a). There is a gradual transition from one colour to another throughout the spectrum and it is a matter of opinion as to exactly where one colour ends, and the next begins.

In Fig. 1.1(b) the amount of light reflected at each wavelength by a particular red surface is plotted as a percentage of the amount of light falling on the surface at each wavelength. The curve thus obtained is called the spectral reflectance curve of the sample, and provides a detailed description of the colour properties of the surface. In the case of this red colour it is clear that about 55 per cent of the red light is reflected, 40 per cent of the orange, 20 per cent of the yellow,

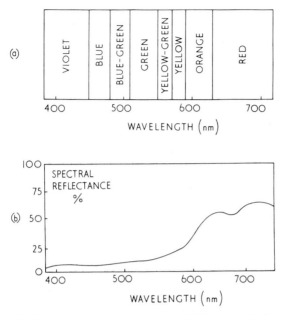

Fig. 1.1 (a) The distribution of colours in the spectrum. (b) The spectral reflectance curve of a red colour.

15 per cent of the yellow-green, 10 per cent of the green, 10 per cent of the blue-green, 5 per cent of the blue, and 5 per cent of the violet. And these reflectances result in the particular red colour of this surface, actually that of a red tomato.

Now suppose we take a colour photograph of a scene containing this particular tomato. We shall reproduce it as a patch of colour, either on a transparency or on paper, and it is obvious that if our patch of colour has the same spectral reflectance curve as the original tomato, then it can produce the same effect; for, physically, the two colours will be identical. And since they are physically identical they will look alike in identical circumstances. Thus if the original and the reproduction are viewed in the same surrounds first in sunlight, then in electric filament light, and then in sodium light, they will always look alike, although of course they will both change colour as the illuminant is changed. Moreover, they will look alike in colour to animals and to colour-blind persons.

1.3 The micro-dispersion method of colour photography

Such colour reproduction would be spectrally correct but can only be achieved in practice by methods that are far too inconvenient for general use. There are two methods that have been suggested and they are both photographic: the micro-dispersion method, and the Lippmann method. The former is shown diagrammatically in Fig. 1.2. The camera lens focuses the image on a coarse grating, consisting of parallel slits, alternately opaque and transparent, about 1/300th of an inch apart. A large plano-convex field lens then collects the light from all the slits and passes it through a narrow-angle prism. Lenses on both sides of the prism focus images of the slits on a photographic plate, and the image of each slit is drawn out into a small spectrum by the prism. Thus the light from each part of the picture is spread out into a spectrum and hence the spectral reflectance curve of every part of the picture is recorded on the plate. The plate is then developed and fixed in the normal way and a positive print made on another plate (or alternatively the original plate can be reversed), and the positive thus obtained is replaced in the plane of the spectra in exact registration. By passing white light through the system in the reverse direction (from right to left in the diagram), and by using the camera lens as a projection lens, a colour reproduction is obtained in which each part of the picture has the same spectral reflectance curve as that of the original.

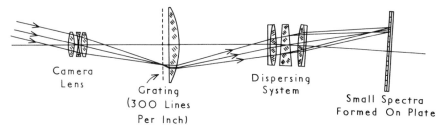

Fig. 1.2. The micro-dispersion method of colour photography (diagrammatic only).

However, the difficulties of the method will at once be appreciated. The more important are: the equipment required is bulky and costly, the grating reduces the amount of light, and an extremely fine-grain (and therefore slow) emulsion has to be used in order to record the minute spectra. But the method is of interest in that it provides colour reproduction which is spectrally correct.

1.4 The Lippmann method

The other method of colour photography that can give spectrally correct colour reproduction is one of the most fascinating photographic inventions ever made. In 1891 Professor Gabriel Lippmann of Paris, by special techniques, made a photographic emulsion with grains of almost unbelievable fineness, 0.01 to 0.04 μm in diameter. This emulsion he coated on plates, which he exposed in an ordinary camera, except that the emulsion side of the plate was turned away from the lens, and a layer of mercury was poured against it, as shown in Fig. 1.3(a). The emulsion-mercury interface then acted as a mirror, and the reflected and oncoming waves interfered with one another to produce standing waves in the emulsion. This standing wave pattern was duly recorded in the emulsion as latent image, and, upon development, parallel plates of silver were produced, the distance between successive plates being equal to half the wavelength of the light used in making the exposure. Thus in Fig 1.3(a), the beam perpendicular to the plate represents green light, and the oblique beam, red light. Since red light is of longer wavelength than green light, the plates of silver are more widely spaced for the oblique beam than for the perpendicular beam. The emulsions were made sensitive throughout the spectrum by the use of a sensitizing dye (Eder, 1945).

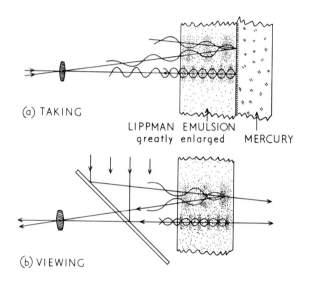

(a) TAKING

LIPPMAN EMULSION
greatly enlarged   MERCURY

(b) VIEWING

Fig. 1.3. The Lippmann method of colour photography (diagrammatic only).

After processing the plate to a negative, it is viewed by reflected light as shown in Fig. 1.3(b). There is no need to make a positive by reversing the plate, since the developed silver layers of the negative give a positive image when viewed by reflected light. This positive image, moreover, is coloured, for the plates of silver will strongly reflect light of half-wavelength equal to the distance between the plates, and weakly, or not at all, light of other wavelengths. Hence all spectral colours, and in fact all other colours also, are reproduced with spectrally correct colour rendering.

Professor Lippmann and other later workers have produced many beautiful colour photographs by this method, and it is probably the most elegant method that will ever be devised. Its disadvantages, however, are of a severe nature. First, the Lippmann emulsions, because of their extremely fine grain, are extremely slow, and exposures of several minutes are necessary to make a Lippmann colour photograph even in bright sunlight. It is impossible to use a fast emulsion because the interference pattern that has to be recorded is smaller than the grain-size of fast emulsions. Secondly, the necessity for viewing the results by reflected light means that it is difficult to project Lippmann colour photographs on to a screen with adequate light; and even when viewed directly by reflected light the angle of viewing is critical.

1.5 Use of identical dyes

In some circumstances it is possible to reproduce the spectral reflectance curves by using the same dyes as were present in the original objects. A textile manufacturer, when trying to reproduce a given colour on an undyed fabric, will achieve spectrally correct colour reproductions if he uses the same dyes in the same amounts as were used on the pattern. In this book, however, we will generally understand the phrase *colour reproduction* to refer to making pictures of original scenes, and the use of identical dyes is then usually possible only in the special case of copying an existing colour photograph or print by means of a process that uses the same dyes or inks (this is discussed in Section 15.7).

1.6 A simplified approach

In view of the difficulties inherent in the micro-dispersion and Lippmann methods of colour photography, it is not surprising that they have never become popularly used; and, indeed, were it not for the fact that when the human eye views colours it simplifies their complexity somewhat, none of the present-day methods of colour reproduction would work.

The rest of the book, therefore, is devoted to describing the principles and methods of achieving colour reproduction by an approach that is basically much more simple: instead of all the colours of the spectrum being dealt with wavelength by wavelength, their effects are considered in a few groups only, as is the case with the human eye.

Although this approach leads to methods of colour reproduction in photography, television, and printing, that are highly successful in practice, we shall see that a proper understanding of them does sometimes involve some quite complicated considerations. It is therefore suggested that the general reader may prefer to omit Chapters 8, 9, 15, 16, 17, and 22, at the first reading.

REFERENCES

Houston, R. A., *Light and Colour*, p. 5, Longmans Green & Co., London (1923).
Eder, J. M., *History of Photography*, Columbia University Press, New York, pp. 668–670 (1945).

GENERAL REFERENCES

Evans, R. M., *An Introduction to Color*, Wiley, New York (1948).
Friedman, J. S., *History of Color Photography*, Chapter 3, American Photographic Publishing Co., Boston (1944).
Le Grand, Y., *Light, Colour, and Vision*, 2nd Edn., Chapman & Hall, London (1967).
Murray, H. D., *Colour in Theory and Practice*, Chapman & Hall, London (1952).
Smith, R. C., *Colour Photography*, p. 157 (April, 1962).
Smith, R. C., *Brit. J. Phot.*, **114**, 1122 (1967).
Wright, W. D., *The Measurement of Colour*, 4th Edn., Hilger, London (1969).

CHAPTER 2

Trichromatic Colour Reproduction and the Additive Principle

1. Introduction – *2*. Maxwell's method – *3*. The physiology of human colour vision – *4*. Spectral sensitivity curves of the retina – *5*. Unwanted stimulations

2.1 Introduction

DURING the seventeenth and eighteenth centuries, the idea that there is something of a *triple* nature in colour steadily grew, and by 1722 Jakob Christoffel LeBlon was using a form of three-colour, or *trichromatic*, printing (Weale, 1957; Wall, 1925; Birren, 1981). By 1807 Thomas Young was instrumental in gaining general acceptance for the view that it is the retina of the human eye that is responsible for this triple feature of colour, and in 1861 James Clerk Maxwell produced the first trichromatic colour *photograph*, not, curiously enough, for its own sake, but as an illustration of the triple nature of colour vision (Maxwell, 1858–1862).

2.2 Maxwell's method

Maxwell's method is fundamental to all modern processes of colour reproduction. He took three photographs—one through a red filter, one through a green filter, and the third through a blue filter, and made three positive slides from the negatives thus obtained. The three slides were placed in three separate projectors, which were arranged to project the three images in register on a white screen, as shown in Fig. 2.1. On placing a red filter in the projector containing the slide made from the negative taken through the red filter, and green and blue filters respectively in those containing the slides made from the negatives taken through the green and blue filters, a colour reproduction was obtained upon the screen. Physically, all the colours on the screen were mixtures of red, green, and blue light only, but, to the eye, white, yellow, orange, mauve, and in fact a whole range of both pale and vivid colours, were seen in addition to red, green, and blue.

Today, colour reproductions, whether in photography, television, or printing, may seem to have little to do with Maxwell's method. But the *principle* of his method, reproduction of all colours by mixtures, in varying amounts, of beams of red, green, and blue light, is retained almost universally; and with modern resources the method itself (triple projection) can produce results of very high quality.

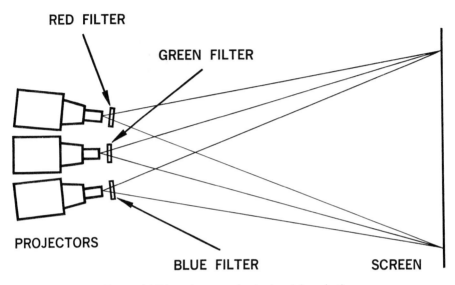

RED FILTER

GREEN FILTER

PROJECTORS

BLUE FILTER   **SCREEN**

Fig. 2.1. Additive colour reproduction by triple projection

For many years it was a historical puzzle as to how Maxwell was able to take pictures through red and green filters when his photographic material was only sensitive to blue light. The ability to extend the sensitivity into the greenish and reddish parts of the spectrum depended on the discovery by Vogel of suitable *sensitizing dyes* some 25 years later. This puzzle was solved by Evans (Evans, 1961), who showed that Maxwell's green filter transmitted just enough blue-green light to enable a record to be obtained, and that the red record was actually produced by ultra-violet radiation which was transmitted through the red filter. Evans finally showed that many red dyes, such as were probably used in the tartan bow that formed the subject of Maxwell's picture, reflect, not only red light, but also ultra-violet radiation. Hence Maxwell's three pictures were obtained with blue, blue-green, and ultra-violet radiation, and yet produced a tolerably acceptable reproduction. So, as has been the case with other great men, Maxwell on this occasion was right for the wrong reasons!

In Chapter 7 the principles of trichromatic colour reproduction will be derived from the experimental facts of colour matching; this approach, though

II

rigorous, is a little intricate. Therefore in this chapter, as an introduction, we shall adopt a different procedure: we shall take the probable basis of human colour vision as a framework, within which we shall be able to see quite quickly, in general terms, both why trichromatic colour reproduction is successful and what its limitations are. The application of trichromatic principles to colour reproduction does not depend, however, on any particular physiological theory, but rather on the physiological fact that a very wide range of intermediate colours can be produced by mixing beams of red, green, and blue light. This mixing can take place either directly, or by using three dyes or pigments: yellow to absorb blue light, magenta to absorb green light, and blue-green or *cyan* to absorb red light.

2.3 The physiology of human colour vision

The human retina contains two main types of light-sensitive cell, known as rods and cones. By 1900 it was well-established that the colourless vision that occurs at very low levels of illumination, such as weak moonlight or starlight, depends on the bleaching of a photo-sensitive substance called *visual purple* contained in the rods. It was therefore natural to ascribe *colour* vision, with its triple nature, to the bleaching of three different photo-sensitive substances contained in the cones. However, the evidence to support this view decisively has not been easy to come by, partly perhaps because the cones, being less sensitive than the rods, and also being far less numerous, offer much less photo-sensitive material to be found. But, in various animals, photo-sensitive pigments have been discovered that absorb in different sections of the visible spectrum, as would be required for a system of colour vision based on such pigments (Dartnall and Lythgoe, 1965); and measurements have been made showing that, when irradiated with strong light, the colour of the light reflected by the human retina back through the pupil of the eye changes in the way to be expected if three such pigments were being bleached (Rushton, 1957 and 1958; Weale, 1959; Ripps and Weale, 1963; Brown and Wald, 1964; Mitchell and Rushton, 1971), one pigment absorbing reddish light, another greenish light, and a third bluish light.

Microscopic studies of individual cones in retinas obtained shortly after death indicate that these pigments are situated in separate cones (Marks, Dobelle, and MacNichol, 1964; Brown and Wald, 1964; Bowmaker and Dartnall, 1980). After the light has been absorbed by the pigments, electrical signals are generated in the form of nerve impulses; it is these signals that convey to the brain the colour and other information concerning the image of the outside world formed on the retina. The nerve fibres along which these signals travel have, in the region where they are joined to the rods and cones, many intricate inter-connections with one another, and it seems unlikely that the colour information consists simply of three signals representing the absorptions of light by the three pigments. There is evidence that in some animals the signals transmitted are analogous to the luminance and colour-difference signals used in colour television (Svaetichin and MacNichol, 1958; De Valois, 1970; MacNichol, 1964), and this also appears to be so in human vision (Hurvich and Jameson, 1957; Hunt, 1982). But once the light has been absorbed in the pigments in the retina, the processing of the subsequent signals will be the same if the absorptions are the same (assuming that the viewing conditions are the same). The key to the success or otherwise of trichromatic colour reproduction must therefore, in the first instance, be sought in the absorptions in the retinal pigments.

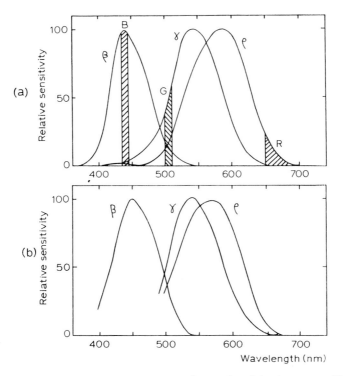

Fig. 2.2. (a) The probable sensitivity curves, β, γ, and ϱ of the three types of light receptor believed to be responsible for colour vision as determined by indirect methods, together with the spectral quality of the three best lights, R, G, and B for additive colour reproduction. (b) Spectral sensitivity curves typical of those found from bleaching experiments on pigments in the human retina.

2.4 Spectral sensitivity curves of the retina

To understand the retinal absorption stage in any detail, we need to know how the ability of each of the three pigments to absorb light and be bleached (and generate a visual response) varies throughout the spectrum. That is, we need to know the spectral sensitivity curve of each of the three pigment-mechanisms. Indirect methods of measuring them have for many years provided what have been believed to be good approximations to these curves, as shown in Fig. 2.2(a) (Wright, 1946; Thomson and Wright, 1953; Stiles, 1978). Direct measurements, made by the method of evaluating the light reflected back through the pupil of the eye, have given similar results as shown in Fig. 2.2(b). This type of measurement is greatly complicated in the blue part of the spectrum because some of the products of the bleaching absorb light in variable amounts but do not generate corresponding colour responses; for this reason the results for the blue-absorbing pigments are less certain than those for the red- and green-absorbing pigments, and in some of the investigations results for the blue-absorbing pigment are not given. The set of curves that best represent the *action spectra* (the visual response to light of different wavelengths), for light incident on the cornea of the human eye, is still a debated subject. Some such sets of curves peak at wavelengths of about 440, 535, and 565 nm; the set shown in Fig. 2.2(a) peak at wavelengths of about 440, 545, and 580 nm, which are close to those advocated by other studies including that by Estevez (Estevez, 1979). The different sets of curves, however, are sufficiently similar for the discussion in this chapter to be equally valid for all.

It will be seen at once that the sensitivity curves overlap to a considerable extent, one type covering chiefly the red, orange, and yellow parts of the spectrum, another the orange, yellow, green, and blue-green parts, and the third the blue-green, blue, and violet parts. We shall regard these three types of sensitivity as belonging to three different types of cone, ϱ, γ, and β, respectively. (If some cones, by having more than one pigment present, had sensitivity curves that were mixtures of those shown, the ensuing arguments would not be affected (Hunt, 1952 and 1959) but studies of the retina show that each cone contains only one of the three types of pigment (Marks, Dobelle, and MacNichol, 1964).)

Considering now Maxwell's method, it is clear that, if our reproduction is to be correct, when taking our three separation negatives, our photographic film should analyse the scene in the same way as the eye. The spectral sensitivities therefore, of the three film-filter combinations should be the same as those of Fig. 2.2. This presents no insuperable difficulties, and can be well enough approximated to, by using panchromatic films and suitable filters. It should be noted that, owing to the broad nature of the ϱ and γ curves, the red filter will in fact look orange, and the green filter paler than a spectral green.

With the spectral-sensitivities of our film-filter combinations the same as those of Fig. 2.2, the amounts of photographic image at any point on our negatives will be functions of the responses of the ϱ, γ, and β cones for the corresponding point at the scene; hence, in our positives the transmission at each point will be proportional to the ϱ, γ, and β responses, it being assumed that the photographic steps are so arranged that the transmissions of the positives bear the correct relation to the exposures received by the negatives.

If the colour filters used in the three projectors were of such colours that the red light stimulated only the ϱ-cones, the green, only the γ-cones, and the blue,

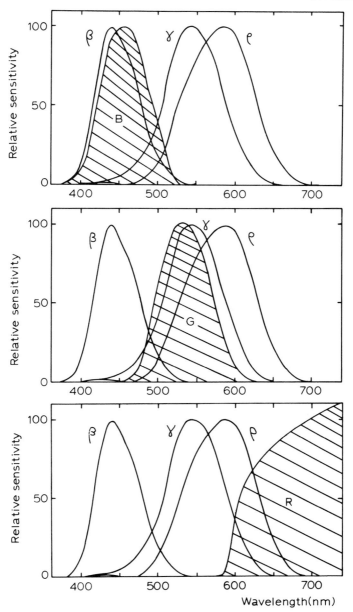

Fig. 2.3. The ϱ, γ, β sensitivity curves of the eye, and the spectral powers of light transmitted by red, green, and blue filters typical of those used in additive colour reproduction (shaded areas, R, G, and B).

15

only the β-cones, correct colour reproduction would result, because each point on the screen would give rise to the same ϱ, γ, and β responses in the retina as those to which the corresponding point in the original scene gave rise. But, unfortunately, the *if* with which this paragraph began is impossible to achieve.

2.5 Unwanted stimulations

From Fig. 2.2(a) it is clear that while a red filter that transmitted only light of wavelength longer than about 650 nm would result in stimulation of only the ϱ-cones, and a blue filter that transmitted only light of wavelength about 450 nm would result chiefly in stimulation of the β-cones and hardly at all the γ- and ϱ-cones, there is no region of the spectrum to which the γ-cones alone are sensitive, and hence no filter can be found to transmit light that stimulates the γ-cones only. The best that can be done is to choose a filter that transmits a narrow band of light in the green part of the spectrum at a wavelength of about 510 nm as shown at G in Fig. 2.2(a); while the bands of light passed by the red and blue filters are those marked R and B.

The effect of this action of the green filter is that wherever, on the screen, the green projector is shining there is an unwanted excess of ϱ- and β- response, and this excess will of course be most noticeable when the γ-response is large, as in the case of greens, which will become paler, and least noticeable in the case of reds and blues where the γ-response is small. Whites would have a medium excess of ϱ- and β-response which would give them a magenta tinge, but this could be overcome by adjusting the relative intensities of the three projectors so that whites looked white, and this would also partly correct all pale colours; vivid colours, however, would then be incorrect in hue and relative intensities, but this is usually much more tolerable than a colour tinge in white and greys.

If the filters used in the three projectors had transmissions as shown at R, G, and B in Fig. 2.2(a) most of the light emitted by the lamps in the projectors would be wasted by being absorbed by the filters, since each filter only transmits a very narrow band of the spectrum. In order to throw more light on the screen from each projector, and hence to produce a brighter picture, filters having broader transmission bands (such as those used for Fig. 2.3) are always used in practice, and this inevitably results in further inaccuracies of colour rendering, since the light from each projector will give even more of the unwanted cone responses than in the case of the filters of Fig. 2.2(a). When the red, green, and blue lights are produced, not by filtering white lights, but by the excitation of phosphors, as is customary in colour television, the situation, as shown in Fig. 2.4, is much the same as in Fig. 2.3.

It is thus clear that the inability of any beams of red, green, and blue light to stimulate the retinal cones separately introduces a basic complication into the whole of trichromatic colour reproduction. If the ϱ and β curves did not overlap in the blue-green part of the spectrum, then green light could be found that stimulated the γ-cones on their own; but since the ϱ and β curves do overlap

16

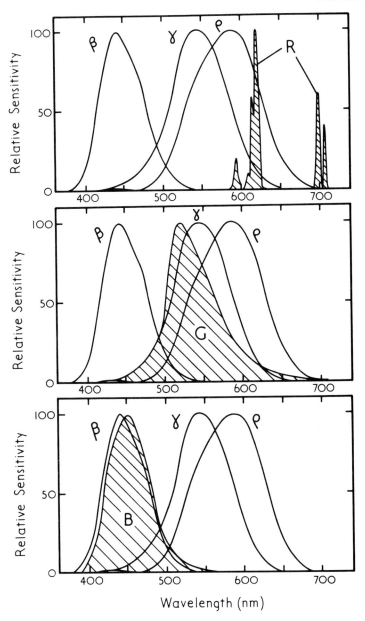

Fig. 2.4. The ϱ, γ, β sensitivity curves of the eye, and spectral power emission curves of red, green, and blue phosphors typical of those used in colour television (shaded areas).

17

appreciably the γ-cones cannot be stimulated on their own. For colour vision, this overlapping provides the basis for good detection of changes in hue throughout the spectrum. But, for colour reproduction, it means that simple trichromatic methods cannot achieve correct colour reproduction of all colours. The difficulty cannot be avoided, because it stems from the basic nature of human colour vision; the result is *unwanted stimulations* in reproduction systems.

The position then becomes as shown in Fig. 2.5. If some particular part of the original gives rise to responses ϱ_O, γ_O, and β_O, and if the strengths R, G, and B of the red, green, and blue beams composing this part of the reproduction are proportional to these responses, then the reproduction is spoiled because the red beam gives rise to an unwanted γ-response, γ_R, the green beam gives rise to unwanted ϱ- and β-responses, ϱ_G and β_G, and the blue beam gives rise to unwanted ϱ- and γ-responses, ϱ_B and γ_B.

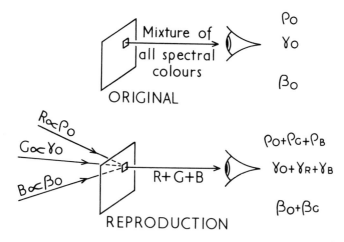

Fig. 2.5. Diagrammatic representation of why a three-colour reproduction of an original scene is inaccurate.

The effect of this may be appreciated in the following way. White and grey sensations may be thought of as corresponding to the values of the responses ϱ, γ, and β being equal, and the colourfulness of a colour as depending upon the ratio of the largest of the three responses to the other two. Thus for a vivid red colour ϱ_O would be several times as large as γ_O or β_O, and it follows, of course, that R will be several times as large as G and B. Therefore the unwanted response γ_R will be larger than the unwanted responses ϱ_G, β_G, and ϱ_B, γ_B. Hence the main effect of the unwanted responses will be to increase γ relative to ϱ and β. Thus the ratio of ϱ to γ will be reduced and the corresponding colour sensation will become less vividly red. The same effect will operate with, for instance, a vivid green colour. G will be large, and hence ϱ_G and β_G will be the main unwanted responses. Hence the ratio of γ to ϱ and β will decrease and the colour will become less vividly green. The argument is quite general, and the effect illustrated in Fig. 2.5 will be to make all colours less colourful. Thus the overlapping of the cone sensitivity curves, as shown in Fig. 2.2, resulting as it does in the inability to stimulate each type of cone on its own, is the reason why correct colour reproduction by simple trichromatic methods is impossible to achieve for all colours.

We shall consider the questions of the importance of this defect and the ways in which its effects can be reduced in later chapters, where, incidentally, we shall see that it is not necessarily the best arrangement to have the spectral sensitivity curves of the film-filter combinations the same as those of the ϱ, γ, and β curves. However, the defects are often unnoticeable in practice, and the pictures that can be obtained by trichromatic reproduction are often extremely pleasing.

REFERENCES

Birren, F., *Color Res. Appl.*, **6,** 85 (1981).
Bowmaker, J. K., and Darnall, H. J. A., *J. Physiol.*, **298,** 131 and 501 (1980).
Brown, P. K., and Wald, G., *Science*, **144,** 45 (1964).
Dartnall, H. J. A., and Lythgoe, J. N., *Vision Research*, **5,** 81 (1965).
DeValois, R. L., *Proc. 1st A.I.C. Congress*, Stockholm, p. 29 (1970)
Estevez, O., Ph.D. Thesis, University of Amsterdam (1979).
Evans, R. M., *J. Phot. Sci.*, **9,** 243 (1961).
Hunt, R. W. G., *J. Opt. Soc. Amer.*, **42,** 198 (1952).
Hunt, R. W. G., *Nature*, **183,** 1601 (1959).
Hunt, R. W. G., *Color. Res. Appl.*, **7,** 95 (1982).
Hurvich, L. M., and Jameson, D., *Psychol. Rev.*, **64,** 384 (1957).
MacNichol, E. F., *Vision Research*, **4,** 119 (1964).
Marks, W. B., Dobelle, W. H., and MacNichol, E. F., *Science*, **143,** 1181 (1964).
Maxwell, J. C., *Proc. Roy. Inst.*, **3,** 370 (1858–62).
Mitchell, D. E., and Rushton, W. A. H., *Vision Research*, **11,** 1045 (1971).
Ripps, H., and Weale, R. A., *Vision Research*, **3,** 531 (1963).
Rushton, W. A. H., *Visual Problems of Colour*, p. 73, N.P.L. Symposium, H.M.S.O., London (1957).
Rushton, W. A. H., *Nature*, **182,** 690 (1958).
Stiles, W. S., *Mechanisms of colour vision*, Academic Press, London (1978).
Svaetichin, G., and MacNichol, E. F., *Ann. N.Y. Acad. Sci.*, **74,** 385 (1958).
Thomson, L. C., and Wright, W. D., *J. Opt. Soc. Amer.*, **43,** 890 (1953).
Wall, E. J., *The History of Three-Color Photography*, p. 1, American Photographic Publishing Co., Boston (1925).
Weale, R. A., *Nature*, **179,** 648 (1957).
Weale, R. A., *Optica Acta*, **6,** 158 (1959).
Wright, W. D., *Researches on Normal and Defective Colour Vision*, Chapters 21 and 30, Kimpton, London (1946).

GENERAL REFERENCES

Boynton, R. M., *Human Color Vision*, Holt and Rinehart-Winston, New York (1979).
Coe, B. W., *Colour photography, the first hundred years*, Ash and Grant, London (1978).
Evans, R. M., Hanson, W. T., and Brewer, W. L., *Principles of Color Photography*, Wiley, New York (1953).
Wall, E. J., *The History of Three-Color Photography*, American Photographic Publishing Co., Boston (1925).

CHAPTER 3

Additive Methods

1. Introduction – *2*. The successive frame method – *3*. The mosaic method –
4. The lenticular method – *5*. The virtual image method – *6*. The diffraction
method – *7*. Errors in additive methods

3.1 Introduction

IT will be appreciated that Maxwell's method of colour photography by triple
projection offers more hope of practical usefulness than the micro-dispersion
and Lippmann methods. Triple projection is used now in television when the
picture is required to be viewed by large audiences. The most common arrangement is for three projection-type television display tubes to project red, green,
and blue images on to a specially shaped viewing screen. The images are
produced initially on small phosphor-coated screens that produce either red, or
green, or blue light, when scanned by an electron beam. A lens system of very
high optical efficiency then produces a magnified image of each of the three small
screens on the viewing screen. One example of this type of system is the Advent
Videobeam (Federman and Pomicter, 1977). For general use, however, the
practical difficulty of getting and retaining exact registration of the three images
over the whole of the picture area, and the expense of triplicating the projection
apparatus, have militated against Maxwell's original method. But red, green,
and blue beams of light can be mixed in other ways, and, leaving aside for the
moment the use of dyes, pigments, or inks (the so-called *subtractive* processes),
there are five other ways of mixing the red, green, and blue beams: the successive
frame method, the mosaic method, the lenticular method, the virtual image
method, and the diffraction method.

These five methods, together with the triple projection method, are usually
called *additive* methods, since all the colours are produced by adding beams of red,
green, and blue light to one another in varying proportions.

3.2 The successive frame method

This method, applicable only to cinematography and television, depends for its
success on the fact that, if beams of red, green, and blue light fall on the retina in

20

quick succession, the individual colours are not seen and the colour sensation is the same as that produced by triple projection. In cinematography, therefore, a filter wheel containing successive segments of red, green, and blue filters is rotated in front of the camera lens, in synchronism with the shutter, so that pictures are taken first through one filter, then through another, and then through the third. The film is processed in the usual way to give a black-and-white positive and this is projected with a similar filter wheel rotating in front of the projection lens and this filter wheel is in synchronism with the shutter in the projector; thus every time a positive, made from a negative taken through the red filter, is in the gate, a red filter is over the projection lens, and similarly for the green and blue. In a colour television system the same method can be applied with a red, green, and blue filter wheel rotating in front of the television camera lens, and a similar wheel rotating in synchronism in front of the television receiver tube.

Unfortunately, in such a system, the blue filter is always much darker than the red and green filters, and the eye can detect the darker filter as a brightness flicker even when the speed of the filter wheel is fast enough to remove all colour flicker. The speed of rotation necessary to lose all sense of flicker depends somewhat on the intensity of the light, but for a stationary scene it is usually about 50 rotations per second giving 150 red, green, and blue 'fields' per second. Even at this speed, a moving object can result in objectionable colour fringing (colour 'break-up'), particularly if it is a highly coloured object, so that it is mainly presented in only one of the three colour-pictures. The Columbia system of colour television which, in 1950, was standardized by the Federal Communications Commission (F.C.C.) for use in the U.S.A., was a successive frame method of this type working at 144 fields per second. But, owing to defence requirements at that time, the system was never widely used commercially, and this was perhaps as well because the system was not *compatible*, that is, the pictures transmitted in the system could not be received as black-and-white pictures on existing black-and-white television sets; furthermore the rotating disc tends to be too awkward to fit in neatly, and is liable to mechanical failure caused by wear. Subsequently the F.C.C. set aside their 1950 decision and standardized a compatible system in 1953 for use in the U.S.A. (Law, 1977).

In photography, the successive frame method has never achieved commercial success because of the high rates of projection (and consequent high consumption of film) necessary to avoid flicker and colour fringing of moving objects.

3.3 The mosaic method

The simplest and most successful additive method has been that used by the mosaic processes. If a very fine mesh of red, green, and blue squares is viewed at a distance, the individual colours are not seen; instead, a single uniform colour appears, the nature of which depends upon the relative amounts of light passing through the three types of square. Hence, if much more light passes through the red and green, than through the blue squares, the same colour is seen as when red and green are mixed by projection, that is, yellow. It is, therefore, possible to produce a colour photograph like Maxwell's by taking a black-and-white photograph through a mosaic of red, green, and blue squares, reversing the negative to a positive or printing a positive from it, and then viewing it through the mosaic in register with the squares on the photograph. Physiologically the success of the method depends upon the fact that the cones of the retina themselves constitute a mosaic, and, if the image of the photographic mosaic on the retina is fine compared with the retinal mosaic, then the three colours will be as effectively mixed as is the case with triple projection.

In photography, the mosaic processes have had a long and distinguished career. The Autochrome plate, which consisted of a random mosaic of red, green, and blue dyed starch grains with the interstices filled with carbon black, came on the market in 1907 and was still a commercial success in the early 1930s. The Agfacolor mosaic process employed a random mosaic of stained resin grains and was rather more transparent. Successful processes employing mosaics of regular areas of red, green, and blue were Finlay and Thames (1906, and later revived in England as the Johnson screen plate), and Dufaycolor (1908) in which the mesh eventually (about 1935) reached the astonishing fineness of a million squares to the square-inch and resulted in considerable commercial success. Regular mosaics proved more satisfactory than random mosaics because, in the latter, random clumpings of elements of the same colour gave an increase in apparent mottle.

The system of 'instant' movies introduced in 1977 by Polaroid as *Polavision* (Land, 1977), instead of using a mosaic, used a mesh of very fine stripes, there being 1500 red, green, and blue triads per inch (about 60 per mm). In 1983 a similar system was introduced for 'instant' slides on 35 mm film as *Polachrome*: in this case there are 1000 triads per inch (about 40 per mm), which, although less than the number used for the Super 8 format movie film, is sufficient to give very sharp results in the 35 mm slides. These systems produce their positive images by the migration of the unexposed silver halide grains to a receiver layer containing nuclei. In the Polachrome system the negative layer is removed after processing (which takes about one minute). The maximum transmission is rather low (about 20 per cent) because of the presence of the filter stripes. A high contrast is used to obtain adequate strength in the colours.

The main interest in mosaic processes is now in connection with colour television, where mosaic dot cathode-ray tubes are very widely used. These tubes, instead of being covered with a uniform coating of phosphor, have a regular mosaic of areas of three different phosphors, one of which fluoresces red, another green, and the third blue. If an electron beam in the tube scanned this regular mosaic one line at a time, it might be thought that a filter wheel in front of the camera could rotate at such a speed that, whenever the electron beam was exciting a red phosphor dot, the red filter was over the camera lens, and similarly for the green and the blue. A simple calculation, however, shows that this is impracticable. For if we had a television system of 500 lines, with, say 500 dots of each of the three phosphors in each line, the filter wheel would have to rotate 500×500 revolutions for each picture, and at 25 pictures per second this requires a rotation of 6 250 000 revolutions per second, which is clearly impossible; amongst other difficulties, no wheel could be made that would not fly apart at such speeds.

To overcome this difficulty it would be necessary in some way to arrange that when, say, the red filter was over the camera lens, the electron beam, as it scanned the mosaic of phosphors, only fell on the red phosphor, the green and blue phosphors being missed; but as soon as the green filter came over the camera, the red and blue phosphors would have to be missed and only the green phosphor irradiated, and similarly for the blue.

One way of achieving this is shown in Fig. 3.1. The red, green, and blue phosphors are deposited on the tube screen, as though they had been fired from three positions R, G, and B respectively, through a metal plate containing a large number of small holes situated just behind the tube screen. Each hole is like a pin-hole camera, and for each hole a triad of red, green, and blue phosphor dots is situated on the screen. The plate with the holes in it is called a *shadow-mask*, and

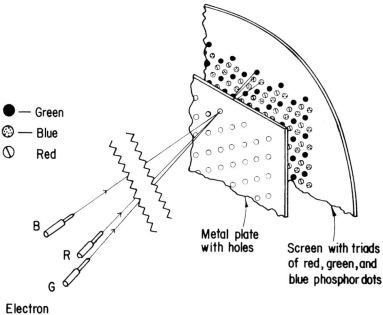

● — Green
◉ — Blue
◑ Red

B
R
G

Electron
Guns

Metal plate
with holes

Screen with triads
of red, green, and
blue phosphor dots

Fig. 3.1. Diagrammatic representation of the shadow-mask type of colour television tube.

23

hence this type of tube is called a *shadow-mask tube*. Three electron guns are used in the tube and they fire from three positions R, G, and B. It could then be arranged that whenever the red filter was over the camera lens, only the gun at position R fired, and hence, only the dots of red phosphor would be irradiated. As soon as the green filter replaced the red filter over the camera lens the gun at R would stop firing and that at G start, so that only the dots of green phosphor would be irradiated; and similarly for the blue. Thus with this arrangement it would be possible to have the filter wheel running only at the usual speeds necessary to overcome flicker and colour fringing on moving objects. A variation of this arrangement would be to have only one electron gun, and to arrange that the electron beam, after passing through each hole, was steered by electrostatic or electromagnetic focusing on to the appropriate phosphor in synchronism with the camera filter-wheel.

Although the successive frame type of camera could, in this way, be used with mosaic television tubes, in practice, because of compatability and other considerations, present colour television systems use cameras in which, by means of semi-reflecting mirrors, three separate but identical images are formed by a single lens, and these pass through red, green, and blue filters on to the light-sensitive surfaces of three television camera tubes; thus the red, green, and blue images are available all the time. All three electron guns in the receiver then fire continuously, the broadcast signals controlling their intensities and hence producing the required colours. The methods of transmitting and receiving such *simultaneous* colour pictures, together with some further discussion of camera arrangements, will be found in Chapters 19 and 20.

It should be noted that in these television systems it is *not* necessary for the lines of dot-triads to coincide with the picture scanning-lines; it is only required that there be a sufficient number of dots to build up the picture without significant loss of detail. Exact registration of the three guns, the apertures in the metal plate, and the phosphor dots is vital for correct colour; but the exact position of the picture lines on the tube is not important.

Just as the mosaic processes have been the most successful in additive colour photography, so far they have also been the most successful in colour television. (See Plate 18, page 288. For location of Plates, see end of Index.)

3.4 The lenticular method

An elegant variation of the mosaic method is the lenticular method, which, as *Kodacolor*,[1] had a successful life in commercial colour photography, and which is also feasible for colour television. The principle, when used in photography, is illustrated in Fig. 3.2. A film is used that has minute furrows, or lenticulations, embossed on one side, and a photographic emulsion coated on the other side. The film is used in the camera with the lenticulations towards the camera lens, as

---

[1] The old *Kodacolor* process 1928–35, not to be confused with the present *Kodacolor* process, which is a subtractive system.

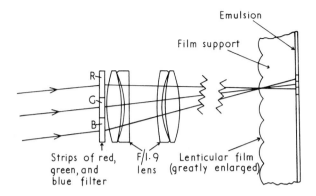

Fig. 3.2. The Lenticular method of additive colour photography (diagrammatic only).

shown in the figure, and the radius of curvature of the lenticulations is made so that they focus an image of the camera lens on the emulsion layer. The camera lens itself is covered with red, green, and blue filters in strips, parellel to the direction of the lenticulations, so that images of the filters are formed in strips along the entire length of each lenticulation. The result is that the picture is divided up into triads of strips, so that the strips opposite the bottom of each lenticulation are exposed only by light that has passed through the red filter, those opposite the middle of each lenticulation only by green light, and those opposite the top only by blue light. The film is then processed to a positive (or to a negative and then printed in register on to positive lenticular stock) and projected with the same, or a similar, array of red, green, and blue filter strips over the projection lens. The lenticulations then ensure that the white projection light passing through the bottom strip of each lenticulation emerges only through the red filter strip on the lens, that through the middle strip, only through the green strip on the lens, and that through the top strip only through the blue strip on the lens. Then the projection lens, by focusing the red, green, and blue beams on to the same point on the screen, additively mixes them, just as effectively as in the case of triple projection. The old *Kodacolor* process, using 22 lenticulations per millimetre, ran for a number of years with considerable success, and the Eastman Kodak Company in America in 1951 offered an improved version for 35 mm professional use (*British Journal of Photography*, 1951). Subsequently, lenticular film was tried out as a medium for recording colour television programmes, where the speed and simplicity of the black-and-white processing is an important asset (Evans and Smith, 1956; Brown, Combs and Smith, 1956; Crane and Evans, 1958; Duke, 1963). It should be noted that, in lenticular systems, proper colour synthesis only requires that the image be accurately located relative to the lenticulations; location of the entire film relative to the filter strips on the projection lens is not critical.

In colour television, experimental colour tubes have been made employing similar principles but using electronic 'lenses' (Dressler, 1953), and considerable commercial success has been achieved with the *Trinitron* tube. In the Trinitron tube the red, green, and blue phosphors are in stripes, and a modified type of shadow-mask is used in which the plate contains slits instead of round holes. The general arrangement of the Trinitron is as shown diagrammatically in Fig. 3.3. The electrons from a single gun are split into three nearly parallel beams; these beams are then made to converge so as to cross one another at the plate with the slits. The geometry of the tube is so arranged that the electrons from one beam can only reach the red phosphor stripes, those from another beam only the green, and those from the third only the blue. The broadcast signals modulate the

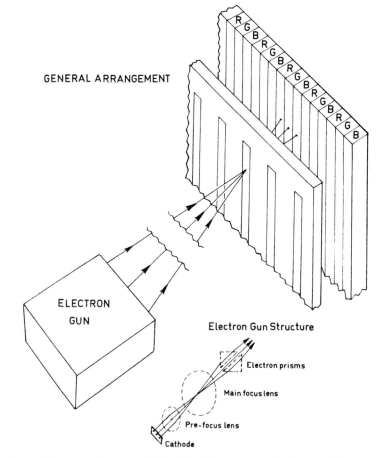

Fig. 3.3. Diagrammatic representation of the Trinitron type of colour television tube.

26

strengths of the three electron beams, to result in the required amount of red, green, and blue light being produced. When a Trinitron tube is viewed at a normal distance, the stripes of phosphor are not resolved by the eye, and hence the red, green, and blue light is effectively mixed as required.

3.5 The virtual-image method

In this method virtual images of ordinary black-and-white positives are seen through red, green, and blue filters, and are superimposed by means of semi-reflecting mirrors. Such devices were used in the early days of colour photography when they were known as Ives *Chromoscopes* or *Kromskops* (Smith, 1967). Their modern equivalent is the *trinoscope* used in colour television (to be described in Section 21.2).

3.6 The diffraction method

In this method, the three images are superimposed, but are modulated by three diffraction gratings. The images are illuminated by white light and different orientations or frequencies of the gratings are used to enable red, green, and blue filters to be inserted in a part of the optical system where the beams from the three images are separated by diffraction. Reduction of this method to practice is difficult because optical alignment is very critical and dirt or blemishes are made very conspicuous (Kurtz, Eisen, and Higgins, 1971).

3.7 Errors in additive methods

As far as errors in additive colour reproductions are concerned, the simplest approach is the one adopted in the previous chapter: to regard any deviation of the camera sensitivity curves from the ϱ-, γ-, β-curves as leading to error, and to regard the optimum red, green, and blue beams as those that most nearly stimulate the ϱ-, γ-, β-types of cone separately. In the case of photographic processes in which the same mosaic is used in the camera as is used for viewing the final result, it is clear that to adopt the ϱ-, γ-, and β-sensitivity curves results in red, green, and blue beams composed of very broad spectral bands and therefore very far from optimum. For this reason, in such processes, filters having narrower transmission bands (and therefore also sharper sensitivity curves) were used with some overall advantage (Sproson, 1949). The whole subject of errors in additive reproduction will be treated more rigorously in Chapter 7; very pleasing additive pictures can be obtained in colour television.

REFERENCES

Brown, W. R. J., Combs, C. S., and Smith, R. B., *J. Soc. Mot. Pic. Tel. Eng.*, **65**, 648 (1956).
British Journal of Photography, **98**, 456 (1951).
Crane, E. M., and Evans, C. H., *J. Soc. Mot. Pic. Tel. Eng.*, **65**, 13 (1958).

Dressler, R., *Proc. I.R.E.*, **41,** 851 (1953).
Duke, V. J., *J. Soc. Mot. Pic. Tel. Eng.*, **72,** 711 (1963).
Evans, C. H., and Smith, R. B., *J. Soc. Mot. Pic. Tel. Eng.*, **65,** 365 (1956).
Federman, F., and Pomicter, D., *J. Roy. Television Soc.*, **16,** vii (May–June, 1977).
Kurtz, C. N., Eisen, F. C., and Higgins, G. C., *Phot. Sci. Eng.*, **15,** 343 (1971).
Law, H. B., *J. Soc. Mot. Pic. Tel. Eng.*, **86,** 214 (1977).
Land, E. H., *Phot. Sci. Eng.*, **21,** 225 (1977).
Smith, R. C., *Brit. J. Phot.*, **114,** 1122 (1967).
Sproson, W. N., *Phot. J.*, **89B,** 108 (1949).

GENERAL REFERENCES

Cornwell-Clyne, A., *Colour Cinematography*, Chapman & Hall, London (1951).
Happé, B., *Brit. Kinematog. Sound Tel.*, **67,** 58 (1985).
Koshofer, G., *Brit. J. Phot.*, **113,** 562 and 824 (1966).

CHAPTER 4

The Subtractive Principle

4.1 Introduction

IN additive methods of colour reproduction, all the colours are produced by the adding or blending together in different proportions of the light from three primary colours, a red, a green, and a blue. Subtractive colour reproduction, at first sight, seems to be quite different, because all the colours are produced by different proportions of three entirely different colours, a cyan (or blue-green), a magenta, and a yellow. In point of fact, however, the subtractive and the additive methods differ only in manner, and not in principle.

In photography, additive methods suffer from two disadvantages. First, somewhere in the display system there must be red, green, and blue filters; these may be in the three projectors as in the triple-projection method, or in the filter wheel as in the successive frame method, or in the three strips across the projection lens as in the lenticular method, or in the minute patches as in the mosaic methods. But wherever such filters occur there is an inevitable loss of light. Therefore, as compared with black-and-white, a projected additive photograph means either a dimmer picture of the same size, or a smaller picture of the same intensity, or a higher wattage projection lamp and hence a more elaborate cooling system in the projector; all this is unwelcome. And as far as reflection prints are concerned, no additive method has been devised; for only the mosaic processes are possible, and in these the mosaic of red, green, and blue patches, even without any image behind them, will blend to give grey instead of white, and it is, therefore, impossible to reproduce whites at all.

The second disadvantage of the additive methods is that they all require either some special equipment, such as triple projectors, or some sort of mosaic which results in a loss of definition.

The attraction of the subtractive principle, which was first described by du Hauron in 1862, is that it overcomes all these difficulties; it is therefore widely used not only in photography but also in the printing industry. Projected pictures are as bright as in black-and-white, reflection prints can be made with good whites, and ordinary cameras and projectors can be used. The disadvantages in photography are that subtractive materials are more complicated, with the result that the costs are greater, and the processing may involve the return of the film to the manufacturers or to a specialized laboratory; this means delay and removes some of the fascination from amateur photography. In colour television the subtractive principle has less advantage because the additive system, using phosphors emitting red, green, and blue light, does not waste light as is the case with the red, green, and blue filters used with beams of white light in additive colour photography; and definition is limited by the lines.

4.2 The subtractive principle

White light contains all the colours of the spectrum. But we may regard the spectrum as consisting of three main parts; first, that containing light of wavelengths over about 580 nm, which contains all the reddish part; secondly, that containing light of wavelengths between about 490 and about 580 nm, which contains all the greenish part; and thirdly, that containing light of wavelengths less than 490 nm, which contains all the bluish part. If we looked at the blended light from each of these three parts of the spectrum, we would simply see three colours, red, green, and blue. It therefore follows that, when the light from a projector falls on a white screen, or when daylight falls on a piece of white paper, we may regard that light as being an additive mixture of a beam of red light, a beam of green light, and a beam of blue light. In order, therefore, to produce a wide range of colours in a beam of white light, all that is required is some means of varying the proportions of the reddish, greenish, and bluish parts, independently (du Hauron, 1869).

In Fig. 4.1(a) there is plotted against wavelength the percentage transmission at each wavelength of a yellow dye at four different concentrations. It is seen that, for all concentrations, the transmission in the reddish part of the spectrum is high, in fact nearly 100 per cent; in the greenish part, the variation of transmission with concentration is not large but, in the bluish part of the spectrum, the transmission depends very markedly on the concentration of the dye. It is therefore clear that, if we insert in a projector a slide on which we can vary the concentration of a yellow dye, as this concentration is altered so the amount of bluish light falling on the screen is altered nearly independently of the amount of greenish and reddish light falling on the screen. Similarly, the amount of bluish light reflected from a piece of white paper, viewed in daylight, would be altered by the concentration of a yellow dye on its surface.

In Fig. 4.1(b) similar spectral-transmission curves are shown for a magenta dye at different concentrations. It is clear that the main effect of altering the concentration of the magenta dye is to vary the transmission in the greenish part of the spectrum. It is true that the transmissions in the bluish and reddish parts of the spectrum do alter also, but they do so to a smaller extent. Finally, in Fig. 4.1(c) similar curves are shown for a cyan dye at different concentrations, and the main effect of varying the concentration is to alter the transmission in the reddish part of the spectrum, and to a less extent the transmissions in the greenish and bluish parts of the spectrum.

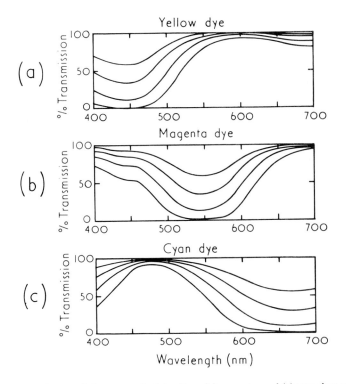

Fig. 4.1. Spectral transmission curves for (a) yellow, (b) magenta, and (c) cyan dyes at four different concentrations.

If, therefore, we now have a slide in a projector, or a surface layer on a piece of white paper, on which we can vary at will the concentrations of a cyan, a magenta, and a yellow dye, we have the means of varying the intensities of the reddish, greenish, and bluish parts of the white light, and, therefore, we can produce a very wide range of colours at different intensities. This is the subtractive principle, and it is clear that, although the colours of the dyes used are cyan, magenta, and yellow, this is merely incidental to the fact that it is dyes of these colours that correspond respectively to a red-absorbing, a green absorbing, and a blue-absorbing substance. In the printing trade the colours of the inks are

31

still sometimes referred to as 'blue', 'red', and 'yellow', but their functions are still to act as absorbers of red, green, and blue light, respectively.

All that is necessary to produce a subtractive colour reproduction is to be able to control the concentrations of the three dyes independently at each point on the transparency or piece of paper. Assuming that this can be done, the successive stages of subtractive colour photography are as follows:

(1) Black-and-white records of the original scene are made by means of red, green, and blue light (as in the additive system).

(2) Dye-image positives are made from the records, the red record giving a cyan image, the green record a magenta image, and the blue record a yellow image.

(3) When superimposed in register, the three dye images are viewed in white light. (See Plate 32, pages 512, 513.)

In colour films and papers the registration of the three images is made automatic by coating layers of photographic emulsions on top of one another in the form of *integral tri-packs* on the same piece of film or paper base, and then processing the material in such a way as to produce the cyan, magenta, and yellow dye images in the appropriate layers. The layers are made extremely thin, so as to minimize any loss of sharpness caused by the upper layers diffusing the light used for exposing the lower layers. A typical thickness for a complete package of image layers in colour film is about 5 to 10 μm, or about one tenth of the thickness of a human hair. This is discussed further in Chapters 12 and 18.

4.3 Defects of the subtractive principle

On many occasions, subtractive colour photographs produced by some of the modern commercial processes are so pleasing to the eye that the impression is made that almost perfect colour rendering has been achieved. In point of fact, however, as far as colour rendering is concerned, all subtractive processes suffer not only from the unwanted stimulations inherent in the additive method (that were described in Chapter 2), but also from some further defects of their own.

In Fig. 4.2 are shown again the probable sensitivity curves of the three types of light-sensitive cell, or cone, ϱ, γ, and β, operating in the human retina. As with the additive process, in order to obtain correct colour reproduction, we may regard it, for the moment, as necessary that the effective sensitivities of the three photographic emulsions used to record the three images should be the same as the three curves of Fig. 4.2. This can be done, but it is further required that the cyan dye controls a band of wavelengths to which only the ϱ-cones respond, the magenta dye a band to which only the γ-cones respond, and the yellow dye a band to which only the β-cones respond. In Fig. 4.2 the approximate bands of wavelength controlled by the cyan, magenta, and yellow dyes of Fig. 4.1 are shown, and it is clear that, as with the additive process, the ϱ-, γ-, and β-responses

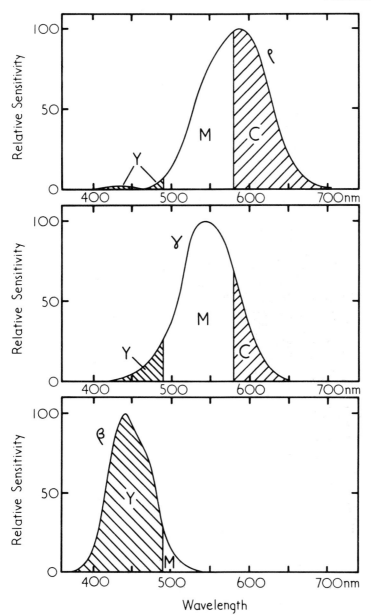

Fig. 4.2. The areas labelled C, M, and Y, show the magnitudes of the retinal responses controlled by the absorptions of cyan, magenta, and yellow dyes respectively. In an ideal system the absorption of each dye would control one retinal response only.

33

THE SUBTRACTIVE PRINCIPLE

are not independently controlled by the three dyes, and perfect colour reproduction does not, therefore, occur. With the additive system, the red, green, and blue lights need not consist of such broad bands of wavelengths as those controlled by the subtractive dyes, and, because of this, additive systems are theoretically superior in this respect; but, in practice, there is not much difference, because of the need to produce adequate light levels in additive systems.

A further disadvantage of the subtractive system, as can be seen from Fig. 4.1, is that cyan, magenta, and yellow dyes have appreciable absorptions in parts of the spectrum where they should have 100 per cent transmission. These

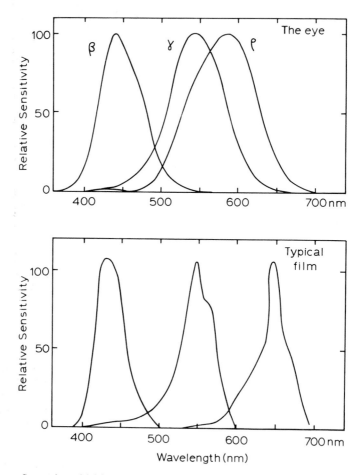

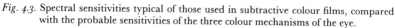
Fig. 4.3. Spectral sensitivities typical of those used in subtractive colour films, compared with the probable sensitivities of the three colour mechanisms of the eye.

34

unwanted absorptions result in colours being reproduced considerably darker than in the original scene unless corrections are made.

Various measures can be adopted in an endeavour to minimize the fundamental shortcomings of subtractive colour processes, and these are often very effective. In the first place, the processes can be operated so that the reproduction is more contrasty than the original. This results in a general improvement in the colourfulness of all colours, but at the expense of tone reproduction. Correct tone reproduction, however, is not always essential for a pleasing picture, and indeed in some cases an increase of contrast improves the tone rendering. This is particularly true of scenes for which the lighting is rather flat, and with some modern subtractive processes a combination of flat lighting and a high-contrast process yields very pleasing results. (In Chapter 6 we shall see that in the case of reproductions seen against dark surrounds, as with projected transparencies or motion-picture films, the pictures must be made more contrasty than when seen against surrounds of average luminance; this is necessary in order that the dark parts of the scene appear dark enough, because dark surrounds have a greater lightening effect on dark than on light parts of the picture, see Fig. 5.2; however, the dark surrounds can also reduce the colourfulness of colours, if they are of large area and near the edge of the surround; see Section 6.7.)

Another beneficial measure is to use sensitivity curves for the three images that are more widely separated along the wavelength axis than those of Fig. 4.2 and typical sets of such curves are shown in Fig. 4.3. This again improves the colourfulness of most colours, but sometimes introduces errors in hue and lightness. (This topic is discussed in more detail in Section 9.5.)

Another measure is to make each dye-image dependent on more than one of the three exposures in such a way as to increase colourfulness: this can be done by means of *inter-image effects* or *masking* (to be described in Chapter 15).

It will be realized that these three expedients cannot correct for the fundamental limitations of the process, which spring from the nature of the colour mechanism of the eye and the shape of the spectral absorption curves of the best available cyan, magenta, and yellow dyes. What is claimed for modern subtractive processes is that they produce pleasing colour pictures, and that the inevitable inaccuracies are balanced in such a way as to be least noticeable.

REFERENCE

Du Hauron, L. Ducos, *The Photographic News*, **13,** 319 (1869).

GENERAL REFERENCE

Evans, R. M., *Eye, Film, and Camera in Color Photography*, Wiley, New York (1959).

CHAPTER 5

Visual Appreciation

1. Introduction – *2.* The basis of judgement – *3.* Variations of hue – *4.* Variations of lightness – *5.* Variations of colourfulness – *6.* Priorities – *7.* Factors affecting apparent colour balance – *8.* Integrating to grey – *9.* The perception of depth

5.1 Introduction

S OMETIMES the imperfections inherent in trichromatic colour photography and television are apparent to the user; some particular colour may appear obviously incorrect in the reproduction. One example of this is the tendency for very pale colours, especially if they are brightly lit, to lose their colour altogether and appear to be white. Another, in colour photography, is the tendency for certain flowers to be reproduced pink instead of blue although every other colour in the picture appears satisfactory. The reason for this last defect is that certain blue flowers, in addition to reflecting strongly in the blue part of the spectrum and thus appearing blue to the eye, also strongly reflect light in the extreme red end of the spectrum; light of this latter wavelength scarcely affects the eye because the sensitivity of the human retina in this part of the spectrum is very low, but the colour film usually has a high sensitivity in the extreme red, as shown in Fig. 4.3, with the result that the film records the redness of the flower more strongly than its blueness. If the sensitivity of the colour film were curtailed in this part of the spectrum, so that it was more nearly like that of the eye, these particular colours would be reproduced more satisfactorily. But such an arrangement would reduce the vividness of most red colours. (In those cases where it is imperative to obtain the correct flower colour a filter-pack consisting of Kodak *Wratten* filters 66, 85B, two CC50M, and CC20M is appropriate for typical daylight colour films, but requires an increase in exposure of about six stops (Reed, 1965).)

In spite of the inability of simple trichromatic methods to render all colours colorimetrically correct, it is often the exception, rather than the rule, for the result to *look* incorrect. Moreover, measurements tell us that the main defect is that colours are rendered insufficiently vivid (because of the unwanted eye-stimulations, see Fig. 2.3) and, in the case of subtractive reproductions, too dark (because of the unwanted dye absorptions, see Fig. 4.1); and yet the user often feels that, far from the colours being too pale and too dark, there is rather a tendency for them to appear, if anything, too vivid and too bright, and the process is accused of exaggerating the colours. In short, measurement defines the short-comings of the process, but when the photographer exposes a colour film or the viewer looks at his colour television, the defects often seem to have disappeared.

To explain this apparent anomaly it has to be remembered that, although colours may be conveniently defined in terms of physical quantities such as spectral transmission and reflection curves, they are perceived as sensations in the mind. We must therefore consider the psychological as well as the physical side of the story.

5.2 The basis of judgement

Let us examine the basis on which colours are criticized in a reproduction. It is only on a few occasions (such as the first viewing of an 'instant' type of print) that the reproduction and the original are side by side; more usually the reproduction is seen at a different place or time, and the time interval may vary from a few hours to several weeks or even months or years. The human memory therefore plays an important part. It might be thought, then, that the process involved in appraising colours in a reproduction consists of making mental comparisons between the sensation produced in the mind by the reproduction, and a recollection from the memory of the colour sensation produced by the original object at the time when the picture was taken. It is, however, a fact that the average person generally feels competent to appraise the colours in pictures taken by people other than himself, of objects that he has never seen, at times when he was not present. This implies that colours in a reproduction are not generally appraised by comparing them either with the original objects, nor even with some mental recollection of them. By what means are they then judged?

There seems no alternative to the idea that the basis of judgement is usually a comparison between the colour sensations aroused by the reproduction, and a mental recollection of the colour sensations previously experienced when looking at objects similar to the ones being appraised (Bartleson, 1959 and 1960).

Let us take an example. Green grass is a fairly common component in colour pictures; when trying to assess whether it is correctly reproduced or not, we make a mental comparison between the colour sensation we experience when looking at the reproduction, and our impression of what green grass usually looks like. Measurement tells us that trichromatic colour processes are particularly bad at reproducing green colours, and that the result will tend to be both darker and less vivid than it should be. Why is it then, that the green colour we see in the picture for green grass seems perfectly satisfactory? The answer is that green grass in original scenes can be any of a wide range of colours. The apparent colour varies with the type of grass, the dampness of the soil, the direction, colour, and

37

intensity of the lighting, the clarity of the atmosphere, the time of year, and even with the colour and size of the objects surrounding the grass. In short, our standard of comparison, a recollection of the usual colour of green grass, is a somewhat vague one, and, consequently, provided that the reproduction of the green grass is included somewhere in the range of colour sensations produced by actual samples of green grass, we are satisfied. These variations in the colours of natural objects are thus of considerable importance. But they are not the only factors that result in a wide range of colour sensations being associated with a given type of object.

Looking at colours is one of a large number of complex activities for which we use our bodies, with scarcely a thought for the processes involved. At first sight it might be thought that our memories could easily supply us with a fairly definite *average* 'green grass' sensation, so that our judgement of the colour reproduction would be clear and fairly precise. But such is far from being the case.

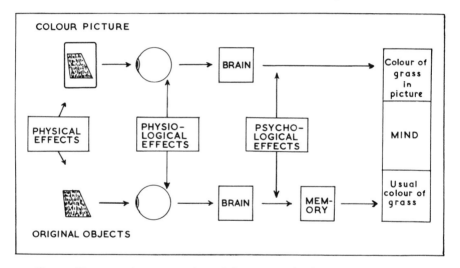

Fig. 5.1. Diagrammatic representations of the processes involved in viewing a colour reproduction of grass (upper line), and (lower line) in viewing original grass. The physical, physiological, and psychological effects differ in the two cases, and also from one original area of grass to another. The standard conception in the memory, therefore, of the usual colour of grass is sufficiently vague that the reduction in colourfulness of grass green inherent in many colour reproductions often passes unnoticed, and this applies to many other colours also.

In Fig. 5.1 an attempt has been made to indicate diagrammatically the factors that complicate the issues. The upper line represents the processes involved in viewing a colour picture of the grass. The light from the reproduction enters the observer's eye, resulting in messages being sent to the brain, and these messages are interpreted by the mind as the colour sensations corresponding to the colour of the grass in the picture. The lower line shows a similar sequence for original grass, the colour sensations of which have been stored in the memory.

The colour sensations produced by the reproduction of the grass will depend upon the physical composition of the light by which it is illuminated, upon the physiological state of adaptation of the eye when viewing it, and upon any psychological effects that the picture as a whole or in part may have on the observer. Similarly, the colour sensation corresponding to any original grass will be affected by the physical, physiological, and psychological conditions under which it is seen, which will not only differ from those obtaining when the reproduction is viewed, but will also vary from day to day for any one area of grass and from one area of grass to another.

It will be clear from the above that the final comparison, shown diagrammatically on the extreme right of Fig. 5.1, between the colour of the grass in the picture and our impression of the usual colour of grass, is complicated at every step by extraneous effects, which not only result in the impressions we have of the usual colours of objects being vague but also prevent the comparison itself being made with any precision. We shall now discuss some of these effects in more detail, for variations in hue, in lightness (by which is meant relative brightness, as discussed further in Section 7.2), and in colourfulness.

5.3 Variations of hue

To take the case of green grass again, the hue will depend to a considerable extent on the type of grass, some grasses being a yellowish-green, and others almost a bluish-green; new spring grass is usually rather yellower than older grass. This applies also to foliage, and here the range of hues is even greater; bright yellow-green in spring, green in summer, yellow in early autumn and red or brown in late autumn. Fruits vary tremendously in hue as they ripen, the unripe fruit generally being green, the hue gradually changing to yellow, orange, or red, as the fruit reaches maturity. Skin-colour, even of so-called 'white' races, varies from light pink or almost white to various shades of brown according to the type of skin and the amount of sun-tan.

The apparent hue of any object is likely to vary, moreover, with the colour of the background against which it is seen. Even the green of a tree, for instance, will apparently change in hue if seen first against a blue sky and then against brown earth.

The colour of the light illuminating a scene will also result in variations in hue. If the sun is low in the sky, for instance, the light will be yellowish, and, although the eye compensates for this physiologically, such compensation is only partial, and objects look yellowish when illuminated by low-altitude sunlight. Furthermore, it is well known that some colours change quite markedly in appearance when taken from daylight into artificial light. This is particularly noticeable with certain mauve colours which tend to look much redder in tungsten-filament lighting.

39

5.4 Variations of lightness

The apparent lightness of surface colours is affected by various factors. For example, haze in the atmosphere can lighten dark objects and darken light objects if they are relatively distant, and dust on surfaces can act in the same way. (See Plate 3, pages 48, 49, and Plate 6, pages 80, 81.)

As with hue, so also with lightness, the background can play an important part; a dark background makes colours appear lighter and a light background makes them appear darker. This effect is illustrated in Fig. 5.2, where the two grey patches reflect exactly the same amount of light, but the patch with the dark surround appears lighter than the patch with the light surround.

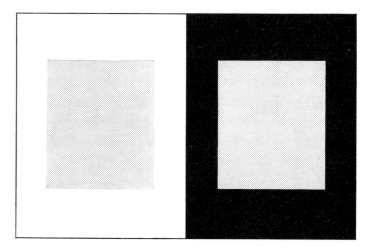

Fig. 5.2. The two grey squares reflect exactly the same amount of light, but the black surrounding the one makes it look lighter than the other which is surrounded by white.

5.5 Variations of colourfulness

The colourfulness of colours seen by the eye is subject to even more variation than hue or lightness.

It is common experience that a scene that looks a little dull and drab when viewed under an overcast sky, becomes apparently much more vivid when the sun comes out. The colours all seem to become more colourful. This is principally caused by two factors. First, the general level of illumination is raised, and secondly the lighting becomes directional instead of diffuse.

To the camera, a variation in the illumination level employed makes no difference to the result, provided that it is adequately compensated by altering the iris diaphragm or the exposure time or both. But the eye, of course, cannot adjust its exposure time, and the control on the iris of the eye is limited to a range of about 8:1 in intensity, so that a major portion of the adjustment in the eye is effected by changes in the actual sensitivities of the different mechanisms of the retina. We are all familiar with the difficulty of seeing our way about in a dimly lit room if we have just been in brilliant sunlight. After some minutes, however, the sensitivity of the retina has risen sufficiently for us to see quite well. In the eye, these changes in sensitivity, known as *adaptation*, are accompanied by quite marked changes in colour vision, which result in colours appearing pale when the intensity is low, and vivid when the intensity is high. An extreme example of this is the appearance of colours by moonlight, when their colourfulnesses are reduced almost to the extent of being indistinguishable from greys. Even at dusk colours are very drab, and the effect is by no means absent at higher illumination levels. For this reason colour reproductions usually look better the more intensely they are illuminated (Bartleson, 1965).

The second effect arises from the fact that all surfaces, whether coloured or not, reflect from their topmost layer a certain proportion of the incident light which is added to that reflected from the body of the surface. This light reflected from the topmost layer is the same colour as the illuminant, and therefore when a coloured surface is viewed in white light, some of the white light is added to the coloured light reflected from the body of the surface and the colourfulness is therefore reduced. Most surfaces exhibit some degree of gloss, and this means that, if the lighting is directional, the white light reflected from the topmost layer of the surface will be confined chiefly to a single direction, and will only rarely enter the eye, so that the coloured light, which is reflected diffusely from the body of the surface, usually enters the eye alone, and no loss of colourfulness takes place. If, on the other hand, the lighting is diffuse, no matter what the direction of viewing, the coloured light diffusely reflected by the body of the surface will always be mixed with some white light specularly reflected from the topmost layer. This is explained diagrammatically in Fig. 5.3. (See Plate 4, page 49.)

As far as adaptation is concerned the variation of colourfulness with light and dark adaptation is an unalterable fact which usually acts adversely, since many colour pictures depict brightly-lit outdoor scenes and are viewed at lower levels of illumination. Occasionally, however, the effect operates to the advantage of the picture. For instance, a colour photograph may have been taken on a very dull day using an exposure sufficiently prolonged to give a correctly exposed result. It may then be viewed by brighter light and a gain in colourfulness obtained. To take an extreme case, a colour photograph may be taken by moonlight. The eye at this low illumination can only just perceive colours and they appear of very low colourfulness indeed. But by giving the colour photograph a sufficiently long exposure (about a million times that given in sunlight) all the colours are reproduced exactly the same as if the colour photograph had been taken in sunlight. Hence, on viewing the photograph with a bright light, the sensations caused by the photograph are very much more colourful than those evoked by the original.

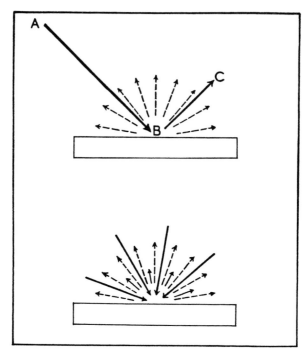

Fig. 5.3. Diagrammatic explanation of the effect of diffuse lighting on colourfulness. Full lines indicate white light, dotted lines indicate coloured light. In the upper diagram the surface receives strongly directional lighting AB, and, when viewed in any direction other than CB, the coloured light diffusely reflected by the surface is seen alone. In the lower diagram the surface is diffusely illuminated, and, no matter what the direction of viewing, the coloured light diffusely reflected by the surface is always mixed with some white light specularly reflected from the topmost layer of the surface. (See Plate 4, page 49.)

Atmospheric haze will reduce the colourfulness of distant objects in that it absorbs part of the coloured light coming from them and superimposes white light instead. The clarity of the atmosphere can make a tremendous difference to the colourfulness of distant objects. (See Plates 2 and 3, pages 48, 49.)

Just as dust affects the lightness of surface colours, so also it affects their apparent colourfulness. Again it absorbs some of the coloured light and superimposes white light so that it always results in a loss of colourfulness. Dust, of course, soon collects on out-door objects in a hot dry spell of weather, but is quickly removed by a shower of rain. Some coloured surfaces show very remarkable increases in colourfulness when they are wetted, even in the absence of dust.

The colour of the lighting can also have an effect on colourfulness in spite of the considerable physiological compensation that the eye makes for differences in illuminant colour. Tungsten-filament lighting and daylight differ considerably in

42

blue content, and this affects the apparent colourfulness of blue and yellow colours.

The colours of blue skies vary considerably in colourfulness according to the state of the atmosphere and the position of the sun, the range extending from white near the sun, especially in the presence of haze, to deep blue opposite to the sun, especially in very clear weather. This variation also makes the blueness of areas of water very variable in colourfulness. (See Plate 1, page 48.)

5.6 Priorities

Since most of the effects described above, such as those caused by atmospheric haze, variation in illumination from directional to diffuse, presence of dust or water on surfaces, adaptation, etc., produce changes in the *colourfulness* of colours, while only a few effects produce changes in hue, we would expect our mental standards of hue to be more precise than those of colourfulness. Thus a pale red tomato, for example, is more acceptable in a reproduction than an orange or a magenta one. Correctness of hue would, therefore, seem to be more important than correctness of colourfulness. Moreover, the variations in colourfulness that occur in natural colours are generally similar to those produced by adding white light uniformly over the whole field of view. Hence, if, in a reproduction, all colours are reduced in colourfulness proportionately, one would expect the result to look more natural than if colours of different hue and colourfulness were reduced in colourfulness to different extents, so that some colours shone out like spectral colours while others were extremely pale. Lightness can probably be regarded as intermediate in importance between hue and colourfulness, and brightness (the *absolute* level of the light response) as of similar importance as colourfulness.

The above considerations would seem to suggest that, as far as colour is concerned, the requirements for a successful colour reproduction are, in order of importance:

(1) Correctness of hue,
(2) Correctness of lightness (tone reproduction),
(3) Colourfulnesses proportional to those in the original,
(4) Colourfulnesses and brightnesses similar to those in the original.

The importance of the first requirement is illustrated by the prime importance of overall colour *balance* in colour reproduction. When a picture becomes unacceptable because of a general excess of magenta, for instance, it is the violent change of hue undergone by pale colours which is the most objectionable feature. Amongst these pale colours, Caucasian skin is one of the most critical in many scenes, and it has been found that this colour tends to be the anchor point for colour balance for many film systems; this can result in greys being reproduced slightly yellowish in low-colourfulness systems, and slightly bluish in high-colourfulness systems, in order that in both cases the skin colour is optimum.

The second requirement usually has a marked effect on the extent to which the reproduction looks natural: too much contrast results in a gaudy, exaggerated, appearance, too little in a smoky or hazy result. This subject is discussed in detail in Chapter 6.

The third requirement is sometimes violated by blue skies in colour photographs: the ultra-violet sensitivity of most photographic materials can render them more colourful than other colours.

The fourth requirement is usually unattainable whenever the original scene is illuminated at a very high level, as occurs, for instance, with bright sunlight; at the lower levels of illumination typical when pictures are viewed indoors, the average and maximum colourfulnesses and brightnesses are generally reduced appreciably.

We can now summarize the situation. For fundamental and unavoidable reasons, simple trichromatic methods cannot result in colorimetrically correct colour reproduction, and the errors inherent in most systems are considerable when measured physically. But when the colour of an object in a colour picture is appraised by an observer, it will generally look acceptable, provided it falls somewhere within the range of colours which that object customarily exhibits in everyday life. Practically all colours met with in everyday experience are subject to wide variations in hue, lightness, and colourfulness, and this means that no precise colour standards of familiar objects can possibly be carried in the memory. In particular, the variations in colourfulness are very great and this obscures the unavoidable tendency of all processes to produce losses in colourfulness. Futhermore, physiological and psychological effects make it extremely difficult to compare sensations produced by original and reproduction colours with any precision. Hence, for general pictorial work, the tolerances are large.

There are, however, some objects whose colours are particularly critical. Thus the reproduction tolerances for human skin, and for most foodstuffs, are smaller than average; and, in advertising work, manufacturers are often very concerned that their products and packages be depicted with little or no apparent errors of colour reproduction, and high standards of accuracy are then required. (See Plate 5, page 80.)

5.7 Factors affecting apparent colour balance

It is clear from the above discussion that colorimetrically correct results are not necessary for a colour reproduction to be acceptable. In fact, some workers have reported that optimum reproduction of some well-known colours, such as skin, is achieved when a definite difference exists between the original and reproduction colours (MacAdam, 1951; Bartleson and Bray, 1962).

There is one property of the appearance of original scenes, however, that remains remarkably constant, and that is their overall colour balance (Evans, 1943). This is partly because of the physiological adaptation of the eye to the prevailing illuminant; but, as has already been mentioned, this effect is only partial, and the consistency of colour balance is also partly caused by the ability of observers subconsciously to discount the colour of an illuminant when looking at objects in its light: this psychological effect takes place more or less instantaneously, whereas the physiological adaptation may take several minutes to complete.

It might, therefore, be thought that consistency of colour balance would also extend to reproductions: unfortunately this is not so. One reason for this is that colour reproductions do not generally fill more than a small part of an observer's field of view, and this reduces both the physiological and the psychological adjustments that take place. Secondly, both these adjustments affect all areas and all tones of a scene almost equally, but reproductions may have deviations that vary according to their geometrical or tonal position in the picture. The avoidance of local areas of colour balance different from that of the whole picture is therefore essential for proper reproduction in pictures: this means that colour photographic materials must be coated and processed very uniformly; colour television cameras and receivers must be completely free of local variations; and half-tone reproductions must be printed uniformly all over. Consistency of colour balance at all tone levels from white to black is equally important, and this requires close control of the relative response characteristics of the red, green, and blue 'channels' of the reproduction system.

Assuming that consistency of the reproduction has been achieved with both area and tone-level, correct colour balance is possible but not automatic. Let us consider, by way of example, the case when transparencies are being projected by tungsten light in a darkened room, the film system being one where optimum colour balance corresponds to greys appearing neutral, that is without any hue bias.

The situation is summarized in Fig. 5.4. In the upper left-hand part of the diagram a neutral grey scale is illuminated with tungsten light, which, being yellowish, results in yellowish stimuli reaching the eye. But the prevailing illuminant being yellowish, the sensitivity of the eye to yellow light is relatively decreased and this compensates for the yellowness of the stimuli entering the eye, thus giving rise to a neutral grey impression in the mind. The reproduction is therefore also required to give a neutral grey impression to the mind, and the eye, seeing a screen lit by tungsten light in a darkened room will be partially, but not quite wholly, adapted to tungsten light. It is therefore necessary that the neutral grey scale be reproduced on the transparency as slightly blue in order to overcome the slightly yellowish appearance of the tungsten light on the screen.

Considering now the bottom half of the figure, when the neutral grey scale is illuminated by daylight, which is bluish, the eye will receive bluish stimuli, but will have decreased sensitivity to blue so that a neutral grey impression will still be produced in the mind. But the projection conditions have not changed, so that the neutral scale must again be reproduced in the transparency as slightly blue. It is therefore clear that, when the photograph is taken in tungsten light, *yellowish* light must produce a slightly blue result, whereas when it is taken in daylight, *bluish* light must produce the same slightly blue result. It is therefore clear that

45

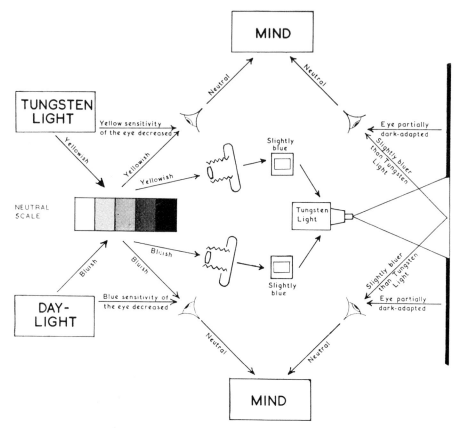

Fig. 5.4. The effects of colour adaptation on colour reproduction.

different films must be used for the two situations, a film with a faster blue layer for tungsten light, and a film with a slower blue layer for daylight, or, alternatively, appropriate filters must be used over the camera lens. Kodak films balanced for daylight use are known as *Daylight type*; those balanced for tungsten light of colour temperature[1] 3400 to 3500 K as *Type A*, and for tungsten light of colour temperature 3100 to 3200 K as *Type B*; those balanced for clear flash lamps as *Type F*; another type, known as *Type G*, is designed to give a balance that is a compromise for daylight and 3200 K tungsten light, the spectral sensitivity of the red layer in this case being shifted towards that of the green layer to reduce the effect of this change in illuminant. In negative-positive colour photographic systems considerable compensation for illuminant colour can be made at the printing stage. (See Plate 20, pages 288, 289).

[1] Colour temperature is defined in Section 10.2.

46

In colour television, similar compensation for the colour of the light illuminating the scene must also be made, and this can be done by altering the relative amplifications of the signals from the red, green, and blue images in the television camera, so that they are always equal to one another for white (and grey) objects in the scene.

If two illuminants of markedly differing colour, such as daylight and tungsten light, are mixed in the same scene, the result in the colour reproduction is usually very unpleasant, and mixed lighting must, therefore, usually be avoided. The appearance of the original scene is often quite tolerable and this seems to be because the light sources are usually visible and subconscious allowance is made for their colour effects, whereas in the colour picture the sources are nearly always excluded; also many colour reproduction systems operate with ultra-violet and infra-red sensitivities greater than those of the eye: hence, in the picture, the difference in colour between the two sources may be increased relative to other colour differences in the scene.

5.8 Integrating to grey

Correct colour balance is particularly necessary in the case of reflection colour prints, because their surroundings provide a reference balance against which they can easily be compared. Hence special techniques are usually needed when making reflection prints and one of the most successful, known as 'integrating to grey', was first suggested by Evans (Evans, 1951). Evans argued that, because any colour reproduction will tend to adapt the eye towards its average colour balance, it would be an advantage if the light from the print, when integrated, appeared grey in colour, because then the eye would be adapted to it immediately. Accordingly, in printing colour negatives, it is often arranged that the cyan, magenta, and yellow layers of the print material receive exposures inversely proportional to the red, green, and blue transmissions of the negative, respectively. In this way, prints that approximately 'integrate to grey' can be made rapidly by automatic printing methods to a consistently high standard of acceptance, the proportion requiring special treatment because of unusual subject matter being surprisingly small. This method of printing has been applied in various ways (Bartleson and Huboi, 1956; Hunt, 1960). This subject is dealt with more fully in Chapter 16. Automatic methods of adjusting the colour balance of television cameras have also been devised (Pearson and Ray, 1978).

5.9 The perception of depth

Our two eyes working together give us *stereoscopic* vision, or perception of depth. For certain tasks, like threading a needle, or putting a nut on a bolt, this ability to perceive depth is extremely useful. But the vast majority of pictures are of objects a metre or more away from us, and then stereoscopic vision contributes little, or – beyond about five metres – nothing, to our perception of depth. Depth is then interpreted from clues such as shadows, perspective, obscuration of objects behind others, relative sizes of objects, and parallax, all of which are monoscopic in nature. This is illustrated by the fact that the interpretation of depth from a monoscopic picture reveals certain rules by which the brain works. A strong rule is that light is generally assumed to be falling on the object from the direction of the top of a picture: thus, turning a picture upside down can reverse all the directions of perceived depth. But this rule can be suppressed in the case of areas that look like a human face: a picture of a plaster cast of a face, even when orientated so that the direction of the light is from the top of the picture, can appear to have the shape of a face, and not that of a cast of a face, so that the direction of perceived depth is opposite from reality and corresponds to a direction of light from the bottom of the picture. (See Plate 8, page 81.)

The realistic rendering of depth in most monoscopic pictures makes the added complexity of producing stereoscopic colour pictures rather unattractive in terms of both inconvenience and high cost: for each stereoscopic reproduction, two (or more) different images must be recorded, and special viewing devices provided. Moreover, stereoscopic *still* pictures (as distinct from *movie* films) suffer from their lack of motion much more than is the case with monoscopic still pictures: stereoscopic stills appear to freeze reality into collections of statues and sculptures having a strangely unreal appearance. For these reasons, stereoscopic pictures have never achieved more than a partial and temporary acclaim, in contrast to monoscopic pictures whose popularity has increased tremendously. For an example of a monoscopic picture portraying a good illusion of depth see Plate 34, page 544.

REFERENCES

Bartleson, C. J., *Phot. Sci. Eng.*, **3**, 114 (1959).
Bartleson, C. J., *J. Opt. Soc. Amer.*, **50**, 73 (1960).
Bartleson, C. J., *Phot. Sci. Eng.*, **9**, 174 and 179 (1965).
Bartleson, C. J., and Bray, C. P., *Phot. Sci. Eng.*, **6**, 19 (1962).
Bartleson, C. J., and Huboi, R. W., *J. Soc. Mot. Pic. Tel. Eng.*, **65**, 205 (1956).
Evans, R. M., *J. Opt. Soc. Amer.*, **33**, 579 (1943).
Evans, R. M., *U.S. Patent 2,571,697. British Patent 660,099* (1951).
Hunt, R. W. G., *J. Phot. Sci.*, **8**, 186 and 212 (1960).
MacAdam, D. L., *J. Soc. Mot. Pic. Tel. Eng.*, **56**, 502 (1951).
Reed, P., *Alpine Garden Society Bulletin*, p. 143 (June, 1965).
Pearson, D. E., and Ray, A. K., *Color. Res. Appl.*, **3**, 117 (1978).

GENERAL REFERENCES

Evans, R. M., *Eye, Film, and Camera in Color Photography*, Wiley: New York (1959).

Plate 1
Most colours can exhibit quite large changes in colourfulness, and this results in the tolerances for colourfulness usually being considerable (see Section 5.5). Blue sky, for instance, is usually pale near the horizon and most colourful at high elevations (in parts of the sky well away from the sun); in this example, the colourfulness of the sky increases steadily towards the top of the picture. Reproduced from a 5 × 4 inch *Ektachrome* transparency on a Crosfield *Magnascan* scanner.

Plates 2 & 3

Haze, mist, smoke, or dust in the atmosphere usually superimposes white light on objects and thus reduces colourfulness. In Plate 2 (*on the left*) the progressive loss of colourfulness through increasing amounts of steam can be seen. In Plate 3 (*above*), objects near the camera are seen through only short lengths of the atmosphere, but more distant objects are considerably reduced in colourfulness because of haze (see Section 5.5). Flare in lenses, and from screens used for projection, or from the surface of reflection prints, has similar effects on colourfulness. The superimposition of white light, whether by atmospheric effects or by flare, also has a profound effect on tone reproduction: the effect appears greatest in dark colours, because the addition of a given amount of flare light represents a greater percentage increase in luminance for dark colours than for light colours, and the response of the eye is approximately proportional to the percentage change. The effects caused by the superimposition of white light can be counteracted approximately by adjusting the black level in television (see Sections 19.13 and 23.14) and this is often done both on the camera and on the receiver. In photography, the shapes of the characteristic curves of films and papers are usually adjusted to counteract the effects of flare caused by typical lenses and by screens or reflection print materials (see Chapter 6). Plate 2 reproduced from a 5 × 4 inch *Ektachrome* transparency, and Plate 3 reproduced from a 36 × 24 mm *Kodachrome* transparency, on a Crosfield *Magnascan* scanner.

Plate 4

The light reflected from the topmost layer of most objects is usually white, and, in the case of matte objects, this reduces the colourfulness of the colour for all directions of viewing; for glossy objects, the white light is reflected as a mirror-image of the light source, and hence does not reduce the colourfulness for other directions of viewing (see Fig. 5.3). Most areas of glossy surfaces can therefore usually be more colourful than matte surfaces (unless the latter have surface textures capable of very low reflectances as occur, for instance, with wool and velvet). In this example, the glossy red is clearly more colourful than the matte red. Reproduced from a 5 × 4 inch *Ektachrome* transparency on a Crosfield *Magnascan* scanner.

CHAPTER 6

Tone Reproduction

6.1 Introduction

IN some situations it may be desirable for a picture to present an appearance quite different from that of the original scene that it represents: for instance, it is usually more convenient to shoot night scenes in normal high levels of illumination, and then to make prints that look as though the picture had been shot at a low level of illumination. Usually, however, systems of colour reproduction are used in order to produce pictures that have an appearance approximating that of the original scene; the purpose of this chapter is to examine the tone characteristics that colour systems must have in order to achieve this result. (Hunt, 1969.)

6.2 Identical viewing conditions

If a picture is viewed under the same conditions as those of the original scene, then it would be expected that its tones should be physically the same as those of the original: that is, a scale of grey surfaces in the scene should be reproduced with the same luminances in the picture. The extensive studies by Bartleson and Breneman on brightness scaling in pictures have shown that this is indeed so (Bartleson and Breneman, 1967; Bartleson, 1975) and hence in this case the tone reproduction of the system should ideally be as shown in Fig. 6.1: the relationship between log luminance in the reproduction and log luminance in the scene should be a straight line passing through the origin at 45 degrees; in this figure, log reproduced luminance is plotted downwards, to be consistent with later figures in

49

which optical density, which is equal to $\log_{10}(100/T)$ where T is the percentage transmittance or reflectance, is always plotted upwards.

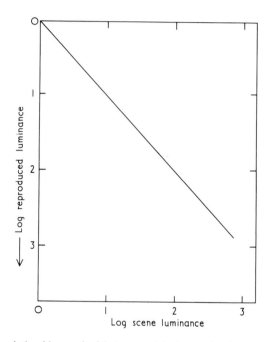

Fig. 6.1. The relationship required between original scene luminance and reproduced luminance when identical viewing conditions are involved.

6.3 Characteristic curves

Graphs in which the log of the reproduced luminance is plotted against the log of the original luminance are very useful in colour reproduction, and, because optical density is widely used for the former and log exposure for the latter, such graphs are often referred to as D—$\log H$ curves, or *characteristic curves*. Such curves often have an approximately straight-line section in the middle with curved sections of lower gradient on either side, as shown for the curve marked 'Actual system' in Fig. 6.4: the low-density curved section is often referred to as the *toe*, and the high-density curved section as the *shoulder*, of the curve. The slope, or gradient, of the straight-line section is called the *gamma*, so that a gamma of 1.0 indicates a slope of 45°, a gamma of 2.0 a slope of about 63°, and a gamma of 0.5 a slope of about 27°. It is also convenient to be able to refer to the slope at any point on the curve, and this should strictly be called the *point-gamma*; but in this book the term gamma will be used in a general sense that includes point-gamma.

6.4 Different luminance levels

If the picture is viewed in conditions that are the same as for the original scene except for an overall shift in the level of illumination, then the work of Bartleson and Breneman has shown that the relationship of Fig. 6.1 should be replaced by that of Fig. 6.2. In this figure a straight line at 45 degrees still represents the relationship, but relative luminances are used instead of absolute luminances: thus, as abscissa, log scene luminance relative to white is plotted instead of log scene luminance; and, as ordinate, reproduction density is plotted instead of log reproduced luminance. Measuring scene luminances relative to white requires the adoption of a suitably defined white as reference. (See Plate 7, pages 80, 81.)

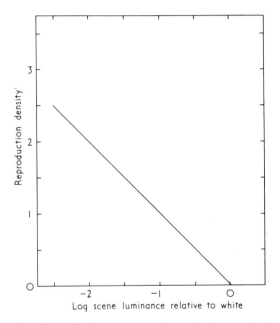

Fig. 6.2. The relationship required between original scene luminance relative to white, and density seen in the reproduction, when the only difference in viewing conditions is an overall shift in luminance level.

The visual justification of Fig. 6.2 is the fact that the eye recognizes objects, not so much by their brightnesses, as by their lightnesses, and, whereas brightnesses are related to log luminances, lightnesses are related to *relative* log luminances. Thus the sensation of whiteness, for instance, is normally evoked whenever an object has a luminance about three times as high as the average luminance of the scene, irrespective of what the absolute value of that average is, over a very wide range of values. Actually, this factor of about three is affected to some extent by the average luminance level, because the apparent contrast of a given scene (or reproduction) does increase a little as the illumination level increases. Hence, if a scene illuminated by bright sunlight at about 50 000 lux were reproduced and viewed at about 1000 lux the illusion of bright sunlight in the reproduction would be improved by an increase in the gamma of the picture; on the other hand, a moonlit scene illuminated at 0.01 lux reproduced and viewed at 1000 lux would call for a reduction in the gamma. However, the change in contrast of the visual response over large ranges of luminance is fairly modest: thus from indoor daylight levels of about 1000 lux, the apparent contrast varies (Hunt, 1965a) by only about +7 per cent upwards to bright sunlight, and −7 per cent downwards to poor artificial light at about 20 lux. For most purposes these changes are small enough to be neglected, and therefore, to a good approximation, in matters of tone reproduction, we may usually omit consideration of the *absolute* log luminances, and consider only the relation between the *relative* log luminances in the original and reproduction; hence the adoption of measures of relative log luminance in Fig. 6.2 is satisfactory.

The scales of density and of log scene luminance relative to white are positioned in Fig. 6.2 so that the corresponding values of the two quantities are equal (apart from a difference in sign). Thus, for example, a grey in the scene that reflects only 10 per cent as much light as the reference white (−1.0 on the scale of log scene luminance relative to white) would have a density in the reproduction of 1.0, and the reference white in the scene would be reproduced with zero density.

The type of reproduction that most closely involves the viewing situation being considered at present (in which the only difference between original and reproduction conditions is an overall shift in the luminance levels) is that commonly used for reflection prints. However, original scenes are not usually confined by a definite border, whereas reflection prints, like all other reproductions, always do have borders; but it is often the case that the average luminance of the areas surrounding the print will be similar to that of the print itself, and this is also the normal situation in original scenes, the average luminance of the part of the scene photographed usually being similar to that of the surrounding areas. The effect of the border is therefore usually small and the characteristic of Fig. 6.2 is therefore still applicable. If the colour of the illuminants for the scene and for the reproduction are different, the same arguments are broadly true as far as tone reproduction is concerned, the added complications being mainly confined to questions of colour appearance (which are discussed in Chapter 11).

To achieve the characteristics of Fig. 6.2 in an actual reflection print system it is necessary for the photographic materials to take account of flare wherever it occurs. Flare always has the effect of lowering gamma: camera flare and viewing flare mostly in the dark tones, and printer flare mostly in the light tones when, as is usual, the print is made from a photographic negative. The effects of viewing flare are particularly large: thus, in typical rooms, the light reflected from the topmost surface of the print, even if it is highly glossy, usually limits the maximum density that can be seen to about 1.6 and causes severe reduction of gamma at densities just below this figure (Hunt, 1965a; Carnahan, 1955). The consequence of the presence of all three sources of flare light, at typical levels, is that, to achieve the characteristic of Fig. 6.2, the photographic materials would have to provide a characteristic that, when measured without flare (Hunt, 1968), is as shown in Fig. 6.3 by the curve marked 'To offset flare'; the 45 degree line of Fig. 6.2, representing the final relationship required when the surround luminance has the same average as that of the picture, is shown for comparison marked 'Average surround'. The curve marked 'To offset flare' has been calculated to allow for: a camera flare luminance of 0.4 per cent of scene-white image-luminance (a figure typical of good quality cameras); a printer flare luminance of 9 per cent (based on practical tests) of the luminance of scene-white reproduced on a negative (therefore as a darkish area) of gamma 0.67; and a viewing flare corresponding to a reflectance of 2.7 per cent on the reproduction, giving a maximum final density of 1.57 (which is representative of average room viewing conditions (Hunt, 1965a), see Section 13.10).

It is seen that the effect of flare is to raise the gamma required by the

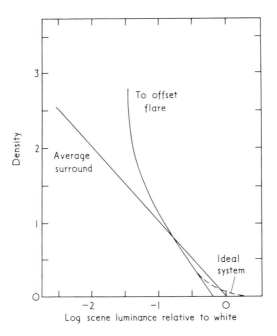

Fig. 6.3. The curve marked 'To offset flare' shows the characteristic that a photographic system must have in order to achieve the final relationship of Fig. 6.2 (shown again in this figure marked 'Average surround') in the presence of typical amounts of camera flare, printer flare, and viewing flare. The broken curve marked 'Ideal system' shows how the gamma has to be reduced at low densities to accommodate extreme highlights.

53

photographic system, and to do so more at the high densities than at the low densities, so that the required characteristic becomes curved.

However, if a photographic material were made having the characteristic shown in Fig. 6.3 ('To offset flare' curve) it would be found that any parts of the scene having luminances greater than that of the reference white (such as specular reflections, see Plate 17, page 241) would have exposures to the right of where the 45 degree line cuts the axis; hence, although they may be properly recorded on the negative, they would all be reproduced at the same (zero) density and would thus be completely lacking in any modulation, appearing as uniform 'holes' in the picture; such areas would appear extremely unnatural and unpleasant. Moreover, even with parts of the scene not having a luminance greater than that of the reference white, even a slight increase over precisely the right exposure level in the printer would have the same effect. For these reasons (and incidentally it is impossible to manufacture photographic materials with characteristic curves having such abrupt intercepts with the zero-density line) an ideal reflection-print system would have a gradually decreasing gamma at low densities as shown in Fig. 6.3 by the broken curve marked 'Ideal system'. In Fig. 6.4 the curve marked 'Ideal system' is the same as the 'To offset flare' curve of Fig.

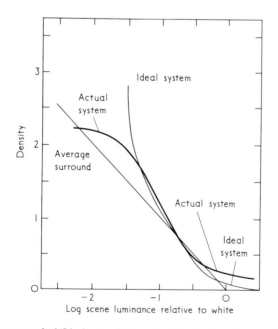

Fig. 6.4. The curve marked 'Ideal system' follows the 'To offset flare' curve of Fig. 6.3 except at the low density end in order to accommodate extreme highlights. The curve marked 'Actual system' shows the characteristics of a commercially successful system used for the production of colour reflection prints.

6.3 except that it follows the broken curve at low densities. Also shown in Fig. 6.4 is a curve marked 'Actual system', which is the characteristic of a commercially-successful system used for the large-scale production of colour reflection prints for the amateur market: it is seen to follow the ideal curve quite closely, except at the low density end where it is slightly too high because of the stain or fog level of the paper, and at the high density end where it is too low because of the impossibility of achieving the infinitely high gamma called for by the ideal curve.

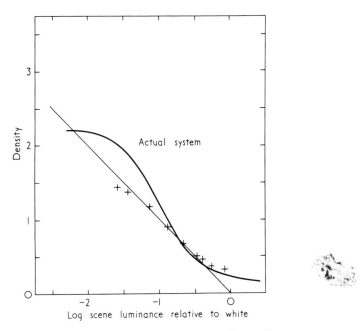

Fig. 6.5. The densities actually seen by an observer (crosses) looking in an ordinary room at a colour reflection-print reproduction of a nine-step grey scale photographed in bright sunlight on a film-paper system having the same characteristics as shown in Fig. 6.4 ('Actual system' curve) and shown again in this figure.

In Fig 6.5 are compared the curve of the actual reflection-print system (replotted from Fig. 6.4) and nine points representing the densities and relative log scene luminances of the nine steps of a grey scale, photographed on the system in bright sunlight (a portrait also being included in the scene to provide a guide to correct exposure levels). The densities and log scene luminances were measured using a telephotometer situated in a typical observer viewing position in a typical room, and from the camera position in the actual scene (Hunt, Pitt and Ward, 1969). It is clear from Fig. 6.5 that, although the characteristic of the actual photographic system is very curved, with a maximum gamma well in excess of unity, yet, over most of their density range, the points lie near the straight line of unity gamma passing through the origin. The system thus achieves the reproduction of tones at about the same density in the reproduction as in the scene (regarding log luminance relative to white as a measure of scene density); deviations only occur below a density of about 0.4 where very light greys and whites are reproduced too dark, and above about 1.4 where very dark greys and blacks are reproduced too light.

6.5 Different surround conditions

If we consider now the viewing of pictures on transparent film, as opposed to reflection prints, we have rather a different situation. Whether the transparency is of the cut-sheet type and is viewed on an illuminated opal, or whether it is a slide or motion-picture film and is viewed by projection, the average luminance of the surround is now normally appreciably less than that of the picture (and incidentally any effects of this may be enhanced by the fact that most reproductions are viewed at reduced angular magnifications).

If the reproduction is in the form of a transparency of cut-sheet size and is viewed on an illuminated opal viewer which it completely covers, then the surround will consist of the rest of the room in which the viewing is taking place; the luminances of objects in the room are usually lower than those in the

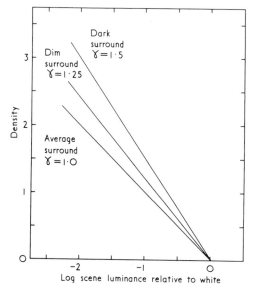

Fig. 6.6. The relationship required between original scene luminance relative to white, and density seen in the reproduction, when the reproduction has an average surround (as in the viewing of reflection prints), a dim surround (as with cut-sheet transparencies viewed on an illuminated opal in room-light), and a dark surround (as with transparencies projected in a dark room).

56

transparency and we may call this a 'dim surround'. If the film is projected in a dark room, the surround is normally of a very much lower luminance than that of the picture, and we may call this a 'dark surround'. The effects of these dim and dark surrounds are to make the pictures appear lighter than would be the case with a surround of the same luminance (see Fig. 5.2). But this lightening occurs to a greater extent in dark areas than in light areas of the picture (Breneman, 1962); hence the surround lowers the apparent gamma. The work of Bartleson and Breneman has shown how this phenomenon affects the gammas required by photographic systems in various viewing conditions: with a dim surround, in order to produce tone reproduction that appears correct, it is necessary to increase the effective objective gamma of the photographic system to about 1.25 (as shown in Fig. 6.6); with a 'dark surround' the apparent lightening of the dark areas of the picture is even more marked, so that to produce tone reproduction that appears correct now requires a gamma of about 1.5 (as also shown in Fig. 6.6). Only if the reproduction is seen with an 'average surround' will a gamma of 1.0 be appropriate, and this condition generally obtains only when reflection prints are being viewed. (The values of 1.25 and 1.5 for dim and dark surrounds, respectively, are taken as representative for fairly high densities; for lower densities values of up to about ten per cent lower may be more appropriate.)

In order to achieve the gammas of 1.25 and 1.5 shown in Fig. 6.6 it is necessary for the photographic materials to have even higher gammas, in order to overcome the effects of flare light in the camera and in the viewing situation (and in the printing step, if any).

In the case of films projected with a dark surround, the curve of Fig. 6.7, marked 'To offset flare', is obtained. The exact shape of this curve depends upon the amounts of flare involved. The curve shown relates to the same camera flare as considered earlier (0.4 per cent of scene-white image-luminance) which is typical of good quality cameras. No printer flare has been included. The amount of viewing flare depends very much on the projection conditions: under the very best conditions the flare light, when projecting film of normal density, is equal to about 0.1 per cent of the open-gate screen luminance (Hunt, 1965a; Estes, 1953); but, under average conditions, it might amount to about 0.6 per cent, and this is the figure that has been used in deriving the curve of Fig. 6.7. It is seen that the gamma of the film now has to be increased at high densities by gradually increasing amounts so that the required relationship becomes curved. If the flare were greater than the value chosen, the curvature would be greater and would extend to lower density levels. If printing flare had been included in calculating the curve of Fig. 6.7, its main effect would have been to require a further increase in gamma, mainly at low densities, similar to that shown in Fig. 6.3, in negative-positive systems; or mainly at high densities in positive-positive systems (see Section 13.3).

As in the case of reflection-print systems, if a film system were used having the abrupt intercept with the zero-density line shown in Fig. 6.7, very unpleasant 'holes' in the picture would occur whenever scene luminances exceeded that of

57

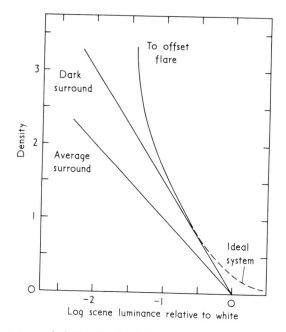

Fig. 6.7. The curve marked 'To offset flare' shows the characteristic that a photographic system must have in order to achieve the final relationship of Fig. 6.6 for dark surround viewing conditions (shown again in this figure), in the presence of typical amounts of camera flare and viewing flare. The broken curve marked 'Ideal system' shows how the gamma has to be reduced at low densities to accommodate extreme highlights.

the reference white. An ideal curve for the system must therefore have a gradually decreasing gamma at the light end of its scale as shown by the broken curve marked 'Ideal system'. In Fig. 6.8 the curve marked 'Ideal system' is the same as the 'To offset flare' curve of Fig. 6.7 except that it follows the broken curve at low densities. Also shown in Fig. 6.8 is a curve marked 'Actual film' and this is the characteristic of a commercially successful reversal[1] film used for the large scale production of transparencies for the amateur market: it is seen to follow the ideal curve quite closely except at the low density end where it is slightly too high because of the stain or fog of the film, and at the high density end where it is too low because of the impossibility of achieving the infinitely high gamma called for by the ideal curve.

In Fig. 6.9 are compared the curve of the actual reversal film (replotted from Fig. 6.8) and nine points representing the densities and relative log scene luminances of the nine steps of the same grey scale as used for Fig. 6.5 but this

[1] A reversal film is one that gives positive images directly, unlike a negative film from which positive images have to be obtained by using a printing step.

time photographed on this reversal film in bright sunlight (with a portrait to define correct exposure levels as before); the densities and log scene luminances were measured using the same telephotometer situated in a typical observer viewing position in the projection room, and from the camera position in the actual scene (Hunt, Pitt and Ward, 1969). It is clear that again, although the characteristic of the actual photographic system is very curved, with a maximum gamma well in excess of 1.5, yet the points lie near the straight line of 1.5 gamma over most of their density range. Moreover, if, again, log scene luminance relative to white is regarded as a measure of scene density, then the reproduced density is equal to 1.5 times the scene density over most of the density range; deviations only occur below a density of about 0.7 where very light greys and whites are reproduced too dark, and above about 2.2 where very dark greys and blacks are reproduced too light.

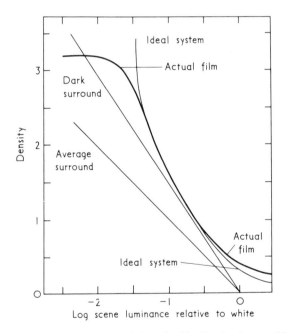

Fig. 6.8. The curve marked 'Ideal system' follows the 'To offset flare' curve of Fig. 6.7 except at the low density end in order to accommodate extreme highlights. The curve marked 'Actual film' shows the characteristic of a commercially successful reversal film used for the production of colour transparencies intended for projection.

In the case of cut-sheet films viewed on illuminators, the dim surround requires a gamma of about 1.25, and the densities of the same nine-step grey scale reproduced on cut-sheet film, and measured by means of the telephotometer from a typical observer viewing position, are shown in Fig. 6.10 (the 'square' points); the results for the reflection print (crosses) and projected transparency (circles) are also shown for comparison.

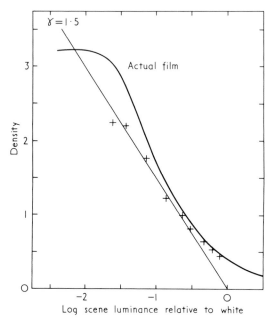

Fig. 6.9. The densities actually seen by the audience (crosses) of the projected colour-transparency reproduction of the nine-step grey scale photographed in bright sunlight on a film having the same characteristic as shown in Fig. 6.8 ('Actual film' curve) and shown again in this figure.

6.6 Complications with solid objects

The results of Fig. 6.10 show that most of the steps of the grey scale in the original scene are reproduced with densities that are the same as those of the original in the case of the reflection print, and increased by a factor of 1.25 in the case of the cut-sheet film on an illuminator, or of 1.5 in the case of the projected transparency. It must be pointed out, however, that this finding is dependent on the particular choice of reference white. Furthermore a flat grey scale is a very artificial scene, and real objects are usually solid, so that their luminance varies as the incident light strikes them at different angles. The simple equivalence of densities shown in Fig 6.10 is therefore complicated in practice, and hence the print or transparency that best depicts a particular scene may have to lighten or darken particular objects in order to effect the best overall compromise. This can be done by careful control of lighting (Evans and Klute, 1944); shadow areas often have to be lightened.

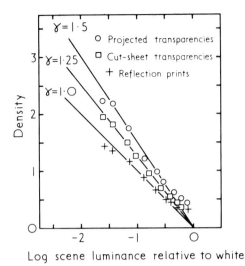

Log scene luminance relative to white

Fig. 6.10. The displayed density of the nine-step grey scale plotted against the log luminances of its steps relative to white. Reflection print systems have a displayed gamma of 1.0, cut-sheet transparency systems a displayed gamma of 1.25, and projected transparency systems a displayed gamma of 1.5.

6.7 Comparisons of transparencies and reflection prints

In Fig. 6.11 the results of Figs. 6.5 and 6.9 have been plotted together using a scale of transmission density reduced by a factor of 1/1.5. It is seen that the two sets of results are in close agreement but that the transparency reproduces both the lightest and the darkest steps of the grey scale with higher gamma. The greater compression of the lightest and darkest tones in reflection prints, as compared to transparencies, is a well-known effect and contributes to their somewhat lower quality (see Chapter 13).

6.8 Colourfulness

Since colourfulness is increased by increasing the gamma (see Section 7.9) it is clear that, other things being equal, projected transparencies and cut-sheet films would show higher colourfulness than reflection prints, because of their use of an effective gamma of 1.5 or 1.25 instead of 1.0. This is in broad agreement with practical experience. But other factors are also at work: for example, interreflections in the image-layers of reflection prints between the base and the top-surface effectively degrade the dyes, thus tending to reduce the colourfulness of reflection prints still further (see Section 13.9); however, colourfulness is increased physio-

61

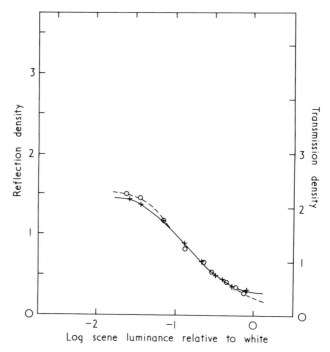

Fig. 6.11. The crosses from Fig. 6.5 together with points (circles) corresponding to the crosses of Fig. 6.9 but plotted on a density scale reduced by a factor of 1/1.5. If the alteration in density scale correctly allows for the effect of the dark surround in the case of the projected transparency, then the near coincidence of the crosses and the circles indicates similarity of apparent tone reproduction in the projected transparency and in the reflection print; however, the latter has lower gamma at both the very light and the very dark ends of the scale.

logically as the general illumination level is raised (Hunt, 1952, 1953, 1965b), and reflection prints can often be illuminated at higher levels than is feasible in the case of projected images of transparencies, and hence some increases in colourfulness can be obtained; transparencies can also suffer some loss of colourfulness because of their dark surrounds, particularly in the case of large areas near the edges of the picture (Hunt, 1950; Rowe, 1972; Pitt and Winter, 1974; Bartleson, 1977; Breneman, 1977).

6.9 Exposure latitude

If the viewing flare light occurring in the projection of transparencies remains of constant magnitude in spite of changes in the density of the transparency, then the characteristic curve effective on the screen will remain substantially constant with variations in exposure level of the film because by far the greater part of the flare is viewing flare. However, part of the viewing flare in projected transparencies derives from light from the screen being reflected by the room back to the screen; hence the viewing flare will decrease somewhat as transparencies become more dense and the effective gamma will rise at high densities; furthermore, light parts of the scene will be reproduced away from the low-gamma curved portion of the film characteristic: hence dark transparencies will have higher effective gammas. Conversely light transparencies will have lower effective gammas. It would therefore be expected that the exposure latitude of transparency materials would be limited by under-exposure resulting in excessive apparent gamma, and over-exposure in insufficient apparent gamma: this is exactly what is found in practice. However, with under-exposed transparencies, insufficient maximum density in the film may result in insufficient gamma in the shadows, and this, when combined with excessive gamma in the rest of the scale, presents a particularly unpleasant appearance. (High density slides may also be unsatisfactory merely because of inadequate screen luminance, of course.)

In reflection prints, the exposure latitude at the camera will be limited by the film, and in the case of colour negative films their long straight characteristic curves (see Section 14.15) usually provide far more camera exposure-latitude than is possible with the curved characteristics required by direct reversal films; however, when printing negatives, the latitude of the printing exposure is limited: in light prints, by insufficient gamma (caused by the curve of the system at low densities); in dark prints, by insufficient density range being available to depict an average scene adequately; and, generally, by the need for objects to be reproduced with approximately the same density as they had in the scene, because reflection prints are seen in the presence of other natural objects.

6.10 Tone reproduction in duplicating

If a transparency, exposed on a film system having as characteristic the 'Actual film' curve of Fig. 6.8, is duplicated by contact printing on to the same film, the resultant characteristic curve is as shown by the broken curve in Fig. 6.12 (obtained by using the densities of the full curve as abscissa values to obtain a new set of densities which are plotted against the original log relative scene luminances). It is assumed that there is negligible printer flare in contact printing (which is probably correct); however, even in optical printing, the amount of printer flare would probably be small compared to the amount required to offset the large increase in the gamma of the transfer function shown by the broken curve of Fig. 6.12. It is clear that the basic gamma of 1.5 of the film system (with even higher gammas at high densities to offset flare), together with the curved low-density portion, have resulted in a much more bowed characteristic for the duplicate system than for the original transparency system. Duplicates made on such a system exhibit excess shadow density (blocked-up shadows) and insuffi-

63

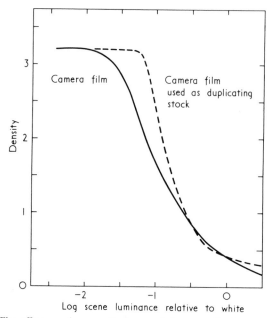

Fig. 6.12. The effect on the overall characteristic curve (broken curve) of using, for duplicating, a film with the 'Actual film' characteristic curve of Fig. 6.8 (shown again in this figure as 'Camera film' curve).

cient high-light contrast (burnt-out highlights) as shown in Fig. 6.13. Similar effects occur, although to a less extent, in the case of cut-sheet film, as shown in Fig. 6.14

The situation can be improved by deliberately adding extra 'printer flare', a technique known as *flashing* (Lighton, 1967; Doody, Lawton, and Perry, 1978). The amount of correction that it is possible to introduce by flashing, however, is limited by the necessity of maintaining an adequate maximum density in the copy.

Ideally, from the point of view of tone reproduction, a duplicating film should have a characteristic curve with a gamma of 1.0 at all densities. In practice it is impossible to avoid low gamma 'toes' and 'shoulders' at the ends of the characteristic curves of photographic films, but in Fig. 6.15 the curve for a film intended for duplicating cut-sheet transparencies is shown, together with the curve of an actual film intended for original cut-sheet transparencies (marked 'Camera film' in Fig. 6.15). It is seen that the duplicating film is a much closer approximation to the ideal $\gamma = 1$ characteristic than the camera film; the slightly higher gamma (as compared to 1) of the duplicating film at high densities provides some correction for printer flare if optical printing (or enlarging) is used. If a duplicating film of the type shown in Fig. 6.15 is not available, a camera type of film can be used and the tone reproduction corrected by using appropriate masks (see Section 15.2 and Chapter 27) or similar devices on scanners (see Chapter 29).

6.11 Tone reproduction in television

The viewing conditions for television are usually similar to those for viewing cut-sheet transparencies on an illuminated opal: in both cases a dim surround is involved. A gamma of about 1.25 is therefore appropriate, and in colour television this is commonly achieved by transmitting signals of gamma 1/2.2 and displaying them on receivers having gammas of about 2.8 (see Section 19.13).

When film is used to originate television signals it is therefore necessary for the gamma to be reduced from that of the film to 1/2.2 (see Section 23.14).

The presence of flare at various stages complicates the situation, and the effects of flare are often reduced by electronic adjustments of the black level of the picture (DeMarsh, 1972).

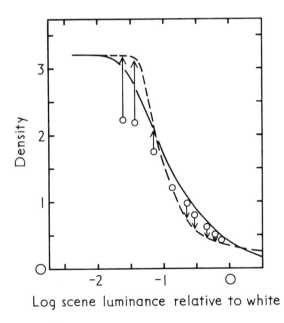

Fig. 6.13. Sensitometric curve of projection-transparency film (full line) compared to the curve resulting from the use of this type of film for duplicating with no flare (broken line). The circles show the results obtained in practice for an original transparency with camera and viewing flare (reproduced from Fig. 6.9): the arrow heads show how their densities are distorted if this type of film is used for duplicating.

65

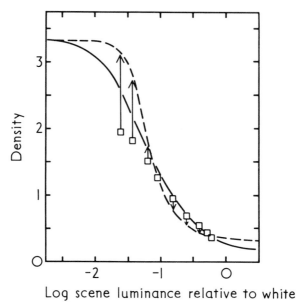

Log scene luminance relative to white

Fig. 6.14. Same as Fig. 6.13 but using throughout a film intended for the production of cut-sheet transparencies.

6.12 Lighting geometry

The geometrical arrangement of the lighting in scenes can greatly affect the apparent tone reproduction in pictures (see Plate 8, page 81); this calls for careful control of the lighting of the scene, and, in film and television studios, elaborate systems of lights are often used. Apparent contrast is affected not only by the positions, but also by the sizes, of light sources. The use of small compact lights results in shadows having sharp edges, whereas larger sources give softer shadows that make pictures look less contrasty. The soft edges that occur when objects are very out-of-focus in pictures can also affect apparent contrast, and this sometimes results in out-of-focus patches of colour in pictures appearing unusually luminous. (See Plate 14, page 177.)

6.13 Conclusions

If a picture is viewed under conditions such that its average luminance and that of its surround are similar to the average luminance of the original scene, then the tone reproduction of the system used to produce the picture should ideally have a gamma of unity.

If the picture is viewed under conditions in which the average luminance of the picture and that of its surround are equal to one another but different from the

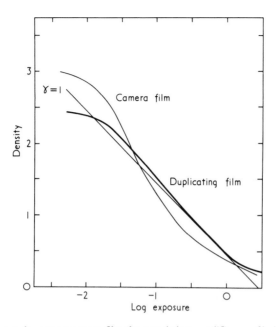

Fig. 6.15. A cut-sheet transparency-film characteristic curve ('Camera film' curve) and that of a duplicating film of gamma approximately 1.0 designed for making cut-sheet duplicates.

average luminance for the original scene, then a gamma of unity is still usually a good enough approximation except for extreme conditions, such as moonlight.

To achieve a gamma of unity in a system, however, it is necessary to take account of flare wherever it occurs: consequently, the photographic materials used in such systems must result in overall characteristic curves whose gammas increase substantially as the exposure approaches the end of the tone scale. Furthermore, the need to provide exposure latitude may also require departures from the concept of a gamma of unity. By allowing for these factors, ideal characteristic curves for the photographic systems can be constructed.

If the picture is viewed in a dark surround, as with projection in a darkened room, the dark surround has the subjective effect of 'subtracting grey' from the picture, but it does so to a greater extent in dark areas than in light areas: hence the dark surround lowers the apparent gamma. Since original scenes are not normally viewed with dark surrounds, it is in this case necessary for the reproduction to operate at a gamma of about 1.5, instead of unity, so as to off-set this effect. Allowing for this factor, and, as before, for flare and latitude, an ideal characteristic curve for a photographic system intended for dark-surround viewing can be constructed. If the picture is viewed with a dim surround, as is the case for cut-sheet transparencies viewed on an illuminated opal, and for television, the gamma required is 1.25 instead of 1.5.

The high gamma required by the dark surround, and the curvature required by flare and latitude considerations, make systems intended for dark-surround projection very unsuitable for duplicating purposes.

Lighting geometry also affects the apparent tone reproduction in pictures.

REFERENCES

Bartleson, C. J., *J. Soc. Mot. Pic. Tel. Eng.*, **84,** 613 (1975).

Bartleson, C. J., Ph.D. Thesis, The City University (1977).

Bartleson, C. J., and Breneman, E. J., *Phot. Sci. Eng.*, **11,** 254 (1967) and *J. Opt. Soc. Amer.*, **57,** 953 (1967).

Breneman, E. J., *Phot. Sci. Eng.*, **6,** 172 (1962)

Breneman, E. J., *J. Opt. Soc. Amer.*, **67,** 657 (1977).

Carnahan, W. H., *Phot. Eng.*, **6,** 237 (1955).

DeMarsh, L. E., *J. Soc., Mot. Pic. Tel. Eng.*, **81,** 784 (1972).

Doody, W. G., Lawton, J. K., and Perry, R. S., *J. Soc. Mot. Pic. Tel. Eng.*, **87,** 373 (1978).

Estes, R. L., *J. Soc. Mot. Pic. Tel. Eng.*, **61,** 257 (1953).

Evans, R. M., and Klute, J., *J. Opt. Soc. Amer.*, **34,** 533 (1944).

Hunt, R. W. G., *J. Opt. Soc. Amer.*, **40,** 362 (1950). See Figs. 3 and 4.

Hunt, R. W. G., *J. Opt. Soc. Amer.*, **42,** 190 (1952).

Hunt, R. W. G., *J. Opt. Soc. Amer.*, **43,** 479 (1953).

Hunt, R. W. G., *J. Phot. Sci.*, **13,** 108 (1965a).

Hunt, R. W. G., *J. Opt. Soc. Amer.*, **55,** 1540 (1965b).

Hunt, R. W. G., in Institute of Printing Conference Proceedings, *Duplication and Conversion of Colour Transparencies*, pp. 5–18 (1968).

Hunt, R. W. G., Pitt, I. T., and Ward, P. C., *J. Phot. Sci.*, **17,** 198 (1969).

Hunt, R. W. G., *Brit. Kinematog. Sound Tel.*, **51,** 268 (1969).

Lighton, C., *Phot. J.*, **107,** 157 (1967).

Pitt, I. T., and Winter, L. M., *J. Opt. Soc. Amer.*, **64,** 1328 (1974).

Rowe, S. C. H., Ph.D. Thesis, The City University (1972).

CHAPTER 7

The Colour Triangle

7.1 Introduction

I N Chapters 2 and 4 the errors of trichromatic colour reproduction by both the additive and the subtractive principles were considered in terms of the assumed sensitivity curves of the three types, ϱ, γ, and β, of retinal cone. This approach to the subject has the advantage of being direct and simple, but it requires some assumptions concerning the eye, and is of a qualitative, rather than a quantitative nature. For a quantitative approach it is necessary first to consider in some detail the phenomenon of trichromatic colour matching. Before we do this, however, a brief consideration of some aspects of colour terminology is advisable (CIE International Lighting Vocabulary, 1970).

7.2 Colour terminology

It is generally agreed that colours have three main perceptual attributes. The most obvious is *hue*, denoting whether the colour appears red, orange, yellow, green, blue, or purple (or some mixture of neighbouring pairs in this list). *Colourfulness* denotes the extent to which the hue is apparent; colourfulness is thus zero for whites, greys, and blacks, is low for pastel colours, and is (normally) high for the colours of the spectrum. *Brightness* denotes the extent to which an area appears to exhibit light; brightness is thus, usually: extremely high for the sun, very high for many other sources of light, high for whites and yellows, medium for greys and browns, and low for blacks. (In the past *luminosity* was sometimes used for this attribute instead of brightness.)

Objects viewed in a high level of illumination generally look brighter than when viewed in a low level, even when the observer is fully adapted to each level; but, in the very important task of recognizing objects, their brightnesses relative to one another are given great attention. Thus, a piece of grey paper seen in sunlight is much brighter than when seen by indoor daylight on a dull day; but it still looks grey, not black, on the dull day because its brightness is judged relative to other objects in the scene. This concept of relative brightness is so important that the term *lightness* is reserved for it, and is defined as the brightness of an area judged relative to the brightness of a similarly illuminated area that appears to be white (or highly transmitting, in the case of transparent objects).

When changes are made in the illumination level to which an observer is adapted, changes occur, not only in the brightnesses of objects, but also in their colourfulnesses. Thus, a scene that looks very colourful in bright sunlight, looks less so in dull cloudy daylight, much less so at twilight, and at moonlight levels looks almost devoid of colourfulness altogether. But, in the important task of recognizing objects, their relative colourfulnesses are given great attention. Thus, a red tomato, seen in sunlight, is much more colourful than when seen indoors on a dull day; but we still perceive it as red, not pink, on the dull day, because its colourfulness is judged relative to other objects in the scene. This attribute of relative colourfulness is called *chroma*; it is defined as the colourfulness of an area judged in proportion to the brightness of a similarly illuminated area that appears to be white (or highly transmitting). Thus, although the tomato on the dull day has a low colourfulness, it is also evident that neighbouring whites also have a low brightness; and this enables the observer to attribute the low colourfulness to the low level of illumination, and not to some change in the object, which is therefore seen as having constant chroma.

It is also possible to judge colourfulness in proportion to the brightness of the object itself (rather than to that of a white), and, when this is done, the relative colourfulness is termed *saturation*. If the level of illumination on an object varies over its surface, because of the three-dimensional shape of the object or because of shadows, it may be difficult to judge what the brightness of a similarly illumin-ated white would be except for a few areas. This means that the perception of lightness and chroma is difficult in many areas, but it is still possible to judge the colour of the object in all its areas in terms of its hue and saturation. Thus, in the case of the tomato, its hue, lightness, and chroma may be judged for the part that is normal to the incident light; but its hue and saturation may be judged for every part of its surface. Hence, when judging the uniformity of the colour of the tomato, hue and saturation are likely to provide a more useful basis than hue, lightness, and chroma.

In the case of light sources it is not possible to make any judgements relative to 'similarly illuminated areas', and therefore lightness and chroma have no relevance; the attributes of light sources are therefore restricted to *hue, colourful-ness, brightness*, and *saturation*.

Before the distinction between colourfulness, chroma, and saturation was recognized as requiring different terms, all three attributes were often referred to as saturation (Hunt, 1977 and 1978). The fourth edition of the International Lighting Vocabulary (CIE, in preparation) recognizes the distinctions between the three attributes, and also allows the use of *chromaticness* as an alternative to colourfulness. The adjectives *bright* and *dim* are used in connection with brightness; *light* and *dark* with lightness; and *strong* and *weak* with chroma.

In colour, both the response of the observer (the subject) and the physical nature of the stimulus (the object) are important, and it is necessary to distinguish clearly between these *subjective* and *objective* aspects of colour. The terms hue, colourfulness, brightness, lightness, chroma, and saturation, as described above, are clearly all subjective terms. Objective terms denote quantities obtained with measuring instruments and, unlike subjective attributes, these quantities are unaffected by changes in the adaptation of the observer. It is desirable to measure quantities that correlate with the subjective attributes defined above, and the relevant objective terms are as follows.

Subjective Terms	Objective Terms	Objective Symbol
Hue	Dominant wavelength	λ_d
	CIE hue-angle	h_{uv} or h_{ab}
Brightness	Luminance	L
Colourfulness (chromaticness)	—	—
Lightness	Luminance factor	β
	CIE lightness	L^*
Chroma	CIE chroma	C^*_{uv} or C^*_{ab}
Saturation	Purity	p
	CIE saturation	s_{uv}
Hue and saturation	Chromaticity	x,y or u',v'

For some of the subjective attributes, two objective terms have been listed; in all these cases the second term (which has the prefix CIE) denotes a measure that is correlated with the subjective attribute more uniformly than is the case for the first term; these more uniform correlates will be considered in detail in Chapter 8. There is no objective term listed to correlate with colourfulness, because no such measure has yet been devised.

Dominant wavelength is defined as the wavelength of the monochromatic stimulus that, when additively mixed in suitable proportions with a specified achromatic stimulus (a white or grey), matches the colour stimulus considered. In the case of purple stimuli, the complementary wavelength, λ_c, has to be used, and this is the wavelength of the monochromatic stimulus that, when additively mixed in suitable proportions with the colour stimulus considered, matches the specified achromatic stimulus. *Luminance* is the luminous intensity in a given direction per unit projected area (the units used are given in Appendix 3). *Luminance factor* is the ratio of the luminance of a colour to that of a perfectly reflecting or transmitting diffuser identically illuminated (by perfect is meant that the diffuser is uniform, isotropic, and does not absorb any light). The more specific terms *reflectance factor* and *transmittance factor* are used for reflecting and transmitting samples, respectively, but in these cases the ratio is that of the light reflected or transmitted by the sample within a defined cone to that by the perfect diffuser within the same cone; if the cone is a hemisphere, the ratio is denoted as the *reflectance* or the *transmittance*; if the cone is very small, these measures are the same as the luminance factor. *Absorptance* is equal to unity minus the transmittance or reflectance. *Opacity* is equal to the reciprocal of the transmittance or reflectance. *Purity* is a measure of the proportions of the amounts of the monochromatic stimulus (or, for purples, of a red and violet spectral mixture) and of the specified achromatic stimulus that, when additively mixed, match the colour stimulus considered. *Chromaticity* is denoted by the proportions of the amounts of three colour-matching stimuli needed to match a colour.

There are occasions when it is extremely important to distinguish between the objective and subjective attributes of colours; but in many contexts only general matters are in view and strict observance of the above terminology would be pedantic. In these cases the subjective terms are used in this book, because they are more widely understood.

In the the Munsell system of colour specification, the variables, although *objective* (they are not affected by the observer's adaptation), have been scaled to be as uniform as possible *subjectively*. Thus, *Munsell Hue* represents perceptually equal hue differences by nearly equal increments; *Munsell Value* does the same for lightness differences, and *Munsell Chroma* for chroma differences. The sizes of the units are such that 1 unit of Munsell Value is roughly equivalent to 2 units of Munsell Chroma, and to 3 units of Munsell Hue for samples of Munsell Chroma 5 (Newhall, 1940). Colour terminology is considered further in Chapter 8.

7.3 Trichromatic matching

If, as in Fig. 7.1 arrangements are made whereby colours can be compared in appearance with an additive mixture of red, green, and blue light, it is found that by varying the relative and absolute amounts of the red, green, and blue lights it is always possible to make the appearance of the mixture identical to that of any chosen colour. A few colours of very high purity appear to be exceptions to this rule, but, as we shall shortly see, they can also be matched by using a special technique.

This phenomenon of trichromatic matching is easily explained in terms of a trichromatic theory of colour vision. For, if all colours are analysed by the retina into only three different types of response, ϱ, γ, and β (proportional, presumably to absorptions in three different photo-sensitive pigments), the eye will be able to detect no difference between two stimuli that give rise to the same ϱ-, the same γ-, and the same β-signal, no matter how different the two stimuli may be in spectral composition. In fact the spectral difference between two matching stimuli can be quite startling as shown in Fig. 7.2, where a stimulus consisting of power throughout the whole of the spectrum is matched by light from three narrow bands of the spectrum only. But both these power distributions give rise to identical ϱ-, γ-, and β-responses, so that the two stimuli are indistinguishable to the eye. Such stimuli, which are spectrally different but visually identical, are known as *metameric pairs*, or *metamers*.

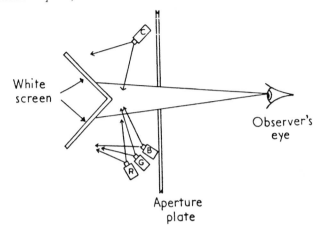

Fig. 7.1. The principle of trichromatic matching. The upper white screen is illuminated only by light of the test colour C. The lower white screen is illuminated only by a mixture of red, green, and blue light from the three projectors R, G, and B. By adjusting the amounts of red, green, and blue light in the mixture, it can be made identical in appearance to the colour C.

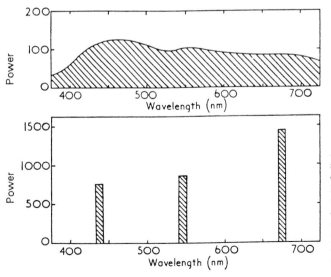

Fig. 7.2. The upper curve shows the spectral power distribution of a white light; the power distribution shown in the lower diagram refers to a light which, to the eye, appears exactly the same colour. (Note the different power scales used.)

73

With some colours of very high purity, and most colours of the spectrum, it is found that, although the red, green, and blue mixture can reproduce the same hue and brightness, the saturation of the mixture is never quite sufficient to match these colours. The case is particularly marked for the blue-green colours of the spectrum. A mixture consisting of blue and green only can be made to reproduce the correct hue and brightness, but the mixture is always paler than the test colour.

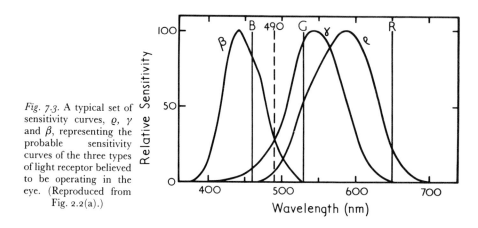

Fig. 7.3. A typical set of sensitivity curves, ϱ, γ and β, representing the probable sensitivity curves of the three types of light receptor believed to be operating in the eye. (Reproduced from Fig. 2.2(a).)

At first sight, it might be thought that this fact invalidates trichromatic theories of colour vision, but it is in fact predicted by them. In Fig. 7.3 the assumed sensitivity curves of the three retinal receptors ϱ, γ, and β, are reproduced from Fig. 2.2(a). Let us consider the case where our red, green, and blue beams of light, R, G, and B, are the most saturated available, that is monochromatic light, of wavelengths 650 nm for the red, R, 530 nm for the green, G, and 460 nm for the blue, B, and suppose that we are trying to match monochromatic light of wavelength 490 nm (a blue-green). It is clear that the red stimulus will give only a ϱ-response, and no γ- or β-response. The green stimulus will give chiefly a γ-response, but also a considerable ϱ-response, although very little β-response. The blue stimulus will give mainly a β-response, together with a small γ-response, but very little ϱ-response. An arbitrary scale has been included in Fig. 7.3 so that the responses given by unit power of each of the three stimuli, R, G, and B, can be tabulated from Fig 7.3 thus:

For unit power of R $\quad \varrho=24 \quad \gamma=0 \quad \beta=0$
,, ,, ,, ,, G $\quad \varrho=56 \quad \gamma=92 \quad \beta=0$
,, ,, ,, ,, B $\quad \varrho=0 \quad \gamma=9 \quad \beta=78$

Now for unit power of light of wavelength 490 nm, from Fig. 7.3:
$$\varrho=8 \qquad \gamma=26 \qquad \beta=26$$

74

and therefore in order to match it with a mixture of R, G, and B it is necessary to choose amounts of them that give rise to the same response magnitudes.

Since only the B stimulus produces any β-response it is clear that we must have a third of a unit of B in order to produce the required β-response. Thus we have:

$$\text{For } \tfrac{1}{3} \text{ unit of B} \qquad \varrho=0 \qquad \gamma=3 \qquad \beta=26$$

We need more γ-response, and since R produces none, we must produce it by means of G. We therefore need a γ-response of 23 units which will be given by $\tfrac{1}{4}$ of a unit of G. Thus we have:

$$\text{For } \tfrac{1}{4} \text{ unit of G} \qquad \varrho=14 \qquad \gamma=23 \qquad \beta=0$$

Hence for $\tfrac{1}{3}$ unit of B together with $\tfrac{1}{4}$ unit of G:

$$\varrho=14 \qquad \gamma=26 \qquad \beta=26$$

It is clear that, although we have the correct amounts of γ and β, we have almost twice as much ϱ-response as we should have, and we have not yet added any R. To do so would merely increase the ϱ-response still further, without altering the γ- and β-responses, thus making the mixture even less like the test colour of wavelength 490 nm. It is therefore clear that no mixture of lights of wavelengths 650, 530, and 460 nm can ever be made to match light of wavelength 490 nm, and this is borne out by experiment. The fact, however, is predicted by the curves of Fig. 7.3, which are the quantitative expressions of trichromatic theories of colour vision, and the theories are in no way invalidated. It is because the ϱ- and β-curves of Fig. 7.3 overlap, thus making it impossible to stimulate the γ-cones on their own, that saturated blue-green colours cannot be matched by additive mixtures of red, green, and blue light.

If monochromatic lights of different wavelengths are chosen for the three matching stimuli, or if matching stimuli each comprised of broad spectral bands of light are used, there will always be some saturated colours that cannot be matched; occasionally light of wavelength 490 nm may be matched, as when for instance one of the three matching stimuli consists of light of this wavelength, but in this case there will be other spectral colours that cannot be matched.

To revert, however, to our stimuli R, G, and B of wavelengths 650, 530, and 460 nm, it is clear that if, instead of adding some of the stimulus R to the mixture of $\tfrac{1}{3}$ unit of B and $\tfrac{1}{4}$ unit of G, we add it in the right amount to the test colour of wavelength 490 nm, we can then obtain a colour match. Thus we already have 14 ϱ-response in the mixture, but only 8 ϱ-response in the test colour. So the addition of 6 ϱ-response to the test colour will make it match the mixture of G and B. This amount of ϱ-response is given by $\tfrac{1}{4}$ unit of R thus:

$$\text{For } \tfrac{1}{4} \text{ unit of R} \qquad \varrho=6 \qquad \gamma=0 \qquad \beta=0$$

We, therefore, now have the situation that:

Unit power of 490 nm $+\tfrac{1}{4}$ unit of R is matched by $\tfrac{1}{4}$ unit of G $+\tfrac{1}{3}$ unit of B.

It is customary to regard this addition of one of the three matching stimuli to the

75

test colour, instead of to the mixture, as a negative quantity of R and we therefore write:

Unit power of 490 nm is matched by $\frac{1}{4}$ unit $G + \frac{1}{3}$ unit of $B - \frac{1}{4}$ unit of R

Using this concept of negative amounts, it is possible to match *all* colours by choosing suitable proportions of three matching stimuli that are additively mixed. The negative amounts arise because of the *unwanted stimulations* produced by the stimuli R, G, and B (see Section 2.5).

7.4 Colour-matching functions

Since, as we have just seen, it is possible to match all colours by means of additive mixtures of three matching stimuli, it is possible to match all the colours of the spectrum. When this has been done the results are often presented as three curves, as shown in Fig. 7.4, in which the amount of R, the amount of G, and the amount of B needed to match unit power of each wavelength of the spectrum, are plotted against wavelength. As would be expected, the maximum of each curve is in a region of the spectrum where the colour is similar to the matching stimulus in question, and it will also be noted that all three curves have negative portions, the largest being that of the R stimulus in the blue-green parts of the spectrum. The units used for the amounts of R, G, and B are not always power units, since it is generally more convenient to use arbitrary units such that some specified white

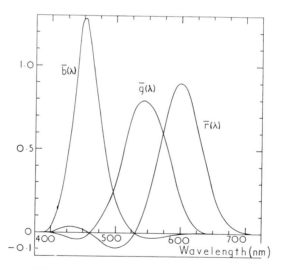

Fig 7.4. Colour-matching functions showing the amounts $\bar{r}(\lambda), \bar{g}(\lambda),$ and $\bar{b}(\lambda)$ of red, green, and blue light required to match unit power of each wavelength of the spectrum, using matching stimuli of wavelengths 650, 530, and 460 nm respectively. The amounts of the stimuli are measured in arbitrary units chosen so that equal amounts of the three stimuli are needed to match the equi-energy white source, S_E.

stimulus is matched by equal amounts of the three matching stimuli. In Fig. 7.4 the white stimulus used for defining the units is one in which the amount of power per unit wavelength is constant throughout the spectrum; this hypothetical white source is of some importance in colorimetry and is known as the *equi-energy source*, with the abbreviation S_E. If a different white had been used for defining the units, or if the units had been defined photometrically (as for instance, in candelas per square metre or lux) or if units of power had been used, the curves would not have differed in shape but only in height, all the ordinates of any one curve being multiplied by the same factor, but the three factors being different for the three curves.

The ordinates of these *colour-mixture curves*, or *colour-matching functions*, are generally denoted by the symbols $\bar{r}(\lambda), \bar{g}(\lambda), \bar{b}(\lambda)$, so that the interpretation of the curves may be written thus:

Unit power of light of wavelength λ is matched by
$\bar{r}(\lambda)$ units of R $+\bar{g}(\lambda)$ units of G $+\bar{b}(\lambda)$ units of B.

It is convenient to abbreviate this to:

$$1.0(\lambda) \equiv \bar{r}(\lambda)\ (R) + \bar{g}(\lambda)\ (G) + \bar{b}(\lambda)\ (B)$$

where the equivalent sign ($\equiv$) is used to mean that the equation represents an equivalence of colours to the eye, and the symbols in brackets do not represent *quantities*, but merely indicate to which stimuli the coefficients (1.0, $\bar{r}(\lambda)$, etc.) refer.

Experiment shows that these equations usually obey the ordinary rules of algebra so that, for instance, if k is a constant:

$$k(\lambda) \equiv k\bar{r}(\lambda)\ (R) + k\bar{g}(\lambda)\ (G) + k\bar{b}(\lambda)\ (B)$$

Moreover if we have k_1 power units of wavelength λ_1 and k_2 power units of wavelength λ_2, represented by the equations

$$k_1(\lambda_1) \equiv k_1\bar{r}_1(R) + k_1\bar{g}_1(G) + k_1\bar{b}_1(B)$$
$$k_2(\lambda_2) \equiv k_2\bar{r}_2(R) + k_2\bar{g}_2(G) + k_2\bar{b}_2(B)$$

then experiment shows that:

$$k_1(\lambda_1) + k_2(\lambda_2) \equiv (k_1\bar{r}_1 + k_2\bar{r}_2)\ (R) + (k_1\bar{g}_1 + k_2\bar{g}_2)\ (G) + (k_1\bar{b}_1 + k_2\bar{b}_2)\ (B)$$

This additive property of equations can be extended to any number of wavelengths so that generally we can write:

$$k_1(\lambda_1) + k_2(\lambda_2) + \ldots \equiv (k_1\bar{r}_1 + k_2\bar{r}_2 + \ldots)(R) + (k_1\bar{g}_1 + k_2\bar{g}_2 + \ldots)\ (G)$$
$$+ (k_1\bar{b}_1 + k_2\bar{b}_2 + \ldots)\ (B)$$

But all colours, whether saturated or pale, light or dark, and of whatever hue, consist of mixtures of spectral colours in varying amounts. Therefore, if we know the spectral power distribution of any stimulus, with the aid of the curves of Fig. 7.4 (or tabulated values of $\bar{r}(\lambda)$, $\bar{g}(\lambda)$, and $\bar{b}(\lambda)$) it is possible to calculate the amounts of R, G, and B necessary to match it, and the results of such calculations agree with the values obtained when the stimulus is matched experimentally.

Suppose then that we have some stimulus, C, represented by some spectral power distribution curve $E(\lambda)$. By means of the curves of Fig. 7.4 we can calculate the amounts of R, G, and B needed to match it, and obtain the equation:

$$(C) \equiv R_C(R) + G_C(G) + B_C(B)$$

where

$$R_C = E_1 \bar{r}_1 + E_2 \bar{r}_2 + \ldots \ldots + E_n \bar{r}_n$$

there being similar expressions for G_C and B_C, and the numerals 1 to n indicating a series of wavelengths equally spaced at a convenient interval throughout the entire visible spectrum. These amounts R_C, G_C and B_C are known as *tristimulus values*.

Suppose now that in an additive colour photographic system the colours of the red, green, and blue lights forming the final picture are the same R, G, and B as we have been considering, namely monochromatic lights of wavelength 650, 530, and 460 nm respectively. And suppose that the three filter-emulsion combinations on which the image of the original scene is focused, have spectral sensitivity curves exactly the same as those of Fig. 7.4 (we shall consider the practical realization of the negative parts of these curves presently). Considering the red negative only, the exposure N_R will be proportional to

$$E_1 \bar{r}_1 + E_2 \bar{r}_2 + E_3 \bar{r}_3 + \ldots \ldots + E_n \bar{r}_n$$

Assuming that the photographic system is linear, the amount of latent image, I_R, formed in the negative, will be proportional to N_R, and the transmission T_R on the positive will be proportional to I_R and therefore also to N_R. But the expression for N_R is exactly the same as that for R_C, the amount of the stimulus R needed to match the colour C. Therefore the projection of the positive with transmission T_R will automatically result in kR_C, of stimulus R being projected on the screen, and k will be constant over the picture area, so that over the entire picture area the amount of R will be proportional to the amount needed to match the colours of the original scene. Similarly the transmission of the green positive T_G will be proportional to G_C, the amount of stimulus G needed to match the colours of the original, and the transmission of the blue positive T_B will be proportional to B_C, the amount of stimulus B needed to match the colours of the original. Hence any colour in the original is represented by:

$$(C) \equiv R_C(R) + G_C(G) + B_C(B)$$

and in the reproduction by:

$$(C') \equiv kR_C(R) + kG_C(G) + kB_C(B)$$

It is thus only necessary to make $k = 1$, that is, to adjust the overall intensity of the picture so that the luminance of the reproduction is the same as that of the original, to obtain exact[1] colour reproduction of all the colours in the original scene (Hardy and Wurzburg, 1937; Harrison and Horner, 1937).

---

[1] By *exact* it is simply meant here that the reproduction colours all have the same tristimulus values and absolute luminances as the original colours. For a discussion as to whether this is the most *desirable* state of affairs reference should be made to Chapter 11.

The conditions that must be satisfied in order to obtain this result are:

(1) The spectral sensitivity curves of the filter-emulsion combinations, used to record the three negatives, must be identical with the colour-matching functions of the three stimuli used to form the final picture;

(2) The transmissions on the positives must be proportional to the exposures at the corresponding points on the negatives;

(3) The overall luminance must be adjusted so that the picture has the same luminance as the original.

Conditions 2 and 3 may not be too difficult to realize, but the negative portions of the colour-matching functions make condition 1 almost impossible to achieve in a photographic system. What is required is a filter-emulsion combination wherein exposure by light in some parts of the spectrum bleaches the latent image formed by light in other parts of the spectrum. Thus in the case of the 'red' negative, for instance, exposure to light from the blue-green part of the spectrum (wavelengths 460 to 530 nm) must *reduce* the amount of latent image formed by light in the red (and blue) parts of the spectrum. By using two *toe-recording* photographic negatives, one for the positive parts of the curve, and the other for the negative part, and binding up a photographic positive made from the latter in exact registration with the former, the result can, at least in theory, be achieved (see MacAdam 1938, page 405). In practice, however, the method is cumbersome and only approximate, and, as far as photography is concerned, no convenient method of introducing the negative parts of the sensitivity curves has yet been found. In colour television, however, the problem is a good deal simpler, at least, in principle. Thus if one television camera-tube had a spectral sensitivity identical to the positive parts of the red colour-matching functions, and another tube a spectral sensitivity corresponding to its negative parts, as shown in Fig. 7.5, then by subtracting the second from the first a red signal would be derived based on a composite spectral sensitivity curve equivalent to the complete matching function. Similar arrangements could be made for the green and blue signals, using a total of six different tubes in all. This arrangement is very cumbersome, however, and it is more convenient in practice to use only three all-positive, curves and to obtain the correct signals by means of a technique known as *matrixing*. If to the red matching function a small fraction of the green matching function is added, a composite all-positive curve can be obtained. If the signal obtained from a tube having such a sensitivity, then has subtracted from it the same small fraction of the green signal, the final signal will be based on the true red matching function. When this technique is applied to obtain signals based on all three colour-matching functions, the correct fractions to be subtracted involve three simultaneous equations, and it is these equations that matrixing is usually intended to imitate. The result can be a television system having exactly the required sensitivity curves of Fig. 7.4. (This is discussed more fully in Section 19.12.)

If then, we have such a colour television system, employing these sensitivity

79

curves, and red, green, and blue lights on the viewing tube that are identical with R, G, and B (that is monochromatic lights of wavelengths 650, 530, and 460 nm respectively), we have in effect fulfilled condition 1. Assuming that conditions 2 and 3 are also effected, we should then have exact colour reproduction. But let us consider the case of a blue-green light, of wavelength 490 nm, for example, in the original scene. Fig. 7.4 tells us that it will require a negative amount of red in its match so that we can write:

$$k(\lambda_{\mathrm{BG}}) \equiv -R_{\mathrm{BG}}(\mathrm{R}) + G_{\mathrm{BG}}(\mathrm{G}) + B_{\mathrm{BG}}(\mathrm{B})$$

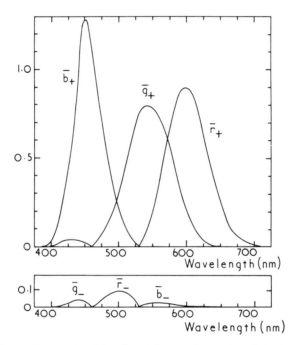

Fig. 7.5. The positive portions (above), and the negative portions (below), of the colour-matching functions of Fig. 7.4.

Now, all the colours on our television tube are formed by mixtures of the stimuli R, G, and B, but these mixtures can now only be all positive mixtures. For the meaning to be attached to the symbol $-R_{\mathrm{BG}}$ is that this amount of stimulus R must be added to the blue-green light; but of course the blue-green light is part of the original scene and, even if we could add red light to it, once it had been added, we would no longer have the same colour. It is clear, therefore, that, although the algebra tells us that we have exact colour reproduction, in fact whenever the electronic signals call for one or more of the three stimuli R, G, or B to be present in a negative amount, the television tube is unable to oblige. This defect only

Plate 5
Good reproduction of Caucasian skin colours requires careful control of hue: although real Caucasian skin varies towards yellowish hues with sun-tan and towards more magenta hues with flushed areas, the subtle changes of hue in these directions from one area to another are important, hence the tolerances may not be large; in the green and purple directions the tolerances are usually quite small (see Section 5.6). Reproduced from a 36 × 24 mm *Kodachrome* transparency on a Crosfield *Magnascan* scanner.

Plates 6 & 7
The relative luminances at which particular objects must be reproduced are often not very
critical, because their level of illumination generally varies considerably with angle, amount of
shadow, or distance from a source of light. Thus, in Plate 6 (*on the left*), the folds on the purple
dress exhibit considerable variation in relative luminance, as is also the case with various parts
of the engine (see Section 6.6). However, the general way in which relative luminances are
reproduced (tone reproduction) is very important in pictures, otherwise either a harsh crude
effect, or a misty dull effect, can result. Relative luminances in pictures can best be measured
relative to that of a reference white (see Section 6.4), but its choice may not be easy because
whites can cover quite a range of luminances, as can be seen from Plate 7 (*above*). Plate 6
reproduced from a 5 × 4 inch *Ektachrome* transparency, and Plate 7 reproduced from a 2¼ inch
square *Ektachrome* transparency, on a Crosfield *Magnascan* scanner.

Plate 8

The perception of roundness and depth in two-dimensional pictures depends largely on the geometry of the lighting (see Sections 5.9 and 6.12). In these three pictures, although the same person, camera, and film were used, the different lighting has produced very different results. Proper control of lighting is thus extremely important. In film and television studios, elaborate systems of lights are often used. Reproduced from $2\frac{1}{4}$ inch square *Ektachrome* transparencies on a Crosfield *Magnascan* scanner.

occurs when the final signals called for are negative. A colour in the original scene can contain light from the blue-green part of the spectrum, for instance, and still be reproduced exactly provided that it also contains light from some other part of the spectrum that makes the resultant quantity of R either zero or positive. Thus a colour consisting of a mixture of light from the blue-green and yellow-green parts of the spectrum would give rise to the following situation;

$$k_1(\lambda_{BG}) \equiv -R_{BG}(R) + G_{BG}(G) + B_{BG}(B)$$
$$k_2(\lambda_{YG}) \equiv R_{YG}(R) + G_{YG}(G) - B_{YG}(B)$$
$$k_1(\lambda_{BG}) + k_2(\lambda_{YG}) \equiv (R_{YG} - R_{BG})(R) + (G_{BG} + G_{YG})(G) + (B_{BG} - B_{YG})(B)$$

Then provided that $R_{YG} - R_{BG}$ and $B_{BG} - B_{YG}$ are both positive, this mixture will be reproduced perfectly correctly on our television screen, despite the fact that neither part of the mixture can be reproduced correctly on its own.

Hence, we can say that, with this television system, all colours will be reproduced exactly, except for those very saturated colours that cannot be matched by an all-positive mixture of the red, green, and blue lights used in the television tube.

7.5 The colour triangle

The consideration of many of these questions is greatly facilitated by the use of the *colour triangle*, which may be thought of as a kind of colour map, in which all colours are represented in a systematic way by points in a triangle.

Given three defined matching stimuli R, G, and B, which could, for instance, be our three monochromatic lights of wavelengths 650, 530, and 460 nm, the amounts of these three stimuli needed to match any colour enable it to be related systematically to all other colours. Thus the equation:

$$k(C) \equiv R_C(R) + G_C(G) + B_C(B)$$

represents k units of the colour C. Now the *amount*, k, of the colour C can be regarded as a physical or photometric quantity, measured, for instance, in power units (such as ergs per square centimetre and per second) or in photometric units (such as candelas per square metre), or, for transmitting and reflecting objects, as the luminance factor. The *colour*, red or yellow, vivid or pale etc., is governed largely by the ratio of the three quantities R_C, G_C, and B_C to one another. It is therefore customary to divide the equation by the sum of the three quantities to give:

$$\frac{k}{R_C + G_C + B_C}(C) \equiv r(R) + g(G) + b(B)$$

where

$$r = R_C/(R_C + G_C + B_C)$$
$$g = G_C/(R_C + G_C + B_C)$$
$$b = B_C/(R_C + G_C + B_C)$$

Since the *amount* of C can be specified separately, we may write the equation without specifying it, using the proportional sign thus:

$$(C) \propto r(R)+g(G)+b(B).$$

r, g, and b are known as *chromaticity co-ordinates*. It is clear that the sum $r+g+b$ is always equal to unity, so that if r and g are known, b can always be deduced from:

$$b=1-r-g$$

We can therefore plot r and g and obtain a *chromaticity diagram* on which all colours are represented. In Fig. 7.6 this has been done with g as ordinate and r as abscissa, the curved line representing the locus of the spectral colours, and the point W, the particular white that was used to define the units of the three stimuli. The matching stimuli themselves and the white, W, have the following values of r and g:

(R)	$r=1$	$g=0$
(G)	$r=0$	$g=1$
(B)	$r=0$	$g=0$
(W)	$r=0.333$	$g=0.333$

and hence occupy the corners and the centre of the triangle as shown.

Since r, g, and b will always have the same signs as R_C, G_C, and B_C, it is inevitable that the value of r is negative in the blue-green part of the spectrum, as shown in the figure.

7.6 The centre of gravity law

Suppose we have two colours, C_1 and C_2, whose positions on the colour triangle are known. It is important to know where the point C_3, representing a mixture of given quantities of C_1 and C_2, will be situated. If C_1 and C_2 are represented by:

$$(C_1) \propto r_1(R)+g_1(G)+b_1(B) \quad \text{where } r_1+g_1+b_1=1$$
$$(C_2) \propto r_2(R)+g_2(G)+b_2(B) \quad \text{where } r_2+g_2+b_2=1$$

and the quantities in the mixture are m_1 units of C_1 and m_2 units of C_2, we have to proceed as follows: m_1 and m_2 are usually given in photometric units (usually units of luminance), so that we have to know the photometric values (for example in candelas per square metre) of the units in which the amounts of R, G, and B are being measured. These values will not necessarily be the same for the three stimuli; they will generally be different, for instance, when the units are defined by stipulating that equal amounts of R, G, and B match a particular white. Let the three photometric values or luminances be denoted by L_R, L_G, and L_B. Then with the amount C_1 measured in photometric units, we may write:

$$L_1(C_1) \equiv r_1(R)+g_1(G)+b_1(B)$$

where $L_1=L_R r_1+L_G g_1+L_B b_1$. Hence 1 photometric unit of C_1 is represented by the equation:

$$1.0(C_1) \equiv \frac{r_1}{L_1}(R)+\frac{g_1}{L_1}(G)+\frac{b_1}{L_1}(B)$$

and therefore m_1 photmetric units by:

$$m_1(C_1) \equiv \frac{m_1}{L_1} r_1(R) + \frac{m_1}{L_1} g_1(G) + \frac{m_1}{L_1} b_1(B)$$

Similarly m_2 photometric units of C_2 are represented by:

$$m_2(C_2) \equiv \frac{m_2}{L_2} r_2(R) + \frac{m_2}{L_2} g_2(G) + \frac{m_2}{L_2} b_2(B)$$

where $L_2 = L_R r_2 + L_G g_2 + L_B b_2$. Therefore the mixture is represented by:

$$m_1(C_1) + m_2(C_2) \equiv \left(\frac{m_1}{L_1} r_1 + \frac{m_2}{L_2} r_2\right)(R) + \left(\frac{m_1}{L_1} g_1 + \frac{m_2}{L_2} g_2\right)(G) + \left(\frac{m_1}{L_1} b_1 + \frac{m_2}{L_2} b_2\right)(B)$$

The new values of r and g are obtained by dividing this equation by the sum of the coefficients (R), (G), and (B), and this sum reduces to $m_1/L_1 + m_2/L_2$. We therefore obtain:

$$(C_3) \propto r_3(R) + g_3(G) + b_3(B)$$

where

$$r_3 = \left(\frac{m_1}{L_1} r_1 + \frac{m_2}{L_2} r_2\right) \bigg/ \left(\frac{m_1}{L_1} + \frac{m_2}{L_2}\right)$$

and similar expressions for g_3 and b_3. The geometrical interpretation of this formula is very simple indeed: C_3 always lies on the line joining C_1 and C_2 and divides it in the inverse ratio:

$$\frac{m_2}{L_2} \bigg/ \frac{m_1}{L_1}$$

as shown in Fig. 7.6. C_3 is in fact at the centre of gravity of weights m_1/L_1 placed at C_1 and m_2/L_2 placed at C_2, hence $C_1 C_3/C_2 C_3 = (m_2/L_2)/(m_1/L_1)$, and this rule of colour mixture is often referred to as the *centre of gravity law*.

It has a number of important consequences in the colour triangle. First, since the spectral locus maintains a convex curvature throughout, all mixtures of spectral colours, and therefore *all* colours, must lie either within or upon the spectral locus, but never outside it; thus, by joining the two ends of the spectral locus with a straight line, the area containing *all* colours is enclosed. Points lying outside this area have one of their three values more negative than is ever required in colour matching.

Secondly, when white light is added in gradually increasing amounts to any colour of the spectrum the position of the point representing the resultant mixture gradually moves in from the spectral locus, along a straight line, towards the white point. Thus, on the colour triangle, the straight line joining the white point to any spectral colour represents colours of constant dominant wavelength[1] but of varying purity, the purest colours lying near or upon the spectral locus, and the least pure colours lying near the white point, with intermediate colours in between. The hues are distributed around the spectral locus in accordance with

[1] MacAdam (1950, 1951) and others have shown that these lines represent colours that are only approximately constant in *hue*.

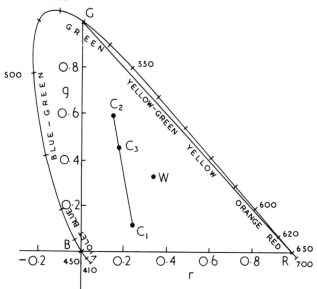

Fig. 7.6. The colour triangle for matching stimuli of wavelengths 650, 530, and 460 nm, showing the locus of spectral colours and the white point W. The units are such that equal quantities of the three stimuli are needed to match the equi-energy white.

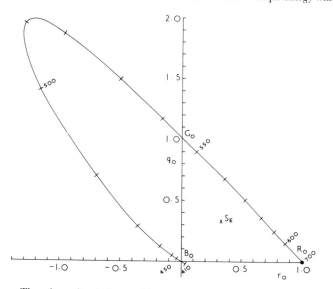

Fig. 7.7. The colour triangle for matching stimuli of wavelengths 700, 546.1, and 435.8 nm, the units being such that equal quantities of the three stimuli are needed to match the equi-energy white, S_E.

Fig. 1.1(a), and have been marked in Fig. 7.6. The purest magentas and purples lie along the line joining the ends of the spectral locus.

The colour triangle now makes clear the limitations of our colour television system. All colours represented by points lying within the triangle R, G, B will be reproduced exactly. But points lying outside, requiring, as they do, one of the amounts to be negative, cannot be reproduced exactly; the reproduction colour will move in to the edge of the triangle. The most severe limitation is in the case of blue-green colours of high purity.

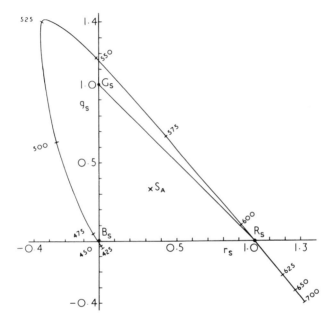

Fig. 7.8. The colour triangle for matching stimuli obtained by isolating three spectral bands from a tungsten filament lamp at a colour temperature of 2856 K (S_A). The spectral bands were: 700 to 580 nm (R_S), 580 to 490 nm (G_S), and 490 to 400 nm (B_S). The units are such that equal quantities of the matching stimuli are needed to match the colour of the light from the lamp (S_A).

7.7 Other colour triangles

If, instead of using matching stimuli of wavelengths 650, 530, and 460 nm, we had used three other lights, R_O, G_O, and B_O say, we would have obtained a colour triangle similar to that of Fig. 7.6, but with R_O, G_O, and B_O at the corners of the triangle and the other colours somewhat shifted in position. Thus, Fig. 7.7 shows the triangle for matching stimuli R_O, G_O, and B_O of wavelengths 700, 546.1, and 435.8 nm, and Fig. 7.8 that for matching stimuli R_S, G_S, and B_S consisting of bands of wavelengths from 700 to 580 nm, from 580 to 490 nm, and from 490 to 400 nm isolated from a tungsten filament lamp operating at a colour temperature (this term is defined in Section 10.2) of 2856 K.

It is obviously desirable that a triangle should be chosen for which, if possible, equal distances in any part of the triangle represent equal colour differences. Actually no triangle can be found that is perfect in this respect, but that shown in Fig. 7.9 is at least approximately uniform in colour differences. It will be seen to be different from those of Figs. 7.6, 7.7 and 7.8 in that the apexes U', V', and W' lie outside the spectral locus. A full description and explanation of this triangle will be deferred to the next chapter; for the moment it is sufficient to

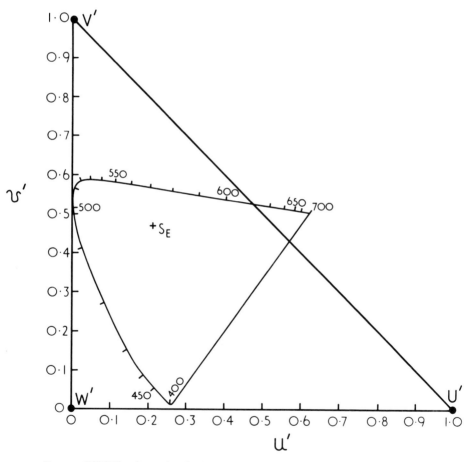

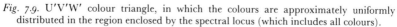

Fig. 7.9. U'V'W' colour triangle, in which the colours are approximately uniformly distributed in the region enclosed by the spectral locus (which includes all colours).

86

remark that it has the same properties as the other triangles. The point representing the mixture of any two colours lies on the line joining the points representing the constituent parts of the mixture, and divides that line in the inverse ratio $(m_2/L_2)/(m_1/L_1)$ as before.

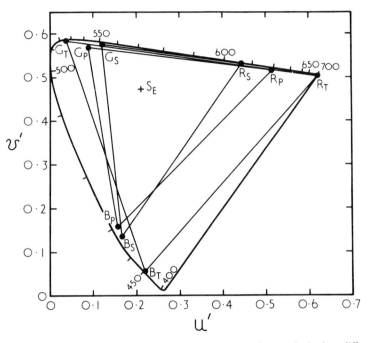

Fig. 7.10. Triangles showing the gamut of colours that can be matched when different matching stimuli are used. $R_TG_TB_T$: theoretical additive stimuli, monochromatic wavelengths 700, 525, and 450 nm. $R_PG_PB_P$: practicable additive stimuli, filters of Fig. 7.11(a) illuminated by light of Fig. 7.11(b). $R_SG_SB_S$: the stimuli of Fig. 7.8, being 'theoretical subtractive' stimuli.

7.8 Additive colour reproduction

With the aid of the colour triangle the limitations of additive colour reproduction can now be seen more rigorously. We have already seen that, in a system in which the sensitivity curves have the required negative parts, the only limitation is that colours lying outside the triangle formed by the reproduction stimuli will move to the edge of this triangle. It is therefore obviously desirable that this triangle should cover as much as possible of the domain of all colours. In Fig. 7.10 the triangle R_T, G_T, B_T, shows about the best that can be done. If G_T were moved to slightly shorter wavelengths the saturated blue-greens would improve at the

87

expense of the saturated yellows, but the latter are far more numerous in nature than the former, so that the position of G_T is probably near the optimum. These stimuli, being on the spectral locus, consist of monochromatic lights, and their wavelengths are approximately 700, 525, and 450 nm.

It is always difficult to obtain very bright beams of purely monochromatic light, and therefore most additive colour photographic systems used reproduction stimuli consisting of red, green, and blue filters, made of glass or dyed gelatin, with transmission curves similar to those shown in Fig. 7.11(a), illuminated by a tungsten light source having a spectral power distribution curve similar to that shown in Fig. 7.11(b). The positions of such stimuli on the triangle are shown by the points R_P, G_P, and B_P, and it is seen that the limitation in the blue-green and magenta directions is now much more marked. In television systems the red is generally worse than R_P, the green slightly worse than G_P, and the blue similar to B_P, (see Fig. 7.17). (The stimuli R_S, G_S, B_S are used in Section 9.2.)

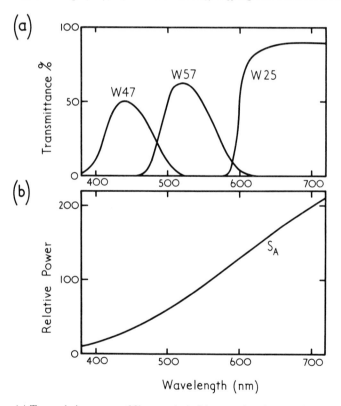

Fig. 7.11. (a) Transmission curves of filters typical of those used as the reproduction stimuli in additive colour photography. (b) Spectral power distribution curve of a tungsten filament lamp of colour temperature 2856 K, often used to illuminate these filters.

As has already been mentioned, in colour photography the negative parts of the sensitivity curves are very difficult to realize, so that in practice they have to be ignored. Each set of reproduction stimuli will have slightly different colour-matching functions; and it is clear that the further the spectral locus lies outside the reproduction triangle, the greater will be the negative portions of these functions, and thus also in the case of the sensitivity curves (which must be identical to them), and hence the greater the errors introduced in ignoring them. Thus in colour photography the choice of reproduction stimuli is doubly important. Not only does a small reproduction triangle limit the gamut of reproducible colours, but it also magnifies the consequences of ignoring the negative parts of the corresponding colour-matching functions. And the absence of the negative sensitivities in colour photography could result in practically all colours being incorrectly reproduced, since one of the three matching stimuli is negative for almost every wavelength of the spectrum. The only exception is in the red, orange, and yellow parts of the spectrum, and these are often the colours reproduced best.

The importance of different types of departure of the sensitivity curves from the theoretical colour-matching functions has been the subject of several studies (Evans, Hanson, and Brewer, 1953, chapter 13; MacAdam, 1953; Neugebauer, 1956; Gosling and Yule, 1960; see Section 9.5).

7.9 The Ives-Abney-Yule compromise

We have seen in Chapter 5 that a common way in which colours vary in real life is by a uniform addition of white to all colours, such as occurs in a hazy atmosphere. For this reason, errors in colour reproduction that are equivalent to the addition of a little white to all colours, are not very noticeable. In Fig. 7.12, three points P_1, P_2, and P_3 are shown, at the apexes of a triangle that just includes the domain of all real colours. If stimuli plotting in such positions were available (which, of course, they are not) all colours would be matched by all positive mixtures of them, and there would be no negative portions to the colour-matching functions. It is in fact quite easy to calculate what the colour-matching functions would be, and, being all positive, the spectral sensitivity curves of our reproduction system could be quite easily matched to them. If then for our reproduction stimuli, we chose three colours Q_1, Q_2, and Q_3, lying on lines joining P_1, P_2, and P_3 to the point S representing a white stimulus, then Q_1, Q_2, and Q_3 could be considered as being mixtures of P_1, P_2, and P_3 with the white light S. The use of them, therefore, as reproduction stimuli instead of P_1, P_2, and P_3 would merely add white light to the scene, and hence not produce errors of a very noticeable character: the dominant wavelengths would be correct, and the purities would be uniformly reduced. This approach is known as the Ives, Abney, and Yule compromise (see MacAdam, 1938, page 415). The colour-matching functions that would correspond to the 'stimuli' P_1, P_2, and P_3 are shown in Fig. 7.13, and a colour reproduction system in which these curves were used for the camera sensitivities,

89

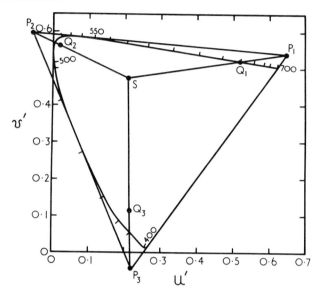

Fig. 7.12. The Ives-Abney-Yule compromise. By using spectral sensitivity curves corresponding to the 'super-saturated' stimuli P_1, P_2, P_3 and real reproduction stimuli Q_1, Q_2, Q_3, the reproduction errors for all colours will be confined to a slight admixture of white.

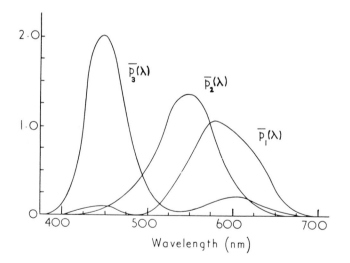

Fig. 7.13. Colour-matching functions for the stimuli P_1, P_2, P_3, of Fig. 7.12. Since the spectral locus lies entirely within the triangle P_1, P_2, P_3, these curves have no negative portions.

90

and the stimuli Q_1, Q_2, and Q_3 were used as reproduction stimuli, should produce errors that are entirely confined to the addition of some white to all colours (although black would still be obtainable, of course, by having zero amounts of Q_1, Q_2, and Q_3). This type of error is represented in the colour triangle by the shifting of points inwards towards the white point as shown in Fig. 7.14. Such shifts are typical of the way in which the purities of colours in real scenes frequently change, and hence would often be acceptable.

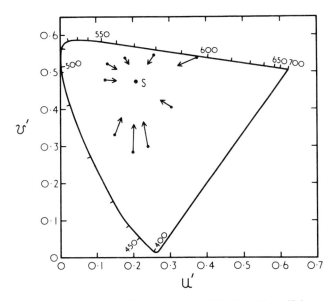

Fig. 7.14. Errors typical of those resulting from the use of the Ives-Abney-Yule compromise shown in Fig. 7.12.

If it were possible to combine the Ives-Abney-Yule compromise with an increase in the gamma of the system, some of the lost colour purity could be made up. In Fig. 7.15 an increase in gamma from 1.0 to 1.5 is shown to increase colour purity by about the same amount as is lost in Fig. 7.14. Although (as discussed in Chapter 6) systems intended for viewing with dark surrounds, in order that the tones look correct, do have gammas of about 1.5, if the dark surround reduces apparent colour saturation again, the net change will be reduced. However, if flat lighting can be used (as in studio work), it may be possible to raise the gamma above that required for correct apparent tone reproduction and thus ensure a net increase in colour saturation.

The Ives-Abney-Yule compromise has the merit of avoiding errors of dominant wavelength, to which the eye is very sensitive, but the overlapping red and green sensitivity curves often result in a loss of efficiency in using the exposing

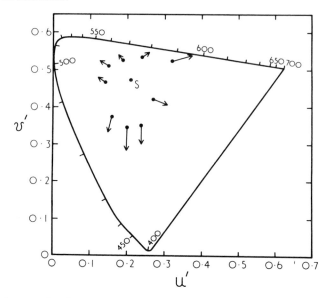

Fig. 7.15. Increases in colour purity typical of those that can be obtained by increasing tone contrast from 1.0 to 1.5.

light. Practical systems of photography and television therefore usually have red and green curves more widely separated along the wavelength axis (see Sections 9.5 and 19.12), and these also increase colour purity (but cause errors of dominant wavelength, unless negative parts of the curves can be used).

7.10 Colour gamuts of reflecting and transmitting colours

When considering the gamut of colours that a system can reproduce, it is useful to bear in mind that there are theoretical limits to the chromaticities that (non-fluorescent) coloured surfaces or filters can attain for any given total reflectance or transmittance. These limits have been worked out by MacAdam (MacAdam, 1935) and are shown in Fig. 7.16. They refer to colours having spectral luminance factors that are either unity or zero at all wavelengths, known as *optimal colours*. Real surface and transmitting colours that do not fluoresce are even more restricted in chromaticity than indicated by Fig. 7.16 (because of having spectral luminance factors intermediate between unity and zero), and surface colours are further restricted because of light reflected from their top-most surfaces (see Figs. 5.3 and 13.4), which can reduce the purities of dark colours very greatly.

In Fig. 7.17 are compared the chromaticity gamut of typical pigments, dyes and inks, occurring in everyday experience (Pointer, 1980; Wintringham, 1951), and the chromaticity gamut covered by phosphors typical of those used in domestic television receivers (B.R.E.M.A., 1969). The two gamuts are shown to

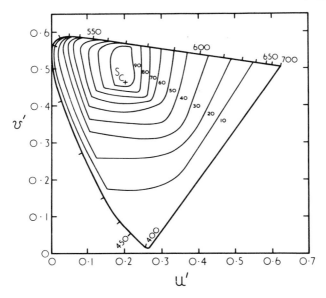

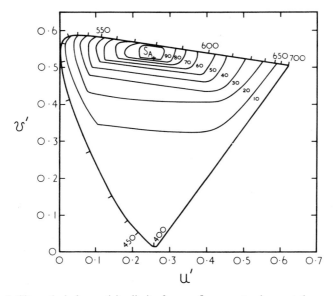

Fig. 7.16. Theoretical chromaticity limits for non-fluorescent colours at the percentage reflectances (or transmittances) shown, for S_A tungsten light (below) and for S_C daylight (above).

93

be similar, except for a slight deficiency of the television gamut in the blue-green and magenta directions: this could cause difficulties with exceptionally vivid green, turquoise, red, and mauve colours.

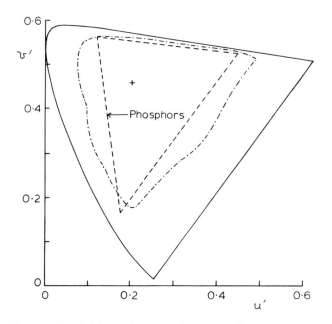

Fig. 7.17. The area bounded by the inner curved (dot-dash) line represents the gamut of chromaticities occurring in pigments, dyes, and inks illuminated by daylight (after Pointer, 1980); the triangle (broken lines) shows the limit of the chromaticities that can be reproduced by phosphors typical of those used in domestic television receivers (BREMA, 1969).

7.11 Two-colour reproductions

If all the colours in a scene were such that the points representing them in the colour triangle lay on a straight line, then it would be possible to obtain an exact colour reproduction by means of a two-colour additive system.

Some interesting effects are obtained in two-colour additive systems if red and white are used as the mixture colours (Cornwell-Clyne, 1951, p. 261; Land, 1959). Because the eye has a strong tendency to discount the overall pinkish colour balance, the white light appears cyan and the reproduction exhibits various cyan as well as pink colours. Judd and others have shown that a number of other effects occur and give rise to quite a wide range of hues being perceived (Belsey, 1964; Judd, 1960; Rushton, 1961; Pearson and Rubinstein, 1970 and

1971); but it has been shown that, contrary to some claims, these reproductions are not independent of the contrasts of the two images (Wilson and Brocklebank, 1960 and 1961).

A remarkable property of these projections is that if the two images are in good registration quite acceptable colour reproduction can be obtained for some subject matter. If the registration is slightly out, the appearance is slightly impaired, but if the registration is grossly out, the observer sees only reds, pinks, and whites. It is clear from these registration effects that more is involved than a general adaptation to the pink colour of the light, for this would be largely independent of registration; what apparently happens is that if the registration is good enough for the two images to give the appearance of a single meaningful scene, then the visual mechanism instantaneously largely discounts the average pink colour, and discerns the objects in the scene as though they were illuminated with a whitish light.

REFERENCES

Belsey, R., *J. Opt. Soc. Amer.*, **54,** 529 (1964).
B.R.E.M.A., P.A.L. Working Party, *Radio and Electronic Engineer*, **38,** 201 (1969).
CIE, *International Lighting Vocabulary*, 3rd Edn., C.I.E. Paris (1970).
Cornwell-Clyne, A., *Colour Cinematography*, Chapman & Hall, London, p. 261 (1951).
Evans, R. M., Hanson, W. T., and Brewer, W. L., *Principles of Color Photography*, Wiley, New York (1953).
Gosling, J. W., and Yule, J. A. C., *Proc. Tech. Assoc. Graphic Arts*, **12,** 157 (1960).
Hardy, A. C., and Wurzburg, F. L., *J. Opt. Soc. Amer.*, **27,** 227 (1937).
Harrison, G. B., and Horner, R. G., *Phot. J.*, **77,** 706 (1937).
Hunt, R. W. G., *Color Res. Appl.*, **2,** 55 and 109 (1977).
Hunt, R. W. G., *Color Res. Appl.*, **3,** 79 (1978).
Judd, D. B., *J. Opt. Soc. Amer.*, **50,** 254 (1960).
Land, E. H., *Proc. Nat. Acad. Sci.*, **45,** 115 and 636 (1959).
MacAdam, D. L., *J. Opt. Soc. Amer.*, **25,** 361 (1935).
MacAdam, D. L., *J. Opt. Soc. Amer.*, **28,** 399 (1938).
MacAdam, D. L., *J. Opt. Soc. Amer.*, **40,** 589 (1950).
MacAdam, D. L., *J. Opt. Soc. Amer.*, **41,** 615 (1951).
MacAdam, D. L., *J. Opt. Soc. Amer.*, **43,** 533 (1953).
Neugebauer, H. E. J., *J. Opt. Soc. Amer.*, **46,** 821 (1956).
Newhall, S. M., *J. Opt. Soc. Amer.*, **30,** 617 (1940).
Pearson, D. E., and Rubinstein, C. B., *J. Opt. Soc. Amer.*, **60,** 1398 (1970).
Pearson, D. E., and Rubinstein, C. B., *J. Soc. Mot. Pic. Tel. Eng.*, **80,** 15 (1971).
Pointer, M. R., *Color Res. Appl.*, **5,** 145 (1980).
Rushton, W. A. H., *Nature*, **189,** 440 (1961).
Wilson, M. H., and Brocklebank, R. W., *J. Phot. Sci.*, **8,** 141 (1960).
Wilson, M. H., and Brocklebank, R. W., *Contemporary Physics*, **3,** 19 (1961).
Wintringham, W. T., *Proc. I.R.E.*, **39,** 1135 (1951).

CHAPTER 8

Colour Standards and Calculations

1. Introduction – *2.* Standard illuminants – *3.* The standard observers – *4.* Colour transformations – *5.* Properties of the XYZ system – *6.* Uniform chromaticity diagrams – *7.* Nomograms – *8.* Uniform colour spaces – *9.* Subjective effects – *10.* Haploscopic matching – *11.* Subjective colour scaling – *12.* Physical colour standards – *13.* Whiteness

8.1 Introduction

IN the previous chapter, various aspects of colour reproduction were considered in a quantitative way with the aid of the colour triangle. We saw, however, that there was no one unique triangle, but several different triangles that could be used. Certain standards have been set up internationally in order to simplify the intercomparison of colour data, and in this chapter we shall briefly review these standards, and describe methods for calculating data from them.

8.2 Standard illuminants

We have already mentioned in Section 5.3 the well-known fact that certain colours exhibit marked changes in appearances as they are viewed in illuminants of different colour. It is therefore clear that an essential step in specifying colour is accurate definition of the illuminants involved. In 1931, in order to simplify the problem, the CIE (Commission Internationale de l'Eclairage) recommended the use of three standard illuminants, A, B, and C, whose spectral power distribution curves are as shown in Fig. 8.1. Standard Illuminant A (S_A) consists of a tungsten filament lamp operating at a colour temperature of 2856 K,[1] while standard illuminants B and C (S_B and S_C) consist of S_A together with certain liquid filters,

---

[1] Equivalent to 2583 degrees Centigrade or Celsius (°C). This colour temperature is usually achieved at a filament temperature of about 2530°C. See Section 10.2.

as shown in Table 8.1. S_A is intended to be representative of tungsten filament lighting, S_B representative of sunlight, and S_C representative of light from an overcast sky. (Billmeyer and Gerrity, 1983).

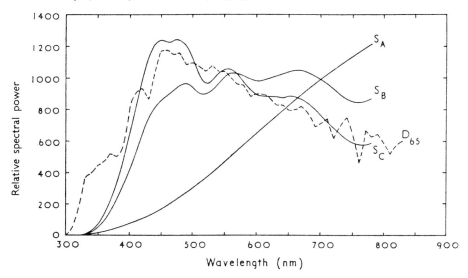

Fig. 8.1. Relative spectral powers of CIE standard illuminants A, B, and C. S_A is representative of tungsten filament lamps. Within the range of wavelengths from 400 to 700 nm, S_B approximates sunlight, and S_C approximates light from an overcast sky. The relative spectral power of CIE standard illuminant D_{65} (representing typical average daylight) is shown by the broken line, and it is seen that S_C is seriously deficient in power at wavelengths below 400 nm.

But, although S_B and S_C represent the spectral power distribution of daylight fairly well over most of the spectrum, they are seriously deficient at wavelengths below 400 nm; this makes them unsuitable for use with samples that absorb power of these wavelengths and then re-emit it by fluorescence at longer wavelengths. The increasingly widespread use of dyes and pigments that fluoresce, as a means of producing brilliant whites for instance, has led to the standardization by the CIE of a series of power distributions representing daylight at all wavelengths between 300 and 830 nm. One of these distributions is shown in Fig. 8.1. In Fig. 8.2 this distribution is shown again together with two others: that labelled 65 represents a standard daylight for general use; that labelled 55 represents a yellower daylight such as may be provided by sun-light with sky-light; and that labelled 75 represents a bluer daylight such as may be provided by a north sky. It should be noted that these standard daylights are defined as spectral power distributions, whereas S_A, S_B, and S_C are defined as actual physical sources: the former are more useful for calculations, the latter for viewing. However, tables of the spectral power distributions of S_A, S_B, and S_C are

97

also available, and the distributions of Fig. 8.2 can be provided approximately by actual sources. The spectral power distributions shown in Fig. 8.2 were founded on measurements made in several different locations in the world (Judd, MacAdam, and Wyszecki, 1964).

These standard illuminants are only intended to be representative of ranges of illuminants, so that any actual sample of sunlight, for instance, might well be redder or bluer than D_{55}, according to the solar altitude, weather conditions, and so on. In addition to these sources the hypothetical equi-energy illuminant E (S_E), consisting of equal power per unit wavelength throughout the visible spectrum, is often referred to in colorimetry.

There are as yet no standard illuminants for the different types of fluorescent lamp, but, in Appendix 2, spectral power distributions are given that are representative of several types of lamp commonly used in practice. The spectral power distributions for standard illuminants A, B, C, D_{50}, D_{55}, D_{65}, and D_{75}, are also given in Appendix 2. The D_{50} illuminant is similar to D_{55}, but slightly yellower; it is used as a standard in the printing industry.

TABLE 8.1

FILTERS FOR USE WITH STANDARD ILLUMINANT A, IN ORDER TO CONVERT IT TO STANDARD ILLUMINANTS B AND C

Each filter consists of two solutions, each one centimetre in thickness and contained in a double cell made of colourless optical glass.

Chemical	Quantities	
	For S_B	For S_C
Copper Sulphate ($CuSO_4.5H_2O$)	2.452 gm.	3.412 gm.
Mannite ($C_6H_8(OH)_6$)	2.452 gm.	3.412 gm.
Pyridine (C_5H_5N)	30.0 c.c.	30.0 c.c.
Water (distilled) to make	1000 c.c.	1000 c.c.
Cobalt Ammonium Sulphate		
($CoSO_4.(NH_4)_2SO_4.6H_2O$)	21.710 gm.	30.580 gm.
Copper Sulphate ($CuSO_4.5H_2O$)	16.110 gm.	22.520 gm.
Sulphuric Acid (Sp.Gr. 1.835)	10.0 c.c.	10.0 c.c.
Water (distilled) to make	1000 c.c.	1000 c.c.

These illuminants are expressed in the CIE System as follows:

$$(S_A) \propto 0.44758(X)+0.40744(Y)+0.14498(Z)$$
$$(S_B) \propto 0.34842(X)+0.35161(Y)+0.29997(Z)$$
$$(S_C) \propto 0.31006(X)+0.31616(Y)+0.37378(Z)$$

8.3 The standard observers

We have seen in previous chapters that a given colour C can be matched by an additive mixture of suitable amounts R_C, G_C, and B_C of three matching stimuli R, G, and B. When different observers match the same colour, using the same

98

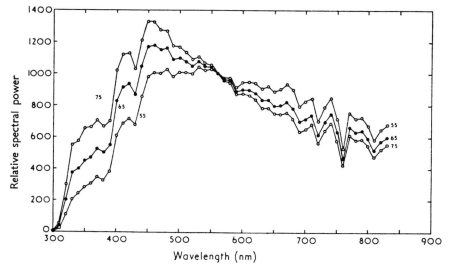

Fig. 8.2. Relative spectral powers of CIE standard distributions D_{55}, representing typical sun-light with sky-light; D_{65} representing typical average daylight; and D_{75} representing typical 'north-sky' light.

matching stimuli, however, it is found that there are slight differences in the amounts that they require to effect a match. Some of these differences are random, and disappear if the results of several matches by each observer are averaged. But there remain real differences, which must be attributed to differences in the colour vision of the individual observers. Some observers are very different from the average and these are classed as colour defective (or 'colour blind'), but the results of most observers are scattered over only a limited range. In 1931 the CIE defined a *Standard Observer* by averaging the results from investigations by W. D. Wright and J. Guild on the colour matching in a 2° field of 17 non-colour-defective observers, and by K. S. Gibson and E. P. T. Tyndall on the relative luminances of the colours of the spectrum, averaged for about a hundred observers. These standard-observer data consist of colour-matching functions for stimuli of wavelengths 700 (R_O), 546.1 (G_O), and 435.8 nm (B_O), with units such that equal amounts of the three stimuli are required to match light from the equi-energy illuminant S_E (Fig. 8.3).

With the aid of these curves, given the spectral power distribution curve of any colour, it is possible to calculate (by the method described in Section 7.4) the amounts of the three stimuli required by the standard observer to match that colour in a 2° field; and this constitutes an exact specification of the colour, which has international significance. Moreover, a calculated specification of this type is derived from purely physical data (the spectral power distribution curve of the colour) without any further colour matching being necessary.

99

The colour triangle corresponding to these stimuli and units is that shown in Fig. 7.7, and the position of any colour in that triangle is calculable from the amounts R, G, and B of the three stimuli by the usual formulae:

$$r = R/(R+G+B)$$
$$g = G/(R+G+B)$$

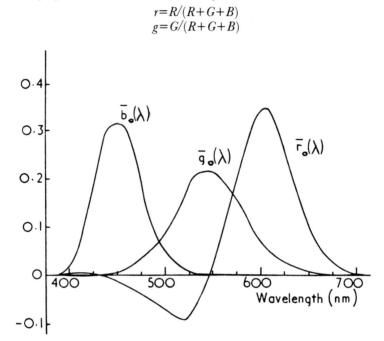

Fig. 8.3. The colour-matching functions for the 2° Standard Observer, using matching stimuli of wavelengths 700, 546.1, and 435.8 nm, with units such that equal quantities of the three matching stimuli are needed to match the equi-energy white, S_E.

The standard observer data adopted in 1931 by the CIE has stood the test of time remarkably well, and has provided a system of colour specification that has been widely and successfully used. But slight errors have occasionally seemed detectable in two respects: the values adopted for the relative luminances of spectral colours; and the effect of the angular size of the field of view. A very thorough redetermination of colour matching data was therefore carried out at the National Physical Laboratory at Teddington (Stiles and Burch, 1958; Stiles, 1955) using both a 2° matching field (the same as that used in establishing the 1931 data) and also a 10° matching field. The results with the 2° field were in close agreement with the 1931 CIE data except that they confirmed earlier suspicions that at the extreme violet end of the spectrum this data ascribed too little luminance to the spectral colours. Although these discrepancies are quite large, in that at some wavelengths the correct values are several times larger than the standardized values, colour specifications are not improved appreciably in practice by revising the data to correct these faults. The reason for this is that at the wavelengths concerned the luminance is so low that the contribution of either the incorrect or the correct values is very small for the vast majority of colours.

Comparison of the results for the 2° and 10° field-size measurements showed that significant differences did occur between them, and 10° Standard Observer data has been adopted for use when large field sizes are involved (CIE, 1960 and 1964). However, in colour reproductions, the interest generally lies much more in patches of colour of about 2° angular size than 10°, and the 1931 CIE data may therefore be used with a fair degree of confidence.

8.4 Colour transformations

If Fig 7.7 is compared with the colour triangles shown in some of the other figures of the previous chapter it will be seen that the spectral locus has an unusually large bulge into the negative r region. This is because the wavelength 546.1 is rather a yellow green. For this and other reasons, the CIE defined three new stimuli X, Y, and Z in terms of which standard-observer results could be expressed. It is possible to calculate the amounts of X, Y, and Z needed to match any colour from the amounts of R, G, and B, of a red, green, blue system, needed to match it, provided that *transformation equations* relating the two systems are known. Thus if we have:

$$C(C) \equiv R(R) + G(G) + B(B)$$

and we know that:

$$1.0(R) \equiv A_1(X) + A_2(Y) + A_3(Z)$$
$$1.0(G) \equiv A_4(X) + A_5(Y) + A_6(Z)$$
$$1.0(B) \equiv A_7(X) + A_8(Y) + A_9(Z)$$

we can substitute for (R), (G), and (B) and obtain:

$$C(C) \equiv X(X) + Y(Y) + Z(Z)$$

where

$$X = A_1 R + A_4 G + A_7 B$$
$$Y = A_2 R + A_5 G + A_8 B$$
$$Z = A_3 R + A_6 G + A_9 B$$

The position of C in the XYZ triangle can then be calculated by obtaining:

$$x = X/(X+Y+Z)$$
$$y = Y/(X+Y+Z)$$

The transformation equations relating the two systems thus contain the coefficients A_1 to A_9, but, as can be seen above, they can be contained either in three equations representing colour matches (as for 1.0 (R), 1.0 (G), and 1.0 (B)) or in three ordinary algebraic equations (as for X, Y, and Z). The former type of equation has sometimes been used without including a distinguishing notation for equations representing colour matches. Unfortunately, the meaning of a given set of equations is quite different according to which type is being used, so that great care has to be taken in interpreting such equations in the literature. However, if one set of equations is known, they can be written in the other form by inspection using the above example as a guide.

Sets of equations of both types can be solved as three simultaneous equations to obtain the reverse transformation equations:

$$1.0 \ (X) \equiv B_1(R) + B_2(G) + B_3(B)$$
$$1.0 \ (Y) \equiv B_4(R) + B_5(G) + B_6(B)$$
$$1.0 \ (Z) \equiv B_7(R) + B_8(G) + B_9(B)$$

$$R = B_1 X + B_4 Y + B_7 Z$$
$$G = B_2 X + B_5 Y + B_8 Z$$
$$B = B_3 X + B_6 Y + B_9 Z$$

These considerations apply not only to the relationship between the stimuli R, G, B and X, Y, Z, but equally to the relationship between any two sets of colour-matching stimuli, the values of the coefficients A_1 to A_9 and B_1 to B_9 depending on

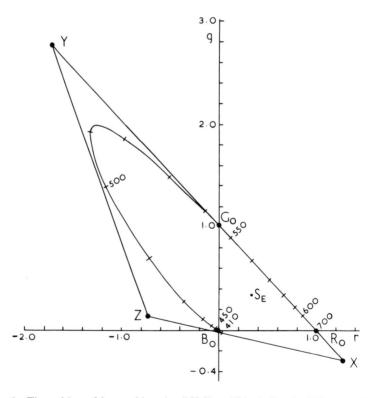

Fig. 8.4. The positions of the matching stimuli X, Y, and Z in the Standard Observer colour triangle $R_O \ G_O \ B_O$ reproduced from Fig. 7.7.

102

the particular stimuli involved and the units adopted for them. In Appendix I, the application of matrix algebra to these relationships is described.

In the case of the CIE stimuli, X, Y, Z, their amounts, X, Y, Z, are related to the amounts, R_O, G_O, B_O, of the stimuli, R_O, G_O, B_O, by the equations:

$$X = 0.49R_O \qquad +0.31G_O \qquad +0.20B_O$$
$$Y = 0.17697R_O \qquad +0.81240G_O \qquad +0.01063B_O$$
$$Z = 0.00R_O \qquad +0.01G_O \qquad +0.99B_O$$

The corresponding set of equations relating the stimuli are:

$$1.0(X) \equiv 2.3646(R_O) - 0.5151(G_O) + 0.0052(B_O)$$
$$1.0(Y) \equiv -0.8965(R_O) + 1.4264(G_O) - 0.0144(B_O)$$
$$1.0(Z) \equiv -0.4681(R_O) + 0.0887(G_O) + 1.0092(B_O)$$

The stimuli X_{10}, Y_{10}, Z_{10} used for the $10°$ standard observer are slightly different, being related to R_O, G_O, B_O by a similar set of equations having slightly different values.

The positions of X, Y, and Z in the $R_O G_O B_O$ colour triangle are given by these last equations by dividing each by the sum of the coefficients and are as shown in Fig. 8.4. It is seen that they lie outside the spectral locus, and that therefore the negative amounts in their specifications are greater than those required in matching even spectral colours. This means that a colorimeter employing these matching stimuli X, Y, and Z, must be arranged so that increasing the amount of X, for example, not only adds some colour (X_+) to the

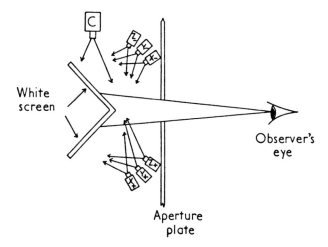

Fig. 8.5. Showing diagrammatically how matching stimuli having negative coefficients greater than those of pure spectral colours can be used. Light from the X_- projector is of a different colour from that of the X_+ projector, but the amounts of light from these two projectors always vary in the same proportion; and similarly for the other two pairs of projectors (Hunt, 1954).

mixture but also adds a proportional amount of another colour (X_-) to the test colour, and similarly for Y and Z, as shown diagrammatically in Fig. 8.5. Only in this way can negative amounts greater than those required by spectral colours be realized. Such a colorimeter can be made quite simply, however, but in point of fact the more conventional type (as shown in Fig. 7.1) is generally used, and the results then transformed into the XYZ system algebraically. It is frequently desirable to transform not the actual values R, G, and B to X, Y, and Z, but only the proportional values r, g, and b, to x, y, and z, where $r+g+b=1$ and $x+y+z=1$. When this is required it is convenient to have the transformation equations in the form:

$$x=\frac{a_1 r + a_2 g + a_3}{a_7 r + a_8 g + a_9}$$

$$y=\frac{a_4 r + a_5 g + a_6}{a_7 r + a_8 g + a_9}$$

$$z=1-x-y$$

The values of the coefficients in these equations are related to those of the equations given above as follows:

$$
\begin{aligned}
a_1 &= A_1 - A_7 & a_4 &= A_2 - A_8 \\
a_2 &= A_4 - A_7 & a_5 &= A_5 - A_8 \\
a_3 &= A_7 & a_6 &= A_8 \\
a_7 &= A_1 + A_2 + A_3 - A_7 - A_8 - A_9 \\
a_8 &= A_4 + A_5 + A_6 - A_7 - A_8 - A_9 \\
a_9 &= A_7 + A_8 + A_9
\end{aligned}
$$

And similar expressions in terms of B_1 to B_9 give the coefficients of the reverse equations

$$r=\frac{b_1 x + b_2 y + b_3}{b_7 x + b_8 y + b_9}$$

$$g=\frac{b_4 x + b_5 y + b_6}{b_7 x + b_8 y + b_9}$$

$$b=1-r-g$$

The following relationships are also useful:

$$
\begin{aligned}
A_1 &= a_1 + a_3 & A_2 &= a_4 + a_6 & A_3 &= a_7 + a_9 - a_1 - a_3 - a_4 - a_6 \\
A_4 &= a_2 + a_3 & A_5 &= a_5 + a_6 & A_6 &= a_8 + a_9 - a_2 - a_3 - a_5 - a_6 \\
A_7 &= a_3 & A_8 &= a_6 & A_9 &= a_9 - a_3 - a_6
\end{aligned}
$$

Similar expressions relate B_1 to B_9 with b_1 to b_9.

The way in which the positions of the points in the colour triangle are altered by transformations from one set of matching stimuli to another can be expressed very simply. All colour triangles based on the 2° standard observer data of Fig. 8.3 are *projective transformations* of the triangle of Fig. 7.7 (and Fig. 8.4); similarly all colour triangles based on the 10° standard observer data are projective transformations of one another. The colour-matching functions for the two standard observers are shown in Fig. 8.6.

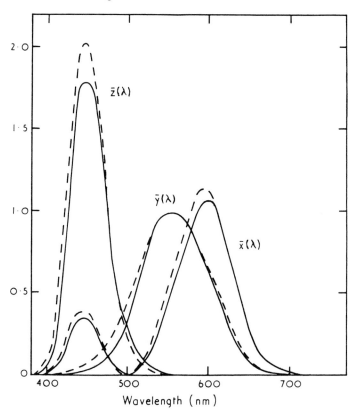

Fig. 8.6. The colour-matching functions for the CIE matching stimuli X, Y, and Z. Full lines: for the 2° Standard Observer, using X, Y, Z; broken lines: for the 10° Standard Observer, using X_{10}, Y_{10}, Z_{10}.

A projective transformation of a triangle is such that it could be obtained by taking a plane and a point, P, suitably situated in space relative to the triangle, and transferring each point of the triangle by means of straight lines drawn through the point, P, until they meet the plane. This is shown in Fig. 8.7. The point M is situated in the RGB triangle. The point P and the plane containing the XYZ triangle have been suitably placed in space. The position of the colour that plots at M in the RGB triangle is then given by N in the XYZ triangle where N is the point of intersection of the line MP with the plane of the XYZ triangle.

That this is indeed the geometrical interpretation of the transformation equations is easily proved as follows.

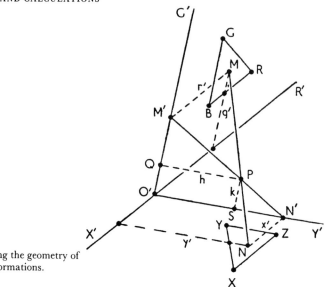

Fig. 8.7. Diagram showing the geometry of projective transformations.

In Fig. 8.7 let R'O'X' be the line of intersection of the RGB and XYZ planes, and let O' be such that PO' is at right angles to R'X'. Let O'G' and O'Y' also be at right angles to R'X' and in the RGB and XYZ planes respectively. Draw MM' at right angles to O'G', and NN' at right angles to O'Y'. Draw PQ parallel to O'Y' and PS parallel to O'G'. Using the small letters on the figure to represent the distances adjacent to them, from similar triangles we have:

$$\frac{x'}{r'}=\frac{PN'}{PM'}=\frac{k}{g'-k} \quad \therefore \quad x'=\frac{kr'}{g'-k}$$

$$\frac{y'}{g'}=\frac{O'N'}{O'M'}=\frac{h}{g'-k} \quad \therefore \quad y'=\frac{hg'}{g'-k}$$

But r' and g' are co-ordinates of M using axes O'R' and O'G', and x' and y' are co-ordinates of N using axes O'X' and O'Y'. They will therefore be related to r and g, and x and y by equations of the type:

$$r'=c_1r-c_2g+c_3$$
$$g'=c_2r+c_1g+c_4$$
$$x'=c_5x-c_6y+c_7$$
$$y'=c_6x+c_5y+c_8$$

On substituting in the equations for x' and y' (above), equations of the form:

$$x=\frac{a_1r+a_2g+a_3}{a_7r+a_8g+a_9}$$

$$y=\frac{a_4r+a_5g+a_6}{a_7r+a_8g+a_9}$$

are obtained. Hence the projective transformation is the geometrical equivalent of the colorimetric transformation equations.

8.5 Properties of the XYZ system

It will be recalled that since the stimuli X, Y, and Z lie outside the spectral locus they consist of light added, not only to the comparison mixture, but also to the test colour C (Fig. 8.5). In the XYZ system it has been arranged that, although the *colour* of the light added by the X stimulus to the test colour (X_-) is different from that added to the comparison beam (X_+), the *luminance* is the same. Similarly, in the case of the Z stimulus, the luminances of the two parts of the stimulus are the same. It therefore follows that all the luminance of the test colour has to be balanced by the Y stimulus. Thus, variation of the amounts of X and Z affect the *colour* of the match, but leave any difference in luminance unchanged. This is an advantage. For supposing that we had two colours C_1 and C_2, whose colour-matching data were known. In the RGB system we have:

$$k_1(C_1) \equiv R_1(R) + G_1(G) + B_1(B)$$
$$k_2(C_2) \equiv R_2(R) + G_2(G) + B_2(B)$$

Now if we want to compare the luminances of these two colours it is necessary to convert the units used for R, G, and B into luminance units. Suppose that the factors for doing this are L_R, L_G, and L_B. Then the luminances are:

$$L_1 = L_R R_1 + L_G G_1 + L_B B_1$$
$$L_2 = L_R R_2 + L_G G_2 + L_B B_2$$

But in the XYZ system if:

$$k_1(C_1) \equiv X_1(X) + Y_1(Y) + Z_1(Z)$$
$$k_2(C_2) \equiv X_2(X) + Y_2(Y) + Z_2(Z)$$

then the luminances are:

$$L_1 = L_X X_1 + L_Y Y_1 + L_Z Z_1$$
$$L_2 = L_X X_2 + L_Y Y_2 + L_Z Z_2$$

But, since X and Z do not affect the luminance of the match, L_X and L_Z are zero, and hence these expressions reduce to:

$$L_1 = L_Y Y_1$$
$$L_2 = L_Y Y_2$$

which are much simpler expressions. Moreover, when it is only required to compare the luminance of one colour with that of another we have:

$$L_1/L_2 = Y_1/Y_2$$

which is an extremely simple relationship. The luminances, L_R, L_G, L_B, of the units used in the R_O, G_O, B_O system are in the ratios 1.0000 to 4.5907 to 0.0601; the co-efficients of R_O, G_O, and B_O in the equation for Y given earlier (0.17697, 0.81240, and 0.01063) are in the same ratios, and this ensures that Y is proportional to luminance.

107

The XYZ system is used very widely for colorimetric specifications. The colour matching functions $\bar{x}(\lambda), \bar{y}(\lambda)$, and $\bar{z}(\lambda)$ for the system are shown in Fig. 8.6 for both the 2° and the 10° standard observers. It will be noted that there are no negative portions to the curves, because, as can be seen from Fig. 8.4, no part of the spectral locus lies outside the triangle XYZ and therefore every colour of the spectrum can be matched by an all positive mixture of X, Y, and Z. InFig. 8.8 the XYZ colour triangles based on the 2° and 10° standard observers are shown, y being plotted against x, where:

$$y = Y/(X+Y+Z)$$
$$x = X/(X+Y+Z)$$

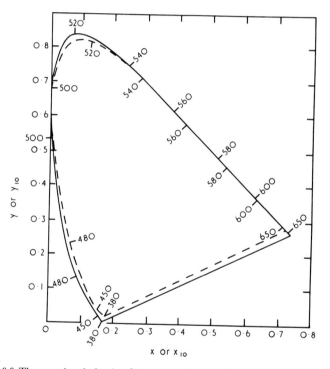

Fig. 8.8. The x, y triangle for the 2° Standard Observer (full line) and for the 10° Standard Observer (broken line) using x_{10}, y_{10}.

(For the 10° standard observer, the stimuli X_{10}, Y_{10}, Z_{10}, are used throughout.) Values of the colour matching functions, $\bar{x}(\lambda), \bar{y}(\lambda), \bar{z})(\lambda)$, and of the chromaticity co-ordinates for spectral colours, $x(\lambda), y(\lambda), z(\lambda)$, are given in Appendix 2 for the 2° standard observer.

In order to obtain the XYZ specification of a colour it can either be matched on a colorimeter as in Fig. 8.5, that gives the direct answer (Hunt, 1954); or, which is more usual, it can be matched on a conventional red, green, and blue colorimeter and the results transformed by means of transformation equations; or, which is more usual still, its XYZ specifications can be calculated from its spectral power curve. To this end, values $\bar{x}(\lambda)$, $\bar{y}(\lambda)$, and $\bar{z}(\lambda)$ are available in standard works on colorimetry and the calculation proceeds as follows.

Suppose we have a transparency or surface whose transmission or reflection factor at, say, 400 nm, is t_1, illuminated by a source whose power at that wavelength is E_1. Then the values of X, Y, and Z for that wavelength are given by:

$$X_1 = t_1 E_1 \bar{x}_1$$
$$Y_1 = t_1 E_1 \bar{y}_1$$
$$Z_1 = t_1 E_1 \bar{z}_1$$

where $\bar{x}_1$, $\bar{y}_1$, $\bar{z}_1$, are the values of $\bar{x}(\lambda)$, $\bar{y}(\lambda)$, $\bar{z}(\lambda)$, respectively, at 400 nm. At another wavelength, say 410 nm, we would have:

$$X_2 = t_2 E_2 \bar{x}_2$$
$$Y_2 = t_2 E_2 \bar{y}_2$$
$$Z_2 = t_2 E_2 \bar{z}_2$$

These products must in fact be worked out at regular wavelength intervals (generally every 5 or 10 nm) throughout the entire visible spectrum and then summated thus:

$$X = t_1 E_1 \bar{x}_1 + t_2 E_2 \bar{x}_2 + t_3 E_3 \bar{x}_3 + \ldots = \Sigma t E \bar{x}$$
$$Y = t_1 E_1 \bar{y}_1 + t_2 E_2 \bar{y}_2 + t_3 E_3 \bar{y}_3 + \ldots = \Sigma t E \bar{y}$$
$$Z = t_1 E_1 \bar{z}_1 + t_2 E_2 \bar{z}_2 + t_3 E_3 \bar{z}_3 + \ldots = \Sigma t E \bar{z}$$

In order to reduce the amount of work, tables for $E\bar{x}$, $E\bar{y}$, and $E\bar{z}$ are sometimes provided. For reflecting and transmitting object colours, these summations are normally multiplied by a common factor so that a perfectly reflecting or transmitting colour ($t=1$ at all wavelengths) has a value of $Y=100$. For any other colour the value of Y then gives the *percentage* reflection or transmission directly, since the Y values of any two colours are proportional to their luminances.

The dependence of luminance entirely on the Y stimulus in the XYZ system, means that the $\bar{y}(\lambda)$ colour-matching function represents the relative luminances of the colours of the spectrum; this is an important function in photometry, where it is known as the *spectral luminous efficiency* function, $V(\lambda)$.

The use of the Centre of Gravity Law in the XYZ system is also simplified by the dependence of the luminance on the Y stimulus only. Suppose we wish to determine the chromaticity of a mixture of m_1 photometric units of a colour C_1 and m_2 photometric units of a colour C_2 where the chromaticity co-ordinates of C_1 and C_2 are x_1, y_1, z_1 and x_2, y_2, z_2, respectively. Then from Section 7.6 the mixture is at the position of the centre of gravity of weights;

$$m_1 / (L_X x_1 + L_Y y_1 + L_Z z_1)$$
$$m_2 / (L_X x_2 + L_Y y_2 + L_Z z_2)$$

But these reduce to $m_1 / L_Y y_1$ and $m_2 / L_Y y_2$, because L_X and L_Z are both zero. But the same result is given by weights m_1/y_1 and m_2/y_2, so that these simple expressions can be used.

The fact that the values of Y are proportional to luminance means that there is some connection between them and the perceptual attributes of brightness and lightness. There is, however, no similar connection between the values of X and Z and the perceptual attributes of hue and colourfulness (or one of the relative colourfulnesses, chroma or saturation). Such a connection with hue can be provided by quoting the *dominant wavelength*, λ_d, which was defined in Section 7.2 as the wavelength of the monochromatic stimulus that, when additively mixed in suitable proportions with a specified achromatic stimulus, matches the colour stimulus considered; it can be conveniently determined graphically by drawing, on a chromaticity diagram, a straight line through the point, N, representing the achromatic stimulus, and the point C, representing the colour considered, and producing it to meet the spectral locus at a point, D, and noting the wavelength (in the case of purple stimuli, the value is called the complementary wavelength, λ_c). A similar connection with saturation can be provided by evaluating the ratio of NC to ND, and when this is done on the CIE x,y chromaticity diagram it is called the *excitation purity*, p_e (in this case, for purple stimuli, D is the point on the line joining the points representing the two ends of the spectrum). These measures, λ_d and p_e, are of limited utility because they result in scales that can be quite severely perceptually non-uniform.

8.6 Uniform chromaticity diagrams

The non-uniformities referred to at the end of the previous Section stem, in part, from the fact that, in the CIE x,y chromaticity diagram, colours are by no means uniformly distributed. This is demonstrated in Fig. 8.9 where the short lines in the diagram all represent colour differences that appear of equal magnitude to the eye in a 2° field. It is seen that towards the top of the diagram, where greenish colours are situated, the lines are much longer than towards the bottom left of the diagram, where bluish colours are situated; the maximum difference is in fact as great as twenty times. Incidentally, the distances shown by these lines correspond to three times a just noticeable difference in colour in a 2° field (Wright, 1941).

Although no projective transformation of the diagram can eliminate the differences in the lengths of these lines completely, by choosing a different diagram, they can be very considerably reduced. Fig. 8.10 shows a selection of the lines shown in Fig. 8.9, but on the u',v' diagram that was extensively used in the previous chapter. It is seen that the lines are much more nearly uniform in length, and in fact the maximum difference is now only about four to one, and over much of the diagram is not greater than two to one.

This particular *uniform chromaticity diagram* is a slight modification of one that was proposed by D. L. MacAdam (MacAdam, 1937), and has the advantage that, as in the XYZ system, two of the stimuli (U' and W') affect only the *colour* of

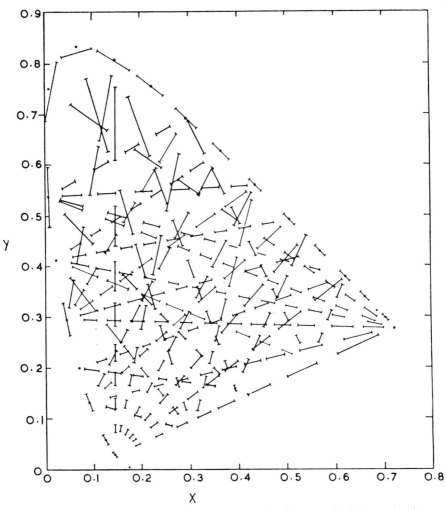

Fig. 8.9. Visually equal chromaticity steps at constant luminance on the CIE x, y triangle (after W. D. Wright).

a match, the *luminance* being affected only by one of the stimuli (V′). The relationship of MacAdam's original diagram to the CIE x, y diagram is given by the conveniently simple equations:

$$u = 4x/(-2x + 12y + 3) \quad x = 3u/(2u - 8v + 4)$$
$$v = 6y/(-2x + 12y + 3) \quad y = 2v/(2u - 8v + 4)$$

The use of this approximately uniform chromaticity diagram was approved by the CIE at its 1959 meeting at Brussels, u being plotted as abcissa and v as

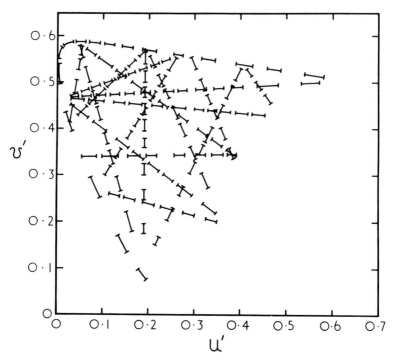

Fig. 8.10. Some of the steps of Fig. 8.9 replotted in the u',v' diagram.

ordinate (CIE, 1971). However, at its meeting in London in 1975, the CIE recommended that this diagram be modified by stretching it in the v direction by a factor of 1.5 (Eastwood, 1975): this change resulted in the colours being distributed in a way that more closely resembles the Munsell system (see Section 7.2), and it also provided an improved basis for a colour-difference formula (to be considered in Section 8.8); the relationship between this modified diagram and the CIE x, y diagram is:

$$u'=4x/(-2x+12y+3) \quad x=9u'/(6u'-16v'+12)$$
$$v'=9y/(-2x+12y+3) \quad y=4v'/(6u'-16v'+12)$$

In this u', v' diagram, u' is plotted as abcissa , and v' as ordinate (as in Fig. 8.10). It is clear from the equations that $u'=u$, and $v'=1.5v$, as required. The various forms of transformation equations connecting the XYZ, the UVW, and the U'V'W' systems, are given in Appendix 2. The complete u', v' diagram is shown in Fig. 7.9, from which it is clear that, for colours in the extreme red corner of the diagram, $u'+v'$ is greater than unity, and hence w' is negative because $u'+v'+w'=1$.

Plate 9

The reproduction of detail, and modulation, in whites and blacks simultaneously is difficult in pictures because of their limited luminance ranges; this is particularly so in reflection prints (see Section 13.10). The difficulty frequently occurs in pictures of weddings, and is best handled by ensuring that the group is uniformly illuminated, as in this example; if the bride is in sunlight and the bridegroom in deep shadow, then a good reflection print may be impossible to achieve, even if the two figures are given different treatment. Reproduced from a 5 × 4 inch *Ektachrome* transparency on a Crosfield *Magnascan* scanner.

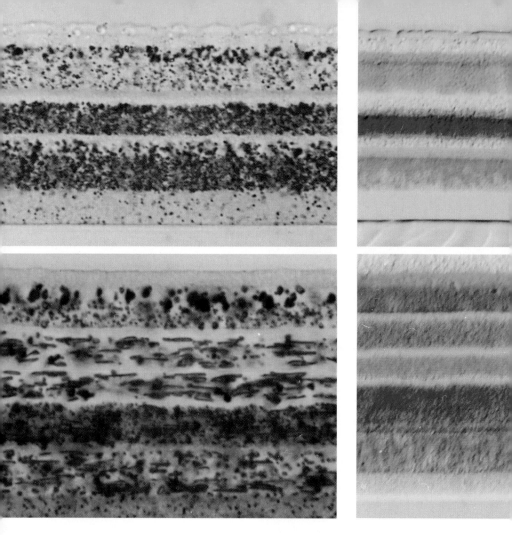

Plate 10

Cross-sections of swollen colour photographic materials magnified 1100 times; the dry thicknesses would be reduced to about one fifth. *Upper left.* A conventional type of colour negative film (*Kodacolor II*) before processing. The blue-sensitive layer is above the yellow filter layer that prevents the lower layers from receiving any blue light. The lowest layer in this example is an anti-halation layer (see Section 18.16) above which can be seen first the red-sensitive layer, and then, after a very thin inter-layer, the green-sensitive layer. The silver halide grains in the layers are clearly visible, but the magnification is not high enough to show the oil globules in which the couplers are incorporated. The pinkish colour of the red layer, and the yellowish colour of the green layer, are caused by the presence of coloured couplers. *Upper centre.* The film shown at top left, after processing. The blue layer has produced a yellow dye, the green layer a magenta dye, and the red layer a cyan dye. It can be seen that each dye layer is made up of two component layers, and careful inspection of the unprocessed film (at upper left) shows that in each case the upper component layer has larger silver halide grains than the lower component layer. Each layer thus consists of a fast component above a slow component, and this improves the speed-grain performance. In this example, the fast components have all been fully developed and their dye clouds have merged to form uniform distributions of dyes; but the slow layers show some granular structures (see Section 18.7). The yellow filter layer, the developed silver, the undeveloped silver halide, and the anti-halation layer, have all been removed by the processing (see Section

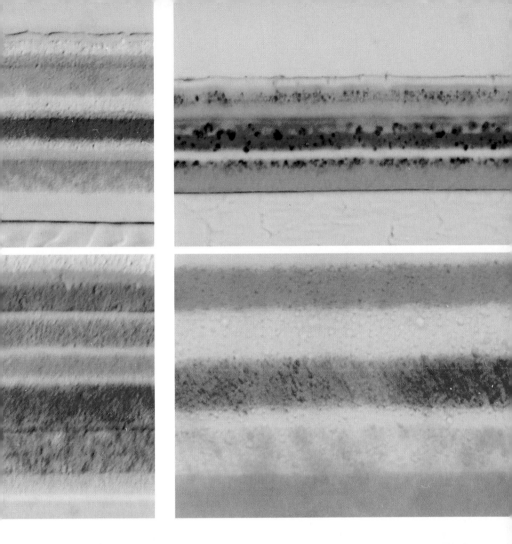

17.9). *Lower left.* A very fast type of colour film (*Kodacolor VR 1000*) before processing. *Lower centre.* The film shown at lower left, after processing. In this type of film, the fast red layer has been coated above the slow green layer to increase the speed. In this example, the yellow filter layer has been omitted, because the emulsions used in the lower layers have enhanced green and red speeds relative to their natural blue speeds; this can be achieved by increasing the surface to volume ratio of the silver halide grains by making them of tabular shape (see Section 18.8). These *T-grains* appear in cross-section as short lines, as can be seen in the unprocessed film (at lower left). The yellow filter layer between the slow red and slow green layers is to adjust colour balance for ease of printing. *Upper right.* An example of colour film processed using couplers in the developers (*Kodachrome 25*). Each layer consists of a fast component above a slow component. It can be seen that the total thickness is less than that of the coupler-incorporated films shown (at left and centre), and this is an advantage for sharpness. (See Section 12.7). *Lower right.* A colour paper after processing (*Ektacolor* paper). The yellow image is at the bottom to minimize the visibility of any roughness of the surface of the paper (see Section 18.17). In some colour films, the magenta image is at the top and the yellow at the bottom to obtain better sharpness (see Sections 12.11 and 18.16). Photomicrographs by courtesy of Frank Judd, Kodak Research Laboratories, Harrow. Reproduced from prints on *Ektacolor* paper by a Crosfield *Magnascan* scanner.

Plate 11

The appearance of processed images on motion-picture films. Each of the three positives on the *right* was made on *Eastman Color Print* film from the adjacent negative on its *left: top left*, camera-original on *Eastman Color Negative* film; *upper left*, intermediate negative made from the camera-original on *Eastman Color Reversal Intermediate* film; *lower left*, intermediate positive made from the camera-original on *Eastman Color Intermediate* film; *bottom left*, intermediate negative made from the intermediate positive on *Eastman Color Intermediate* film; (see Section 12.11). The orange appearance of the films on the left is caused by the presence of coloured couplers (see Section 15.4). In the negatives, both the tones and the colours are reversed. Reproduced on a Crosfield *Magnascan* scanner.

In the U'V'W' system the luminances, $L_{U'}$, $L_{W'}$, of the units used for U' and W' are both zero. The centre of Gravity Law is therefore operated using weights m_1/v'_1 and m_2/v'_2, where v'_1 and v'_2 are the v'-co-ordinates of the two colours C_1 and C_2 being mixed in the proportions m_1 and m_2 photometric units, respectively. Values of $u'(\lambda)$, $v'(\lambda)$, the chromaticity co-ordinates of the spectral colours, are given in Appendix 2.

In colour television a just noticeable difference in chromaticity has been estimated to be about 0.004 in the u, v chromaticity diagram (Jones, 1968), but for matching D_{65} whites on monitors it has been reported that skilled observers can see differences of only about 0.002, so that in this case a tolerance of only about $\pm$ 0.001 is desirable (Knight, 1972).

8.7 Nomograms

A useful device for transferring data from one chromaticity diagram to another is a nomogram. In Fig. 8.11 a nomogram is shown for obtaining the values of u' and v' directly from x and y, or vice versa. It is only necessary to place a straight edge across the nomogram so that it cuts the x and y scales at the values of the point concerned: the corresponding values of u' and v' are then given by the intersection of the straight edge with the u' and v' scales. To obtain the values of x and y corresponding to known values of u' and v' the straight edge is aligned with the values of u' and v' and the values of x and y are then read off directly.

Since a nomogram of this type is so useful, and appears to be little known, it is of interest to record briefly here how it can be derived. In Fig. 8.11 two variables

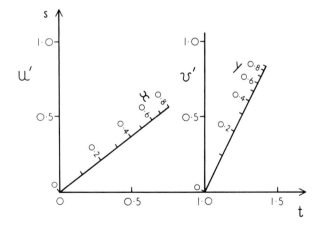

Fig. 8.11. Showing the co-ordinates s and t used in working out nomograms for colorimetric transformations. The nomogram of Fig. 8.12 was worked out in these co-ordinates and then plotted with the s-scale doubled in length in order to give more convenient scales.

s and t are plotted along two axes at right angles. The scale of u', which is uniform, runs along the s axis, and the scale of v', which is also uniform, runs parallel to it, at a distance of 1 unit from it along the t axis. If the equation for x is:

$$x = \frac{a_1 u' + a_2 v' + a_3}{a_7 u' + a_8 v' + a_9}$$

then the x-scale in the nomogram is given by:

$$s = \frac{a_3 - a_9 x}{(a_7 + a_8)x - (a_1 + a_2)}$$

$$t = \frac{a_8 x - a_2}{(a_7 + a_8)x - (a_1 + a_2)}$$

And if the equation for y is:

$$y = \frac{a_4 u' + a_5 v' + a_6}{a_7 u' + a_8 v' + a_9}$$

then the y-scale in the nomogram is given by:

$$s = \frac{a_6 - a_9 y}{(a_7 + a_8)y - (a_4 + a_5)}$$

$$t = \frac{a_8 y - a_5}{(a_7 + a_8)y - (a_4 + a_5)}$$

It sometimes happens, in calculating nomograms, that a scale passes through the infinity point and the useful values of the scale are very awkwardly placed. In these cases it is sometimes profitable to calculate the nomogram for different pairs of variables, such as y, z and v', w'. Alternatively one of the scales can be reversed in direction, for instance by writing $v'' = -v'$ and working out the nomogram for v'', and then using the negative part of the v'' scale as a positive scale for v'. Also it is useful to remember that since, in projective transformations, all straight lines remain straight lines, any projective transformation of the nomogram is permissible, and sometimes a more convenient shape can be arrived at. We have made use of this in Fig. 8.12, in which the nomogram of Fig. 8.11 has been stretched in the vertical direction by a factor of 2 to give more convenient scales.

8.8 Uniform colour spaces

Colours that are seen against surroundings of very much lower luminances usually appear to be self-luminous and without any grey content; they are typically regarded as light sources: the sun, tungsten-filament lamps, traffic-light signals, fluorescent lamps, neon lights, light-emitting diodes, and flames, usually come into this category. Such colours are sometimes referred to as *unrelated colours*. For unrelated colours, the quotation of two chromaticity co-ordinates, together

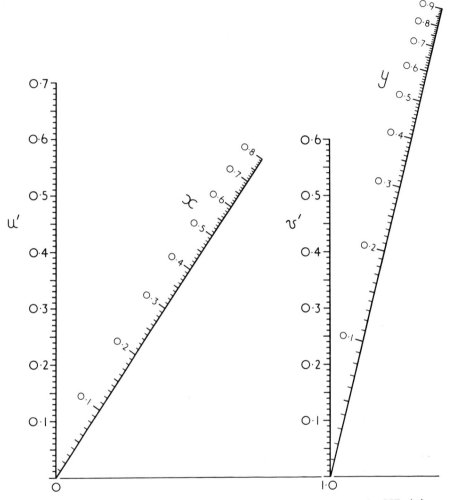

Fig. 8.12. Nomogram for transferring data from the CIE *x,y* diagram to the CIE *u′,v′* diagram and vice-versa. A straight-edge placed across the four scales gives the values of *u′* and *v′* corresponding to those of *x* and *y* and vice-versa.

with a suitable measure of the absolute luminance as the third variable, usually provides a sufficient basis for their colour specification.

But for most colours, the luminances of the surroundings are sufficiently similar for them to appear, not as self-luminous, but as reflecting or transmitting objects, usually having a grey content in their colour appearance. These colours are sometimes referred to as *related colours*. For related colours, the luminance of the colour relative to the average luminance of the surroundings must be considered in addition to two chromaticity co-ordinates and absolute luminance (Hunt, 1977 and 1978). This can be illustrated by the following example. Suppose a colour has a chromaticity $u'=0.20$, $v'=0.47$, and an absolute luminance of 45 cd/m² (for a list of photometric units and their meanings, see Appendix 3): if the surroundings have the same average chromaticity ($u'=0.20$, $v'=0.47$), and an absolute luminance of 10 cd/m², the colour will usually look white; but if the average absolute luminance of the surroundings is 100 cd/m², the colour will usually look grey. In this case, the appearance of the colour (white or grey) is associated much more with the relative luminance than with the absolute luminance (which was constant). It is therefore customary, for related colours, to use, as the three variables, two chromaticity co-ordinates and a measure of *relative* luminance; this relative luminance is usually measured as a percentage of the luminance of some reference white, such as the perfect reflecting diffuser. In the CIE X, Y, Z system, for related colours, the tristimulus values X, Y, Z are usually measured or calculated in such a way as to ensure that $Y=100$ for the perfect diffuser. In the example, if the perfect diffuser had a luminance of 50 cd/m² in the first case, and 500 cd/m² in the second, the two colours might then have been specified as $u'=0.20$, $v'=0.47$, $Y=90$ for the white, and $u'=0.20$, $v'=0.47$, $Y=9$ for the grey. When the absolute luminances of related colours are also of importance, these must be quoted in addition; but, instead of quoting the absolute luminances of individual colours, it is usually more helpful to quote either the luminance of the reference white, or the absolute level of the illumination on the colours.

Although the appearance of related colours is strongly dependent on their relative luminances, a uniform linear scale of relative luminance does not represent a uniform visual scale: for example, the apparent difference between two samples of relative luminances 10 and 15 per cent is much greater than that between two samples of 70 and 75 per cent. To allow for this, a quantity called *CIE 1976 lightness*, L^*, is calculated as

$$L^*=116\,(Y/Y_n)^{1/3}-16$$

where Y_n is the Y tristimulus value of the reference white being used. Equal increments on the L^* scale do represent approximately equal steps in the perceived lightness of related colours. (The L^* scale is, in fact, an approximation to the Munsell Value scale, L^* being equal to approximately 10 times Munsell Value; see Section 14.24). It should be noted that $L^*=100$ for the reference white (for which $Y=Y_n$). For values of Y/Y_n less than 0.008856, L^* is not calculated by the formula given above, but by $L^*=903.3\,(Y/Y_n)$. (L^* is very similar to an earlier CIE correlate of lightness, W^*, recommended by the CIE in 1963. W^* was defined as equal to $25(100Y/Y_n)^{1/3}-17$. When $Y_n=100$, $L^*=W^*+1-0.0086Y^{1/3}$).

It might be thought that u', v', and L^* would be the most appropriate three variables to use for related colours. There are, however, two further factors that need to be taken into account. First, the appearance of a related colour depends not only on the luminance of its surroundings, but also on the chromaticity of its surroundings. Thus if, in our example, the chromaticity of the colours had been, not $u'=0.20$, $v'=0.47$, but $u'=0.24$, $v'=0.52$, then the colours would still have tended to appear white and grey if, again, the surroundings had the same average chromaticity as the colours. The appearance of related colours is, in fact, dependent more on the chromaticity relative to that of the surroundings, than on the absolute chromaticity. It is therefore advantageous to consider, not u' and v', but $u'-u'_n$ and $v'-v'_n$, where u'_n and v'_n are the u', v' values for the reference white. The second factor that needs consideration is that, for a given chromaticity difference, the magnitude of the perceived colour difference depends on the lightness: the lighter the samples, the greater the perceived difference (Wyszecki, 1963; Wyszecki and Wright, 1965). To allow for this effect the following variables can be used:

$$u^*=13L^* \ (u'-u'_n)$$
$$v^*=13L^* \ (v'-v'_n)$$

The multiplication of the chromaticity differences by L^* results in u^* and v^* indicating more nearly the visual significance of those differences. The factor 13 is introduced in order to make equal differences in L^*, u^*, and v^* correspond to roughly equal visual differences.

By plotting u^* and v^* in a horizontal plane, and L^* along a vertical axis, as is

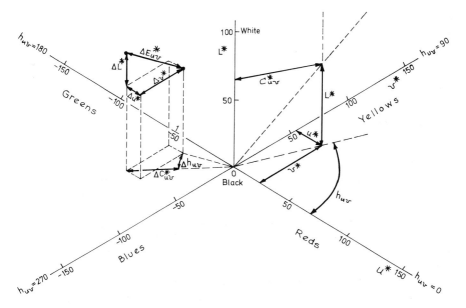

Fig. 8.13. The L^*,u^*,v^* space. The three axes for L^*, u^*, and v^*, are at right-angles to one another. Equal colour differences are represented by approximately equal distances in the space.

done in Fig. 8.13, it is thus possible to construct a three dimensional *colour space*, or a *colour solid* as it is sometimes called, in which equal distances in any direction represent colour differences of roughly equal visual magnitudes. Thus, if two samples have differences in L^*, u^*, v^* equal to ΔL^*, Δu^*, Δv^*, respectively, then the total colour difference, ΔE^*_{uv}, is equal to the distance between the points representing them in the colour solid (as shown in Fig. 8.13), and this distance is given by the formula:

$$\Delta E^*_{uv} = [(\Delta L^*)^2 + (\Delta u^*)^2 + (\Delta v^*)^2]^{\frac{1}{2}}$$

This quantity is called the *CIE 1976 (L*u*v*) colour difference* or *CIELUV colour difference*. (This formula superseded that recommended in 1963, in which $\Delta E = [(\Delta W^*)^2 + (\Delta U^*)^2 + (\Delta V^*)^2]^{\frac{1}{2}}$ where $\Delta U^* = 13W^*(u-u_O)$, and $\Delta V^* = 13W^* (v-v_O)$, u_O and v_O being the values of u and v for the reference white).

This type of space is useful in studies of colour television because, in the associated chromaticity diagrams, additive mixtures of stimuli lie on straight lines joining the points representing those stimuli. The similar earlier CIE U^*, V^*, W^* system was used in one such study (Schroeder, 1979).

In the textile industry, a colour difference formula known as ANLAB(40) was widely used, based on the work of Adams and Miss Nickerson, using variables L, A, B, with one of its scaling factors equal to 40 (McLaren and Coates, 1970 and 1972; McLaren, 1971). This formula makes use of the Munsell Value function (given in Section 14.24) which is of an awkward nature. The CIE has therefore also recommended a similar formula in which cube root relationships (Faulhaber and Witherell, 1971) and the L^* function are used instead of the Munsell Value function, as follows:

$$\Delta E^*_{ab} = [(\Delta L^*)^2 + (\Delta a^*)^2 + (\Delta b^*)^2]^{\frac{1}{2}}$$

where

$$a^* = 500[(X/X_n)^{\frac{1}{3}} - (Y/Y_n)^{\frac{1}{3}}]$$
$$b^* = 200[(Y/Y_n)^{\frac{1}{3}} - (Z/Z_n)^{\frac{1}{3}}]$$

X_n, Y_n, Z_n, being the X, Y, Z tristimulus values for the reference white being used. This quantity, ΔE^*_{ab}, which is called the *CIE 1976 (L*a*b*) colour difference* or *CIELAB colour difference*, does not have an approximately uniform chromaticity diagram associated with it, but a three-dimensional colour space analogous to that shown in Fig. 8.13 can be constructed by plotting a^* and b^* in a horizontal plane, and L^* along a vertical axis. A colour space similar to the L^*, a^*, b^* space has been used to evaluate colour photographic systems (Baumann, 1980).

Colour difference formulae of this type (ΔE^*_{uv} and ΔE^*_{ab}) are intended to apply to cases where the samples are seen side by side with little or no spatial separation between them; if, however, the samples are widely separated, then the weight given to the lightness component in the difference should be reduced (by using $k\Delta L^*$ where k is about 0.5 or less), because differences in lightness are then less visible (Judd and Wyszecki, 1975).

The $L^*u^*v^*$ and $L^*a^*b^*$ colour spaces are only approximately uniform, and their associated colour difference formulae can therefore evaluate colour differences only approximately. More precise spaces and formulae are being sought, but, meanwhile, the $L^*u^*v^*$ space has the advantage of having an associated approximately uniform chromaticity diagram, while the $L^*a^*b^*$ formula is similar to one already widely used in the textile industry.

In the $L^*u^*v^*$ colour space, u^* and v^* are both zero for the reference white. As indicated in Fig. 8.13, increases in u^* represent increases mainly in redness; decreases in u^* represent increases mainly in greenness; increases in v^* represent increases mainly in yellowness; decreases in v^* represent increases mainly in blueness. In the $L^*a^*b^*$ space, a^* and b^* are both zero for the reference white, and a^* is roughly similar to u^*, and b^* to v^* for changes in colour. This approximate coincidence of two of the axes of the colour spaces with the basic hue-pairs, red and green, and yellow and blue, is a useful feature.

It is sometimes desirable to identify components of colour differences in terms of correlates of chroma and of hue. This can be done by using quantities C^*, called *CIE 1976 chroma*, and h, called *CIE 1976 hue-angle*. C^* is calculated as

$$C^*_{uv}=(u^{*2}+v^{*2})^{\frac{1}{2}} \quad \text{or} \quad C^*_{ab}=(a^{*2}+b^{*2})^{\frac{1}{2}}$$

and, in the colour spaces, it is the distance of the colour from the L^* axis (see Fig. 8.13). h is calculated as:

$$h_{uv}=\arctan(v^*/u^*) \quad \text{or} \quad h_{ab}=\arctan(b^*/a^*)$$

(where arctan means: the angle whose tangent is). h_{uv} is the angle, in the colour space, between the plane containing the L^* axis and the colour, and the plane containing the L^* and u^* axes; the way in which the values of h_{uv} are related to the u^*, v^* axes is shown in Fig. 8.13; similar relationships also apply between h_{ab} and the a^*, b^* axes. So that colour differences can be broken up into components of lightness, chroma, and hue, whose squares sum to the square of ΔE^*, a quantity ΔH^*, called *CIE 1976 hue-difference*, can be calculated as:

$$\Delta H^*_{uv}=[(\Delta E^*_{uv})^2-(\Delta L^*)^2-(\Delta C^*_{uv})^2]^{\frac{1}{2}}$$

or

$$\Delta H^*_{ab}=[(\Delta E^*_{ab})^2-(\Delta L^*)^2-(\Delta C^*_{ab})^2]^{\frac{1}{2}}$$

so that

$$\Delta E^*_{uv}=[(\Delta L^*)^2+(\Delta C^*_{uv})^2+(\Delta H^*_{uv})^2]^{\frac{1}{2}}$$

and

$$\Delta E^*_{ab}=[(\Delta L^*)^2+(\Delta C^*_{ab})^2+(\Delta H^*_{ab})^2]^{\frac{1}{2}}$$

For small colour differences away from the L^* axis, $\Delta H^*=C^*\Delta h(\pi/180)$ approximately.

In the L^*,u^*,v^* space, it is possible to obtain a simple correlate of saturation, *CIE 1976 u,v saturation*, s_{uv}. This measure is proportional to the distance on the u', v' chromaticity diagram between the points representing the colour considered and the reference white; thus

$$s_{uv}=13[(u'-u'_n)^2+(v'-v'_n)^2]^{\frac{1}{2}}$$

From this equation it follows that

$$C^*_{uv}=L^*s_{uv}$$

showing that differences in chromaticity between colours and the reference white are reduced, as the lightness is reduced, in the evaluation of this correlate of chroma. It is not possible to calculate a similar simple correlate of saturation in the L^*,a^*,b^* space, because it does not have an associated chromaticity diagram.

In both colour spaces, planes parallel to that containing the axes u^* and v^* or a^* and b^* are planes of constant L^* and contain colours of approximately constant lightness. In the L^*,u^*,v^* space, horizontal planes contain u',v' chromaticity diagrams, in the sense that, on any given plane, if points in the solid are projected on to it by means of straight lines passing through the origin, then the projected points generate a u',v' diagram multiplied by an overall scale factor.

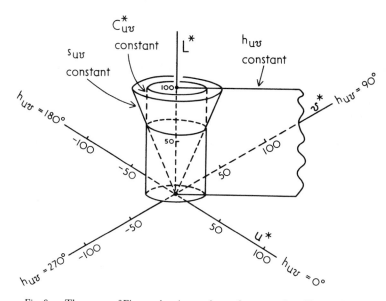

Fig. 8.14. The space of Fig. 13 showing surfaces of constant h_{uv}, C^*_{uv}, and s_{uv}.

The relationships between the correlates of hue, chroma, and saturation in the L^*,u^*,v^* space is illustrated in Fig. 8.14. Vertical planes containing the L^* axis as one edge are planes of constant CIE hue-angle, h_{uv}; cylinders having L^* as their axis are planes of constant CIE chroma, C^*_{uv}; and cones having L^* as their axis and their apexes at the origin are surfaces of constant s_{uv}. The relationships of Fig. 8.14 also apply to the L^*,a^*,b^* space except for the correlate of saturation.

These colour solids are similar in general terms to that of the Munsell Colour System. In the Munsell solid, Munsell Value, which is very similar to L^*, is arranged along a vertical axis with white at the top and black at the bottom; planes of constant Munsell Hue, which are approximately similar to planes of constant h, are arranged to contain the Munsell Value axis as one edge and occupy a series of angular positions round a complete circle; and samples of a given Munsell Chroma, which have an approximately constant value of C^*, are arranged at a constant distance from the Munsell Value axis. (The numerical values of C^* are typically some 5 to 10 times those of Munsell Chroma, depending on the hue and lightness and on which of the two CIE spaces is being used; as already mentioned, the numerical values of L^* are approximately 10 times those of Munsell Value.)

Samples in Munsell colour atlases are usually arranged on pages each of which shows colours all of the same Munsell Hue, those of equal Munsell value being arranged in rows (with the lightest at the top and the darkest at the bottom), and those of equal chroma being in columns (with the least chromatic being closest to the grey axis and the most chromatic farthest away). In Figs. 8.15 and 8.16 are shown the positions, in L^*, C^*_{uv} plots, of the points representing such rows and columns of Munsell colours for Standard Illuminant C for four different hues for the range of samples typically included in the atlases: the points in the plots are in very nearly equally-spaced rows because L^* and Munsell Value are very similar, but they show some deviations from equally spaced columns because the L^*,u^*,v^* space is only a rough approximation to Munsell space. (Deviations of similar size but different in detail occur when the same comparison is made in the L^*,a^*,b^* space.)

Fig. 8.17 shows the positions in the u',v' diagram of colours of Munsell Value 5; this corresponds to a value of L^* of 51.0, and this diagram is therefore a horizontal section through the L^*,u^*,v^* space at this level of L^*. If the Munsell

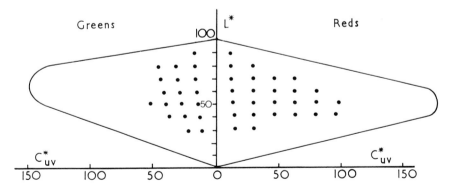

Fig. 8.15. Plot of L^* against C_{uv}, showing the positions of samples in the Munsell atlas having Munsell Values of 3,4,5,6,7,8, and 9, and Munsell Chromas of 2,4,6,8, etc., for Munsell Hues 5R and 5G. The outer lines show the optimal colour limits for non-fluorescent colours. (Standard Illuminant C.)

121

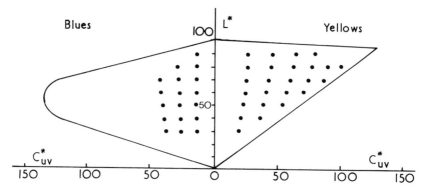

Fig. 8.16. Same as Fig. 8.15 but for Munsell Hues 5Y and 5B.

and L^*,u^*,v^* spaces were identical, the lines in Fig. 8.17 would all be equally spaced concentric circles or straight radii. This is only very approximately the case, and in particular it should be noted that many of the lines of constant Munsell Hue are somewhat curved. The L^*,a^*,b^* space is similarly only an approximation to the Munsell space; compared to the L^*,u^*,v^* space, it approximates the Munsell space more closely for some colours (for example, colours of low chroma), but less well for others (for example, yellow and purple colours of high chroma).

An approximately uniform colour solid makes it possible to represent the optimal colour limits (see Section 7.10) for non-fluorescent colours in a more meaningful way than can be done in a chromaticity diagram. In Figs. 8.15 and 8.16, the outer lines show these limits in the L^*,u^*,v^* space for illuminant S_C for these hues. Other vertical planes containing the L^* axis would show intermediate sections through the solid. It is clear that the maximum value possible for C^*_{uv} ranges from about 120 to 180 according to the hue, and that these maximum values occur at values of L^* of about 90 for yellows, but only about 35 for blues, with intermediate values of L^* for the other hues.

In Appendix 2 a table of values of Y and L^* is given, together with other data and a worked example in which a colour difference is evaluated.

In the case of samples viewed with a dim surround, the resulting drop in apparent contrast (see Section 6.5) could be allowed for by using, instead of $L^*=116(Y/Y_n)^{1/3}-16$, $L^*_{dim}=116(Y/Y_n)^{(1/3.75}-16$; a similar allowance for a dark surround could be made by using $L^*_{dark}=116(Y/Y_n)^{1/4.5}-16$. If L^*_{dark} is used instead of L^* in evaluating u^* and v^*, modified values, u^*_{dark} and v^*_{dark}, would be obtained; if the dark surround produced any desaturating effects on the colours (see Section 6.7) this could be allowed for by reducing appropriately the factor 13 when making the calculations. Similar considerations also apply to dim surround conditions.

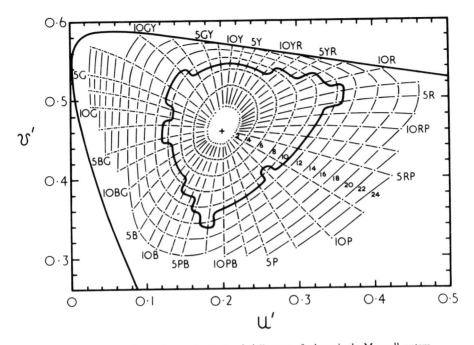

Fig. 8.17. The dots indicate the position in the u',v' diagram of colours in the Munsell system, all of Munsell Value 5, of Munsell Hue 2.5, 5, 7.5, and 10 R (Red), YR (Yellow-Red), Y (Yellow), GY (Green-Yellow), G (Green); BG (Blue-Green), B (Blue), PB (Purple-Blue), P (Purple), and RP (Red-Purple), for Munsell Chromas of 2,4,6,8,10,12, etc., out to the optimal colour limit for non-fluorescent colours. The unbroken irregular line inside the spectral locus shows the gamut of chromaticities for which samples are available in the Munsell Book of Colour (matt sample version) for Munsell Value 5. (Standard Illuminant C.)

8.9 Subjective effects

It is important to recognize the limitations of chromaticity diagrams and colour solids, and their associated parameters, as a means of describing the appearance of colours. Thus, although CIE lightness, L^*, provides a reasonably uniform scale of perceived lightness under average viewing conditions, the actual lightness perceived depends significantly on the actual viewing conditions: thus, a neutral sample having a value of L^* of, say, 50 will appear as a medium grey when seen against a grey background, but as a light grey if the background is black, and as a dark grey if the background is white; this effect is illustrated in Fig. 5.2. Similarly, although CIE chroma, C^*, provides a reasonably uniform scale of perceived chroma under average viewing conditions, the actual chroma perceived depends significantly on the actual viewing conditions: thus a sample having a value of C^* of say, 100, will appear very colourful if the level of illumination is at thousands of lux, of only moderate colourfulness if the level is at tens of lux, and it will approach zero colourfulness if the level is below about a tenth of a lux.

These effects have been studied by several different methods, including haploscopic matching and subjective colour-scaling; we shall consider these two methods in the next two sections.

8.10 Haploscopic matching

The changes in the appearance of colours caused by changes in the lightness and colour of their backgrounds, and in the level of their illumination, occur largely independently in an observer's two eyes; this is presumably because most of the visual adaptation to changes in viewing conditions occurs at the retina of each eye. (The pupil diameters of the two eyes always alter together and they both have the diameter characteristic mainly of the illumination falling on whichever eye happens to be more intensely illuminated; but pupil diameter contributes only a small factor in adaptation phenomena.) This approximate independence of adaptation in the two eyes means that one eye can be adapted to one set of viewing conditions and the other to another set, and then a colour seen by one eye can be matched by varying a colour seen by the other eye until there is equality of appearance. This *haploscopic* matching method was first extensively used by Wright (1934 and 1946) and has since been used by a number of workers including Winch and Young (1951), Burnham, Evans, and Newhall (1952),

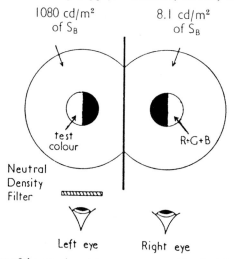

Fig. 8.18. Diagram of the experimental arrangement used in obtaining the results of Fig. 8.19. The right eye always saw the red, green, and blue mixture under the same conditions of adaptation. Changes in appearance of test colours occasioned by changes in the left eye adaptation were measured by matching them with the red, green, and blue mixture seen by the right eye.

124

Burnham (1959), and Hunt (1950, 1952, 1953a, and 1965); and, in a similar method, neighbouring areas of the retina of the same eye were used by MacAdam (1956a and b).

In one haploscopic investigation, the experimental arrangements shown diagrammatically in Fig. 8.18 were used, to obtain the results shown in Fig. 8.19.

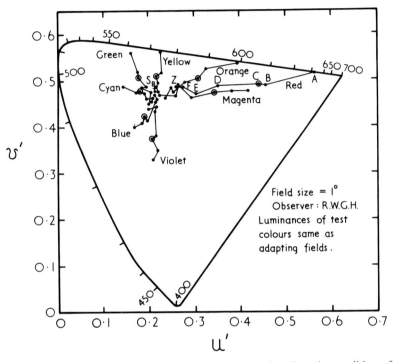

Fig. 8.19. The chromaticities of colours which, when seen under adaptation conditions of Standard Illuminant B at 8.1 cd/m², matched a set of eight test colours seen under various levels of Standard Illuminant B. The points A, B, C, D, E, and F, shows how the appearance of a red colour becomes less saturated as its luminance and that of the adapting field are reduced from 1080 (A), to 65 (B), to 8.1 (C), to 2.7 (D), to 0.32 (E), to 0.075 (F), cd/m²; the point Z refers to the F luminance for the colour but zero luminance for the adapting field. The points marked 'Orange' show similar results for an orange colour; and the points marked with the other colour names show similar results for the other six colours (Hunt, 1952 and 1953a). If the average reflectance of a typical scene is taken as 20 per cent, these luminance levels correspond to the following levels of illumination:

A	16 900 lux	cloudy daylight or operating theatre
B	1020 lux	dull daylight or shop window
C	127 lux	twilight or living room
D	42 lux	twilight or good street lighting
E	5.0 lux	poor street lighting
F	1.2 lux	ten times full moon lighting

Provision was made for the right eye to see a patch of colour composed of a red, green, and blue mixture, the amounts of which the observer could vary at will. Surrounding this patch, and controlling the state of adaptation of the eye, was a surround field, of standard illuminant B, that was kept at a constant luminance of 8.1 cd/m². The left eye viewed a test colour in a central patch, with a surround field of standard illuminant B at various levels (produced by changing the neutral density filter shown in Fig. 8.18), and this controlled the adaptation of the left eye.

The outermost points in Fig. 8.19 show the stimuli necessary to produce in the right eye sensations that matched those produced in the left eye by eight different test colours when surrounded by 1080 cd/m² of standard illuminant B. The stimuli needed by the right eye to match the eight test colours when the left eye was adapted to six lower levels are also shown in Fig. 8.19; at all adaptation levels the luminances of the test colours were the same as that of the adapting field. The results for the seven levels of adaptation for each test colour are connected by straight lines. The lines thus provide a description of the way in which colours appear to become of lower colourfulness as the general illumination level falls. It is seen that the changes in colourfulness with changes of adaptation are very considerable. In Fig. 8.20 similar results are shown for the

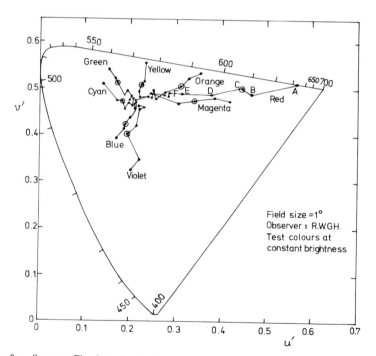

Fig. 8.20. Same as Fig. 8.19, but for test colours at constant brightness (equal to that produced by the right-eye surround).

case where the luminances of the test colours were such that their brightness was the same for all levels of adaptation: it is seen that there is still a large general loss of colourfulness with falling level of adaptation in Fig. 8.20, so that the effect shown in Fig. 8.19 is not caused mainly by the fact that the brightness was falling.

In Fig. 8.21 some results are shown for another set of measurements (Hunt, 1965) made in a manner similar to that shown in Fig. 8.18, but in this case the right eye was adapted to a luminance of 3600 cd/m² (at a colour temperature of 4000 K) and the left eye was allowed an unimpeded view of the outside world. The right-eye adapting luminance was equivalent to a scene of average reflectance (say about 20 per cent) illuminated by sun-light (about 50 000 lux) and this made it possible to match colours seen by the left eye in bright sun-light. Fig. 8.21 shows the combined effect of changing both the colour and the intensity of the adapting light in the left eye: the arrows depict the way in which the appearances of various colours change as the left-eye adaptation is changed from bright sun-light at 50 000 lux to tungsten light at 28 lux, and it is clear that the colours become generally yellower and less colourful. Calculation of values analogous to L^*, as shown in the caption to Fig. 8.21, also indicate that there was a marked drop in brightness when the colours are viewed in the tungsten-light condition. It is thus clear that adaptation in this case has corrected only partially for the yellower and dimmer nature of the tungsten-light condition, and, in addition, a loss of colourfulness of the type shown in Fig. 8.19 has taken place. (In this case, the left-eye adapting illuminant also illuminated the test colours, so that, in going from sunlight to tungsten light, the chromaticities of the test colours would have changed in a manner dependent on their particular spectral reflectance curves).

It must be remembered, when considering the above results, that chromaticity diagrams or solids show relationships between stimuli, and that the corresponding colour sensations vary with the conditions of viewing. Thus Fig. 8.21, although used to describe the effects of altering the viewing conditions, is, none the less, like all chromaticity diagrams, still merely a diagram for plotting stimuli (in this case those needed by the right eye in order to match some stimuli in the left eye) and not a diagram relating sensation magnitudes.

8.11 Subjective colour scaling

Sensations, by their very nature, can only be measured, if at all, by introspection; and, because of the inherent difficulties of this approach, colorimetry has rightly been based entirely on relationships between stimuli, sensations only being classed as equal, or equally different, in some or all respects. Stevens, however, persevered with the method of asking observers to estimate the magnitudes of sensations of brightness and lightness and found that a power-law relationship exists between the luminance or luminance factor of a stimulus and the corresponding estimated subjective magnitude (Stevens and Stevens, 1963). He found, moreover, that a power law exists in a number of other stimulus-sensation relationships, such as loudness, electric shock, weight lifting, heat, and cold. In

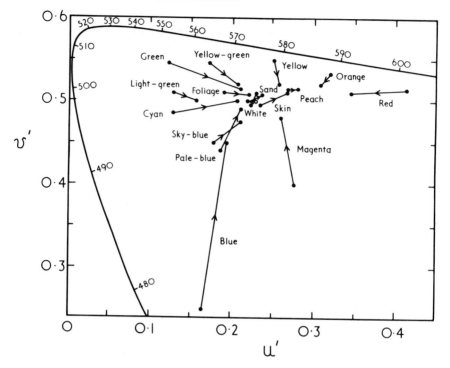

Fig. 8.21. Variation in the appearance of sixteen different test colours as the adapting conditions were changed from bright sun-light at an illumination level of 50 000 lux to tungsten room light at 28 lux; the conditions of reference field were 3600 cd/m² at 4000 K (Hunt, 1965).

The following changes in apparent lightness (evaluated as $116 (Y/Y_n)^{\frac{1}{3}} - 16$) also occured:

Colour	Apparent lightness	
	Bright sun	Tungsten room
White	87	54
Yellow	79	45
Flesh	72	40
Sand	68	36
Orange	61	32
Light-green	61	24
Yellow-green	63	32
Peach	56	35
Sky-blue	56	19
Pale-blue	58	30
Foliage	53	19
Red	44	21
Magenta	40	20
Green	41	17
Cyan	30	11
Blue	28	12

128

the case of luminance, the exact relationship found was that sensation-magnitude is proportional to the cube root of the luminance in the absence of contrast effects; but in their presence it changes more rapidly with luminance or luminance factor, the exact value depending on the conditions, but not normally exceeding a cubic relationship. Others have found similar relationships (Hopkinson, 1956; Padgham and Saunders, 1966), while the writer has derived similar relationships between the physiological response and luminance (Hunt, 1953b). Measures that correlate well with differences in lightness, such as that used in CIE lightness, L^*, also correlate well with experimentally-determined scales of subjectively-estimated lightness-magnitude (Bartleson, 1975).

The extension of these subjective scaling techniques to hue and colourfulness (or chroma or saturation) has also been undertaken. For unrelated colours seen against a dark background, a colour naming technique was used by Kelly to obtain the results shown in Fig. 8.22 (Kelly, 1943); these results would, however, be changed somewhat if the observer were adapted to a particular colour. In most recent investigations of this type, only four hue names are used: red, yellow, green and blue. Violets, purples, mauves, magentas, and pinks, are then described as reddish-blues or bluish-reds, and orange colours as yellowish-reds or reddish-yellows. The wavelengths of the spectrum corresponding to pure red, green, yellow, and blue, vary according to the state of adaptation of the eye, and have been determined for various adaptations (Hurvich and Jameson, 1951; Thomson, 1954).

Using these four fundamental hue names, red, yellow, green, and blue, observers can be asked to scale hue subjectively by expressing it as a certain proportion of two of them: thus an orange might be scaled as 60 per cent red and 40 per cent yellow; or a purple as 80 per cent blue and 20 per cent red. For related colours, lightness can be scaled subjectively by using a suitable range of number such as 100 for white down to zero for black. The third variable for related colours can be scaled in various ways, but colourfulness has been found to be the most satisfactory (Pointer, 1980).

There has been some discussion on the validity of including brown in a colour-scaling system using only four hue names (Rowe, 1973); but it has been shown that browns can be adequately described using red and yellow hue-combinations, provided that they are combined with suitably low estimates of lightness (Bartleson, 1976).

Helson, Judd, and Warren trained a group of observers to scale colours in terms of hue, lightness, and chroma (Helson, Judd, and Warren, 1952). They scaled 60 surface colours illuminated by Standard Illuminant C, and they then scaled the same colours illuminated by Standard Illuminant A. Attempts were then made to express the results in terms of a combination of the objective change in stimulus and the subjective change in observer-adaptation. Similar studies and analyses have also been carried out by Nayatani and his co-workers (Sobagaki, Yamanaka, Takahama, and Nayatani, 1974; Sobagaki, Takahama, Yamanaka, Nishimoto, and Nayatani, 1975).

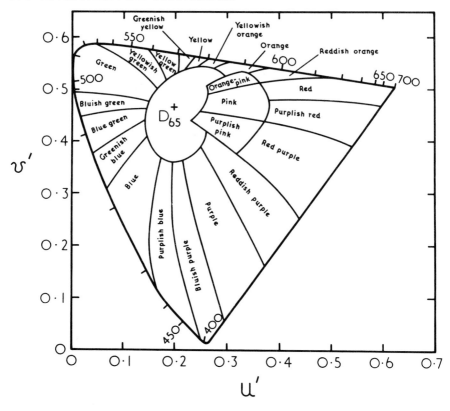

Fig. 8.22 The division of this diagram into a number of regions corresponding to various hues, together with a central region to which no hue name is given, is from Kelly's work on colour designations for lights (Kelly, 1943). It refers to observation of self-luminous areas seen against a dark background.

An alternative, and rather less complicated, approach to the problem is to keep the stimuli constant and to scale the change in their appearance caused by the change in adaptation only. This requires self-luminous stimuli that can be viewed in different surround illuminants in such a way that changing the latter does not affect them objectively: this means that such stimuli must reflect no light from the surround illuminants. Using this method, Rowe scaled stimuli whose luminances were about twice that of the surrounds (Rowe, 1972). Pointer and Ensell have carried out scaling experiments similar to those by Rowe, but using colours whose luminance was half that of the surround (Pointer, Ensell, and Bullock, 1977); the results obtained for Standard Illuminant D_{65} and Standard Illuminant A adapting fields (both at 110 cd/m²) are shown in Fig. 8.23, the colourfulness contours shown being for values of 20, 40, 60, 80, etc.

Grids of lines of the type shown in Fig. 8.23 may be used for predicting changes in colour appearance caused by changes in adaptation. Thus if a colour had a chromaticity in Standard Illuminant D_{65} of $u'=0.35$, $v'=0.45$, its appearance would be about 70 per cent Red, 30 per cent Blue, and colourfulness 80: if this colour was such that its chromaticity under Standard Illuminant A was $u'=0.40$, $v'=0.52$ its appearance would be about 100 per cent Red, and colourful-

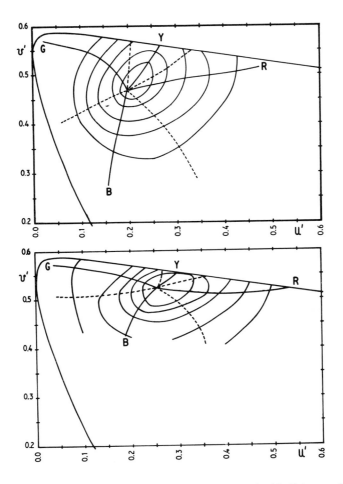

Fig. 8.23. Contours of constant hue and colourfulness as determined by Pointer and Ensell for colours of luminance equal to half that of the surround. Upper: for adaptation to Standard Illuminant D_{65}; lower: for adaptation to Standard Illuminant A. (Pointer, Ensell, and Bullock, 1977). Average results for 5 observers. Adapting fields: luminances: 110 cd/m²; angular subtense 18°. Test colours: luminances: 55 cd/m²; angular subtense 2°.

ness 62; thus the colour had become redder and more colourful. In this way, questions of colour constancy can be answered (Bartleson, 1979).

It must be emphasized that the grids of lines shown in Fig. 8.23 only apply to the conditions used in this particular experiment, and if a different luminance level had been used, or if the luminance level had been different for the two adaptations, or if the colour stimuli had had different luminances relative to the surround, or if they had been seen in a partly dark surround, for instance, then different pairs of grids of lines would probably be required. It has also been shown that adaptation to the colour of an illuminant tends to be incomplete for whites and light greys, complete for medium greys, and in excess for dark greys and near-blacks (Judd, 1940 and 1960).

Evans has investigated the conditions required for the sensation of fluorescence to be experienced (Evans, 1959; Evans and Swenholt, 1967 and 1969; Hunt, 1982). He has shown that there is a transition value of the relative luminance of related colours above which the colour appears first to fluoresce and then at still higher values takes on the appearance of a light source. This transition value may be referred to as the zero-grey point. In colour reproduction, if reflecting objects are reproduced with relative luminances above their zero-grey points they are usually completely unacceptable.

A method for predicting the scaled appearance of surface colours, in terms of their apparent hue, lightness, and chroma, when seen against medium grey backgrounds, is given in Appendix 2. This method is based on a physiologically plausible model of colour vision, and is applicable to illuminants within the range of colours that are normally considered 'white', and to levels of illumination for which colour vision is operating normally. (Hunt, 1982; Hunt and Pointer, 1985.)

8.12 Physical colour standards

As was mentioned in section 7.2, the Munsell Colour System, although objective in nature in that it consists of samples of painted cards, is scaled subjectively in that equal steps on its scales represent as nearly as possible equal subjective colour differences. It is no doubt partly for this reason that it has gained wide recognition as a very useful method of colour specification. One disadvantage to be reckoned with surface colour standards, however, is the difficulty of preventing them from deteriorating with use; this is avoided if transparent glass samples are used, although at the expense of some inconvenience when surface colours are being compared. The Tintometer system uses cyan, magenta, and yellow glasses, that are calibrated on scales of approximately equal visual increments for each colour and are of excellent permanence. Both the Munsell and the Tintometer systems have found wide application, and their usefulness as portable physical sub-standards of colour has been greatly enhanced by the publication of CIE specifications for their samples (see Kelly, Gibson, and Nickerson, 1943, for the Munsell system; see Schofield, 1939, and Haupt, Schlater, and Eckerle, 1972, for the Tintometer system), thus making it possible to transfer results from either system to the CIE system or vice-versa. In the case of the Munsell system a smoothed calibration was also prepared, results in which are referred to as *Munsell Renotation* (Newhall, Nickerson, and Judd, 1943); this is used for specifying the glossy Munsell samples. Another useful system using reflecting samples is the Swedish (NCS) Natural Colour System (Swedish Standards Institution, 1979).

Ceramic tiles can be used as surface colour standards of high permanence, and sets of twelve different colours have been made available by the British Ceramic Research Association (Clarke, 1969; Malkin and Verrill, 1983).

8.13 Whiteness

The CIE has recommended the following formula for providing a measure, W, of the whiteness of samples that are 'called white commercially':

$$W = Y + 800(x_n - x) + 1700(y_n - y)$$

where Y is the tristimulus value of the sample, x and y are its chromaticity coordinates, and x_n and y_n are those of the perfect reflecting diffuser. A measure of the tint, T_W, of the white is given by:

$$T_W = 1000(x_n - x) - 650(y_n - y)$$

The higher the value of W, the greater is the indicated whiteness; the more positive the value of T_W, the greater is the indicated greenishness, and the more negative, the greater the reddishness. Similar formulae are also recommended for use with the 10° Standard Observer.

REFERENCES

Bartleson, C. J., *Color Res. Appl.*, **1**, 181 (1976).
Bartleson, C. J., *Color Res. Appl.*, **4**, 119 and 143 (1979).
Bartleson, C. J., *J. Soc. Mot. Pic. Tel. Eng.*, **84**, 613 (1975).
Baumann, E., *Phot. Sci. Eng.*, **24**, 14 (1980).
Billmeyer, F. W., and Gerrity, E. A., *Color Res. Appl.*, **8**, 90 (1983).
Burnham, R. W., *J. Opt. Soc. Amer.*, **49**, 254 (1959).
Burnham, R. W., Evans, R. M., and Newall, S. M., *J. Opt. Soc. Amer.*, **42**, 597 (1952).
CIE., *Proc. 14th Session (Brussels)*, Vol. A, p. 95 (1960).
CIE., *Proc. 15th Session (Vienna)*, Vol. A, p. 113 (1964).
CIE., Publication No. 15, *Colorimetry* (1971).
Clarke, F. J. J., *Printing Technology*, **13**, 101 (1969).
Eastwood, D., *Die Farbe*, **24**, 97 (1975).
Evans, R. M., *J. Opt. Soc. Amer.*, **49**, 1049 (1959).
Evans, R. M., and Swenholt, B. K., *J. Opt. Soc. Amer.*, **57**, 1319 (1967).
Evans, R. M., and Swenholt, B. K., *J. Opt. Soc. Amer.*, **59**, 628 (1969).
Faulhaber, M. E., and Witherell, P. G., *Applied Optics*, **10**, 950 (1971).
Haupt, G. W., Schlater, J. C., and Eckerle, K. L., N.B.S. Technical Note 16 (1972).
Helson, H., Judd, D. B., and Warren, M. H., *Illum. Eng.*, **47**, 221 (1952).
Hopkinson, R. G., *Nature*, **178**, 1065 (1956).
Hunt, R. W. G., *J. Opt., Soc. Amer.*, **40**, 362 (1950).
Hunt, R. W. G., *J. Opt., Soc. Amer.*, **42**, 190 (1952).

Hunt, R. W. G., *J. Opt., Soc. Amer.*, **43,** 479 (1953a).
Hunt, R. W. G., *J. Phot. Sci.*, **1,** 149 (1953b).
Hunt, R. W. G., *J. Sci. Instrum.*, **31,** 122 (1954).
Hunt, R. W. G., *J. Opt. Soc. Amer.*, **55,** 1540 (1965).
Hunt, R. W. G., *Color Res. Appl.*, **2,** 55 and 109 (1977).
Hunt, R. W. G., *Color Res. Appl.*, **3,** 79 (1978).
Hunt, R. W. G., *Color Res. Appl.*, **7,** 95 (1982).
Hunt, R. W. G., and Pointer, M. R., *Color. Res. Appl.*, **10,** to be published (1985).
Hurvich, L. M., and Jameson, D., *J. Exp.* Psychology, **41,** 455 (1951).
Jones, A. H., *J. Soc. Mot. Pic. Tel. Eng.*, **77,** 108 (1968).
Judd, D. B., *J. Opt. Soc. Amer.*, **30,** 2 (1940).
Judd, D. B., *J. Opt. Soc. Amer.*, **50,** 254 (1960).
Judd, D. B., and Wyszecki G., *Color in Business, Science, and Industry*, p. 317, Wiley, New York (1975).
Judd, D. B., MacAdam, D. L., and Wyszecki, G., *J. Opt. Soc. Amer.*, **54,** 1031 (1964).
Kelly, K. L., *J. Opt. Soc. Amer.*, **33,** 627 (1943).
Kelly, K. L., Gibson, K. S., and Nickerson, D., *J. Opt. Soc. Amer.*, **33,** 355 (1943).
Knight, R., *J. Roy. Television Soc.*, **14,** 39 (1972).
MacAdam, D. L., *J. Opt. Soc. Amer.*, **27,** 294 (1937).
MacAdam, D. L., *J. Opt. Soc. Amer.*, **45,** 500 (1956a).
MacAdam, D. L., *J. Soc. Mot. Pic. Tel. Eng.*, **64,** 455 (1956b).
Malkin, F., and Verrill, J. F., *Proc. CIE 20th Session, Amsterdam,* Paper E37 (1983).
McLaren, K., *J. Soc. Dyers and Colorists,* **87,** 159 (1971).
McLaren, K., and Coates, E., *J. Soc. Dyers and Colorists,* **86,** 354 (1970).
McLaren, K., and Coates, E., *J. Soc. Dyers and Colorists,* **88,** 28 (1972).
Newhall, S. M., Nickerson, D., and Judd, D. B., *J. Opt. Soc. Amer.*, **33,** 385 (1943).
Padgham, C. A., and Saunders, J. E., *Trans. Illum. Eng. Soc.*, **31,** 122 (1966).
Pointer, M. R., *Color Res. Appl.*, **5,** 99 (1980).
Pointer, M. R., Ensell, J. S., and Bullock, L. M., *Color Res. Appl.*, **2,** 131, (1977).
Rowe, S. C. H., Ph.D. Thesis, The City University (1972).
Rowe, S. C. H., *Colour 73*, p. 391, *Proc. 2nd AIC Congress, York 1973,* Hilger, London (1973).
Schofield, R. K., *J. Sci. Instrum.*, **16,** 74 (1939).
Schroeder, A. C., *J. Soc. Mot. Pic. Tel. Eng.*, **88,** 706 (1979).
Sobagaki, H., Yamanaka, T., Takahama, K., and Nayatani, Y., *J. Opt. Soc. Amer.*, **64,** 743 (1974).
Sobagaki, H., Takahama, K., Yamanaka, T., Nishimoto, A., and Nayatani, Y., *Proc. CIE 18th Session, London,* CIE Publication No. 36, p. 192 (1975).
Stevens, S. S., and Stevens, J. C., *J. Opt. Soc. Amer.*, **53,** 375 (1963).
Stiles, W. S., *Phys. Soc. Year Book*, p. 44 (1955).
Stiles, W. S., and Burch, J. M., *Optica Acta*, **6,** 1 (1958).
Swedish Standards Institution, *Colour Atlas*, Swedish Standard SS 01–91–02, Stockholm (1979).
Thomson, L. C., *Optica Acta*, **1,** 93 (1954).
Winch, G. T., and Young, B. M., *G.E.C. Journal*, **18,** 88 (1951).
Wright, W. D., *Proc. Roy. Soc. B.*, **115,** 49 (1934).
Wright, W. D., *Proc. Phys. Soc.*, **53,** 99 (1941).
Wright, W. D., *Researches on Normal and Defective Colour Vision*, Kimpton, London, pp. 209–255 (1946).
Wyszecki, G., *J. Opt. Soc. Amer.*, **53,** 1318 (1963).
Wyszecki, G., and Wright, H., *J. Opt. Soc. Amer.*, **55,** 1166 (1965).

GENERAL REFERENCES

Evans, R. M., *The Perception of Color*, Wiley, New York (1974).
Hardy, A. C., *Handbook of Colorimetry*, The Technology Press, Cambridge, Massachusetts (1963).
Judd, D. B., and Wyszecki, G., *Color in Business, Science, and Industry*, Wiley, New York (1975).
Murray, H. D., *Colour in Theory and Practice*, Chapman & Hall, London (1952).
Wright, W. D., *The Measurement of Colour*, 4th Edn., Hilger, London (1969).

CHAPTER 9

The Colorimetry of Subtractive Systems

9.1 Introduction

SUBTRACTIVE systems of colour reproduction, in which images are formed with cyan, magenta, and yellow dyes or inks, are of great importance because of their widespread use in colour photography and in printing. In this chapter we shall examine the colorimetric properties of such systems, but we shall defer until Chapter 26 a discussion of the added complications that arise when such images are formed by using ink images in the form of dots.

9.2 Subtractive chromaticity gamuts

As shown in Chapter 4, the function of the cyan dye is to absorb red light, that of the magenta dye to absorb green light, and that of the yellow dye to absorb blue light, so that, ideally, the three dyes should have spectral transmission curves as shown by the full lines in Fig. 9.1. These curves are such that at every wavelength two of the dyes have 100 per cent transmission, while only the third dye absorbs. If the absorption bands were narrower than those shown, so that at some wavelengths no light was absorbed, then, no matter how concentrated the deposits of the three dyes, it would not be possible to form black. If, on the other hand, the absorption bands were wider than those shown, the colours would be darker than they need be. The two wavelength values at which the absorption bands change must be at about 500 and 600 nm in order that the colours controlled by the dyes are in fact red, green, and blue, and not, for instance, orange, cyan, and violet, which would limit unnecessarily the range of colours that could be formed. The exact optimum position of these two wavelength values is somewhat indeterminate, but most estimates (see for example Clarkson and Vickerstaff, 1948) give values around 490 and 580 nm and these have been used in Fig. 9.1

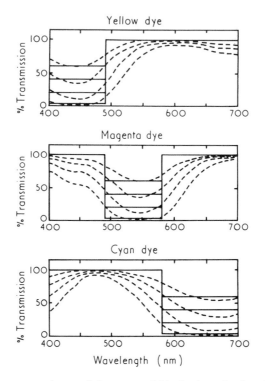

Fig. 9.1. Full lines: spectral transmission curves of 'ideal' subtractive dyes (block dyes) at four different concentrations. Broken lines: spectral transmission curves of dyes typical of those used in practice (reproduced from Fig. 4.1) at four concentrations.

The reproduction stimuli corresponding to these dyes (often referred to as *block dyes*) will be the three blocks of wavelengths 580 to 700 nm, 490 to 580 nm, and 400 to 490 nm, suitably weighted according to the spectral power distribution of the light source used. For a tungsten filament light source operating at a colour temperature of 2856 K (Standard Illuminant A) the reproduction primaries are at R_S, G_S, and B_S in Fig. 7.10. These points are the centres of gravity of the appropriate blocks of wavelengths after weighting for the light source and dividing by the values of v' in the u', v' diagram (see Section 8.6).

The gamut of colours that can be matched with mixtures of the three dyes of Fig. 9.1 is then the triangle R_S, G_S, B_S in Fig. 7.10. It is seen that there are large

136

regions of chromaticity that cannot be matched with these dyes, most notably in the green and magenta parts of the diagram, and it is clear that processes of colour photography that use dyes for forming their colours, that is, subtractive processes, will be unable to match many colours of high saturation. It is important, however, to remember that additive processes are theoretically restricted to a triangle such as R_T, G_T, B_T, and in practice are restricted to one, such as R_P, G_P, B_P, that is not usually very much larger than R_S, G_S, B_S (see Fig. 7.10). It is interesting, in fact, to compare the chromaticity gamuts obtainable using subtractive systems with those obtainable in television. In Fig. 9.2 are shown the gamuts for block dyes used with Standard Illuminant D_{65} (which is the white normally used for European colour television) and the gamut for phosphors representative of those used in typical domestic colour television receivers, having chromaticities:

$u'=0.451$, $v'=0.523$ for red; $u'=0.121$, $v'=0.561$ for green; and $u'=0.175$, $v'=0.158$ for blue (B.R.E.M.A. P.A.L. Working Party, 1969).

It is clear that the two gamuts are very similar.

So far, we have been thinking entirely in terms of dyes having transmission

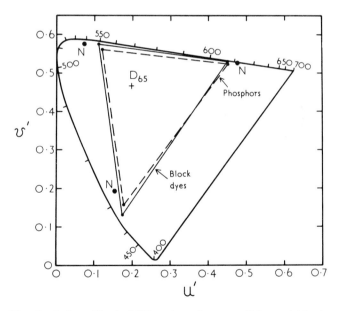

Fig. 9.2. The triangle formed by the full lines shows the gamut of chromaticities that can be reproduced by the block dyes of Fig. 9.1 used with the Standard Illuminant D_{65}; the triangle formed by the broken lines shows the gamut of chromaticities for typical television phosphors used in domestic receivers. (The points marked N show the chromaticities originally chosen for the N.T.S.C. colour television system.)

curves as shown by the full lines of Fig. 9.1. In practice no such dyes exist, and instead use has to be made of dyes having transmission curves of the type shown in Fig. 4.1 (broken lines in Fig. 9.1). They differ in several respects from the 'ideal' curves of Fig. 9.1, but photographically their greatest defect is that they do not transmit 100 per cent of the light in the regions where they are supposed to. These *unwanted absorptions* result in many colours, particularly blues and greens, being reproduced too dark, but the sloping sides of the absorption bands enable a few colours outside the triangle R_S, G_S, B_S to be matched. The reason for this can be seen from Fig. 9.1. As the concentration of the dyes increases the sloping sides of

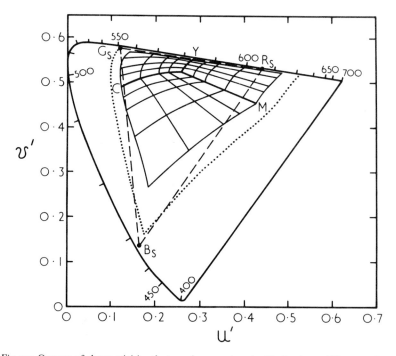

Fig. 9.3. Gamuts of chromaticities that can be reproduced with the dyes of Fig. 9.1 when illuminated by a tungsten filament lamp at a colour temperature of 2856 K. Broken lines: gamut produced by the block dyes (which produce the stimuli R_S, G_S, B_S). Full lines: gamuts produced by various combinations of the four different concentrations of the three typical practical dyes shown in Fig. 9.1. The dotted lines show how the gamut is extended by using, in addition, a fifth concentration, equivalent in the case of each dye to the two highest concentrations combined together; such high concentrations are seldom present in pictures but the chromaticities they produce indicate approximately the limits of a typical modern subtractive film. The five concentrations represent minimum transmissions in the main absorption bands of approximately 56 per cent, 31 per cent, 10 per cent, 1 per cent, and 0.1 per cent, which are equivalent to densities of approximately 0.25, 0.5, 1.0, 2.0, and 3.0.

the absorption bands effectively narrow the red, green, and blue blocks of wavelengths so that the reproduction primaries become slightly more like monochromatic stimuli and hence more saturated. (For a method of analysing the effect of this on colour reproduction, see MacAdam, 1938, page 466.) The gamut of colours obtainable with the real dyes of Fig. 9.1, used with Standard Illuminant A (2856 K), is shown in Fig. 9.3, and is seen to lie partially outside the triangle R_S, G_S, B_S, in the cyan and magenta directions where the triangle R_S, G_S, B_S, is most restricted; flare usually restricts the gamut to that corresponding to maximum densities of about 2.0 (the full lines), but if maximum densities of about 3.0 can be attained the gamut becomes considerably larger (as shown by the dotted line).

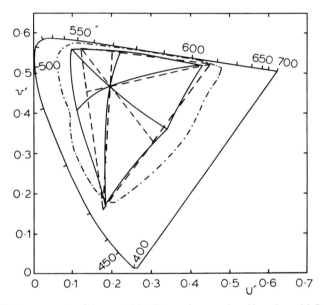

Fig. 9.4. Full lines: gamuts of chromaticities that can be reproduced by using, with Standard Illuminant D_{65}, the dyes of a typical current film, using combinations of concentrations of dye corresponding to densities in the main absorption bands of 2.0. Broken lines: gamuts of chromaticities that can be reproduced by typical television domestic colour receiver phosphors. Dot-dash line: gamut of real colours.

In Fig. 9.4 the gamut obtainable with the real dyes of Fig. 9.1 (considering maximum densities of only 2.0) is compared, using standard illuminant D_{65}, with the television gamut shown in Fig. 9.2. It is clear that the two systems are remarkably similar, both in their gamuts and also in the directions of their principal hue lines; but in the cyan direction the film can produce chromaticities of greater purity, and approaches more nearly the gamut of real colours (shown by the dot-dash line, reproduced from Fig. 7.17). However, the unwanted absorptions of the dyes have a darkening effect on the colours, and this further restricts the gamut of colours that can be reproduced (as will be discussed in the next Section).

9.3 Subtractive gamuts in the colour solid

The restrictions caused by the unwanted absorptions cannot be seen in simple chromaticity plots, such as that of Fig. 9.4, because relative luminance is not shown.

A more comprehensive way of showing the complete gamut is to show the range of chromaticities attainable at various luminances relative to that of a reference white. This has been done by Ohta who has studied the effects of the sizes and widths of the main absorption bands of the dyes as well as those of the unwanted absorptions (Ohta, 1971d, 1972a). For transparencies, he concluded that the peak absorptions should be between 640 and 660 nm for cyans, between 530 and 540 nm for magentas, and at about 430 nm for yellows. The gamut was more sensitive to the position of the magenta peak than to the positions of the cyan and yellow peaks (Ohta, 1971a, 1971b). For reflection prints he showed that the gamut is smaller than that for transparencies using the same dyes, because of the dye broadening effect caused by inter-reflections in the layer (illustrated in Fig. 13.3) and because of white light reflected from the surface of the print (Ohta, 1971c and 1972b).

Another way of showing the complete gamut is to use a three-dimensional colour solid (Yule, Pearson, and Pobboravsky, 1968; Pearson, Pobboravsky, and Yule, 1968), and it is convenient to use for this purpose the CIE L^*, u^*, v^* solid (in relation to a reference white chosen on the grounds of convenience). This is done in Fig. 9.5 where the correlate of chroma C^*_{uv}, is plotted as abscissa, and the correlate of relative luminance, L^*, is plotted as ordinate. The figure is in three sections, each section showing two of the six principle hues: red, green, blue, cyan, magenta, and yellow. If the colour solid is visualized as having the L^* axis vertical, and the u^* and v^* axes lying in a horizontal plane, then Fig. 9.5 shows vertical sections through the solid, all containing the L^* axis, but located at different hue angles. (It is also sometimes helpful to plot v^* against u^* and thus to consider the solid as viewed from the top.) In Fig. 9.5 gamuts are shown for the film dyes and television phosphors of Fig. 9.4. In the case of the television phosphors, the sections are planes through the solid, but in the case of the film dyes, the sections are slightly curved because, as can be seen from Fig. 9.4, for these dyes, as their concentrations are altered, the loci of the six principal hues are not quite straight lines on the chromaticity diagram.

In Fig. 9.5 the L^* axis represents the grey scale in each of the three sections of the figure, with the reference white at the top ($L^*=100$) and black at the bottom ($L^*=0$). Horizontal distance away from the L^* axis represents the degree to which the colour is distinguishable from a grey of the same relative luminance (this is similar to Munsell Chroma). It is thus clear that the colours of highest relative luminance will plot towards the top of each section and those of highest chroma farthest away from the L^* axis. Straight lines radiating from the zero point represent colours of equal chromaticity but different relative luminance, and are of approximately constant saturation.

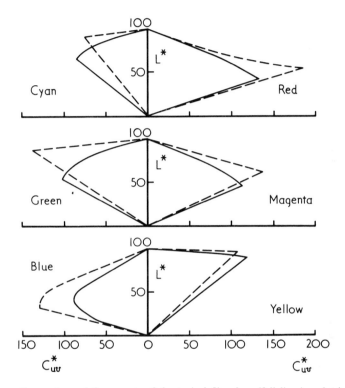

Fig. 9.5. Comparison of the gamuts of the typical film dyes (full lines) and television phosphors (broken lines) of Fig. 9.4, for the six principal hues, on plots of C^*_{uv} against L^*. Film illuminant and television reference white: Standard Illuminant D_{65}. Maximum dye concentrations corresponding to densities of 2.0.

Comparing now the gamuts on Fig. 9.5 (full lines for film, broken lines for television), it can be seen that the conclusions to be drawn are rather different from those from Fig. 9.4. The greater gamut for the film in the cyan direction is still in evidence, but it can now be seen that, although this is effective for values of L^* between 0 and 80, it is the television system that has the greater gamut in the cyan direction when L^* lies between 80 and 100. This is because the unwanted absorptions of the cyan dye prevent cyans of very high relative luminance from

141

being produced in the film. In the red, green, magenta, and blue directions, Fig. 9.4 indicates that the film and television gamuts are rather similar, but Fig. 9.5 makes it clear that, while this is true for colours of low relative luminance (L^* less than about 50 for red, green, and magenta, and less than about 20 for blue), for colours of high relative luminance, the television gamut is, in fact, larger than that for the film. Again, the difference can be attributed to the unwanted absorptions of the film. In the yellow direction, the differences are quite small but with the film having the slightly larger gamut.

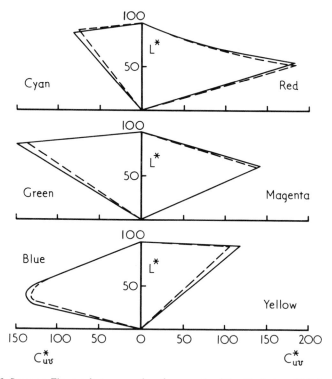

Fig. 9.6. Same as Fig. 9.5 but comparing the gamuts of the block dyes (full lines) and television phosphors (broken lines).

Colour television is thus shown to be capable in most hues of producing colours of higher relative luminance than film, and bright pastel colours are certainly a feature of good colour television. The colours of low relative luminance have more similar gamuts in film and in television, in theory, but, in practice, noise and flare light may restrict these colours more severely in television than in film.

In Fig. 9.6 the same type of plot is used as in Fig. 9.5, but the television gamuts (broken lines) are now compared to those of the block dyes (full lines). It is now seen that the gamuts are very similar, the absence of any unwanted absorptions enabling the colours of high relative luminance to be reproduced by the dyes. When film is used in tele-cine equipment with electronic masking (see Section 23.13) to produce a television display, the effects of the unwanted absorptions are also removed and the gamut then becomes virtually the same as that for the television phosphors (broken lines). If a positive film were made having the unwanted absorptions of its dyes compensated by means of coloured couplers (see Section 15.4), then the gamut attainable would be similar to that of colour television, but it would have to be viewed by a light source of the right colour to offset the colour cast caused by the presence of the coloured couplers.

The above arguments, however, only apply to situations where the television signals or film densities can never result in values of L^* in excess of 100. In television, signal magnitudes usually are restricted to a maximum corresponding to peak white, but, in colour films, peak whites are usually reproduced with a density of about 0.3. Lower densities than this are possible, however, and this means that inter-image effects (see Section 15.5) can result in densities in some layers being less than those for white (even for non-fluorescent colours). This effect does occur in practice especially in reversal films (Clapper, Gendron, and Brownstein, 1973): if inter-image effects were such as to compensate for unwanted absorptions fairly completely, the film gamut would be much more like that of Fig. 9.6 than that of Fig. 9.5, in which case the differences in gamut between film and television in the colour solid would be very small.

9.4 Spectral sensitivities for block dyes

Assuming that we are using the block dyes of Fig. 9.1 with S_A, what must be the spectral sensitivities of the three layers or parts of our process in order that colours be reproduced accurately? Once again they are, of course, colour-matching functions, in this case those corresponding to the primaries R_S, G_S, and B_S. They are shown in Fig. 9.7(a). It is seen that the curves show large negative portions, and this is a direct consequence of the fact that the spectral locus in Fig. 7.8 lies so far outside the triangle R_S, G_S, B_S.

We have already seen (see Section 7.4) that in photographic processes, unless cumbersome and complicated procedures are used, it is impossible for the emulsions to have any negative sensitivities at all. This means that, not only will colours that plot outside the triangle R_S, G_S, B_S be desaturated, but also all colours that plot inside that triangle will be reproduced incorrectly, and, although the most usual fault will be desaturation, errors in hue may also occur, unless, for instance, the Ives-Abney-Yule compromise is used (see Section 7.9).

If it were possible in a photographic system to use emulsions having negative as well as positive portions to their sensitivity curves, correct reproduction of all colours within the triangle R_S, G_S, B_S might still not occur. It would be further necessary that the tone reproduction of each emulsion be correct, and in practice all photographic systems distort the rendering of tones to some extent at the extreme highlight and shadow ends of the scale.

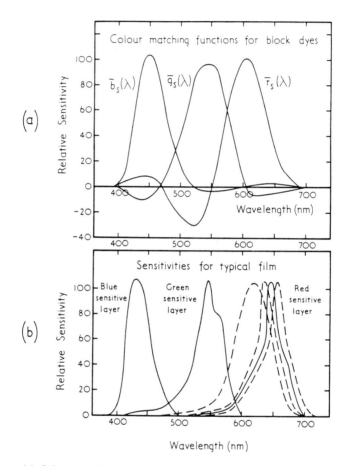

Fig. 9.7 (a). Colour-matching functions for the 'theoretical subtractive' stimuli R_S, G_S, B_S of Figs. 7.8 and 7.10. (*b*) Spectral sensitivities typical of those used in films in practice; the broken lines show alternative choices for the red sensitive layer.

9.5 Spectral sensitivities for real dyes

The block dyes considered in the previous section provide a system in which the chromaticities of the effective reproduction primaries are easily calculated, and they in turn define the appropriate set of colour-matching functions. However, because real dyes do not absorb uniformly in each third of the spectrum, and the chromaticities of their effective primaries therefore vary as the concentrations of the dyes are altered, there is no unique set of theoretically correct colour-matching functions for subtractive systems using real dyes.

In subtractive systems, therefore, accurate colorimetric reproduction of all matchable colours cannot be achieved by simply using a set of theoretically correct spectral sensitivity curves in the camera film. Instead, the approach has to be statistical, and the variables in the system have to be adjusted so that, if accurate colorimetric reproduction is the objective, the departures from it are minimized for ranges of colours typifying those most often encountered in practice, and weighted according to their relative importance and the relative importance of different types of error for them.

The best set of spectral sensitivity curves to choose in practice depends on several requirements, some of which are conflicting. If possible, it is clearly desirable that the curves be a set of colour-matching functions, because only then will colours that look alike in the original always look alike in the reproduction, and that look different in the original always look different in the reproduction. Colour-matching functions with negative portions are not usually realized because most photographic effects, such as inter-image effects, that might be used to generate negative portions, result approximately in the subtraction of log exposures, whereas what is required is the subtraction of exposures. On the other hand, the curves of all-positive sets of colour-matching functions (such as the $\bar{x}(\lambda)$, $\bar{y}(\lambda)$, $\bar{z}(\lambda)$ curves of Fig. 8.6) overlap one another considerably; and this results in high levels of masking or inter-image effects being required to obtain satisfactory colour reproduction (Hanson and Brewer, 1955), and these high levels may be difficult or impossible to achieve. Widely overlapping sensitivity curves also tend to use the light available for exposing layers of films rather inefficiently. For all these reasons, photographic systems are usually optimized with sensitivity curves that depart somewhat from any set of colour-matching functions. This is illustrated in Fig. 9.7(b) where the film sensitivity curves are clearly different from the colour matching functions shown or any possible linear combination of them (which define all other possible sets). By using all-positive sets of curves that overlap one another less, the levels of masking or inter-image effects required are reduced, and the exposing light is used more efficiently. The penalty is that some colours with unusual spectral power distributions that look alike in the original will not look alike in the reproduction and others that look different in the original will look alike in the reproduction; in particular, the large shift in the spectral sensitivity of the red layer to longer wavelengths, to compensate for its lack of the large negative portion, leads to difficulties with some colours. Certain blue flowers are reproduced too pink (see Section 5.1), and some green fabrics grey or even brown; these distortions can be reduced by adopting smaller shifts for the red layer (see the broken lines in Fig. 9.7(b)), but, in the absence of high levels of masking or inter-image effects, this leads to greens being reproduced too yellow, reds too dull, and Caucasian skin colour (which is an extremely important colour in the Western world) being too ashen and grey. The optimum

145

choice of sensitivity curve for the red layer is, therefore, a compromise, and the full line in Fig. 9.7(b) represents a typical choice. What is usually done in practice, therefore, is to use separations of the spectral sensitivity curves along the wavelength axis as a substitute for negative portions, and inter-image effects and masking as means for correcting for deficiencies in the absorption characteristics of the image-dyes (Yule, 1971; Pearson and Yule, 1973).

Colour films having very high photographic speeds usually achieve some of their high sensitivity by the use of sensitivity curves that are somewhat broader than those shown in Fig. 9.7(b); this has the advantage that more of the exposing light is absorbed, but some reductions in the saturations of the reproduced colours usually occur unless corrected by other means.

A quantity known as the *Colorimetric Quality Factor* or *q-factor* has been proposed as a measure of how closely a sensitivity function approximates a colour-matching function (Yule, 1967, p. 138). A limitation of the Colorimetric Quality Factor is that all wavelengths of the spectrum are given equal weight, whereas in practice some must be more important than others, and a satisfactory method of weighting different wavelengths is required.

The interpretation of colorimetric problems in subtractive colour reproduction would be greatly facilitated if it were possible to describe additive primaries that, for real dyes as well as for the 'ideal' or block dyes, represented the colour of the light controlled by each dye. Various attempts have been made to do this and those made by MacAdam, and by Umberger, will now be described.

9.6 MacAdam's analysis

MacAdam has shown (MacAdam, 1938) that, although the chromaticities of the primaries corresponding to sets of typical real dyes vary with dye concentration, it is possible, at least for some dye sets, to find a much more stable set of primaries that correspond to three carefully chosen mixtures of pairs of the three dyes. The implementation of such a system requires a comprehensive system of photographic masking (see Section 15.8), but, if this is used together with the spectral sensitivity curves generated by the primaries, then a closer approximation is made to a theoretically correct system. However, errors still occur, so that a statistical minimizing of the discrepancies is still necessary. (MacAdam optimized for chromaticity only, but optimization for chromaticity and relative luminance is necessary.)

The shapes of colour-matching functions do not change much for small changes in the chromaticities of the primaries, and the differences between the spectral sensitivity curves of typical colour films and *any* reasonable set of colour matching functions is usually quite large, as illustrated in Fig. 9.7. MacAdam (MacAdam, 1966) has reported that colour films can be made with spectral sensitivities approximating the positive portions of colour-matching functions typical of additive and subtractive primaries, and that such films avoid the common tendency for standard films to render certain blue flowers much too pink (as mentioned in Section 5.1). Compared to standard films, such films also exhibit less shift in colour balance with different illuminants, but the change in spectral sensitization does result in some loss of colour saturation of red and blue colours; if the saturation is restored again by some other means, for instance by the use of better dyes or of inter-image effects, then the shifts caused by different illuminants increase again.

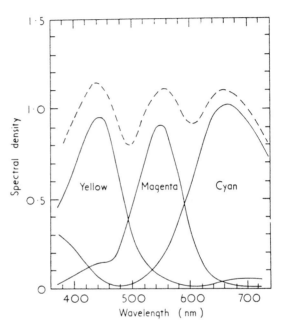

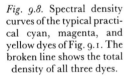

Fig. 9.8. Spectral density curves of the typical practical cyan, magenta, and yellow dyes of Fig. 9.1. The broken line shows the total density of all three dyes.

9.7 Umberger's analysis

In Umberger's analysis (Umberger, 1963), it is convenient to consider the spectral absorption properties of the dyes in terms of their variation of density (rather than transmission) with wavelength; a set of three such spectral density curves is shown in Fig. 9.8. In transparency materials, the density, D_λ, of a dye at any wavelength is usually quite accurately proportional to its concentration, c_1 (this is not so for reflection prints because of the effects of inter-reflections in the dye-layer). For transparencies we may therefore write:

$$D_\lambda = a_{1\lambda}.c_1,$$

where $a_{1\lambda}$ represents the spectral density function of the dye at unit concentration. The spectral transmittance, T_λ, is therefore given by:

$$\log T_\lambda = -D_\lambda = -a_{1\lambda}c_1$$

If now, two further dyes having spectral density functions $a_{2\lambda}$ and $a_{3\lambda}$ are added at concentrations c_2 and c_3 respectively, then the spectral transmittance, T_λ, is given by

$$\log T_\lambda = -a_{1\lambda}c_1 - a_{2\lambda}c_2 - a_{3\lambda}c_3$$

If now a change from c_1 to $c_1{}'$ is made in the concentration of one of the dyes, the transmittance, $T_\lambda{}'$, is given by

$$\log T_\lambda{}' = -a_{1\lambda}c_1{}' - a_{2\lambda}c_2 - a_{3\lambda}c_3$$

These two transmittance curves, together with the spectral power distribution curve, E_λ, of whatever light source is used to illuminate the transparency, will constitute two colours $E_\lambda T_\lambda$ and $E_\lambda T_\lambda{}'$; and the difference between these two expressions,

$$E_\lambda T_\lambda - E_\lambda T_\lambda{}',$$

is the spectral power distribution of a colour, P_1, that when added to $E_\lambda T_\lambda{}'$ produces $E_\lambda T_\lambda$. In other words the colour, P_1, is acting as the additive stimulus corresponding to the dye whose concentration was altered from c_1 to $c_1{}'$.

The problem is to evaluate the expression, $E_\lambda T_\lambda - E_\lambda T_\lambda{}'$, in a useful manner. To do this, we consider the effect of very small changes in c_1; differentiating the expression for $\log T_\lambda$ with respect to c_1, we obtain:

$$\frac{dT_\lambda}{T_\lambda dc_1} = -2.3 a_{1\lambda}$$

where $2.3 = \log_e 10$ approximately.

But for very small changes in c_1, $dT_\lambda = T_\lambda{}' - T_\lambda$. Therefore,

$$E_\lambda T_\lambda - E_\lambda T_\lambda{}' = -E_\lambda dT_\lambda = 2.3 T_\lambda E_\lambda a_{1\lambda} dc_1.$$

This shows that the stimulus P_1 depends on the transmission curve T_λ of the area of film under consideration; and hence P_1 will be dependent on the colour being considered. Subtractive dyes therefore have corresponding primaries that are *unstable* (we have already noted this, and a tendency for the primaries to become more saturated at high concentrations).

If we consider, for a moment, grey colours, then T_λ will be approximately constant throughout the spectrum, and to a first approximation can be replaced by a constant T_n. Then:

$$E_\lambda T_\lambda - E_\lambda T_\lambda{}' = 2.3\ T_n\ E_\lambda a_{1\lambda} dc_1 = k E_\lambda a_{1\lambda} dc_1$$

where k is a constant. The nature of P_1 then becomes very simple: its spectral power distribution is obtained by taking the spectral *density* curve $a_{1\lambda}$ and regarding it as the *transmission* curve of an additive filter; the spectral power distribution thus obtained, $E_\lambda a_{1\lambda}$, is then proportional to the spectral power distribution, $E_\lambda T_\lambda - E_\lambda T_\lambda{}'$, of P_1. (Higher accuracy can be obtained by evaluating $T_\lambda E_\lambda a_{1\lambda}$, if it is desired to avoid the approximation involved in regarding the neutral as non-selective.)

This definition of P_1, though simple, may seem to be of limited application because it applies only to neutral colours and to very small changes in dye concentrations. If, however, T_λ is regarded now, not as the spectral transmission of a patch of uniform colour, but as the integrated transmission of the light from the whole picture area of the transparency, the arguments above remain true, but become more general. For, while few patches of uniform colour in a transparency will be neutral, the integrated light from the whole of the transparency will very often be approximately neutral. The power distribution, $E_\lambda a_{1\lambda}$, can then be regarded as the primary corresponding to the dye for the whole picture. As far as the restriction to very small changes in dye concentration is concerned, the results will provide useful approximations to larger changes, but the instability of the primaries precludes the possibility of finding a primary representative of large changes in a dye concentration.

Power distributions, $E_\lambda a_{2\lambda}$ and $E_\lambda a_{3\lambda}$, can of course be determined to represent primaries P_2 and P_3 representing the dyes having spectral density curves $a_{2\lambda}$ and $a_{3\lambda}$ respectively. The three primaries, P_1, P_2, and P_3, can then be used for various purposes, such as the determination of theoretical sensitivity curves for the set of dyes being used, and the movement of colours on the colour triangle as the concentration of the dyes are varied from neutral. Other sets of primaries can be worked out for non-neutral colours if required, using the expressions: $T_\lambda E_\lambda a_{1\lambda}$, $T_\lambda E_\lambda a_{2\lambda}$, $T_\lambda E_\lambda a_{3\lambda}$.

9.8 Two-colour subtractive systems

If all the colours of a scene could be matched by a mixture of only two dyes, colorimetrically correct two-colour subtractive reproductions could be obtained. The colours in most scenes, however, are not confined even approximately to the above type of restrictions, but two-colour (usually cyan and orange) reproductions can sometimes be surprisingly realistic, and in cinematography have been used commercially (Cornwell-Clyne, 1959, p. 343). Plate 19 (page 288) illustrates the type of colour rendering obtainable: the top right result was obtained with a television gamut restricted to the orange-cyan direction, and can be compared to the normal result at bottom right.

The acceptability of two-colour reproductions is markedly dependent on the subject matter. Indoor scenes are often very realistic, probably because light sources very deficient in blue content, such as candles and yellowish tungsten lamps, are commonly experienced, and the low level of the blue signal tends to reduce vision to nearly two variables. Outdoor scenes, on the other hand, are generally less acceptable, and the inability to render the hue difference between blue sky and green foliage is a serious drawback.

9.9 Subtractive quality

For a more detailed analysis of subtractive systems, including the effects of changes in spectral sensitivities, contrasts, and dye characteristics, on the accuracy of colour reproduction, reference should be made to Chapters 13 and 14 of Evans, Hanson, and Brewer, 1953.

Enough, however, has been said to show that there are fundamental limitations to the fidelity of colour reproduction by subtractive photographic means. Some colours are too saturated to be matched by the dyes available; all colours, whether matchable or not, may be reproduced to some extent erroneously, because of the impossibility of incorporating negative sensitivities in the three emulsions; the tone reproductions of the three emulsions are not usually exactly linear and therefore may introduce errors; and the dyes available absorb in parts of the spectrum where they should have 100 per cent transmission. On the other hand, narrowing and shifting the spectral sensitivity curves can compensate for the lack of negative sensitivities; inter-image effects and masking can greatly reduce the effects of unwanted absorptions; the dyes that we have to use can match a few more colours than can the theoretical dyes; and, as we have seen in Chapter 5, the visual tolerances can be fairly large. In consequence, the practical results can be, and often are, extremely pleasing.

REFERENCES

B.R.E.M.A. P.A.L. Working Party, *Radio Electron. Eng.*, **38,** 201 (1969).
Clapper, F. R., Gendron, R. G., and Brownstein, S. A., *J. Opt. Soc. Amer.*, **63,** 625 (1973).
Clarkson, M. E., and Vickerstaff, T., *Phot. J.*, **88b,** 26 (1948).
Cornwell-Clyne, A., *Colour Cinematography*, p. 343, Chapman & Hall, London (1951).
Evans, R. M., Hanson, W. T., and Brewer, W. L., *Principles of Color Photography*, Wiley, New York (1953).
Hanson, W. T., and Brewer, W. L., *J. Opt. Soc. Amer.*, **45,** 476 (1955).
MacAdam, D. L., *J. Opt. Soc. Amer.*, **28,** 399 (1938).
MacAdam, D. L., *J. Phot. Sci.*, **14,** 229 (1966). See Figs. 34 and 35.
Ohta, N., *Phot. Sci. Eng.*, **15,** 399 (1971a).
Ohta, N., *Phot. Sci. Eng.*, **15,** 416 (1971b).
Ohta, N., *Phot. Sci. Eng.*, **15,** 487 (1971c).
Ohta, N., *Applied Optics*, **10,** 2183 (1971d).
Ohta, N., *J. Opt. Soc. Amer.*, **62,** 129 (1972a).
Ohta, N., *J. Opt. Soc. Amer.*, **62,** 185 (1972b).
Pearson, M. L., Pobboravsky, I., and Yule, J. A. C., *Proc. Tech. Assoc. Graphic Arts*, **20,** 330 (1968).
Pearson, M. L., and Yule, J. A. C., *J. Color and Appearance*, **2,** 30 (1973).
Umberger, J. Q., *Phot. Sci. Eng.*, **7,** 34 (1963).
Yule, J. A. C., Pearson, M. L., and Pobboravsky, I., *Printing Technology*, **12,** 150 (1968).
Yule, J. A. C., Private Communication (1971).
Yule, J. A. C., *Principles of Color Reproduction*, p. 138, Wiley, New York (1967).

CHAPTER 10

Light Sources

10.1 Introduction

W E have already seen that the eye is able to *adapt* to illuminants of different colours. For example in tungsten light, which is deficient in blue light, the eye increases its blue sensitivity. In this way illuminants of different colours result in changes in colour rendering that are much reduced. It was also pointed out in Chapter 5 that cameras must similarly adapt to the colour of the illuminant, since the final picture is generally presented to the eye in such a way that the adaptation that takes place is much less than when the original is viewed. These variations in illuminant colour are of such importance in colour reproduction that we shall devote a chapter to them.

10.2 Tungsten lamps

The most important artificial illuminants are tungsten filament lamps, because of their extremely widespread use. The colour of the light they emit is affected by the colour of the glass used for the envelope, although this is generally very nearly colourless (usually very slightly greenish) and therefore has only a small effect. By far the most important factor determining the spectral power distribution of the light emitted is the temperature at which the filament is operated, and this in turn depends on the resistance of the filament and the voltage applied to the lamp. As the temperature of the filament is raised from room temperature the colours listed in Table 10.1 are produced. The temperatures are listed both in degrees Centigrade or Celsius (°C) and kelvins (K), since the latter figure (which exceeds the former by 273) is generally used for light sources. The temperatures assigned to the colour names are only approximate and the temperature at which the light becomes white depends on the state of adaptation of the observer, and also on the intensity of the light (Hurvich and Jameson, 1951). The maximum temperature obtainable with tungsten filaments is fixed by the melting point of tungsten, which is about 3700 K. Modern tungsten lamps, which run at about 3000 K, give light of a colour that most people describe as white when they are fully adapted to it.

TABLE 10.1

Temperatures of heated objects (incandescent sources)

COLOUR	TEMPERATURE °C	TEMPERATURE K
Extremely dull red	480	753
Very dark red	630	903
Dark red	750	1023
Cherry red	815	1088
Light cherry red	900	1173
Orange red	990	1263
Yellow	1150	1423
Yellow-white	1330	1603

The spectral power distribution of the light emitted by certain sources can be defined very simply; these sources are known as *full radiators* or *black bodies*, and consist of heated enclosures with a small opening through which the light is emitted; this opening must be small in the sense that its area is a small fraction (e.g. one hundredth) of the area of the interior of the enclosure, like the door of a furnace, for example. For full radiators, the spectral power distribution is given by *Planck's Radiation Law:*

$$P(\lambda) = \frac{c_1}{\lambda^5} \cdot \frac{1}{e^{c_2/\lambda T} - 1}$$

where $P(\lambda)$ is the power in watts radiated per square centimetre of surface per micro-metre (μm) wavelength band at wavelength λ; λ is the wavelength in micro-metres, T is the temperature in kelvins; $c_1 = 37\ 418$; $c_2 = 14\ 388$; and $e = 2.718$. When the wavelength is short and the temperature not too high, $e^{c_2/\lambda T}$ becomes very large compared with one, and, to a close approximation:

$$P(\lambda) = c_1/\lambda^5 e^{c_2/\lambda T}$$

This is known as *Wien's Radiation Law*, and for temperatures typical of those used in tungsten filament lamps and for wavelengths in the visible part of the spectrum it is accurate to about 1 per cent.

Tungsten filament lamps are obviously not full radiators in the sense of being heated enclosures with small openings. But the power distributions that they emit are very nearly identical to those emitted by full radiators with temperatures about 50° higher than those of the filaments, so that it has become customary to designate the colour of tungsten filament lamps by quoting these temperatures, which are referred to as *colour temperatures*. Thus a lamp of colour temperature 3000 K, for example, emits light of spectral power distribution almost identical with that of a full radiator operating at this temperature; the actual filament temperature would be about 2950 K but this figure is of little interest and is not generally quoted.

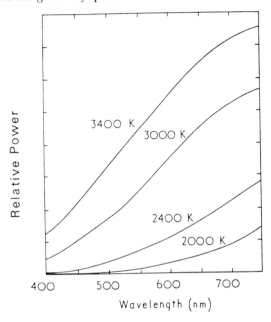

Fig. 10.1. Spectral power distribution curves for four different colour temperatures.

In Fig. 10.1, spectral power distributions for four different colour temperatures are shown (after Murray, 1952). The higher the colour temperature the greater is the efficacy of the lamp, because more visible light (as shown in the figure) and less infra-red light is emitted for a given wattage. Also, high colour temperatures correspond to bluer light and a reduction in the difference between the colour of the light from the lamp and that of daylight. Therefore tungsten lamps are made to operate at the highest possible colour temperatures. For lamps with thick filaments (about a fifth of a millimetre in diameter) colour temperatures of about 3400 K are possible, but with thin filaments (about a fiftieth of a millimetre in diameter) colour temperatures of only about 2500 K can be achieved owing to the fragility of the filament and the serious weakening resulting from any evaporation from it. Thick filaments, of course, are of lower electrical resistance than thin filaments, so that they can only be used for low voltage or high wattage lamps. Thus for a lamp to operate at a colour temperature of 3000 K or over and have a reasonable life, the wattage must be 250 or more, if it is in the 200–250 volt range; but in the 12–24 volt range wattages as low as about 30 can be achieved at these colour temperatures. Conversely, for a given wattage, a low voltage lamp can be operated at a higher colour temperature (and hence at a greater efficacy) than a high voltage lamp. For these reasons, tungsten lamps used in motion-picture and television studios are generally of 100–120 volts and not 200–250 volts. In cine-projectors, where a highly efficient and compact source is required, voltages of 100–120, or even 6–12, are used; for not only are high voltage lamps less efficient, but compact filaments are difficult to make because of a tendency for arc discharges to occur from one filament to another, causing early lamp failure (Aldington, 1954).

Typical efficacies for tungsten filament lamps are about 25 lumens/watt for 3200 K lamps and about 12 lumens/watt for 2650 K lamps.

Failure of tungsten filament lamps is most commonly caused by evaporation of tungsten from the filament: for various reasons this evaporation occurs more markedly at some points than at others so that the filament develops local 'waists' that are thinner than the rest. These waists then become hotter, because of the high resistance caused by the reduced diameter, and the tungsten therefore evaporates at an even greater rate, until finally a break occurs. The tungsten evaporated from the filament is deposited on the inside of the glass wall of the lamp, where it forms a grey or brownish deposit which absorbs light and therefore reduces the light-output.

In *tungsten-halogen* lamps (Zubler and Mosby, 1959; Strange and Stewart, 1963) the blackening caused by the evaporated tungsten is avoided by running the filament in an atmosphere of low-pressure iodine (or other halogen) vapour, and constructing the envelope of the lamp of quartz so that its wall can be maintained above about 250°C in temperature (Levin and Westlund, 1966). When this is done, the tungsten combines with the iodine at the wall to form tungsten iodide:

$$W+I_2 \rightarrow WI_2$$

The tungsten iodide then returns to the neighbourhood of the filament, where, under the influence of its temperature of over 2000°C, the tungsten iodide then dissociates to form tungsten and iodine:

$$WI_2 \rightarrow W+I_2$$

The tungsten is then redeposited on the filament. Unfortunately the tungsten does not go preferentially to the hottest (and therefore thinnest) parts of the filament, so that lamp failure still occurs because of local filament breakage; but such lamps have their average life extended as a result of the iodine cycle, or can be run at a higher colour temperature than ordinary lamps for the same average life. The quartz envelopes of these lamps are maintained at the high temperature required to operate the iodine cycle by making them very compact. The compactness of the lamp is an advantage in that it enables highly efficient light-collecting optical components to be used. Hence the tungsten-halogen lamp has the advantages that it does not blacken with use, and therefore is more efficient and less variable during its life; it can be run at a higher colour temperature, and therefore has a higher efficacy; and it is compact, so that it is convenient and efficient when used with optical components. The iodine vapour does absorb slightly in the yellow-green part of the spectrum so that if too much iodine vapour is included the light has a purplish tinge (Studer and Van Beers, 1964).

In colour photography, when films are designed for use with tungsten light, the colour temperature is usually specified. The photographer frequently finds, however, that the lamps with which he is obliged to illuminate his scene are not of exactly the required colour temperature. But the colour temperature of tungsten lamps can be modified by means of filters, and if all the lamps illuminating a scene are of the same (but wrong) colour temperature, a very convenient method of correction is to put the appropriate filter over the camera lens. If the lamps vary appreciably in colour temperature amongst themselves, some or all of them must be filtered individually, either with or without a filter over the camera lens for general correction.

The colour temperatures of tungsten lamps can be conveniently measured by means of photoelectric colour temperature meters; these instruments compare the intensity of illumination through red and blue filters, and are calibrated in kelvins (Harding, 1952; Palmer, 1965).

10.3 Spectral-power converting filters

When using filters for modifying the colour temperature of the light emitted by lamps, it is convenient to use the reciprocal of the colour temperature, rather than the colour temperature itself. In order to obtain numbers of convenient size, these reciprocals are multiplied by a million and the values thus obtained are called micro-reciprocal degrees, or *mireds*. Thus a colour temperature of 2000 K is equivalent to 500 mireds; 4000 K to 250 mireds. A filter that raised the colour temperature of the light emitted by a source from 2000 K to 4000 K would thus produce a change of -250 mireds. A filter of this type can be designed so that it *always* produces a change of -250 mireds no matter what the original colour temperature of the source. This can be true for filters of all mired-shift values, whether positive or negative, to the same accuracy as that to which Wien's Radiation Law is true. This important property of *spectral-power converting filters* can be proved as follows.

Assuming that the temperatures of the sources and the wavelengths of the spectrum being considered are such that, to a reasonably good approximation, Wien's Radiation Law applies, we have:

$$P(\lambda) = c_1/\lambda^5 e^{c_2/\lambda T}$$

Converting this to logarithms (to the base e), for two temperatures T and T' we have:

$$\log_e P(\lambda) = \log_e c_1 - 5\log_e \lambda - \frac{c_2}{\lambda T}$$

$$\log_e P'(\lambda) = \log_e c_1 - 5\log_e \lambda - \frac{c_2}{\lambda T'}$$

155

Therefore, by subtraction:

$$\log_e P(\lambda) - \log_e P'(\lambda) = \left(\frac{1}{T'} - \frac{1}{T'}\right)\frac{c_2}{\lambda}$$

The expression $\log_e P(\lambda) - \log_e P'(\lambda)$ is the difference (in logarithmic units) between the two power distributions and therefore represents the optical density (to the base e) which a filter must have at each wavelength in order to convert the power distribution from that which is characteristic of a colour temperature T to that characteristic of a colour temperature T'. The way in which the density of this filter varies with wavelength is shown by the above equation to be simply inversely proportional to the wavelength λ, and directly proportional to:

$$\frac{1}{T'} - \frac{1}{T}$$

But this expression is simply one millionth of the mired shift, so that the nature of the filter depends only on the mired shift, and not on the individual colour temperatures T and T'.

The derivation of the spectral density curve required for a filter of a given mired shift, M, is then calculated as follows. The required density (D_e) to the base e is given by:

$$D_e = \left(\frac{1}{T'} - \frac{1}{T}\right)\frac{c_2}{\lambda}$$
$$= 10^{-6}Mc_2/\lambda$$

But the density is usually evaluated to the base 10, so that on this basis the density D is given by:

$$D = 10^{-6}Mc_2/2.303\lambda$$

Inserting the value 14 388 for c_2 we obtain:

$$D = 0.00624M(1/\lambda).$$

Hence if D is plotted against $1/\lambda$, a straight line is obtained, the slope, m, of which is given by:

$$m = 0.00624M$$

Conversely if a filter has a slope of m, the mired shift is given by:

$$M = (160.2)m$$

The slope m is that of the line when plotted against scales such that one unit of density on one axis is the same length as one unit of reciprocal micro-metres on the other axis.

It will be noted that if M is positive, m is also positive, so that for a positive mired shift, that is, a lowering of colour temperature, density increases with reciprocal wavelength, and therefore decreases with wavelength. Conversely, for a negative mired shift, that is, an increase in colour temperature, density increases with wavelength. .

In Fig. 10.2 the spectral densities of two Wratten filters, Numbers 81EF and 82C, are plotted against wavelength (on a reciprocal scale increasing from right to left). If these filters were true power-converting filters, their spectral density curves would be straight lines on this graph. It is clear that over the major part of the visible spectrum the filters do have curves that approximate fairly closely to the theoretical requirements (indicated by the broken lines); the discrepancies at the short wavelength end of the spectrum are quite large, but the wavelength scale is very extended in this region on the reciprocal scale, so that these shortcomings tend to be somewhat over-emphasized. In practice these two filters can be used as power-converting filters quite successfully, and similar filters, of both glass and gelatin, giving various mired shifts, are commercially available.

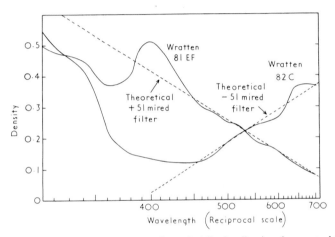

Fig. 10.2 Spectral density curves for two theoretical (broken lines) and two actual (full lines) spectral-power converting filters.

10.4 Daylight

The most important – and the most variable – source of light is daylight. The sun from which all phases of daylight are derived is believed to have a temperature of millions of degrees at its centre, but its surface is much cooler and the colour temperature of the light that it emits is probably between 6000 and 7000 K. The exact determination of this figure is difficult because the light passes through the atmospheres of both the sun and the earth, which are neither neutral nor constant in their spectral absorptions (Moon, 1940; Taylor and Kerr, 1941; Jones and Condit, 1948; Hull, 1954). In Fig. 10.3 a typical spectral power distribution of sunlight as received on the earth's surface is shown as reported by MacAdam (MacAdam, 1958). The curve exhibits a number of undulations, some caused by absorption bands in the solar atmosphere (Fraunhofer lines) and others by absorptions in the terrestrial atmosphere (caused by oxygen and water vapour, for instance).

If the atmosphere is clear and cloudless, the total daylight consists of a mixture of the direct light from the sun together with the diffuse light scattered by the atmosphere. Because light of short wavelengths is scattered much more than light of long wavelengths, this diffuse skylight consists mainly of blue light, and gives rise to the blueness of the clear sky. The diffuse light, however, is not only

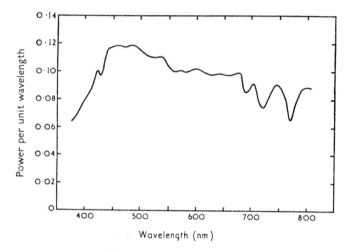

Fig. 10.3. A typical spectral power distribution curve for sunlight as received on the earth's surface (MacAdam, 1958).

scattered downwards to the earth, but also outwards into space, so that there is a net loss of blue light in the combined sunlight and skylight incident on the earth. The sun's surface probably approximates closely in power distribution to a full radiator; but, as seen from the earth's surface, because of the loss of blue light by scattering, and because of the absorptions in the atmosphere of both the earth and the sun, the departures from full radiation are considerable. In Fig. 10.4 are spectral power distributions typical of daylight when the sun is shining and when the conditions are cloudy (Henderson and Hodgkiss, 1963 and 1964). For comparison with the results for the sunny conditions the spectral power distribution of a full radiator at 5630 K is shown by the broken line C, and it is seen that while the general distribution is similar there are some quite appreciable differences. The difference between the colour temperature to which the sunlight now approximates (5630 K) and that of the surface of the sun (6000–7000 K) is a measure of the loss of light of the shorter wavelengths by scattering into space. Compared with the full radiator, daylight is particularly deficient in power at wavelengths below about 430 nm. For comparison with the results for the cloudy

158

conditions the spectral power distribution of a full radiator at 7730 K is shown by the broken line D.

When the weather is cloudy the spectral power distribution of the daylight depends on the height of the clouds. If the clouds are low, then they simply act as a neutral diffusing and absorbing layer that mixes the blue skylight and direct sunlight incident upon them to produce a diffuse light of colour similar to that of the sun and sky together on a clear day. But if the top of the cloud layer is very

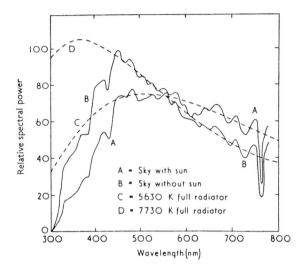

Fig. 10.4. Relative spectral power distribution curves typical of daylight when the sun is shining and when the conditions are cloudy (Henderson and Hodgkiss, 1963 and 1964), together with those of full radiators of about the same colour temperatures.

high, the spectral power distribution on the earth approximates to that of the sun outside the earth's atmosphere. The reason for this is that a very high layer of cloud can catch much of the scattered blue light before it is lost to space and can reflect it back to earth again; that this is possible can be deduced from the fact that at altitudes of 40 000 feet the sky appears quite dark (Harding and Lambert, 1951), indicating that most of the blue light of the sky is scattered at lower altitudes. The colour temperature to which the spectral power distribution for cloudy conditions most closely approximates is about 6500 K, and this in turn approximates to the estimated colour temperature of the sun outside the earth's atmosphere. Fig. 10.5 shows spectral power distributions of daylight for various weather conditions (Condit and Grum, 1964). It has also been shown on theoretical grounds (Middleton, 1954) that the colour of the ground has an appreciable effect on the colour of cloudy daylight, making it greenish over grass, for instance.

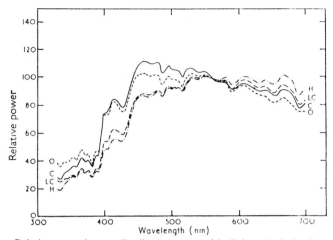

Fig. 10.5. Relative spectral power distribution curves of daylight typical of various weather conditions as received on a nearly vertical surface facing towards the sun. O: Overcast; C: clear; LC: light cloud; H: hazy (Condit and Grum, 1964).

The way in which the colour of the illumination changes as the sun sets depends on the weather conditions. If the sky is cloudless, the increased thickness of the atmosphere through which the rays of the sun must pass before reaching the earth's surface produces the familiar reddening of the light as shown by the spectral power distribution curves of Fig. 10.6 (Condit and Grum, 1964). Thus colour photographs taken in low altitude sunlight often show a pronounced

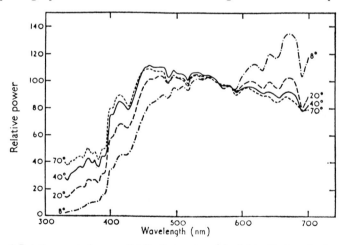

Fig. 10.6. Relative spectral power distribution curves of daylight with clear sky for various solar altitudes, as received on a nearly vertical surface facing towards the sun (Condit and Grum, 1964).

orange cast. But, as sunset is approached, the direct rays from the sun become much weaker so that diffuse light, which is very blue, becomes of more and more importance (see, for instance, Fig. 18 of the paper by Jones and Condit (Jones and Condit, 1948) where sunlight and skylight intensities, and their ratio, are plotted against the solar altitude). Hence the light becomes first redder, and then bluer. With cloud at medium heights, some reddening of the light may be expected at first, but this will soon give way to an increase in blueness as sunset is approached. With high cloud, little or no reddening of the light should occur before sunset. Pitt and Selwyn (1938) found that, except when the direct rays of the sun provided an important part of the general illumination, the colour of the light remained remarkably steady until the rapid increase in blueness took place at sunset. The total variations that can be produced by different phases of daylight are very considerable, as illustrated by Fig. 10.7, where the spectral power distributions are shown for surfaces facing towards clear sun at a solar altitude of 8° and facing away from it at a solar altitude of 30° (Condit and Grum, 1964).

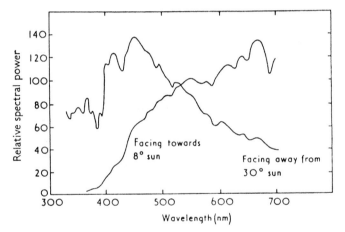

Fig. 10.7. Relative spectral power distribution curves of an extremely bluish (facing away from 30° sun) and an extremely reddish (facing towards 8° sun) sample of daylight.

In Table 10.2 are listed Kodak Colour Compensating Filters that can be used to correct various phases of daylight, so that, as far as colour photography is concerned, the results will always approximate to those that would occur in sunlight on a clear day when the solar altitude is 55°. It must be emphasized that these filter recommendations are only very approximate, and that on individual occasions the required filter may be very different, because of the particular weather conditions prevailing.

Photoelectric colour temperature meters can be used for assessing the

161

relative blue-to-red balance of daylight, but should be calibrated in terms of a suitable range of correcting filters, since the colour temperature readings will be upset by the departures of the power distribution of daylight from those of full radiators.

TABLE 10.2

Kodak Colour Compensating Filters required to correct the colour of various phases of daylight for colour photography. B: Sun behind camera. C: Sun in front of camera.

WEATHER	SUNNY		CLOUDY	
DIRECTION	B	C	B	C
SOLAR ALTITUDE				
10–15°	20B+5C	None	10B+5C	5B+5C
15–20°	10B	,,	10B	5C
20–30°	5B	5Y	5B	None
30–40°	None	10Y	None	,,
40–50°	,,	10Y	,,	,,
50–60°	,,	15Y	,,	5Y

It should be noted that it may not always be desirable from the artistic point of view to correct the colour of the lighting; some distortion in the final colour reproduction may well be useful in creating the right 'mood'; bluish when cloudy, yellowish in low altitude sunlight, for instance.

Standardized spectral power distributions have been drawn up by the CIE to represent daylight for *correlated colour temperatures* (this term is defined in Section 10.11) from 4000 to 25 000 K (Judd, MacAdam, and Wyszecki, 1964), and are known as Standard Illuminants D. Some of these are tabulated in Appendix 2.

10.5 Fluorescent lamps

Fluorescent lamps have spectral power distributions that are mixtures of those of various fluorescent powders and that of the mercury vapour spectrum (Ranby, 1968). Typical examples are shown in Fig. 10.8. The sharp peaks in the spectral power distributions are caused by the presence of the mercury vapour spectrum. The curve labelled WW refers to a warm white tube, while that labelled WWX refers to a warm white de luxe tube which has more light in the far red part of the spectrum but a lower efficacy. The curve labelled N refers to a tube having a correlated colour temperature of about 6500 K, while that labelled AD has a similar correlated colour temperature but has a higher ultra-violet content in order to approximate standard illuminant D_{65} more closely in this part of the spectrum. Fig. 10.9 shows the spectral power distribution of a 'three-band' type of fluorescent lamp; these lamps have phosphors that emit light mainly in fairly narrow bands of light centred at wavelengths of about 435, 545, and 612 nm, and they cause many colours to be perceived with increased saturation (Thornton, 1972). Spectral power distributions of fluorescent lamps vary considerably, and

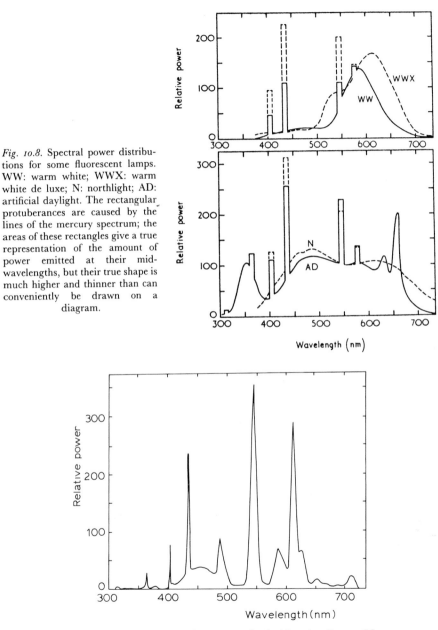

Fig. 10.8. Spectral power distributions for some fluorescent lamps. WW: warm white; WWX: warm white de luxe; N: northlight; AD: artificial daylight. The rectangular protuberances are caused by the lines of the mercury spectrum; the areas of these rectangles give a true representation of the amount of power emitted at their mid-wavelengths, but their true shape is much higher and thinner than can conveniently be drawn on a diagram.

Fig. 10.9. Spectral power distribution for an example of a 'three-band' type of fluorescent lamp.

163

those shown in Figs. 10.8 and 10.9 are only given as a few examples. Further examples are given in tabular form in Appendix 2.

When fluorescent tubes are used with colour films, the mercury line at 546 nm, especially if accompanied by a low level of light in the red part of the spectrum (as in curve WW), often gives results that are unpleasantly greenish. The great variety of film sensitizations, and of fluorescent lamps and their fittings, that are in use makes it impossible to recommend a single filter to correct this greenishness, and for critical results practical tests are always advisable. However, in Table 10.3 some suggestions are given for initial tests. If unusually long exposure times are used, some colour correction may also be necessary for reciprocity failure (see Section 14.3) of the film.

TABLE 10.3

Kodak Wratten filters and Colour Compensating (CC) and Light Balancing filters for various types of colour film.

TYPE OF ILLUMINATION	DAYLIGHT FILMS	TYPE A FILMS	TYPE B FILMS
Daylight (but see Table 10.2)	None	85 or 85C	85B
Blue flashbulbs	None	85	85B
Clear flashbulbs (Aluminium)	80C	81C	81C
Clear flashbulbs (Zirconium)	80D	81C	81C
Electronic flash (Xenon)	None	85	85B
3200 K tungsten	80A	82A	None
3400 K tungsten (photoflood)	80B	None	81A
Fluorescent lamps*:			
Daylight types	CC20M	—	85B+ CC20M
Warm white types	CC30B	—	CC40R
Colour television tubes*	None	85 or 85C	85B

* Note: results obtained with fluorescent tubes and colour television tubes are variable and practical tests are always advisable for critical work.

10.6 Sodium, mercury, and metal-halide lamps

It is well known that discharge lamps using sodium or mercury vapour at low pressures give very poor colour rendering and therefore tend to be used only for street lighting. Such sodium lamps emit almost all their light in the two sodium lines at 589 nm, while the mercury lamps emit virtually no red light. By increasing the pressure of the sodium or mercury vapour in the lamp, the lines are broadened and a continuum throughout the spectrum is added, and high pressure sodium lamps having a correlated colour temperature of about 2000 K are used for street lighting.

By adding metal halides to the vapour, extra lines can be produced, and the colour rendering can then be improved without converting the source into a large area, lower luminance type, as is the case for fluorescent lamps (Beeson and Robinson, 1969; Aldworth, 1971). Metal-halide discharge lamps are therefore useful for flood-lighting sports stadia, because, being compact, they can be used in reflectors so that the light is beamed in the right direction, and the metal-halide additives increase the efficacy of the already efficient discharge type of lamp. Lamps with correlated colour temperatures from 3000 to 6000 K are used, and the efficacy can be as high as nearly 100 lumens per watt (compared to about 25 lumens per watt for a 500 watt 3200 K tungsten lamp, or about 12 lumens per watt for a 40 watt 2650 K tungsten lamp). The power distribution of the lamps, however, is very different from that of full radiators: two examples are shown in Fig. 10.10. Such lamps are particularly useful for the colour televising of sports events (Aldworth and Beeson, 1971; Davies, Jackson, and Rogers, 1972; Ald-

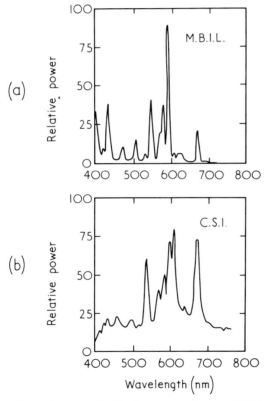

Fig. *10.10.* Spectral power distribution curves for metal-halide discharge lamps: (a) M.B.I.L. type (high-pressure mercury quartz arc tube with metal-halide additives in linear form); (b) C.S.I. type (high-pressure mercury quartz compact-source arc tube in outer bulb). These curves are only examples: many different curves are possible for each type of lamp.

worth, 1975). Metal-halide lamps may show some changes in colour as they warm up, and during their lives (Kaufman and Sauter, 1974), and special precautions may be necessary to avoid flicker (Samuelson, 1977).

10.7 Xenon arcs

Another source providing a mixture of a continuous spectrum and emission of discrete lines is the Xenon arc (Beeson, Bocock, Castellain and Tuck, 1958; Uffers, 1958); a typical spectral power distribution of this source is shown in Fig. 10.11; the exact power distribution depends somewhat on the pressure of the Xenon gas in the lamp, but it is usually fairly similar to that of daylight having a correlated colour temperature of about 6000 K; however, the emission at the red and blue ends of the spectrum is usually rather higher so that the light is very slightly purplish compared to daylight.

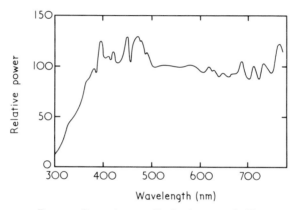

Fig. 10.11. Spectral power distribution curve for Xenon arc.

Xenon lamps are available for running continuously, or with very short pulses of power to give flashes of light of about 1/1000 of a second duration for flash photography. The continuously-run lamps can be used in professional film projectors, for studio lighting, and for flood-lighting, and are particularly useful when light of near-daylight colour is required.

10.8 Carbon arcs

Carbon arcs, operating in air without any glass envelope, have long been used for projecting professional motion pictures. The light produced comes partly from the intensely hot craters of the carbon rods forming the arc, and partly from the combustion of gases between the arcs. The efficacy and colour of the emission are improved by incorporating additives, such as cerium, in the carbon rods, and a typical spectral power distribution is as shown in Fig. 10.12(a) (Dull and Kemp,

1956). Sometimes it is required to supplement studio tungsten lamps with arcs, and in this case arcs are required that emit light having a correlated colour temperature of about 3200 K. The carbon arc can be used for this purpose, too, by suitable choice of additives, a typical spectral power distribution then being as shown in Fig. 10.12(b) (Holloway, Plasket, Dull, and Handley, 1955; Dull and Kemp, 1956).

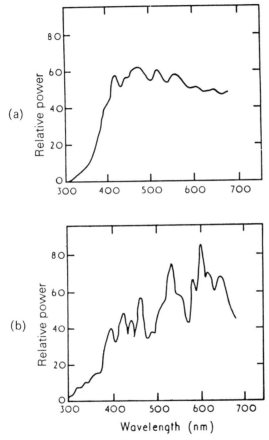

Fig. 10.12. Spectral power distribution curves for carbon arcs. (a) White-flame arc giving light of approximately average daylight quality. (b) Yellow-flame arc giving light having a correlated colour temperature of about 3200 K.

10.9 Photographic flash-bulbs

The ordinary photographic flash-bulb usually consists of a combustible metallic wire, such as aluminium wire, enclosed in a glass envelope containing oxygen.

167

The light emitted is usually similar to that of a full radiator at about 3800 K for an aluminium filling, or about 4000 K for a zirconium filling. Flash-bulbs intended for use with films balanced for use in daylight are usually coated with a lacquer containing a blue dye so as to raise the effective colour temperature to around 5500 K (Keeling, 1969). Electronic flash usually depends on Xenon (see Section 10.7).

10.10 The red-eye effect

It is sometimes found that, in colour photographs taken by means of flash light, the pupils of people's eyes are reproduced red instead of black. This effect is caused by light being reflected by the layers of the eye immediately behind the retina, and since these layers are reddish the reflection has this colour. It is not noticed in everyday life because the amount of light involved is small compared to the general level of illumination. But in flash photography, during the time of the exposure, the flash light produces an illumination level far higher than the ambient light, and the optical system of the eye focuses the reflected light in a fairly narrow beam back towards the flash, so that if the camera lens is close to it, it picks up the reddish light and records the pupils as red instead of black. The effect can only be entirely avoided by having the flash several inches away from the camera lens; but if the ambient lighting is kept high the trouble is alleviated because the pupils of the eyes are then small, thus reducing both the illumination level inside the eye, and the proportion of light reflected.

10.11 Correlated colour temperatures of commonly used light sources

If the relative spectral power distribution of a source is exactly the same as that of a full radiator (black body), then the temperature of the latter is referred to as the *distribution temperature* of the source. Most sources, however, do not duplicate the relative power distribution of a full radiator exactly, but many have the same chromaticity as that of a full radiator; in this case the temperature of the latter is referred to as the *colour temperature*. It is common with other sources of whitish light to quote their *correlated colour temperature*: this is defined as the temperature of the full radiator that produces light most closely matching the particular source. These correlated colour temperatures then provide a useful indication of the relative bluishness or yellowishness of the sources. In Table 10.4 the correlated colour temperatures are given for typical examples of a number of sources commonly used in colour reproduction systems; the corresponding mired values, M, are also given. This mired scale is particularly useful because, over the range of mired values involved, it so happens that equal mired intervals are to a good approximation equivalent to equal colour differences (Knight, 1972). (See Plate 20, pages 288, 289.)

In Fig. 10.13 the chromaticities of full radiators at various colour temperatures are shown on the u',v' chromaticity diagram by a curved line,

TABLE 10.4

Correlated colour temperatures of commonly used light sources

SOURCE	KELVINS	MIREDS
Typical north-sky light	7500	133
Typical average daylight	6500	154
Artificial Daylight fluorescent lamps[1]	6500	154
Xenon (electronic flash or continuous)	6000	167
Typical sunlight plus skylight	5500	182
Blue flash-bulbs	5500	182
Carbon arc (for projectors)	5000	200
Sunlight at solar altitude 20°	4700	213
Cool White fluorescent lamps[2]	4300	233
Sunlight at solar altitude 10°	4000	250
Clear flash-bulbs	3800	263
White fluorescent lamps[3]	3500	286
Photo-flood tungsten lamps	3400	294
Tungsten-halogen lamps	3300	303
Projection tungsten lamps	3200	312
Studio tungsten lamps	3200	312
Warm White fluorescent lamps	3000	333
Floodlighting tungsten lamps	3000	333
Domestic tungsten lamps (100 to 200 W.)	2900	345
Domestic tungsten lamps (40 to 60 W.)	2800	357
Sunlight at sunset	2000	500
Candle flame	1900	526

[1] Sometimes called North-light or Colour Matching lamps.
[2] Sometimes called Daylight lamps.
[3] Sometimes called Natural lamps.

which is known as the *full-radiator locus* (or the *black-body locus*). For sources that do not lie on the full-radiator locus, the correlated colour temperature is calculated as that colour temperature whose chromaticity lies closest to the chromaticity of the source in question on the u,v diagram (this diagram, defined in Section 8.6, is used instead of the $u'v'$ diagram for historical reasons); since the u,v diagram represents equal colour differences by approximately equal distances, this method of calculation gives results reasonably close to those that would be obtained by direct visual comparison by a normal observer. In Fig. 10.14 the part of the full-radiator locus covering the range of colour temperatures of greatest practical importance is shown on a larger scale, together with the chromaticities of some important illuminants.

10.12 Colour rendering of light sources

Sources of different relative spectral power distributions have different *colour rendering* properties: thus sodium lamps which emit almost monochromatic light render colours very poorly. With the advent of fluorescent lamps, in which the

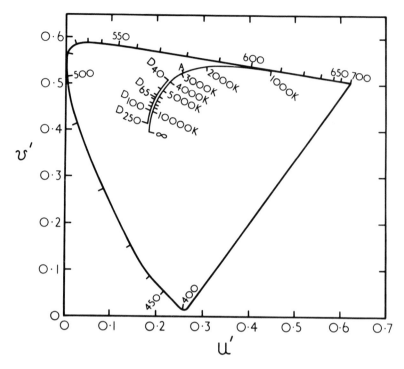

Fig. 10.13. The chromaticities of full radiators in the u',v' diagram. The series D_{40} to D_{250} refers to the CIE standard daylight illuminants having correlated colour temperatures from 4000 K to 25 000 K.

relative spectral power distribution could be varied at will over quite a wide range, it became very desirable to have some means of expressing the degree to which any given source possessed satisfactory colour rendering. To this end, in 1965, the CIE (CIE, 1965) defined a *General Colour Rendering Index*, R_a, as:

$$R_a = 100 - \frac{4.6}{8} (d_1 + d_2 + d_3 + d_4 + d_5 + d_6 + d_7 + d_8)$$

where d_1 is the distance on the u,v chromaticity diagram (multiplied by a factor of 800) between points representing colours having the same spectral reflectance as the Munsell colour 7.5R 6/4 (that is, a Munsell hue of 7.5R, a Munsell value of 6, and a Munsell chroma of 4), when illuminated by the source in question and by the CIE D-illuminant having the chromaticity nearest to it in the u,v diagram (except that for sources of correlated colour temperatures below 5000 K a full-radiator source is used instead of a D-illuminant); d_2, d_3, d_4, d_5, d_6, d_7, and d_8, are similar distances for colours having the same spectral reflectances as the Munsell colours: 5Y6/4, 5GY6/8, 2.5G6/6, 10BG6/4, 5PB6/8, 2.5P6/8 and 10P6/8. The

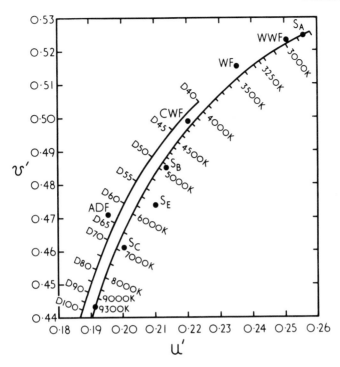

Fig. 10.14. The chromaticities of some important illuminants together with those of full radiators of similar correlated colour temperature. Fluorescent lamps are indicated thus: WWF, warm white; WF, white; CWF, cool white; ADF, artificial daylight. The series D_{40} to D_{100} refers to the CIE standard daylight illuminants having correlated colour temperatures from 4000 K to 10 000 K.

factor of 800 was included so as to make the units similar in size to those used in the U^*, V^*, W^* space: in this space, chromaticities are multiplied by $13W^*$, and, since a Munsell value of 6 corresponds to a W^* value of about 61, $13W^*$ is equal to approximately 800. (The u,v, diagram is defined in Section 8.6, and the U^*, V^*, W^*, space in Section 8.8.)

CIE *Special Colour Rendering Indices* can also be evaluated for individual colours and are given by

$$R_i = 100 - 4.6d_i$$

where d_i is the distance measured in U^*, V^*, W^* space, for the individual colour concerned between the points representing its positions when illuminated by the source being considered and by the nearest D-illuminant (or full radiator for sources below 5000 K).

In a revised method of calculating the General Colour Rendering Index, the

Special Indices are first evaluated for each of the eight Munsell colours in the manner just described, and then averaged (CIE, 1973): this procedure enables any variations in the rendering of the lightnesses of the samples to be included. The revised method also includes a more elaborate method of allowing for the effects of differences between the chromaticities of the source considered and the D-illuminant or full radiator.

Alternatively, the colour rendering of sources can be expressed in terms of the degree to which the percentages of the light they emit in a series of bands throughout the spectrum differ from the percentages present in those bands for the appropriate D-illuminant or full radiator (Crawford, 1963; British Standard 950: 1967). The bands used are given in Table 10.5, together with the amounts of light in each band for illuminants D_{65} and D_{50} (two ultra-violet bands are also

TABLE 10.5

Spectral Band Method of Expressing Colour Rendering

BAND NUMBER	WAVELENGTH RANGE (NM)	BAND VALUE FOR 100 lm OF D65	BAND VALUE FOR 100 lm OF D50
UVI	300–340	11.2 mW	4.7mW
UVII	340–400	43.2 mW	22.4 mW
1	400–455	0.79 lm	0.573 lm
2	455–510	11.2 lm	9.6 lm
3	510–540	23.1 lm	21.8 lm
4	540–590	43.7 lm	44.2 lm
5	590–620	14.4 lm	15.8 lm
6	620–760	6.8 lm	8.01 lm

TABLE 10.6

Calculations of figure of merit for colour rendering by the spectral band method

Spectral band (nm.)	Band luminance for test source	Band luminance for D_{65}	Ratio of band luminance	Single band deviation (per cent)	Double band deviation (per cent)	Excess over tolerance
400–455	0.864	0.79	1.09	+ 9		0
					− 2	0
455–510	9.78	11.2	.87	−13		3
					−13	8
510–540	20.2	23.1	.88	−12		2
					− 1	0
540–590	48.6	43.7	1.11	+11		1
					+ 2	0
590–620	13.4	14.4	.93	− 7		0
					− 1	0
620–760	7.12	6.8	1.05	+ 5		0

Figure of merit = 1024−14 = 1010 or 914−14 = 900.

Sum of excesses $S = 14$

included, because the ultra-violet content of sources is important when samples containing optical brighteners or other fluorescing agents are involved). A figure of demerit, S, can be derived by the method shown in Table 10.6 (in which an artificial Daylight tube is compared to D_{65}), and a figure of merit can be obtained by subtracting S from some suitable number (1024 and 914 have both been used in the literature). The value, S, is the sum of the excesses over the tolerances for the percentage differences in the bands: the tolerances are ± 10 per cent for single bands, and ± 5 per cent for the average percentage difference in all pairs of contiguous bands. Tolerances for illuminants to be used as artificial daylight for the assessment of colour (see Sections 27.9 and 27.10) are sometimes specified in terms of tolerances of ± 15 per cent deviation for the light in single bands and $\pm 7\frac{1}{2}$ per cent deviation for the light in contiguous pairs of bands (with ± 30 per cent for each of the ultra-violet bands; British Standard 950: 1967).

TABLE 10.7

Properties of various lamps

Lamp type	Colour temperature K	Colour rendering index (R_a)	Efficacy lm/W
Tungsten (240 V, 40 W)	2650	100	12
Tungsten (240 V, 500 W)	3200	100	25
Tungsten-halogen	3200	100	25
Daylight (D_{65})	6500	100	—
High pressure Xenon	5290	93	25
Fluorescent:			
Artificial daylight	6500	94	32
3800 de luxe type	3800	92	46
Natural	3800	85	52
Deluxe warm white	3000	80	48
White	3500	56	72
Warm white	3000	54	71
Colour corrected mercury (MBF: high-pressure mercury quartz arc tube in outer bulb with internal fluorescent coating)	3830	44	60
High pressure sodium	2100	21	110
Metal halide (MBI: high-press- ure mercury quartz arc tube with metal-halide additives in clear outer bulb):			
Dy, Na, Tl, In	6430	88	85
Dy, Tl, In	6750	86	85
Sn (+Br)	5010	84	45
Na, Tl, In, Li	4640	69	80–100
Se, Na, Th	4300	66	75–100
Na, Tl, In	4500	64	90–100
Na, Tl, In	5300	62	90–100

In Table 10.7 CIE General Colour Rendering Indices are given for a variety of lamps, together with some other useful data. This table is intended to give the reader a general view of the values of R_a which may be expected for various types of lamp; but the actual figures quoted for R_a, as well as those quoted for colour temperature and efficacy, may differ from those of actual lamps because of various factors, such as the wattage, operating temperature, and life of the lamp. A general trend to be noted is that higher efficacies usually involve lower values of the colour rendering index. (Davies, Jackson, and Rogers, 1972; Moore, Stott, Davies, and Halstead, 1973.)

10.13 Visual clarity

It has been found that, if a type of lamp is used that results in an increase in the perceived saturation of most of the colours in a scene, then, for a given level of illumination, that scene appears to be brighter than with a conventional type of lamp (Thornton, 1972). The three-band type of fluorescent lamp, therefore, results in an apparently higher level of illumination than a conventional lamp of the same light output as measured photometrically. The phenomenon is often referred to as *visual clarity*, and occurs, at least in part, because higher colour-fulnesses normally occur at higher levels of illumination (Boyce and Lynes, 1976; Hunt, 1979).

10.14 Polarization

The light emitted by most light sources is *unpolarized*, by which is meant that the transverse direction in which the light waves vibrate is random. However, in some circumstances the light received by camera lenses is *partially polarized*, by which is meant that the power of the light beam is greater in some directions of transverse vibration than in others. This can occur with specular reflections from non-metallic surfaces; for such surfaces, the light is in fact *fully polarized* (the transverse vibrations all in the same direction) for one particular angle of incidence and reflection, i_p, which is related to the refractive index n, by *Brewster's law*:

$$i_p = \tan n$$

For $n = 1.53$ (a value typical of glass and many plastics) the angle i_p is equal to about $57°$; for $n = 1.33$ (a value typical of water), i_p is equal to about $53°$. The degree of polarization decreases gradually as the angle of incidence and reflection becomes increasingly different from the angle i_p.

When taking pictures through windows it is sometimes possible to reduce unwanted reflections from the glass by using a filter that absorbs light having transverse vibrations in one particular direction, while freely transmitting light having vibrations at right angles to that direction. Such *polarizing filters* are used at the angle that minimizes the unwanted reflections. Unwanted reflections from metallic surfaces can sometimes be similarly reduced, but in this case polarizing filters have to be used over both the camera lens and the light sources. The light from blue skies is partially polarized, especially in directions at right-angles to the sun's rays; hence polarizing filters can be used to darken blue skies relative to other parts of scenes.

In colour measurement, care has to be taken so that partial polarization of the light in colorimeters and spectrophotometers does not cause errors with glossy non-metallic colours.

REFERENCES

Aldington, J. N., *Trans. Illum. Eng. Soc.*, **19**, 319 (1954).
Aldworth, R. C., *Light and Lighting*, **64**, 154 (1971).
Aldworth, R. C., *J. Soc. Mot. Pic. Tel. Eng.*, **84**, 70 (1975).
Aldworth, R. C., and Beeson, E. J. G., *Brit. Kinematog. Sound Tel.*, **53**, 222 (1971).
Beeson, E. J. G., Bocock, W. A., Castellain, A. P., and Tuck, F. A., *Brit. Kinematography*, **32**, 59 (1958).
Beeson, E. J. G., and Robinson, K. G., *Brit. Kinematog. Sound Tel.*, **51**, 212 (1969).
Boyce, P. R., and Lynes, J. A., *Proc. CIE 18th Session, London*, CIE Publication No. 36, p. 290 (1976).
British Standard 950: 1967, Artificial daylight for the assessment of colour. Parts I and II (1967).
CIE Publication No. 13, Method of measuring and specifying colour rendering properties of light sources (1965 and 1973).
Condit, H. R., and Grum, F., *J. Opt. Soc. Amer.*, **54**, 937 (1964).
Crawford, B. H., *Brit. J. Appl. Phys.*, **14**, 319 (1963) and *Trans. Illum. Eng. Soc. (London)*, **28**, 50 (1963).
Davies, I. F., Jackson, M. G. A., and Rogers, B. C., *Lighting Research and Technology*, **4**, 181 (1972).
Dull, R. B., and Kemp, J. G., *J. Soc. Mot. Pic. Tel. Eng.*, **65**, 432 (1956).
Harding, H. G. W., *J. Sci. Instrum.*, **29**, 145 (1952).
Harding, H. G. W., and Lambert, G. E. V., *Nature*, **167**, 436 (1951).
Henderson, S. T., and Hodgkiss, D., *Brit. J. Appl. Phys.*, **14**, 125 (1963).
Henderson, S. T., and Hodgkiss, D., *Brit. J. Appl. Phys.*, **15**, 947 (1964).
Holloway, F. P., Plaskett, C. A., Dull, R. B., and Handley, C. W., *J. Soc. Mot. Pic. Tel. Eng.*, **64**, 657 (1955).
Hull, J. N., *Trans. Illum. Eng. Soc.*, **19**, 21 (1954).
Hunt, R. W. G., *Lighting Research and Technology*, **11**, 175 (1979).
Hurvich, L. M., and Jameson, D., *J. Opt. Soc. Amer.*, **41**, 521, 528, and 787 (1951).
Jones, L. A., and Condit, H. R., *J. Opt. Soc. Amer.*, **38**, 123 (1948).
Judd, D. B., MacAdam, D. L., and Wyszecki, G., *J. Opt. Soc. Amer.*, **54**, 1031 (1964).
Kaufman, A., and Sauter, D., *J. Soc. Mot. Pic. Tel. Eng.*, **83**, 20 (1974).
Keeling, D., *Brit. J. Phot.*, **116**, 329 (1969).
Knight, R. E., *J. Roy. Television Soc.*, **14**, 39 (1972).
Levin, R. E., and Westlund, A. E., *J. Soc. Mot. Pic. Tel. Eng.*, **75**, 589 (1966).
MacAdam, D. L., *J. Opt. Soc. Amer.*, **48**, 832 (1958).
Middleton, W. E. K., *J. Opt. Soc. Amer.*, **44**, 793 (1954).
Moon, P., *J. Franklin Inst.*, **230**, 583 (1940).
Moore, J. R., Stott, P., Davies, I. F., and Halstead, M. B., *Lighting Research and Technology*, **5**, 17 (1973).
Murray, H. D., *Colour in Theory and Practice*, Chapman & Hall, London, p. 205 (1952).
Palmer, D. A., N.P.L. Quarterly, page 2 (July to September, 1965).
Pitt, F. H. G., and Selwyn, E. W. H., *Phot. J.*, **78**, 115 (1938).
Ranby, P. W., *Light and Lighting*, **61**, 227 (1968).
Samuelson, D. W., *Brit. Kinematog. Sound Tel.*, **59**, 314 (1977).
Strange, J. W., and Stewart, J., *Trans. Illum. Eng. Soc.*, **28**, 91 (1963).
Studer, F. J., and Van Beers, R. F., *J. Opt. Soc. Amer.*, **54**, 945 (1964).
Taylor, A. H., and Kerr, G. P., *J. Opt. Soc. Amer.*, **31**, 3 (1941).

Thornton, W. A., *J. Opt. Soc. Amer.*, **62,** 457 (1972).
Uffers, H., *J. Soc. Mot. Pic. Tel. Eng.*, **67,** 389 (1958).
Zubler, E. G., and Mosby, F. A., *Illum. Engr.*, **54,** 734 (1959).

GENERAL REFERENCES

Barrows, W. E., *Light, Photometry, and Illuminating Engineering*, McGraw Hill, New York (1951).
Cayless, M. A., and Marsden, A. M., *Lamps and Lighting*, 3rd Edn., Arnold, London (1983).
Edwards, E. F., and Burgin, R., *Phot. J.*, **106,** 319 (1966).
Henderson, S. T., *Daylight and its Spectrum*, Hilger, London (1970).
Moon, P., *The Scientific Basis of Illuminating Engineering*, McGraw Hill, New York (1936).
Murray, H. D., *Colour in Theory and Practice*, Chapman & Hall, London (1952).
Stiles, W. S., and Wyszecki, G., *Color Science*, Wiley, New York (1967).
Stimson, A., *Photometry and Radiometry for Engineers*, Wiley, New York (1974).
Walsh, J. W. T., *Photometry*, Constable, London (1953).

Plate 12

The reproduction of texture requires good resolution of fine detail, particularly luminance detail. In television, with its limited resolution, the use of a camera with a long focal-length lens can be used to show texture in small sections of the scene, and cameras with shorter focal-length lenses to show more general views (see Section 23.2). Reproduced from a $2\frac{1}{4}$ inch square *Ektachrome* transparency on a Crosfield *Magnascan* scanner.

Plate 13

(*Overleaf*). This example illustrates the remarkably high definition possible with subtractive colour film and modern scanners (see Sections 18.16 and 29.7): reproduced from an area measuring only 16 × 25 mm of a *Kodachrome* transparency on a Crosfield *Magnascan* scanner.

Plate 14
The small diameter of the pupil of the eye, its ability to change its focus rapidly and automatically as a scene is scanned, and the ability of the brain to ignore irrelevant information, tend to give us the impression that all objects are in focus at the same time. In pictures, however, out-of-focus effects are more severe, because camera lenses usually have apertures of larger diameters, and the picture itself is viewed as an object. Out-of-focus objects have 'soft' edges and these can give their colours an enhanced luminosity which is rarely seen in real scenes. In this example, the daffodils in the extreme foreground are very out-of-focus, and seem more 'luminous' than those that are in focus. By a similar process, the control of the sizes of light sources in studios can alter the sharpness of the edges of shadows, and affect the apparent contrast of scenes and their reproductions. (See Section 6.12 and 18.11.) Reproduced from a $2\frac{1}{4}$ inch square *Ektachrome* transparency on a Crosfield *Magnascan* scanner.

CHAPTER 11

Objectives in Colour Reproduction

11.1 Introduction

THE ultimate test of any colour reproduction is the opinion of the person who views it. But opinions differ, and, in cases where dissatisfaction is felt, the viewer often finds great difficulty in saying exactly why he does not like the sensations that he experiences when looking at the picture. Trained observers may feel more competent to name the faults in a reproduction, but training often makes an observer especially sensitive to certain faults that have been prevalent in his experience, while other faults, equally bad to a naïve, but less articulate, observer, he may overlook. A scientific approach to the problem, though difficult, has therefore to be attempted.

11.2 Comparative methods

If it is required to know simply by how much one colour reproduction of some given scene is better than another, a quantitative assessment can be made by recording the independent judgments of a number of observers. Thus if, out of 50 observers, 35 preferred reproduction A, 10 preferred reproduction B, and 5 rated reproductions A and B as being of equal merit, the distribution of the votes can be used as a quantitative measure of the subjective difference between A and B. Similar judgments can then be made between other reproductions, A and C, B and C, A and D, B and D, C and D, and so on, and an order of merit drawn up in which the number of times each reproduction was preferred provides its index of quality. This method of *paired comparisons* is a very powerful tool, and enables a very thorough comparison of a small number of alternative reproductions to be made. For a large number of reproductions, however, it becomes a very time-consuming and laborious undertaking, and in this case the *single-stimulus* method is more practicable.

In the *single-stimulus* method, the alternative reproductions are shown to the observers one at a time, and they are asked to rate them according to some given scale, such as: Acceptable, Doubtful, Not Acceptable; or Excellent, Good, Fair, Poor, Bad (Allnatt, 1965, 1966, 1968; Corbett, 1970). In each case some number of merit points is allocated arbitrarily to each category, such as 1, $\frac{1}{2}$, 0 for the first series, or 4, 3, 2, 1, 0, for the second series, and the total number of points obtained by each reproduction from all the observers is expressed as a percentage of the total number of points that it could have obtained if all observers rated it as high as possible. In this way, merit-percentages are obtained, that provide a quantitative assessment of the reproductions. One difficulty with the single-stimulus method is that the observers may tend to change their standards as the tests proceed, since each picture has to be judged against some mental standard in the observers' minds. But the effects of this difficulty can be greatly reduced by showing the reproductions in random order, and by varying the order for different panels of judges. It is also possible to arrange for observers to scale numerically various perceptual attributes of pictures, such as sharpness, graininess, or contrast, as well as overall quality (Bartleson, 1981).

11.3 Absolute methods

The above methods are useful when a number of existing reproductions have to be compared; and they can also help in answering such questions as 'What are the main faults in this system of colour reproduction?' But to predict *quantitatively* changes of colour that should be made in a system in order to improve it, colorimetry has to be used, although the difficulties are very considerable. However, before colorimetry can be applied it is clearly important first of all to define the objectives of the system. The main purpose of this chapter is to discuss six different types of colour reproduction: *spectral, colorimetric, exact, equivalent, corresponding*, and *preferred*; in different applications the objective could be any one of these six (Hunt, 1970).

11.4 Spectral colour reproduction

If a colour reproduction system is being used for the production of a mail-order catalogue, for example, then it is desirable that the colours of the goods displayed in the catalogue appear the same as those of the actual goods themselves; it is further desirable that this equality of appearance be maintained when the illuminant colour is changed. For instance, if a prospective purchaser is looking at the catalogue in daylight, then the colours should match those of the goods in daylight; but, in addition, if the catalogue and the goods are both taken into electric tungsten filament lighting then the match should still hold good; if the appearance of the goods has changed, that of the catalogue should have done so equally. And the same should be true for all other common illuminants, such as fluorescent lamps.

This requirement (colour-matching independent of the illuminant) can only be met if the spectral reflectance curves of the original and reproduced colours are identical; this is called *spectral colour reproduction* (Hunt, 1970). In colour television, the concept of spectral colour reproduction is also useful, but in this case, since the picture is self-luminous, it has to be defined as equality of relative spectral power distributions. An important feature of equality of spectral reflectance curves, or of relative spectral power distributions, is that it would ensure that the colours matched for all observers (assuming identical viewing conditions), whatever the nature of their colour vision.

Only the Lippman and microdispersion methods (see Sections 1.3 and 1.4) attempt to achieve spectral colour reproduction, and neither is convenient enough to be useful. In photography, sets of cyan, magenta, and yellow dyes are used that cannot achieve spectral colour reproduction except for a few special colours; in printing, much the same is true, although the use of a black ink, in addition to the cyan, magenta, and yellow inks, can provide some help, as can the use of additional coloured inks; in colour television, the spectral emission curves of the phosphors are such that the relative spectral power distributions of the displayed colours are usually markedly different from those of the original colours.

There is, moreover, a sense in which modern systems are becoming worse as far as spectral reproduction is concerned. In order to enlarge the gamut of reproducible colours and improve colour reproduction generally, the dyes used in colour photography are tending to become more spectrally selective, and this means that the more prevalent pale and dull colours will be reproduced with poorer spectral colour reproduction. This is illustrated in Fig. 11.1; amongst real objects, greys tend to be rather non-selective, so that the horizontal line (a) of the figure represents a typical object-colour grey; the broken curve (b) shows how an obsolete colour film used to reproduce this grey, while the continuous curve (c) shows the result for a modern film. It is clear that the modern film has poorer spectral colour reproduction for this colour than the obsolete film; the modern film would thus be more liable to have its matches on this colour upset by illuminant changes or by the spread of characteristics of colour vision amongst observers.

Colours that match one another but differ in spectral composition are known as *metamers*, and it would be useful if the *metamerism* (that is, the extent to which the spectral composition of matching colours differed) could be measured: there is at present no general method of doing this, but, for any pair of illuminants, the

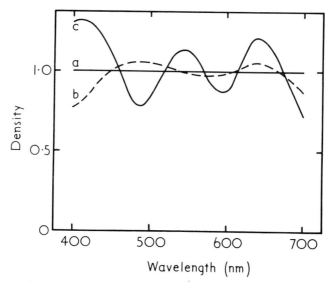

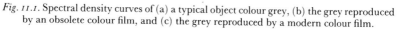

Fig. 11.1. Spectral density curves of (a) a typical object colour grey, (b) the grey reproduced by an obsolete colour film, and (c) the grey reproduced by a modern colour film.

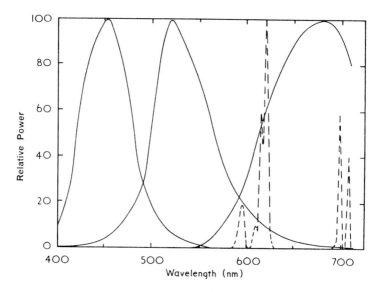

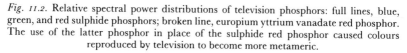

Fig. 11.2. Relative spectral power distributions of television phosphors: full lines, blue, green, and red sulphide phosphors; broken line, europium yttrium vanadate red phosphor. The use of the latter phosphor in place of the sulphide red phosphor caused colours reproduced by television to become more metameric.

180

metamerism can be measured in terms of the extent to which colours that match under one of the illuminants, fail to match under the other. It has, however, been found empirically by Pinney and DeMarsh that the general effects of illuminant metamerism can be minimized in colour photography if the cyan, magenta, and yellow dyes are chosen so that the spectral density minima at about 500 and 600 nm in reproduced greys are made approximately equal (Pinney and DeMarsh, 1963). The same may be true of observer metamerism, but this would have to be checked by suitable techniques, such as computation using the Standard Deviate Observer suggested by Allen (Allen, 1969).

In colour television, spectral colour reproduction has also become worse in recent years with the introduction of the rare-earth red phosphor as shown in Fig. 11.2.

With all the current practicable methods of colour reproduction, whether by photography, by television, or by printing, it is usually impossible to achieve spectral colour reproduction; the only exception is the duplication of an original that itself consists of mixtures of the reproduction dyes, inks, or phosphors. The concept of spectral colour reproduction is, nevertheless, useful, in that it defines the requirement for independence of illuminant colour and of observers' colour vision; and the extent to which any colour reproduction system is sensitive to these factors can be assessed by considering the effects of specified changes in illuminant or observer.

11.5 Colorimetric colour reproduction

Observer metamerism cannot be eliminated in practical situations, but it has been found that, if computations are made with the CIE Standard (2°) Observer data, the results usually accord well with assessments made by (non-colour-defective) real observers. It then becomes possible to define colorimetrically the particular metamer in the reproduction that would match any colour in the original. Such metameric matches are characterized by the original and the reproduction colours having the same CIE chromaticities and relative luminances. This is called *colorimetric* colour reproduction (Clapper and DeMarsh, 1969), which may therefore be defined as reproduction in which the colours have chromaticities and relative luminances equal to those of the original. In the case of reflection prints this would normally imply that the original and reproduction illuminants had the same chromaticities (but their spectral power distributions could be different). The colorimetry is usually carried out relative to a well-lit reference white in the original, and relative to its reproduction in the picture. This procedure makes the relative luminances independent of changes in the intensity of either the original or the reproduction illuminant (or, in television, the luminance of the screen). This is a simplification that has some limitations, which we shall discuss later, but it enables the usual type of colour difference formula to be used. Thus, for daylight viewing of reflection reproductions of scenes lit by daylight, the colorimetric evaluations could be made using a daylight type Standard Illuminant, and departures from colorimetric colour reproduction (Pitt, 1967) could be calculated using the colour difference formula currently recommended by the CIE. However, it must be remembered that, in pictures, the colours of some objects (such as skin, blue sky, grass, foliage, and greys) are more important than others, and errors in some directions (such as hue) are more serious than in others: a distinction therefore has to be made between the *perceptibility* and the *acceptability* of colour differences.

If the appearance of colours were independent of illuminant intensity, then the concept of colorimetric colour reproduction might be applicable to all cases where the original and reproduction illuminants had the same colour (chromaticity co-ordinates); but the appearance of colours certainly is affected, sometimes quite markedly, by the illuminant intensity, which, as can be seen from Table 11.1, can vary very widely, and hence the achievement of colorimetric colour reproduction does not necessarily imply equality of appearance of colours in the original and in the picture; moreover, other factors affecting the appearance of colours are also important.

TABLE 11.1

Typical levels of illumination met with in practice

	TYPICAL DAYLIGHT ILLUMINATION LEVELS			
Bright Sun	50 000	—	100 000	lux
Hazy Sun	25 000	—	50 000	,,
Cloudy Bright	10 000	—	25 000	,,
Cloudy Dull	2000	—	10 000	,,
Very Dull	100	—	2000	,,
Sunset	1	—	100	,,
Full Moon	0.01	—	0.1	,,
Star Light	0.0001	—	0.001	,,
	TYPICAL ARTIFICIAL LIGHT ILLUMINATION LEVELS			
Operating Theatre	5000	—	10 000	lux
Shop Windows	1000	—	5000	,,
Drawing Offices	300	—	500	,,
Offices	200	—	300	,,
Living Rooms	50	—	200	,,
Corridors	50	—	100	,,
Good Street Lighting	20			,,
Poor Street Lighting	0.1			,,

11.6 Exact colour reproduction

If, in addition to the chromaticities and relative luminances being equal, the absolute luminances of the colours in the original and in the picture are also equal, we have a situation in which differences in illuminant intensity (or screen luminance in the case of television) have been eliminated (see Section 7.4): this is

182

called *exact colour reproduction*. Hence the reproduction of a colour in a picture is *exact* if its chromaticity, its relative luminance, and its absolute luminance are the same as those in the original scene. This would result in equality of appearance of the reproduced and original colours providing that the state of adaptation of the eye were the same when viewing the picture as when viewing the original scene; factors that can have an important effect on the adaptation of the eye include the luminance and colour of the surround, the angular subtense, and glare, and only if all these viewing conditions are similar will the adaptation be the same.

Thus, if the reproduction of a certain colour is *exact*, the observer will only see the same colour as he would have done when looking at the original scene, if a number of important conditions are simultaneously met. In general, there would be a difference in colour appearance: if the viewing conditions were not the same for the original object and for the reproduction; or if the observer differed appreciably from the CIE 2° standard observer; and, in practice, it is frequently the case that the spectral power distributions of the illuminants are not quite identical to those assumed for calculating the chromaticities and relative luminances (so that colorimetric errors may be present).

11.7 Equivalent colour reproduction

There are many situations where colorimetric and exact reproduction are known to be erroneous objectives. For instance, if a scene lit by tungsten light is reproduced in a viewing situation in which the ambient lighting is daylight, then colorimetric and exact colour reproduction would both produce results that are too yellow. This situation commonly occurs in colour television: a studio scene lit by tungsten light, if reproduced on a colour receiver with colorimetric or exact colour reproduction, would look too yellow when viewed in ambient daylighting; this is because the eye would be adapted mainly to the daylight, as a result of its larger area, whereas, in the case of the original, the eye would have been adapted to tungsten light and hence would have had its blue sensitivity increased, and its red sensitivity decreased, relative to its green sensitivity. (The optimum colour balance to choose for viewing a colour television display in a variety of ambient illuminant colours is discussed in Section 21.12.)

Because of the effects of the viewing conditions, such as those just described, it is necessary to define a fourth type of objective, *equivalent colour reproduction*; this is defined as reproduction in which the chromaticities, relative luminances, and absolute luminances of the colours are such that, when seen in the picture-viewing conditions, they have the same appearance as the colours in the original scene.

There are at least three types of effect that are of practical importance in this connection: the effects of differences in colour between the original-illuminant and the reproduction-illuminant; the effects of differences in intensity between the two illuminants; and the effects of differences in the surround of the original and of the reproduction. To predict these effects adequately requires more research, but the following examples of results obtained by haploscopic matching (see Section 8.10) illustrate their natures.

In Fig. 11.3 the chromaticities of pairs of equivalent colours (Hunt, 1957) are shown for tungsten light (dots) and daylight (arrow-heads); it is seen that, as expected, stimuli have to be bluer in daylight adaptation to elicit the same sensations as in tungsten-light adaptation. Equations relating equivalent colours for tungsten light and daylight adaptation have been proposed (Burnham, Evans, and Newhall, 1957; Nayatani, Y., Takahama, K., and Sobagaki, H., 1981).

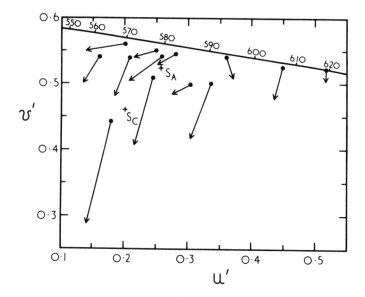

Fig. 11.3. The chromaticities of colours that appear the same in 8.1 cd/m² of Standard Illuminants A (dots) and C (arrow-heads).

In Fig. 8.20 the chromaticities of series of equivalent colours are shown for a series of changes in illuminant intensity (Hunt, 1952 and 1953). It is seen that as the illuminant intensity is decreased there is a gradual decrease in colourfulness. Fig. 13.5 shows the luminances of series of equivalent colours for a series of greys from white to black viewed under a range of illuminant intensities (Hunt, 1965a). It is seen that, as the level of illumination drops, the brightnesses decrease, and there is also a slight reduction in apparent contrast (in this figure this could be an artefact of the scale used as ordinate, but other investigations also support this finding (Bartleson and Breneman, 1967a)).

In Fig. 11.4 an example of the effect of the surround is given; equivalent colours were measured for a grey scale seen first with a grey surround and then with a dark surround; it can be seen that (as discussed in Chapter 6) the dark surround has the effect of decreasing the apparent gamma (Hunt, 1965b). A dark surround may also decrease the apparent colourfulness of colours (Hunt, 1950; Rowe, 1972; Hunt, 1973; Pitt and Winter, 1974; Breneman, 1977). It is interesting to note that stage make-up usually results in an enhancement of tonal and colour differences, and this is presumably necessary in order to overcome reductions in apparent 'gamma' and colour saturation as a result of the stage being seen with the dim or dark surround of the rest of the auditorium.

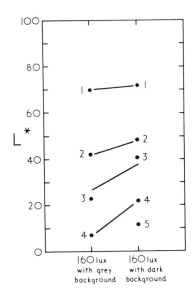

Fig. 11.4. The luminances of greys (plotted on the CIE L^* scale) that, when seen under adaptation conditions of 3600 cd/m² at 4000 K, are equivalent to a series of greys illuminated at 160 lux (of tungsten light) with a grey surround and with a dark surround. The dark surround makes the dark greys appear lighter, and hence reduces the apparent gamma.

In Fig. 11.5 an example is given of the combined effects of changes in illuminant colour (on both the sample and the eye), illuminant intensity, and surround: equivalent colours were measured for the samples in a colour chart for bright sunlight and for the chart illuminated by the light from a tungsten projector so as to give the appearance of a picture projected with a dark surround (Hunt, 1965b). It is clear that the appearance of the colours in the two cases is quite different, the change to the (dimmer) tungsten projector making the colours less colourful with a shift towards yellow-orange in addition. (However, these results are dependent on the particular spectral reflectance curves of the sample

185

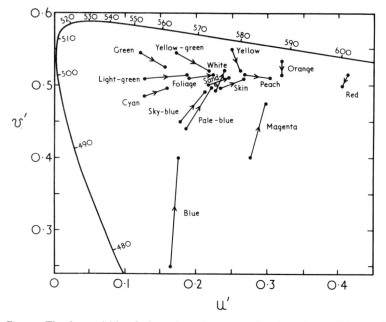

Fig. 11.5. The chromaticities of colours that, when seen under adaptation conditions of 3600 cd/m² at 4000 K, are equivalent to a selection of particular colours viewed first under bright sunlight at 43 000 lux and then under the light from a tungsten projector at 160 lux with a dark surround. The arrowed lines show that the effect of changing to the dimmer tungsten-projector illumination is to make the appearance of the colours less colourful, with a shift in the yellow-orange direction superimposed.

The following changes in apparent lightness (evaluated as $116 \, (Y/Y_n)^{1/3} - 16$) also occurred:

Colour	Apparent lightness	
	Bright sun	Tungsten projector
White	87	64
Yellow	79	75
Flesh	72	53
Sand	68	43
Orange	61	36
Light-green	61	32
Yellow-green	63	36
Peach	56	42
Sky-blue	56	32
Pale-blue	58	35
Foliage	53	28
Red	44	22
Magenta	40	19
Green	41	13
Cyan	30	9
Blue	28	10

186

colours used in the chart: other colours, having the same chromaticities in sunlight but different spectral reflectance curves, would exhibit different shifts in chromaticity when in the projection light, and hence would have different equivalent colours; but the general trends would be the same.)

When equivalent colour reproduction is the objective, it is now possible to take colour motion pictures under all conditions of illumination without the use of supplementary light sources: this is achieved by using a film of speed 160 ASA, a camera with a 240° shutter, and an f/1.2 lens (*Movie Maker Magazine*, 1972). With this system, if the scene is dimly lit, the picture will be dense and will project as a dim reproduction, and hence equivalent colour reproduction will have been achieved, at least as far as brightness is concerned.

11.8 Colorimetric colour reproduction as a practical criterion

It is clear from the foregoing that the effects of various differences in the viewing conditions between the original and the reproduction are complicated and quite large; yet the recognition of the colours of objects in real life can be undertaken with reasonable consistency over a wide variety of illuminant and viewing conditions. This is in large part because, with the vast majority of naturally occurring objects, we are concerned with surface colours (as distinct from self-luminous colours) and recognition is then a relative, rather than an absolute, matter. For instance, if a grey surface is viewed first in bright sunlight at 50 000 lux and then in artificial light at 50 lux, its absolute luminance will have dropped by a factor of 1000 to 1, and although it will certainly look of lower brightness, so will all the other objects in the field of view: because of this, the surface maintains its appearance as a grey, and the apparent amount of greyness remains approximately constant. It is therefore helpful to consider colours relative to all the other objects in the field of view. This approach is simplified by considering colours relative to white, as is done in colorimetric colour reproduction, because in any normal viewing situations there is a fairly small range of luminances that appear white, those above appearing fluorescent or luminous and those below appearing greyish (Evans, 1959).

It has been found (Hunt, Pitt, and Ward, 1969) that, by considering the reproduction of greys relative to white (as is discussed in Chapter 6), it is possible to deal in concepts that avoid the difficulty that in many situations it is impossible for the reproduction to produce the same sensations as the original. For instance, a white object seen in bright sunlight produces a brightness that a reflection print seen in artificial light is completely incapable of matching: but, if the object is reproduced as a white on the reflection print, correct tone reproduction relative to white has been achieved, and the result is found to be fairly satisfactory.

A similar problem arises with respect to colourfulness. If colours of high purity are viewed in sunlight, the sensations they produce are more colourful than any stimuli (even including spectral stimuli) can produce under typical levels of artificial lighting (Hunt, 1953). To conclude that it must be impossible to produce satisfactory colour reproductions of sunlit scenes for viewing in artificial light is contrary to experience.

There are, therefore, very good reasons for considering colour reproduction relative to white, and hence much justification for the measurement of luminances relative to a well-lit reference white. Equality of such relative luminances and equality of chromaticities we have termed *colorimetric colour reproduction*.

Colorimetric colour reproduction is thus perhaps quite a good criterion for reflection prints (assuming they have a surround similar to that of the original) viewed in light of the same colour, but usually of different intensity, as was used for the original. However, it must not be forgotten that, because the apparent brightness and colourfulness vary with the illumination level, it will always be the case that the print will look more like the original if the illumination on the print is adjusted in intensity so as to be closer to that in the original.

11.9 Corresponding colour reproduction

The same problems concerning brightness and colourfulness arise in connection with equivalent colour reproduction. To overcome this difficulty we need the concept of *corresponding colour reproduction*, which is defined as reproduction in which the chromaticities and relative luminances of the colours are such that, when seen in the picture-viewing conditions, they have the same appearance as the colours in the original would have had if they had been illuminated to produce the same average absolute luminance level as that of the reproduction. By eliminating any differences in absolute luminance levels between the original and the reproduction we avoid, as in the case of colorimetric colour reproduction, unrealistic conclusions that pictures of brightly-lit scenes cannot be reproduced for viewing at lower levels of illumination; but by requiring equality of appearance in other respects we can allow for the effects of differences in surround and illuminant colour.

The way in which the above definition enables allowance to be made for the effect of the surround on the reproduction of white, grey, and black colours is illustrated in Fig. 11.6 (reproduced from Chapter 6). The reproduction density required in order to achieve corresponding colour reproduction is plotted as ordinate, and the logarithm of the exposure of the original scene relative to white (Bartleson and Breneman, 1967b) is plotted as abcissa (Hunt, 1969). The requirements for three different surround conditions are shown: average sur-round, such as occurs with reflection prints; dim surround, as for television viewing or viewing sheet transparencies on illuminated opals; and dark sur-round, as for films projected in a dark room. The gamma has to be raised to about 1.25 and 1.5 respectively for the latter two cases because the dim and dark surrounds reduce the apparent gamma. (Ambient lighting in television viewing situations is usually variable, but it has been found that, if the television display has a gamma of about 1.5 in dark surround conditions, then, as the ambient lighting is increased to typical levels, the amount of viewing flare added usually reduces the gamma to about 1.25 as required (Novick, 1969).)

When the gamma of a reproduction is increased, the purity of its colours is also increased, because the higher gamma results in greater ratios between the tristimulus values (MacAdam, 1938). (In these discussions all the tristimulus values are relative to the well-lit reference white.) But, as has already been mentioned, a dark or dim surround may result in a *reduction* in apparent colour saturation for a given purity.

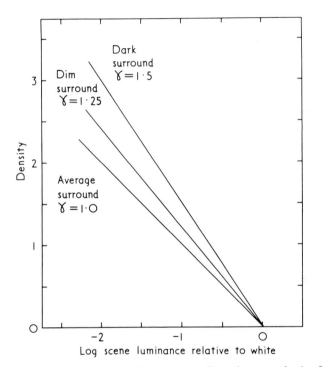

Fig. 11.6. The density required to achieve corresponding colour reproduction for whites, greys, and blacks, is shown for three different surround conditions: average surround, as is common for reflection prints; dim surrounds, for television and for the viewing of sheet-films on illuminated opals; and dark surround, for pictures projected in dark rooms.

To allow for the effects of changes in illuminant colour, Evans (Evans, 1943) proposed that the three tristimulus values (X, Y, and Z) of each original colour (relative to the well-lit reference white) be multiplied by the appropriate one of the three different factors representing the ratios between the tristimulus values of white in the reproduction and in the original. These modified original tristimulus values could then be compared to those for the reproduction (evaluated relative to the well-lit reference white in the reproduction). Although this procedure operates in the right general direction, it can be refined by using, not just three multiplying factors, but three simultaneous equations to provide a full colour transformation with nine constants (Burnham, Evans, and Newhall, 1957). Such transformation equations (whether of the three or nine term type) need to allow for the fact that the colour of the illuminant does not always correspond to the subjective white point. For instance, when transparencies are being projected in a dark room with tungsten light, the colour of the light on the screen appears slightly yellowish, even when the observer is fully adapted to it, and hence the film has to be slightly bluish to produce an apparent white (Hunt, 1965b; see Section 5.7); but adaptation may be complete for medium greys and be in excess for dark greys (Helson, 1938; Judd, 1940).

The concept of corresponding colour reproduction is probably the most appropriate to use generally in colour reproduction problems. It has the same advantage over equivalent colour reproduction as colorimetric colour reproduction has over exact colour reproduction: by relating the colours both in the original and in the reproduction to a reference white, allowance is made for the fact that observers tend to perceive not in isolation but with reference to a framework provided by the environment. For example, when a sunlit scene is projected by tungsten light with a dark surround, equivalent colour reproduction would call for colours of very high purity in order to produce the sensations of high colourfulness experienced in bright sunlight; but the observer knows that, in the somewhat dimmer conditions provided by the projector, all colour sensations are lower in colourfulness, and the picture will look more natural if this is taken into account. More research is required to quantify these effects, and it is important not to forget that (just as with the concept of colorimetric colour reproduction) the picture will tend to look more like the original if the illumination level is adjusted to be closer to that for the original scene. For example, it is well known that raising the screen luminance of a projected colour transparency raises the quality of a picture of a sunlit scene (Bartleson, 1965); and reflection prints of brightly-lit scenes are usually much improved if viewed under strong lighting. Similar effects also occur in television: when the green sulphide phosphor was introduced (see Section 21.17), there was an appreciable reduction in colour purity, and colorimetric colour reproduction deteriorated; but, because the green sulphide phosphor enables pictures of higher absolute luminance to be produced, the loss of purity was offset by a gain in the apparent colourfulness of the colours, and the final effect was that the pictures were improved (Matthews, 1963): in this case corresponding colour reproduction was made worse, but equivalent colour reproduction was made better.

11.10 Preferred colour reproduction

There is a considerable body of evidence that for Caucasian skin colour the above concepts must be supplemented to allow for the fact that a sun-tanned appearance is generally preferred to average real skin colour (MacAdam, 1951; Bartleson and Bray, 1962). There may also be other colours where similar considerations apply: for instance, blue sky and blue water are usually preferred in real life to grey sky and grey water; colour films can have some sensitivity to ultra-violet radiation and hence tend to increase the blueness of sky and water relative to the saturation of the other reproduced colours, but such a tendency, if not overdone, may well be preferred to a more consistent reproduction. It may also be desirable to introduce other distortions of colour rendering to create mood or atmosphere in a picture. These factors may be very important in practice, but it is felt that the concepts of spectral, colorimetric, exact, equivalent, and corresponding colour reproduction, provide a framework that is a necessary preliminary to any discussion of deliberate distortions of colour reproduction. In this context, *preferred colour reproduction* is defined as reproduction in which the colours depart from equality of appearance to those in the original, either absolutely or relative to white, in order to give a more pleasing result to the viewer.

In Figs. 11.7 and 11.8, chromaticities are shown for preferred colour reproduction of blue sky, green grass, and Caucasian skin colours (Hunt, Pitt and Winter, 1974). These results were obtained by making colour photographs of out-door scenes containing well-defined areas of one of these test colours, and then varying the colours of those areas only. This was achieved by using pairs of opaque masks to obscure either the test part of the picture or the rest of it: by making two successive exposures in register in an enlarger, series of reflection prints for each scene were made in which the colour of the area of sky, grass, or skin was varied, but the colour of the rest of the picture was kept constant; and, by using the masks with pairs of slides projected on a screen, a similar result was obtained for transparencies. The colour of the sky, grass, or skin area was varied by covering that area with uniform pale colour-filters when enlarging or while projecting; in this way, these areas retained their inherent variety of tones and colours, and only the overall average colour was altered; neutral filters were used to control the luminances of the two parts of the picture.

Each reflection print and projected picture thus obtained was judged by a panel of observers for the quality of the colour reproduction of its blue sky, green grass, or Caucasian skin. The average chromaticity and relative luminance of each of these colours was then measured and correlated with the observers' judgments to obtain the results shown in Figs. 11.7 and 11.8. For the reflection prints, the judgments were made in typical indoor daylight, and the colorimetry was evaluated for Standard Illuminant C. For the projected transparencies, the colorimetry was evaluated for the actual projector illuminant. The results of a similar investigation on the preferred reproduction of Caucasian skin on a typical television display (Novick, 1972) are also shown in Fig. 11.7. The broken lines in the figures indicate areas of chromaticity giving acceptable colour reproduction. Chromaticities achieved by a negative-positive system of colour photography used for producing reflection prints are shown in Fig. 11.7 ($\times$); and chromati-

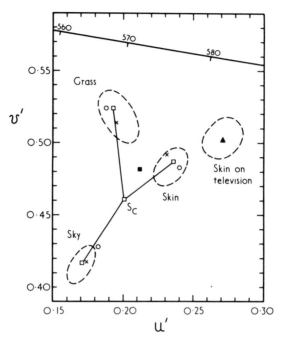

Fig. 11.7. Chromaticities of preferred colour reproductions (☐) for green grass, blue sky, and Caucasian skin colours in reflection prints, together with the chromaticities of typical real samples of these objects (○), and typical reproductions given by a negative photographic system (✕), all for Standard Illuminant C (S_C). Also shown are the chromaticities for the preferred reproduction of Caucasian skin on television viewed in dim ambient tungsten light (▲), together with the associated subjective neutral point (■). The broken lines indicate areas of chromaticity giving acceptable colour reproduction. The percentage relative luminances for the preferred colours were 27, 30, and 39 for grass, sky, and skin, respectively; typical figures for real grass, and skin are about 13 and 35, respectively.

cities achieved by two different reversal films used for producing transparencies are shown in Fig. 11.8 (+ and ✕).

It is clear from Fig. 11.7 that, for the reflection prints, the preferred skin colour (☐) lies, as expected, on the yellowish (sun-tanned) side of typical average real skin (○) (Thomas, 1973), but the difference is small; and the preferred grass colour (☐) lies on the yellowish side of typical average real grass (○) (Thomas, 1973), but again, the difference is small. The chromaticities for real skin and grass lie within the area of acceptable colour reproduction, and this suggests that, for these colours, colorimetric and preferred colour reproduction are similar (although the relative luminances are rather different for grass). But, for the blue sky colour, although the dominant wavelength of the preferred (☐) and real (○) (Hendley and Hecht, 1949) colours are closely similar, the preferred colour has

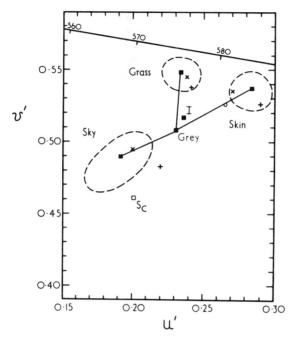

Fig. 11.8. Chromaticities of preferred colour reproduction (■) for green grass, blue sky, and Caucasian skin colours, in transparencies projected with tungsten light, together with the chromaticities of typical reproductions given by two reversal films (+ and ×). The chromaticity of the open-gate light from the projector (I) has a correlated colour temperature of about 3400 K; the subjective neutral point, marked 'Grey', had a correlated colour temperature of about 3700 K. Also shown is the chromaticity of a typical real sample of Caucasian skin illuminated by the light of the projector (○). The broken lines indicate areas of chromaticity giving acceptable colour reproduction. The percentage relative luminances for the preferred colours (expressed relative to normal open-gate luminance) were 6, 16, and 34 for grass, sky, and skin, respectively.

an appreciably higher purity. The preferred skin colour on television (▲) has a dominant wavelength similar to that of real (Illuminant C) skin (○), but is of considerably higher purity; this is partly because the associated subjective neutral point (■) was displaced towards yellow, but perhaps also partly because of a desaturating effect of the dim surround in which the television display was viewed.

The preferred colours shown in Fig. 11.8 for the projected pictures are displaced in an orange direction relative to those for the reflection prints; this is because of visual adaptation to the tungsten light of the projectors: the point marked 'Grey' was the chromaticity that appeared neutral to observers in the viewing conditions used for the projection, and it is shifted in the orange direction

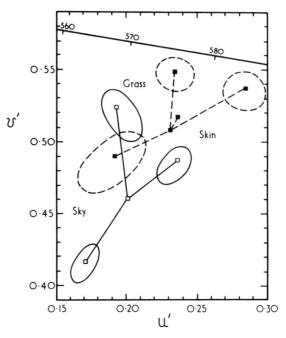

Fig. 11.9. The main results of Fig. 11.7 (full lines) and of Fig. 11.8 (broken lines) shown together.

from points representing daylight (such as S_C). The point marked 'I' shows the chromaticity of the open-gate light from the projectors and its displacement from the 'Grey' point indicates that the adaptation to the projector light was not quite complete. In Fig. 11.9 the main results of Figs. 11.7 and 11.8 are shown together, and the shift to more orange chromaticities in the case of the pictures projected with tungsten light is very apparent. It is interesting also to note that there is a small area of chromaticity that is acceptable as grass in reflection prints seen in daylight and as blue sky in tungsten-light projection. If the chromaticities of the real colours of Fig. 11.7 could be converted into the corresponding chromaticities for the viewing conditions used in obtaining the results of Fig. 11.8, it would be possible to say if the preferred colour reproduction in the projected pictures involved any subjective distortion of the colours. Compared to the chromaticity of a typical sample of real Caucasian skin colour illuminated by the light of the projector (◯), the preferred colour has an appreciably higher purity, and this is perhaps to offset a desaturating effect of the dark surround to the projected picture.

The preferred relative luminances found were, no doubt, affected by the relative luminance corresponding to the apparent white level, and this level may be expected to vary from scene to scene and from area to area within a scene. This makes the specification of preferred relative luminances in pictures a rather complicated matter.

Comparison of the results shown in Figs. 11.7 and 11.8 with those obtained in earlier investigations (MacAdam, 1951; MacAdam, 1954; Bartleson, 1959; Bartleson and Bray, 1962) suggest that the latter may have been influenced to some extent by the generally lower levels of colour purity available in the systems then used. Results for preferred colour reproduction may, therefore, be influenced by the nature of the reproduction system used, and it also seems likely that they may be influenced by the particular form of the viewing conditions, and by cultural, ethnic, and psychological features of the observers: results such as shown in Figs. 11.7 and 11.8 can therefore only be regarded as examples and not as definitive for particular applications.

11.11 Degree of metamerism

In the case of colorimetric or exact colour reproduction, the degree of metamerism can be assessed by a direct comparison of the spectral reflectance (or relative power distribution) of the original and reproduction. But in the cases of equivalent, corresponding, and preferred colour reproduction, the colours in the reproduction must, in general, be physically different from those in the original: hence there must always be some metamerism. However, some reproduction colorants will tend to produce corresponding colours with greater degrees of metamerism than others, and some means of assessing this would be desirable. For this purpose it would probably be good enough to assess the degree of metamerism for an appropriate colorimetric colour reproduction situation, and to regard the results as indicative of the degree of metamerism for the other cases. Thus, for a picture of a sun-lit outdoor scene projected with a dark surround by tungsten light, the degree of metamerism could be assessed by comparing the spectral reflectance curves of various original colours with those of dye-concentration combinations in the film that are metameric matches to them for tungsten light.

11.12 Conclusions

Spectral colour reproduction (equality of spectral reflectances or of relative spectral power distributions), though not attainable in most situations, provides a useful basis for determining the degree of metamerism of reproduction systems.

Colorimetric colour reproduction (equality of chromaticities and relative luminances) is a useful criterion when the original and reproduction have the same viewing conditions and use illuminants of the same colour; this is often roughly true for reflection prints.

Exact colour reproduction (equality of chromaticities, relative luminances, and absolute luminances) ensures equality of appearance for original and reproduction if the viewing conditions are the same for both.

Equivalent colour reproduction (chromaticities, relative luminances, and absolute luminances such as to ensure equality of appearance) can allow for all effects of viewing conditions, but may be an unrealistic criterion if there is an appreciable difference in luminance level between original and reproduction.

Corresponding colour reproduction (chromaticities and relative luminances such as to ensure equality of appearance when the original and reproduction luminance levels are the same) allows for all effects of viewing conditions except absolute luminance levels, and provides a realistic criterion for most situations.

However, for some objects whose colours are well-known, *preferred* colour reproduction may be required, wherein departures from equality of appearance (whether at equal or at different absolute luminance levels) may be required in order to achieve a more pleasing result.

REFERENCES

Allen, E., *Color Engineering*, **7**, 35 (1969).
Allnat, J. W., *Proc. Inst. Elec. Eng.*, **112**, 1819 (1965).
Allnat, J. W., *Proc. Inst. Elec. Eng.*, **113**, 551 (1966).
Allnat, J. W., *Proc. Inst. Elec. Eng.*, **115**, 371 (1968).
Bartleson, C. J., *Phot. Sci. Eng.*, **3**, 114 (1959).
Bartleson, C. J., *Phot. Sci. Eng.*, **9**, 174 (1965).
Bartleson, C. J., *J. Phot. Sci.*, **30**, 33 (1981).
Bartleson, C. J., and Bray, C. P., *Phot. Sci. Eng.*, **6**, 19 (1962).
Bartleson, C. J., and Breneman, E. J., *J. Opt. Soc. Amer.*, **57**, 953 (1967a).
Bartleson, C. J., and Breneman, E. J., *Phot. Sci. Eng.*, **11**, 254 (1967b).
Breneman, E. J., *J. Opt. Soc. Amer.*, **67**, 657 (1977).
Burnham, R. W., Evans, R. M., and Newhall, S., *J. Opt. Soc. Amer.*, **47**, 35 (1957).
Clapper, F. R., and DeMarsh, L. E., Private communication (1969).
Corbett, J. M., *Proc. Inst. Elec. Eng.*, **117**, 512 (1970).
Evans, R. M., *J. Opt. Soc. Amer.*, **33**, 579 (1943).
Evans, R. M., *J. Opt. Soc. Amer.*, **49**, 1049 (1959).
Helson, H., *J. Exp. Psychol.*, **23**, 439 (1938).
Hendley, C. D., and Hecht, S., *J. Opt. Soc. Amer.*, **39**, 870 (1949).
Hunt, R. W. G., *J. Opt. Soc. Amer.*, **40**, 362 (1950). See Figs. 3 and 4.
Hunt, R. W. G., *J. Opt. Soc. Amer.*, **42**, 190 (1952).
Hunt, R. W. G., *J. Opt. Soc. Amer.*, **43**, 479 (1953).
Hunt, R. W. G., *The Reproduction of Colour*, 1st Edn., Fountain Press, p. 196 (1957).
Hunt, R. W. G., *J. Phot. Sci.*, **13**, 108 (1965a).
Hunt, R. W. G., *J. Opt. Soc. Amer.*, **55**, 1540 (1965b).
Hunt, R. W. G., *Brit. Kinematog. Sound Tel.*, **51**, 268 (1969).
Hunt, R. W. G., *J. Phot. Sci.*, **18**, 205 (1970).
Hunt, R. W. G., *Colour 73, Proc. 2nd. A.I.C. Congress*, York, pp. 62–3, Hilger, London (1973).
Hunt, R. W. G., Pitt, I. T., and Ward, P. C., *J. Phot. Sci.*, **17**, 198 (1969).
Hunt, R. W. G., Pitt, I. T., and Winter, L. M., *J. Phot. Sci.*, **22**, 144 (1974).
Judd, D. B., *J. Opt. Soc. Amer.*, **30**, 2 (1940).
MacAdam, D. L., *J. Opt. Soc. Amer.*, **28**, 399 (1938).
MacAdam, D. L., *J. Soc. Mot. Pic. Tel. Eng.*, **56**, 502 (1951).
MacAdam, D. L., *Proc. I.R.E.*, **42**, 166 (1954).

Matthews, J. A., Private communication (1963).
Movie Maker Magazine, **6,** 424 (1972).
Nayatani, Y., Takahama, K., and Sobagaki, H., *Color Res. Appl.*, **6,** 161 (1981).
Novick, S. B., *Brit. Kinematog. Sound Tel.*, **51,** 342 (1969).
Novick, S. B., *Brit. Kinematog. Sound Tel.*, **54,** 130 (1972).
Pinney, J. E., and DeMarsh, L. E., *J. Phot. Sci.*, **11,** 249 (1963).
Pitt, I. T., In *Colour Measurement in Industry*, p. 234, The Colour Group (Great Britain) (1967).
Pitt, I. T., and Winter, L. M., *J. Opt. Soc. Amer.*, **64,** 1328 (1974).
Rowe, S. C. H., Ph.D. Thesis, The City University (1972).
Thomas, W., *S.P.S.E. Handbook of Photographic Science and Engineering*, pp. 441–2, Wiley, New York (1973).

PART TWO
COLOUR PHOTOGRAPHY

CHAPTER 12

Subtractive Methods in Colour Photography

1. Introduction – *2.* Relief images – *3.* Colour development – *4.* Integral tripacks – *5.* Processing with the couplers incorporated in the film – *6.* Reversal processing – *7.* Processing with the couplers in developers – *8.* The philosophy of colour negatives – *9.* Subtractive methods for amateur use in still photography – *10.* Subtractive methods for professional use in still photography – *11.* Subtractive methods for motion picture use

12.1 Introduction

THE basic step in subtractive colour photography is the formation of cyan, magenta, and yellow dye-images. For the dyes to be present as *images*, it is necessary for their concentrations to vary from point to point in the picture area in a manner that is appropriately dependent on the distribution of the colours of the scene. There are several ways of accomplishing this, but the one that has achieved the widest commercial success is known as *colour development*; however, before describing this, it is convenient to consider the *relief image* method, which has had a considerable use in the professional motion picture industry, and is also used on a small scale for professional still photographs.

12.2 Relief images

The principle of the *relief image* method of producing dye images is as follows. The thickness of a gelatin layer is made to vary from point to point in the picture according to the intensity of the exposure. On immersing such a layer in a solution of dye, more dye will be taken up by the areas where the gelatin is thick than where it is thin, and hence a dye image is obtained. The way in which a relief gelatin image is produced from a negative (exposed through a red, green, or blue filter) is shown in Fig. 12.1 which relates to the Kodak *Dye Transfer* process.

Dye Transfer Matrix film, on which the positive dye-images are produced, consists of an ordinary black-and-white emulsion coated on film base, except that the emulsion is unhardened, contains a yellow dye, and is not dye-sensitized (so that it is sensitive only to blue light). The emulsion is exposed through the base, and the yellow dye absorbs the blue light to which alone the emulsion is sensitive. Parts of the emulsion near to the base are easily exposed, because the light can reach them without having to traverse more than a very thin layer of yellow dye. But parts of the emulsion further away from the base can only be exposed with difficulty, for the light must traverse a comparatively thick layer of yellow dye

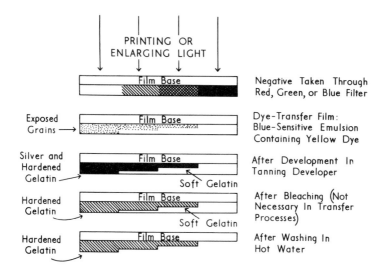

Fig. 12.1. Method of forming a gelatin relief image from a negative.

before it can reach them. The exposure of this film, therefore, occurs most easily near the base, and becomes progressively more and more difficult throughout the emulsion layer. Hence, when exposures of different intensities are made over the area of the film, the latent image formed will vary in depth according to the intensity of the exposure. Heavily exposed areas (such as that on the extreme left in the diagram) will exhibit latent image throughout practically the whole layer, but less intensely exposed areas will have a shallower latent image concentrated near the film base. The image is, therefore, an image in depth or a *relief image.*

What is now required, is to remove all the gelatin that does not contain latent image and to leave the rest; we shall then have a gelatin relief image which can be dyed the appropriate colour. Now gelatin, in its usual state, is soluble in hot water. But by suitable chemical treatment it can be hardened, or tanned, so that it becomes insoluble. If, therefore, we could harden all the gelatin around the latent image and leave the rest unchanged, by washing the film in hot water we could remove the unwanted gelatin and leave the relief gelatin image adhering to the film base. The way in which this can be done is surprisingly simple.

Amongst other ingredients, ordinary photographic developers contain a developing agent and sulphite. The reaction with an exposed silver bromide crystal can then be represented thus:

AgBr+Developing Agent ⟶ Ag+Oxidized developing agent

+sulphite

Ag+Colourless waste product

The developing agent reduces the silver bromide to silver, and, by doing so, it becomes oxidized, whereupon it reacts immediately with the sulphite to form a colourless waste product. To harden the gelatin around the latent image and nowhere else, it is only necessary to use pyrogallol as the developing agent and greatly to reduce the amount of sulphite in the developer. The first part of the reaction then takes place as before, but the oxidized developer, having very little sulphite with which to react, proceeds to react with the gelatin, and in fact hardens it. We thus have:

AgBr+Developing Agent ⟶ Ag+Oxidized developing agent

+Gelatin

Ag+Hardened gelatin

The matrix film is, therefore, processed in a 'pyro' developer with very little sulphite, and then washed with hot water to leave a hardened gelatin relief image. This relief image is then dyed, and since the amount of dye absorbed is proportional to the thickness of the gelatin, the amount of dye present at each point will be a function of the image exposure, as required. If three such dye images are made using cyan, magenta, and yellow dyes, after bleaching away the silver, a subtractive colour photograph can be produced simply by superimposing the three images in register. If a paper print is required, it is possible to transfer the dye from the gelatin relief image to a suitably prepared paper surface as in Fig. 12.2. It is in this way that the Kodak *Dye Transfer* process works.

If a colour transparency or film is required the dye image can be transferred on to a suitably prepared transparent support; this was the principle of the *Technicolor* process. In any transfer process, after the dye image has been transferred, the relief image can be dyed again and a second transfer made on another piece of support. By doing this with each dye, a second copy of the colour photograph is made; and the process can be repeated, more or less indefinitely, so as to obtain a large number of copies. Incidentally, since in transfer processes the dye is always transferred from the gelatin relief image and viewed on another support, it is not necessary to remove the silver from the relief image.

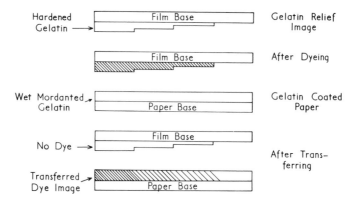

Fig. *12.2.* Method of using a gelatin relief image to transfer a dye image to paper.

12.3 Colour development

The *colour development* method of producing dye images is simpler to operate than the relief image method because it obviates the need for separate developing and dyeing stages. This method is very similar to the tanning development technique in that only a very little sulphite is present in the developer, but, instead of letting the oxidized developer react with the gelatin of the emulsion, a *coupler* is present in the developer, or in the emulsion layer, and this reacts with the oxidized developing agent to form an insoluble dye.

We thus have:

$$AgBr + Developing\ Agent \longrightarrow Ag + Oxidized\ developing\ agent$$
$$+ Coupler$$
$$Ag + Insoluble\ dye$$

This reaction only works satisfactorily with some developing agents, notably paraphenylenediamine and some of its derivatives. It is clear that the dye is formed jointly from the coupler and the oxidized developer, and the colour of the dye formed is determined by the nature of the coupler and the developing agent, although they are themselves usually colourless. The amount of dye formed depends on the amount of oxidized developer available and this in turn depends on the amount of silver that has been developed; thus the amount of dye is related directly to the amount of exposure given at each point, and is therefore laid down as an image and not as a uniform layer. The reaction of the oxidized developer is localized around silver halide crystals containing latent image, so that, on colour development, blobs of insoluble dye are formed only around the crystals developed by the developing agent, and hence the dye image obtained reproduces in a somewhat blurred manner the granular nature of the silver image from which it is derived. If three such dye images are produced, using cyan, magenta, and yellow dyes, by bleaching out the silver and superimposing the images in register, a subtractive colour photograph is obtained. Alternatively the dye images could be transferred to another support, but colour development processes cannot be used to give large numbers of copies in the same way as the relief-image processes.

The great advantage of the colour development technique, however, is that it becomes possible to produce dye images of different colours in different layers of a single film. Colour development is discussed further in Chapter 17.

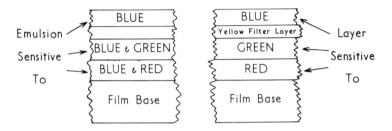

Fig. 12.3. Sensitization of the layers in an integral tripack typical of those used for films of camera speed.

12.4 Integral tripacks

Any process of colour photography that involves taking three pictures one after the other clearly has the severe limitation that only 'still-life', or very slowly moving scenes can be taken. In the mosaic and lenticular additive processes the three pictures were taken at the same time on neighbouring areas of film. In modern subtractive processes of colour photography the three pictures are taken on three emulsions coated one on top of the other, as shown in Fig. 12.3, an arrangement known as an *integral tripack*.

Photographic emulsions are naturally sensitive only to the blue part of the spectrum, and their sensitivity is extended to the green and red parts only by the addition of *sensitizing dyes*. An ordinary, unsensitized, emulsion usually constitutes the top layer in a tripack, and in it is produced a negative that provides the blue record of the scene, but in this case no blue filter is necessary because the emulsion itself responds only to blue light. The bottom layer of the film consists of an emulsion sensitized only to red light. It still has its natural sensitivity to blue light, of course, but this is rendered inoperative by means of a yellow filter layer immediately beneath the top layer. In this bottom layer, therefore, is produced a negative providing the red record of the scene; but once again no red filter is needed, because the yellow filter together with the red sensitizing of the emulsion make the layer sensitive only to red light. Between the yellow filter layer and the bottom layer is an emulsion sensitized to green light only. This sensitizing, together with the yellow filter layer, constitutes a layer sensitive to green light only, and in it is produced a negative providing the green record of the scene, but without using a green filter. (See Plate 10, pages 112, 113.)

It will be clear that with such a three-layer film a single exposure suffices to record the three images required, one being effectively taken through a red filter, another through a green, and the third through a blue. It remains to process the film in such a way that cyan, magenta, and yellow dye-images are formed in these three layers respectively. There are two main methods of achieving this by colour development. In one method the couplers are incorporated in the film; in the other they are in three separate developers.

12.5 Processing with the couplers incorporated in the film

Fig. 12.4 shows in diagrammatic form the way in which an integral tripack material can be processed when the couplers are incorporated in the film. Each of the four large circles depicts a highly magnified cross-section of the three emulsion layers and the yellow filter layer. Each small triangle represents a silver halide crystal, or *grain*, and the triangles with dots in them indicate grains that

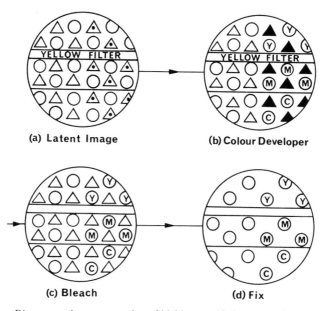

(a) Latent Image

(b) Colour Developer

(c) Bleach

(d) Fix

Fig. 12.4. Diagrammatic representation of highly magnified cross-sections of an integral tripack material with incorporated couplers being processed so as to give a negative image. △: unexposed silver halide grain. △̇: exposed silver halide grain. ▲: developed grain of silver. ○: particle of coupler. Ⓨ: particle of yellow dye. Ⓜ: particle of magenta dye. Ⓒ: particle of cyan dye.

have been exposed and contain latent image, while those without dots indicate grains that have not been exposed and do not contain latent image. It is thus clear, Fig. 12.4 (a), that in this example light has fallen on the right-hand part of the film but not on the left. The circles represent particles of couplers, those in the top (blue-sensitive) layer being capable of forming yellow dye, those in the bottom (red-sensitive) layer being capable of forming cyan dye, and those in the other (green-sensitive) emulsion layer being capable of forming magenta dye. The couplers are prevented from wandering away from their proper layers, either by attaching long molecular chains to them (used first by Agfa in 1936 (Koshofer, 1966)) or by dissolving them in oily solvents and then dispersing them in the form of minute oil globules (used first in Kodak materials (Mees, 1942)): when oil globules are used they are usually of about a tenth of the diameter of the silver halide grains. On immersing the material into a solution containing a suitable developing agent the situation becomes as shown in Fig. 12.4(b). The developing agent converts the silver halide to silver (represented by the black triangles) wherever latent image was present, and, around each grain of silver thus formed, the oxidized developer reacts with coupler to produce dye: yellow (Y) in the top layer, cyan (C) in the bottom layer, and magenta (M) in the other emulsion layer. The dyes are deposited as very small 'clouds' of molecules or globules around each developed grain. It is now necessary to remove the unexposed silver halide from the film, for this would gradually darken as the film was viewed, and this *fixing* is carried out as in black-and-white films by means of a 'hypo' solution; but it is also necessary to remove the silver image, which otherwise would darken the result, and this is conveniently done by converting the silver back to silver ions by means of a suitable bleach, used prior to the fixing stage. The remainder of the process is therefore basically as shown in Fig. 12.4(c) where the bleach converts the silver to silver ions again, and Fig. 12.4(d) where all the silver ions are removed by the fixer. (In some processes the bleaching and fixing steps are combined in a single solution known as a *blix*.) The yellow filter layer generally also disappears at the bleach stage. The unused coupler is harmless and is allowed to remain; in fact in some colour films the unused coupler is actually used to improve the accuracy of the final results, and in these cases it is usually coloured yellow or pink, but normally it is colourless (see Sections 15.4 and 17.4).

The final result in Fig. 12.4 is that on the right-hand side of the film, where the light originally fell, all three dyes, cyan, magenta, and yellow are produced (resulting in a dark area), whereas on the left-hand side, where no light fell, no dyes are produced (resulting in a light area). The result is thus a negative: light becoming dark, and dark becoming light. In addition, colours will be reversed; red becoming cyan, green becoming magenta, and blue becoming yellow, and *vice-versa*. That this is so can be seen by considering the following example. If the light falling on the film had been red, only the bottom layer would have been exposed; hence only cyan dye would have been formed. Conversely if the light falling on the film had been cyan (blue-green) only the blue and green sensitive layers would have been exposed, and hence only yellow and magenta dye would have been formed and the superimposition of these two dyes results in a red colour. To produce a colour positive from such a record, which is called a *colour negative*, it is only necessary to re-photograph or *print* the processed negative on to a similar piece of film or paper: once again, by the same arguments, both the tones and the colours will be reversed and the final result will be in its correct colours. (This is explained more fully in Section 12.8). Systems operating in this way include *Kodacolor*, *Eastman Color*, *Agfacolor*, *Ektacolor*, *Vericolor*, and *Fujicolor*.

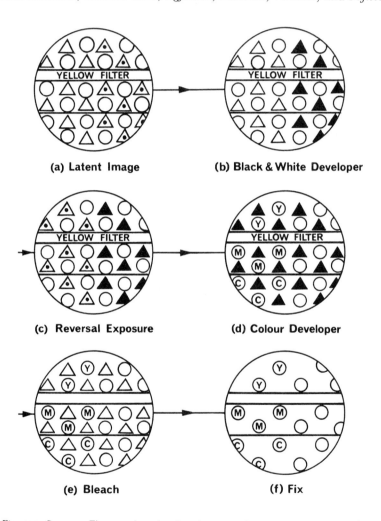

(a) Latent Image   (b) Black & White Developer

(c) Reversal Exposure   (d) Colour Developer

(e) Bleach   (f) Fix

Fig. 12.5. Same as Fig. 12.4 but showing the processing sequence necessary (reversal process) to obtain a positive image directly on the camera film.

12.6 Reversal processing

As in black-and-white photography, if a separate negative stage is not required, the film exposed in the camera can be *reversal processed* to give a positive image directly. The way in which this can be achieved is shown diagrammatically in Fig. 12.5 in which the symbols all have the same meanings as in Fig. 12.4. As in Fig. 12.4 light has fallen on the right-hand part of the film but not on the left, so that the latent image is present only on the right, Fig. 12.5(a). The film is then immersed in an ordinary black-and-white developer, which converts the exposed silver halide to silver, thereby oxidizing the developing agent; but being an ordinary black-and-white type developing agent its oxidized form does not react with the couplers and hence no dye is formed at this stage, Fig. 12.5(b). The next step is to re-expose the film uniformly to a strong white light so that latent image is formed in all the undeveloped silver halide, Fig. 12.5(c); the film then enters a colour developer, which converts this silver halide to silver and the oxidized developer formed in the vicinity of this silver reacts with the couplers to form cyan, magenta, and yellow dye images as before, Fig. 12.5(d). The usual bleaching, Fig. 12.5(e), and fixing, Fig. 12.5(f), stages then remove all the silver, to give, now, all three dyes (a dark area) on the left hand part of the film where the light originally did not fall, and no dyes (a light area) on the right hand part of the film where the light originally did fall. The result is thus a positive, as required. That colours also come out correctly can be seen by the following example. If the right-hand side was exposed only to red light, only the bottom layer would have been exposed, so that, in the black-and-white developer, silver would only have been produced in the bottom layer. The reversal exposure would therefore produce latent image in the top two light-sensitive layers, but not in the bottom layer. On colour development, yellow and magenta dyes, but no cyan, would therefore be formed and hence a red colour produced on the film in the area in which the red light originally exposed it. Past and present films operating in this way include *Ektachrome, Agfachrome, Fujichrome, Perutzcolor, Orwochrome, Ferrania-color, Anscochrome*, and *Gevachrome*. In some processes, chemical fogging is used instead of re-exposure by light.

12.7 Processing with the couplers in developers

In Fig 12.6 a diagrammatic example is given of one way in which direct positive colour images can be obtained using the other main method of colour development: that in which the couplers are in three separate developing solutions. The symbols used are the same as before, and once again light has fallen on the right-hand part of the film and not on the left, Fig. 12.6(a). An ordinary black-and-white developer therefore produces silver images on the right-hand part, Fig. 12.6(b) as before. The film is then re-exposed uniformly, not to white light as in Fig. 12.5, but to red light from the bottom, Fig. 12.6(c). Since only the bottom layer is sensitive to red light, latent image is formed only in this layer so that, on

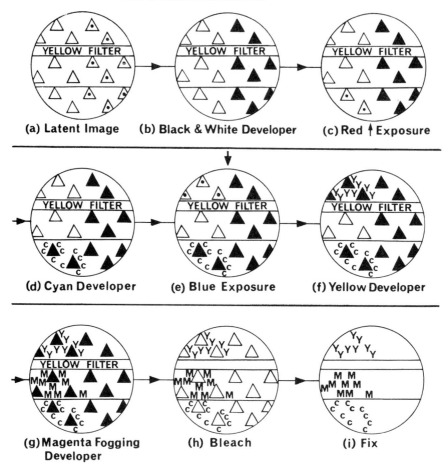

Fig. 12.6. Same as Fig. 12.5 but using the method where the three couplers are in three different developers instead of in three layers of the tripack.

immersing the film in a colour developer containing cyan-forming coupler, cyan dye is formed in the left-hand part of the bottom layer only, Fig. 12.6(d). The film is next exposed to blue light from the top and, since the yellow filter layer protects the bottom two layers from all blue light, latent image is at this stage formed only in the top layer, Fig. 12.6(e), so that on immersing the film in a colour developer containing yellow-forming coupler, yellow dye is formed in the left-hand part of the top layer only, Fig. 12.6(f). It is now required to form magenta dye in the left hand part of the middle layer, and since it is only in this part of the entire film that there is any silver halide left, use is made of a colour developer containing magenta-forming coupler and of such a character that development takes place even when the silver halide has not been exposed, Fig. 12.6(g). The final two stages are the usual bleaching, Fig. 12.6(h), and fixing, Fig. 12.6(i), steps. It will be seen that the final disposition of the dyes is the same in Fig. 12.6 as in Fig. 12.5, so that a positive has been achieved, and by the same arguments as used previously the colours as well as the tones are correctly reproduced.

This method of colour development cannot conveniently be used to obtain colour negative images because it is the reversal exposure step that provides the opportunity for the colour developers to affect each layer in turn independently.

It will be appreciated from Fig. 12.6 that this type of process is of considerable complexity, and is, for this reason, only operated at a few large processing stations. In return for this complexity, however, the process can yield an extremely high resolving power; this is partly because, without any couplers in them, the layers can be made thinner (see Plate 10, pages 112, 113). This type of process can, therefore, be used with advantage for 8 mm cinematography where the frame size is only 4.9×3.7 mm, while for 35 mm use its resolving power is usually ample. Examples of this type of process include *Kodachrome* (introduced in 1935), *Kodachrome II* (1961), *Kodachrome X* (1963), *Ilfochrome* (1948, see Hornsby, 1950); *Kodachrome 25, 40* (Type A), and *64* (1974), and *200* (1986). The original Kodachrome process, which was worked out by two musicians, Leopold Mannes and Leopold Godowsky (Davies, 1936; Matthews, 1955), was in fact rather different and involved a series of differential bleaching steps (Weissberger, 1970); the type of process shown in Fig. 12.6 was introduced in 1938.

In principle it is possible with this successive development type of process to have the red, green, and blue sensitive emulsions present as a mixture in a single layer, but attempts to reduce such a system to practice have encountered various difficulties (Hanson, 1977; Ryan, 1977).

12.8 The philosophy of colour negatives

A colour negative, fundamentally, consists merely of three negative images (exposed by red, green, and blue light) superimposed in register, but differing from one another in such a way that they can be distinguished from one another. Thus three black-and-white silver images superimposed would be of value if they could be stripped apart; this, incidentally, is a practicable form of colour negative and has been tried in the motion-picture industry (Capstaff, 1950). But, of course, there is no need to strip the images apart when they are differently coloured. Thus if one of them is a cyan image, another magenta, and the third yellow, when the colour negative is viewed by red light only the cyan image will be seen, and in green light only the magenta will be seen, and in blue light only the yellow will be seen (to a first approximation). It is customary, but not essential, to make the image that was exposed in the camera by red light a cyan image, that by green light a magenta image, and that by blue light, a yellow image. When this is the case the colour negative, as well as reversing all tones (blacks becoming whites and whites becoming blacks) reverses all colours too, so that reds become blue-greens, blue-greens become reds, greens become magentas, magentas become greens, blues become yellows, and yellows become blues; in other words every colour takes on the complementary hue. The important consideration, however, is not what the negative looks like in white light to the eye, but how efficiently it enables the three superimposed negatives to be distinguished at the printing stage. With the above arrangement, the printing material must produce a red-absorbing (cyan) image when exposed to red light, because light of this colour isolates the negative image that was made by red light in the camera. Similarly the printing material must produce a magenta image when exposed to green light, and a yellow image when exposed to blue light. A print on such a material will reverse not only the tones but also the colours of the negative to give correct tones and hues in the print. (Ashton, 1980.)

In principle, however, it would be just as good if, for instance, in the colour negative the image exposed by the red light in the camera were developed magenta, and that by green light were developed cyan. Then the printing material must be such that green light, which will isolate the negative exposed by red light in the camera, must produce a red-absorbing (cyan) dye. Similarly red light would have to produce a green-absorbing image on the printing material. Furthermore, again in principle, the lights used for printing the negative on to the printing material need not be red, green, and blue. They could, sometimes with advantage, be for instance infra-red, green, and ultra-violet. The only requirement is that the light forming the red-absorbing (cyan) image in the print must isolate the negative that was formed by red light in the camera; and similarly for the green and blue. The intermediate dyes used in the negative and the spectral content of the printing light are entirely a matter of convenience; and of course it is not essential to print with separate beams of, for instance, red, green, and blue light, since, by making the print material sensitive only in the required narrow bands of the spectrum, white light can be used instead. (In practice, however, something might be gained in colour saturation in the print by using a printing light consisting of only narrow spectral bands specially chosen to coincide with the absorption peaks of the dyes used in the colour negative, if this more effectively isolated each negative image and reduced contamination from the other two.)

Some of these principles have found application in colour films used for professional motion picture and aerial survey work (see Sections 12.10 and 12.11).

12.9 Subtractive methods for amateur use in still photography

Both the negative-positive and the reversal versions of subtractive colour photography are widely used by amateur photographers. The colour transparency, so conveniently provided by reversal processing of the film used in the camera, gives results of excellent quality at remarkably low cost, and the system is used very widely indeed. The disadvantages of the colour transparency, however, are first that it requires equipment (and preferably a darkened room) for viewing, and secondly that reflection prints and duplicate transparencies cannot easily be made from original transparencies without some loss of quality.

For these reasons, the negative-positive systems are also widely used: they have the advantages that as many identical prints or transparencies as are required can be made from the same negative, and the printing operation enables corrections for exposure errors (in intensity and in colour) to be made, thus giving a system that can have excellent exposure latitude, and good flexibility as far as illuminant colour is concerned. The disadvantages of the negative-positive system are as follows: first, because of the use of both negative and positive materials and the necessity for the printing operation, the cost of making one picture from a scene is higher than in the case of reversal materials; if more than one picture is required the negative-positive system is usually cheaper, but most amateurs require only one of most of the pictures they expose. The second disadvantage of the negative-positive system is that the quality of the transparencies it produces is not usually quite as good as can be obtained on reversal film (but it may be better than that obtained on copies of reversal-film transparencies).

The choice of system for amateur use therefore depends upon whether prints or transparencies are the prime requirement; whether only one, or more than one, copy is required; and whether the cheapest system is desirable. In practice, most amateurs use reversal film for transparencies, and the negative-positive system for reflection prints.

The history of the popular amateur colour transparency for still photography started with the introduction of 35 mm *Kodachrome* film in 1936. The history of the colour negative also extends back to the 1930s, when it was introduced for use both in mosaic additive processes (Harrison and Spencer, 1937), and also in integral tripack subtractive materials (Berger, 1950; Koshofer, 1966). But it was not until the beginning of the next decade that a colour negative system intended mainly for amateur snapshot use was placed on the market in the form of *Kodacolor* in 1942 (Mees, 1942). At first, the unexposed areas of *Kodacolor* negatives were clear, as is the case for black-and-white negatives, but in 1944 a silver mask was introduced in order to improve colour reproduction (Evans, Hanson, and Brewer, 1953), and the negatives then looked greyish (see Section 15.2). In 1949 the silver mask was removed and coloured couplers (see Section 15.4) were used, with the result that the negatives became orange in the unexposed areas; a further change to a slightly different orange colour was made in 1955 when the same film was sold for both daylight and clear flash use; this film possessed increased latitude so as to accommodate both types of illuminant without reducing the permissible margin of exposure error for the user. Subsequently, flashbulbs covered with blue filter material, so as to make the colour of the light emitted by them similar to that of daylight, became standard for most amateur colour photography. In 1963 the Kodak *Instamatic* system with ASA64 speed films was introduced, and in 1972 *Kodacolor II* film for the *Pocket Instamatic* programme, this film being sharper and of finer grain (and being of a paler orange colour). *Kodacolor* films of higher quality still were introduced, in 1982, for use with the Kodak disc cameras, and, in 1983, for general use as *Kodacolor VR 100, 200, 400,* and *1000* films, the numerals indicating the film speeds. The problems involved in printing amateurs' colour negatives are discussed in Chapter 16.

12.10 Subtractive methods for professional use in still photography

Many of the same considerations apply in professional as in amateur subtractive colour photography, but generally speaking the emphasis in professional work is more on quality than on cost. For this reason reversal films in large formats (for example, 4×5 in.) are quite commonly used in order to obtain extremely high definition and to facilitate retouching. Professionals also require more often to be able to process their own films and coupler-incorporated reversal films such as *Ektachrome* (introduced in 1946) are therefore needed.

The dye transfer system is important for professional users because, in spite of the high cost of operating it, its greater flexibility, arising from the independent handling of the three coloured images, is often very useful.

Professional photographers sometimes need to give very long or very short exposure times; and for this reason certain films are made specially for particular ranges of exposure times, since it is often impossible to make a film in which the contrasts of the three layers are equal at all exposure times. Thus Kodak *Vericolor S* film, for instance, is intended for exposure times of 1/10 sec. and shorter, while Kodak *Vericolor L* film is for times from 1/10 to 60 secs.

A particular branch of professional photography is aerial survey work, and special films have been made for this purpose. In one of these films the three layers, instead of being sensitive to the red, green, and blue parts of the spectrum and yielding cyan, magenta, and yellow images respectively, are made sensitive to the green, red, and infra-red parts of the spectrum yielding yellow, magenta, and cyan dyes respectively (Tarkington and Sorem, 1963). When this film records green vegetation whose coloration is caused by chlorophyll, on reversal processing, a red or magenta result is obtained because, in addition to its green reflection, chlorophyll reflects strongly in the infra-red; green paints, however, do not usually have this property and reproduce as blues. The film therefore distinguishes very sharply between vegetation and green paints: it was therefore introduced (in 1942) to overcome camouflage in aerial reconnaisance work. This type of film can also be used in aerial survey work (Smith, 1968) to detect the distribution of certain types of trees in a forest (Spencer, 1947) and is also very valuable for recording underwater detail (Mott, 1966).

Films of this type that have no layer sensitive to blue light, can be coated with their magenta-forming layers on top, in order to improve sharpness (see Section 12.11). It is also possible to coat this layer on top, even when one of the layers is blue-sensitive, if the speed of the blue-sensitive layer is very much greater than that of the blue sensitivities of the other layers; a yellow filter can then be coated, above all the layers, that is transparent enough to blue light to allow exposure of the blue-sensitive layer, but dense enough to prevent exposure of the other layers by blue light (Fritz, 1971; Moser and Fritz, 1975).

12.11 Subtractive methods for motion-picture use

For amateur motion pictures, where the desire to have copies made is even smaller than in amateur still photography, the reversal types of film are used almost exclusively; *Kodachrome* was introduced for this purpose in 16 mm and 8 mm sizes in 1935, and in the Super 8 size in 1965.

With the introduction in 1971 of the Kodak XL cameras for Super 8 size film, it became possible to take motion pictures under any lighting conditions without the use of any supplementary lights. This was achieved by using a camera with an f/1.2 lens and a 240° shutter (giving exposure for two-thirds of the running time), and a 160 ASA speed type A reversal film of the *Ektachrome* type; the system is used without a filter in tungsten lighting and with an amber filter in daylight (*Movie Maker Magazine*, 1972).

In professional motion picture work, both reversal and negative camera films are used. Fig. 12.7 summarizes the different methods employed. (See Plate 11, page 113.)

Since more than one copy is normally required in professional work, the direct use of reversal film is not usually possible. The next simplest system is the straight negative-positive combination shown in Fig. 12.7(a), using for instance, *Eastman Color Negative* and *Print* films (or, in the past, *Eastman Color Negative* film with the *Technicolor* process for the positive prints); this system can give pictures of excellent quality (Hanson, 1952; Hanson and Kisner, 1953; Dundon and Zwick, 1959; Kisner, 1962; Koshofer, 1967; Kennel, Sehlin, Reinking, Spakowsky, and Whittier, 1982).

It is interesting to note that in *Eastman Color Print* film the conventional order of the layers, shown in Fig. 12.3, is not followed. Because each emulsion layer in a colour film diffuses the light somewhat, the top layer is usually the sharpest and the bottom layer the least sharp. Therefore, to obtain pictures of maximum sharpness it is desirable to have the magenta layer, whose image is the most strongly visible of the three, at the top, and the yellow layer, whose image is the least visible of the three, at the bottom. This can be done if emulsions can be made that are not sensitive to blue light. This is possible by using silver-chloride or silver-chloro-bromide emulsions, which have their natural sensitivity in the ultra-violet instead of in the blue. These emulsions cannot usually be used for camera films because of their rather low sensitivity, but for print films they are fast enough, and *Eastman Color Print* film has the magenta layer on top, the cyan layer in the middle, and the yellow layer at the bottom. (The cyan and magenta layers cannot be made with no sensitivity at all to blue light, but, by making the yellow layer much faster than the other two, their speed to blue light is negligible.)

Although the system shown in Fig. 12.7(a) is used whenever possible, because of its quality and cost advantages, there are a number of reasons why extra stages are sometimes advisable between the camera film and the release print (Gale and Kisner, 1960). First, many motion-picture productions require special effects, such as dissolves and wipes, and these require intermediate steps if they are to be carried out conveniently. Secondly, intermediate records are useful as an insurance against loss or damage to the original camera film. Thirdly, intermediate records are useful for exporting to foreign countries for local release-printing. Fourthly, intermediate records facilitate any change of size or format between the camera film and the release prints. Fifthly, in the case of a reversal camera original, cheaper prints can sometimes be made from an intermediate record than from the camera film.

Fig. 12.7(b) shows a system in which the colour negative is duplicated by printing it through red, green, and blue filters on to suitable black-and-white films. These black-and-white films used, at one time, to be printed on to a special *internegative* film having false colour sensitization, the blue layer giving the

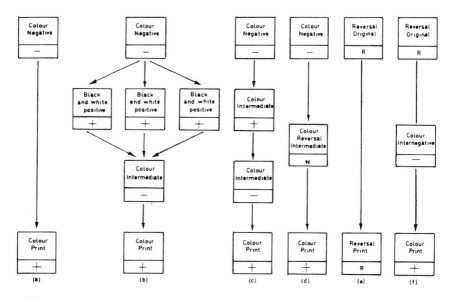

Fig. 12.7. Different methods of producing release prints in colour in professional motion picture work. Films bearing the following different types of image are indicated thus:

    − Negative image.
    + Positive image printed from a negative.
    R Positive image obtained by reversal processing.
    N Negative image obtained by reversal processing.

magenta image, the green layer the cyan image, and the red layer the yellow image (Anderson, Groet, Horton, and Zwick, 1953). This arrangement was chosen so that the magenta image (the most important for sharpness) was at the top, and the yellow image (the least important for sharpness) was at the bottom, of the tripack. This film was superseded, however, by an *intermediate* film having the conventional layer order (using absorbing dyes in the three layers to obtain good definition) known as *Eastman Color Intermediate* film (Bello, Groet, Hanson, Osborne, and Zwick, 1957). By making this intermediate film have a closely controlled gamma of 1.0, the system shown in Fig. 12.7(c) provides an alternative method which is much simpler to operate, because it avoids the necessity for printing three separate films in register; however, the use of four colour films in cascade makes it difficult to avoid some loss in quality. A simpler system is achieved by using a reversal intermediate film such as *Eastman Color Reversal Intermediate* film (Deer and Donlon, 1970) as shown in Fig. 12.7(d).

Reversal camera films are sometimes used when, for reasons of economy or portability, a 16 mm camera has to be employed. When such reversal camera films are used with reversal print films, the arrangement is known as a *reversal-reversal* or *positive-positive* system. Special films for operating this system (shown in Fig. 12.7 (e)) have been made available, such as *Ektachrome Commercial, Ektachrome E.R.*, and *Ektachrome Reversal Print* films (Groet, Liberman, and Richey, 1959; Groet, Murray, and Osborne, 1960; Beilfuss, Thomas, and Zuidema, 1966) and *Eastman Reversal Color Print* film (Thomas, Rees, and Lovick, 1965; Wall and Zuidema, 1966). Release prints can be made more cheaply, however, from a negative than from a reversal film, and so if the number of copies required is sufficient, it becomes more economical as well as more flexible, to use an internegative film, such as *Eastman Color Internegative* film as shown in Fig. 12.7 (f): this film is of low contrast and has conventional layer order and sensitizing (Zwick, Bello, and Osborne, 1956) and is printed on to ordinary *Eastman Color Print* film. All the negative, intermediate, and internegative films mentioned employ coloured couplers: the function of coloured couplers will be explained more fully later (see Section 15.4), but in these films they provide a means of correcting for the unwanted absorptions of cyan and magenta dyes, their use being particularly important in all systems where the dyes are used in several successive stages.

REFERENCES

Anderson, C., Groet, N. H., Horton, C. H., and Zwick, D. M., *J. Soc. Mot. Tel. Eng.*, **60**, 217 (1953).
Ashton, G. W., *Brit. J. Phot.*, **127**, 169 (1980).
Beilfuss, H. R., Thomas, D. S., and Zuidema, J. W., *J. Soc. Mot. Pic. Tel. Eng.*, **75**, 344 (1966).
Bello, H. J., Groet, N. H., Hanson, W. T., Osborne, C. E., and Zwick, D. M., *J. Soc. Mot. Pic. Tel. Eng.*, **66**, 205 (1957).
Berger, H., *Agfacolor*, W. Girardet, Wuppertal-Elberfeld, p. 32 (1950).
Capstaff, J. G., *J. Soc. Mot. Pic. Tel. Eng.*, **54**, 445 (1950).
Davies, E. R., *Phot. J.*, **76**, 248 (1936).

Deer, P. L., and Donlon, J. H., *J. Soc. Mot. Pic. Tel. Eng.*, **79**, 1009 (1970).
Dundon, M. L., and Zwick, D. M., *J. Soc. Mot. Pic. Tel. Eng.*, **68**, 735 (1959).
Evans, R. M., Hanson, W. T., and Brewer, W. L., *Principles of Color Photography*, Wiley, New York, p. 307 (1953).
Fritz, N. L., *Proc. Am. Soc. Photogram.*, Workshop on Color Aerial Photography and Plant Sciences and Related Areas (March 1971).
Gale, R. O., and Kisner, W. I., *J. Soc. Mot. Pic. Tel. Eng.*, **69**, 874 (1960).
Groet, N. H., Liberman, M., and Richey, F., *J. Soc. Mot. Pic. Tel. Eng.*, **68**, 8 (1959).
Groet, N. H., Murray, T. J., and Osborne, C. E., *J. Soc. Mot. Pic. Tel. Eng.*, **69**, 815 (1960).
Hanson, W. T., *J. Soc. Mot. Pic. Tel. Eng.*, **58**, 223 (1952).
Hanson, W. T., *Phot. Sci. Eng.*, **21**, 293 (1977).
Hanson, W. T., and Kisner, W. I., *J. Soc. Mot. Pic. Tel. Eng.*, **61**, 667 (1953).
Harrison, G. B., and Spencer, D. A., *Phot. J.*, **77**, 250 (1937).
Hornsby, K. M., *Brit. J. Phot.*, **97**, 132 (1950).
Koshofer, G., *Brit. J. Phot.*, **113**, 644 (1966).
Koshofer, G., *Brit. J. Phot.*, **114**, 1125 (1967).
Kisner, W. I., *J. Soc. Mot. Pic. Tel. Eng.*, **71**, 776 and 779 (1962).
Kennel, G. L., Sehlin, R. C., Reinking, F. R., Spakowsky, S. W., and Whittier, G. L., *J. Soc. Mot. Pic. Tel. Eng.*, **91**, 922 (1982).
Matthews, G. E., *P.S.A. Journal*, **21**, 33 (1955).
Mees, C. E. K., *Phot. J.*, **82**, 300 (1942).
Moser, J. S., and Fritz, N. L., *Phot. Sci. Eng.*, **19**, 243 (1975).
Mott, P. G., *Photogrammetric Record*, **5**, 221 (1966).
Movie Maker Magazine, **6**, 424 (1972).
Ryan, R. T., *A History of Motion Picture Colour Technology*, pp. 216–220, Focal Press, London (1977).
Smith, J. T., *Manual of Aerial Color Photography*, American Society of Photogrammetry, Virginia (1968).
Spencer, D. A., *J. Roy. Soc. Arts*, **95**, 675 (1947).
Tarkington, R. G., and Sorem, A. L., *Photogramm. Engng.*, **29**, 88 (1963).
Thomas, D. S., Rees, H. L., and Lovick, R. C., *J. Soc. Mot. Pic. Tel. Eng.*, **74**, 671 (1965).
Wall, C. M., and Zuidema, J. W., *J. Soc. Mot. Pic. Tel. Eng.*, **75**, 345 (1966).
Weissberger, A., *American Scientist*, **58**, 648 (1970).
Zwick, D. M., Bello, H. J., and Osborne, C. E., *J. Soc. Mot. Pic. Tel. Eng.*, **65**, 426 (1956).

GENERAL REFERENCES

Berger, H., *Agfacolor*, W. Girardet, Wuppertal-Elberfeld (1950).
Bomback, E. S., *Manual of Colour Photography*, 2nd Edn., Fountain Press, London (1972).
Collins, R. B., *Phot. J.*, **100**, 173 (1960).
Cornwell-Clyne, A., *Colour Cinematography*, Chapman & Hall, London (1951).
Evans, R. M., *Eye, Film, and Camera in Color Photography*, Wiley, New York (1959).
Evans, R. M., Hanson, W. T., and Brewer, W. L., *Principles of Color Photography*, Wiley, New York (1953).
Hanson, W. T., *J. Soc. Mot. Pic. Tel. Eng.*, **89**, 528 (1980).
Hanson, W. T., *J. Soc. Mot. Pic. Tel. Eng.*, **90**, 791 (1981).
Happé, L. B., *Brit. Kinematog. Sound Tel.*, **67**, 58, 179, 242, and 418 (1985).
Koshofer, G., *Brit. J. Phot.*, **112**, 780 (1965); **113**, 562, 606, 644, 738, 824, 920 (1966), and **114**, 128 and 1125 (1967).
Mees, C. E. K., *From Dry Plates to Ektachrome Film*, Chapter 17, Ziff-Davis, New York, (1961).
Ryan, R. T., *A History of Motion Picture Colour Technology*, Focal Press, London (1977).

CHAPTER 13

Reflection Prints in Colour

1. Introduction – *2*. Direct reflection-print systems – *3*. Reversal-reversal (positive-positive) systems – *4*. Negative-positive systems – *5*. Internegative systems – *6*. Printing from electronic images – *7*. Basic difficulties in reflection prints – *8*. Effect of surround – *9*. Inter-reflections in the image layer – *10*. Luminance ranges – *11*. Luminance levels

13.1 Introduction

A LTHOUGH in cinematography the transparency is the ultimate requirement, in still photography reflection prints are usually preferred. Transparencies can justifiably claim the advantages of being inexpensive and of providing excellent photographic quality, but they suffer from the disadvantage of being inconvenient to view; for best results, a projector must be provided, a room must be darkened, a screen has to be set up, and a source of electricity provided; none of this is necessary for reflection prints.

13.2 Direct reflection-print systems

At first sight it might seem that the simplest way of producing reflection prints would be just to coat a reversal tripack material, such as *Kodachrome* or *Ektachrome*, on a paper support, instead of on the usual transparent film base, and to use it in the camera. But this suffers from a number of serious difficulties. Such a system gives laterally reversed, or mirror-image, pictures of the real world. This can be overcome by using a prism or mirror over the camera lens, but this complicates the camera. In the original *Polacolor* process, the image, immediately after development, is soluble, and this enables it to transfer to a receiving sheet, placed in contact with it, where it becomes insoluble again. Since the receiving sheet is placed face to face with the material exposed in the camera, the transferred image is laterally reversed with respect to the camera image, and is therefore correct with respect to the original scene. The development and transference of the image take place actually in the camera and the user can see the completed print within one minute of making the exposure (see Section 17.10). In the new *Polacolor* process the receiving sheet is integral with the exposed material and lateral reversal is corrected by means of a mirror contained in the camera (Land, 1972).

But whether by mirror or by transfer, two disadvantages of direct reflection-print systems may be noted. First, no enlargement occurs, so that a compromise choice has to be made between large cameras and small prints; secondly, accurate determination of exposure is necessary because the tolerances for image density and colour balance are smaller for reflection prints (which can be compared with other objects in the field of view surrounding them) than for transparencies projected in a dark room.

13.3 Reversal-reversal (positive-positive) systems

If the production of reflection prints directly in the camera presents difficulties, it might be thought that the most attractive alternative would be to provide means whereby reflection prints could be made from transparencies. An enlarging stage could easily be incorporated so as to produce large prints from small cameras; correction for variations in the density or colour of the transparencies could be provided so as to produce prints of correct exposure and colour balance; and lateral reversal is of course avoided merely by turning the transparency over before printing it. Selection of those transparencies from which prints are especially wanted is rendered easy by virtue of the fact that colour transparencies consist of positive, and not negative, images. It is not surprising, therefore, that the first colour reflection prints offered to the public on a wide commercial scale used colour transparencies as intermediates, both in the U.S.A. (in 1941) and in England (in 1954). It is interesting to note that some of these prints have been made by processes in which the dye images, instead of being formed in the layers by colour development, are formed by image-wise destruction of dyes that are already present in the layers: the *Ilford Colour Prints* introduced in 1954 were of this type, and the *Cilchrome*, later *Cibachrome*, prints introduced in 1964 also use this system; the chemical basis of these systems is described in Section 17.10. Papers on which positive transparencies can be printed directly to give positive images are known as *reversal colour papers* and the system as the *reversal-reversal* or *positive-positive* system. Several such papers are commercially available (for instance *Ektachrome* paper), and print services depending on this system are offered by some photographic manufacturers and by some photofinishing laboratories. (For reversal-reversal film systems, see Section 12.11.)

13.4 Negative-positive systems

There are, however, several advantages in making *negatives* instead of positives on camera films when requiring reflection prints. Since colour negatives reverse all the tones and colours of the original scene, they are obviously not intended to depict its appearance, and hence their characteristics can be adjusted solely to obtain the highest quality, at the greatest convenience, on the final print. Thus negative materials can be made of low contrast (a high contrast print material being used to give the required overall contrast to the system); and their low contrast enables them to possess good exposure latitude without over-exposure resulting in very high densities (which tend to be difficult to print). They can also incorporate couplers that are coloured, to correct for the unwanted absorptions of the dyes in the system (see Section 15.4). Finally, negative photographic materials are, at least in principle, inherently easier to manufacture and process than reversal materials. Colour negative films are therefore very widely used as camera materials when reflection prints are required.

13.5 Internegative systems

Another method of making prints which, like the reversal-reversal system, employs a transparency as the starting point is the *internegative* system. In this system, a colour negative is made from the transparency and this is then printed on to the same type of colour paper as is used for the ordinary negative-positive system.

The internegative stage provides an opportunity for introducing colour correction and this is important because the use of cyan, magenta, and yellow dye-sets three times (in the transparency, in the internegative, and in the print material) means that the effects of their deficiencies, such as unwanted absorptions, are much more noticeable; the use of coloured couplers at some stage in the system is therefore desirable, and they can be incorporated in the internegative film.

Furthermore, by making the internegative film have a higher contrast at the high-exposure end of its scale than at its low-exposure end (see Fig. 14.10), some correction can also be made for the tendency for light transparencies to be of lower contrast than dark transparencies. This tendency arises from the fact that films intended for the production of transparencies must have a gamma that increases with density, so that the tone reproduction of the pictures is subjectively satisfactory (see Section 6.5).

Thus the internegative system offers more scope for colour and tone correction than the reversal-reversal system, but the necessity for producing the internegative makes it more expensive (unless the cost of the internegative can be spread over more than one print), it is more complicated to operate, and it may involve loss of definition unless special precautions are taken.

13.6 Printing from electronic images

If it is required to obtain reflection prints from pictures that are in the form of electronic signals, these signals can be used to obtain images on cathode-ray tubes (or on some other television type of display device) which can then be printed optically on to a suitable colour photographic paper. An alternative method, used in the *Mavigraph* printer in connection with the *Mavica* electronic camera, is to use the electronic signals to control the heat output of a special head that traverses a coloured sheet of material causing evaporation of a pigment on to a receiving paper; separate coloured sheets, and transfer steps, have to be used for the cyan, magenta, and yellow pigments (Sony, 1982). (See also Section 16.15.)

13.7 Basic difficulties in reflection prints

Whether made by a direct, a reversal-reversal, a negative-positive, an inter-negative, or an electronic system, for reflection prints in colour to be successful, three difficulties additional to those inherent in transparencies have to be accommodated. These difficulties are: the effect of the surround; the effect of inter-reflections in the image layer; and the limited tone range caused by light reflected by the top-most surface of the prints.

13.8 Effect of surround

The first difficulty is that, when a reflection print is viewed, other objects in the field of view provide a reference framework of lightness and colour balance against which any deviations in these respects in the print are easily noticed. But when a transparency is projected in a dark room the viewer has much less of a reference framework and hence variations in the lightness and colour balance of transparencies often pass undetected. The tolerances in density and colour balance when making reflection prints to a given standard of acceptability are therefore smaller than in the case of transparencies, and special printing techniques are required: these have been successfully established and will be described in Chapter 16.

The effect of surrounds on apparent contrast and colourfulness has been discussed in Chapter 6.

13.9 Inter-reflections in the image layer

The second difficulty encountered in making reflection prints arises from the fact that the dyes composing the image are situated in a layer that is bounded at one side by the diffusing paper surface, and at the other by the gelatin-air interface. This means that the light by which the images in a reflection print are viewed passes not just once each way through the image layer, but, on the average, several times. The reason for this can be seen from Fig. 13.1. Gelatin, which is used as the vehicle for photographic emulsions, has a refractive index of about 1.5. This means that light reflected from the paper base beneath the gelatin can only escape from the gelatin if it emerges within an angle of about 40° from the perpendicular to the gelatin surface. Light reflected outside this cone is totally internally reflected back on to the paper where it is rediffused for a second attempt at escaping. Light reflected at the critical angle, which for gelatin is

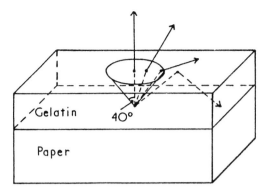

Fig. 13.1. Gelatin has a refractive index of 1.5, so that only light reaching its surface within a cone of semi-angle 40° is able to escape, the rest being totally internally reflected.

about 40°, emerges along the actual surface of the gelatin. The paper reflects light in all directions, and the fraction of the rays that emerge at the first attempt is less than half the total, actually only 38.6 per cent (Williams and Clapper, 1953). Thus 38.6 per cent of the incident rays emerge at the first attempt, the rest travelling back to the paper and up again, thus traversing the layer four times instead of twice and magnifying the density by a factor of two. Again only 38.6 per cent of the rays escape, and the remainder have to traverse the layer a further twice, six times in all, before another attempt is made, and so on. It is clear, therefore, that reflection densities are not simply equal to twice the transmission densities but to more than twice. The magnitude of the increase in density, however, is dependent on the transmission density of the layer; for if the density is very high the contributions of the rays that have traversed the layer four or more times will be quite small, but, if the density is low, rays which have traversed four, six, or even eight or more times will still have a noticeable effect. Suppose, for instance, the layer transmitted only one-tenth of the light, so that its transmission density was 1.0 (optical density is defined as $\log_{10}(100/T)$ where T is the percentage transmittance). The light that had passed twice through the layer would be reduced to only one-hundredth of its original intensity, equivalent to a density of the layer of 2.0; but light that had passed four times through the layer would be reduced to one ten-thousandth, which is fairly negligible compared to a hundredth. But if the layer had a transmission density of only 0.15, then the intensities would be reduced by effective densities of 0.3 after two passes, 0.6 after four, 0.9 after six, and so on, and the corresponding intensities of one-half, one-quarter, one-eighth, and so on, do not fall quickly to a negligible proportion of the total reflection. The result, therefore, is that low transmission densities are magnified by factors considerably in excess of the simple doubling that might be expected, but that as the density of the layer increases the increase in the factor gradually reduces. This is shown in Fig. 13.2.

This behaviour of absorptions in reflection print layers has two important consequences in colour reflection prints. In the first place, it means that, in areas intended to be white, very great care must be taken to minimize any traces of residual dyes, because, as can be seen from Fig. 13.2, very low transmission densities are magnified by factors as large as five or more. Secondly, because the

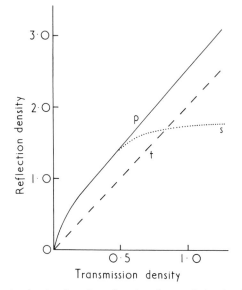

Fig. 13.2. Reflection density plotted as a function of transmission density: (p) when the dye layer is in optical contact with a diffusely reflecting white layer, as in a reflection print; (t) when the optical contact is broken and the reflection density is simply twice the transmission density; (s) the way in which surface reflections modify curve (p) in typical room viewing conditions.

unwanted absorptions of cyan, magenta, and yellow dyes are of lower density than the wanted absorptions, they will be increased in density more than the wanted absorptions; this means that the unwanted absorptions have a stronger effect in reflection than in transmission materials. The effect also results in an effective broadening of the spectral absorption curve of a dye as shown in Fig. 13.3 (Evans, Hanson, and Brewer, 1953). Once again low densities have been raised by a greater factor than high densities so that although the required peak density is obtained in a reflection image by having slightly less than half the amount of dye present, as compared with the amount required in a transmission image, the absorptions at wavelengths off the peak are increased more and hence the effective absorption of the dye becomes broader. Reflection prints, therefore, even if possessing dyes chemically identical to those used in transmission materials, have colorants that are broader in their absorption bands and worse in

224

their unwanted absorptions, and these effects tend to reduce colour saturation and lightness.

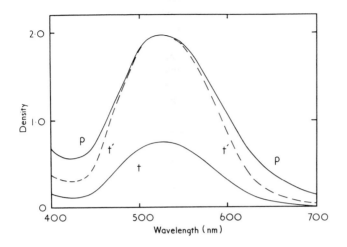

Fig. 13.3. Spectral absorption curves of the same dye: (t) in a transparency material; (p) in a reflection print material; and (t'), curve (t) multiplied by 2.61 to make the peak density equal to that of curve (p). Curve (p) is broader and has higher unwanted absorptions than curve (t').

13.10 Luminance ranges

The third difficulty encountered in reflection prints arises from the existence of the top-most surface of the image layer; this surface inevitably reflects some light, and this light, because it has not traversed the image layer, simply acts as a whitish flare, desaturating the picture, its greatest apparent effect being in dark areas. Of even more importance than the desaturating effect, however, is the limitation that this surface reflection has on the range of tones that can be reproduced in reflection prints.

Jones and Condit (1941) have shown that the range of luminances in outdoor scenes varies from about 1.45 to 2.8 log units, with an average of 2.2 log units (that is, from 28 to 1 to 630 to 1 with an average of 158 to 1, in arithmetic units). Transparency materials usually have maximum densities of over 3.0, corresponding to a tone-range of 1000 to 1, and this suggests that they should in theory be able to handle the maximum luminance range of outdoor scenes satisfactorily. However, measurements of typical maximum and minimum luminances actually present in projected pictures shows the range to be only about 2.1 log units or 126 to 1 (Estes, 1953; Hunt, 1965a). The difference is made up of several factors. First, transparency films all exhibit some absorption in fully exposed areas and this 'stain' often amounts to 0.2 or more log units. Secondly, it is impossible to make photographic materials with characteristics that remain linear at the ends of their exposure scales, so that for whites to be reproduced with adequate tonal modulation their density has to be higher than that of the stain; this can account for a further loss of about 0.2 log units. Thirdly, if the maximum density of a slide is about 3.0, a fairly typical figure, and the luminance caused by ambient light is 3.0 log units less than the open-gate screen luminance at the same point (a figure representative of good projection conditions), then the minimum luminance for a black would be equal to twice that of the ambient luminance, or 0.3 log units above that corresponding to the maximum density of the film. Fourthly, because of vignetting in camera lenses and fall-off in the luminance provided by projectors towards the corners of the pictures, typical maximum luminances in pictures are further reduced by about 0.2 log units, because only in some cases will they occur in the centre of the picture where the maximum luminance is available. These factors, amounting to 0.9 log units, reduce the tone range from the 3.0 log units of the maximum density of the films to 2.1 log units in picture elements on the screen.

However, the effect of the dark surrounds in reducing the apparent contrast of projected transparencies (see Section 6.5) is such that this range of reproduction luminances, 2.1 log units, is equivalent to only about 1.4 log units in scene luminance, and this is appreciably less than the 2.2 log units for an average outdoor scene.

But, when similar measurements are made on the range of tones available between typical maximum and minimum luminances in reflection prints viewed in average conditions, the range is even smaller than for transparencies.

It is possible to make colour photographic papers that, when processed to give highly glossy surfaces, possess blacks that when measured in good reflection densitometers have maximum densities of 2.4 or more. However, as with colour transparencies, the luminance range is reduced by several factors. First, the minimum density or 'stain' of colour photographic papers usually amounts to about 0.1 log unit; secondly, another 0.2 log unit is normally lost because of the need for reproducing whites at a high enough density for adequate tonal modulation to be achieved. Vignetting is usually only a small factor in negative-positive print systems because the negative tends to have light corners and this helps to correct fall-off of luminance towards the corners of the picture on enlargement: in reversal-reversal systems vignetting may add a further important loss. The factors so far considered amount to a total loss of range of about 0.3 log units, so that it might be expected that 2.1 log units should remain. But measurements show that the actual tones perceivable often run from about 0.3 to a maximum of only 1.55 giving a range of only about 1.25 log units (Hunt, 1965a). A further large loss of 0.85 has therefore occurred, and this is the result of the light reflected from the topmost surface of the print. Similar results for black-and-white photographic papers have also been reported (Carnahan, 1955).

The reason why a black, that a densitometer can measure as having a maximum density of 2.40, appears to have a maximum density of only about 1.55

when seen under typical viewing conditions in a room can be explained by reference to Fig. 13.4. The top half of the figure illustrates the situation in the densitometer. The print is illuminated by a beam of light perpendicular to the surface and the reflection from the glossy top surface of the print is mainly back along the same path; the photocell, which views the print from an angle of 45°, therefore picks up only the light reflected diffusely by the print after traversing the image layer. The area of the inside of the densitometer that is specularly reflected

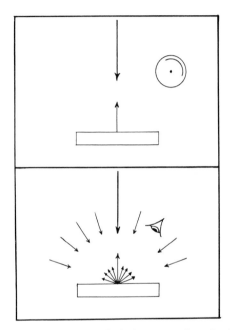

Fig. 13.4. Top: when illuminated perpendicularly, no specular reflection from a reflecting sample reaches a photo-cell placed at 45°. Bottom: when viewed in a light room, no matter where the eye is placed, some specularly reflected light is seen, and this limits the maximum density that can be seen on reflection prints in typical viewing conditions.

along the direction of the photocell is arranged to be of extremely low luminance; hence the effect of the light reflected by the topmost surface of the paper is virtually eliminated. The lower half of the figure illustrates the conditions when the print is viewed in a room: although most of the light may come from one (or a few) directions, as indicated by the heavy vertical arrow, some light will fall on the print from all other directions since the room, in general, has significant luminances in all its parts. Hence, no matter where the eye is placed, specular reflection of the light coming from some part of the room will enter the eye and prevent it from seeing the true maximum density of the paper. It is found that

1.55 represents about the maximum density that can be seen in ordinary rooms, and the effect of this on the relationship between reflection and transmission densities is shown by the dotted line in Fig. 13.2. The value of 1.55 is not affected by the maximum density evaluated in a densitometer, provided this figure is not less than about 2.0. (It should not be construed from this that the maximum densities of papers are unimportant so long as they are above 2.0: changes in the maximum amount of dye available in a colour print can affect the rendering of saturated colours by virtue of changes in the sloping sides of its spectral absorption curve, even though the appearance of blacks as viewed in rooms is unaltered.) The figure of 1.55 applies to rooms with rather directional light sources, such as are provided by tungsten lamps: in diffusely lit rooms, it may be only about 1.40. In highly unusual viewing situations, such as viewing the print by a shaft of sunlight in a coal-cellar, or, more realistically, holding prints in a beam from a projector shining in an otherwise dark room, figures higher than 1.75, and even approaching the densitometer figure, may be achieved.

For ordinary conditions the range of luminances available in reflection colour prints can therefore be taken as 1.55 minus 0.30 (to allow for the losses mentioned earlier), or a range of no more than 1.25 log units, or 18 to 1. This range is thus less than available in projected transparencies (2.1 log units in a dark surround, equivalent to an effective range of about 1.4 log units in an average surround) and both these ranges are much less than that of average out-door scenes (2.2 log units). Hence, although densities below those needed for reproducing modulated whites will also be used, some compression of whites or blacks or both must take place, but more so in reflection prints than in transparencies; this is illustrated in Fig. 6.11. (When transparencies are projected in conditions of considerable ambient illumination, and when reflection prints are made on very matt paper, such as newsprint, the ranges of luminance are decreased even further.) When photographing scenes with the intention of producing colour reflection prints the successful photographer is therefore careful to compose or arrange the scene so as to keep all important subject matter within a limited luminance range; in general this means that heavy shadows must not be allowed to fall on important parts of the scene. Even when this is done, difficulties remain with some types of scene: for instance, in wedding pictures, while a transparency may handle the white of the bride's dress, and the black of the bridegroom's suit without difficulty, in a reflection print it is not easy to avoid either loss of detail in the former by flat highlights or in the latter by blocked-up shadows. (See Plate 9, page 112.)

It is sometimes useful to lighten heavy shadows in out-door scenes by using *fill-in flash*; that is, a (blue-coated) flash-bulb or a Xenon flash can be fired at the time of exposure from a position near the camera. Similar results can also be obtained by using white or metallic reflectors suitably placed. In these ways the lighting can be adjusted to suit the system. (See Plate 33, page 513.)

13.11 Luminance levels

The apparent brightness of any given luminance depends markedly on the adaptation conditions of the observer's eye. Thus the motor-car headlamp that dazzles painfully after dark, appears to be a more modest source in bright sunlight: the luminance has remained constant, but the brightness has changed enormously.

In comparing the luminance ranges available in colour transparencies and prints, some attention must also be paid, therefore, to the absolute luminances involved and to the adaptation conditions. Projectors that provide about 100 lux on the screen produce whites of luminances comparable with those on reflection prints illuminated at about 50 lux: these levels are roughly equivalent to those obtained with average domestic projectors and average tungsten-lit rooms in houses. Projectors giving 500 lux are sometimes used and are becoming more common, and living rooms lit at considerably more than 50 lux are certainly encountered; but to consider the luminance of the whites in the two cases as similar seems broadly justifiable. What then of the brightnesses? In the case of the projected slide the surround is dark and this has the effect of increasing the brightness, or 'subtracting grey' from the picture, whereas no such effect usually occurs when viewing prints. The result is that the whites on projected transparencies look much whiter than those on reflection prints; moreover, as we have seen, the transparencies also have a greater luminance range, and this is usually sufficient to make the blacks look blacker as well. The advantage here, therefore, lies entirely with the transparency, the whites are brighter and the blacks blacker. But if colour reflection prints are viewed in bright sunlight the illumination level can reach 50 000 lux, and this leaves even a powerful projector far behind. But what happens to the brightness? Some indication can be obtained by using the haploscopic matching technique (as described in Section 8.10), some results of which will now be described (Hunt, 1965a).

In the instrument used, the observer's left eye had an unobstructed view of the scene, while his right eye viewed a white field of $15°$ subtense, and 3600 cd/m^2 luminance with a centre spot of subtense $1\frac{1}{2}°$ which could be adjusted in colour and luminance to produce a wide range of brightnesses. A wide range of brightnesses seen by the left eye could therefore be matched by the centre spot in the right eye, and the luminance of the centre spot necessary to do this was used as a measure of the brightnesses seen in the left eye. Observations of this type were made on a white, and four grey, squares of a chart containing a neutral scale of six squares, and eighteen coloured squares. The results are plotted in Fig. 13.5. The luminances of the central spot seen by the right eye are plotted on a log scale which has been spaced so as to represent Munsell value intervals uniformly. The luminance corresponding to the right-eye surround luminance has been set arbitrarily at the point on the scale representing Munsell value 6. The dots marked 1 all refer to the white square on the chart, those marked 2, 3, 4, 5 refer to progressively darker greys. The left-eye viewing conditions used are marked along the top and the results obtained for each are plotted vertically below them. The spacing of the viewing conditions along the abscissa is entirely arbitrary, except that they are progressive in luminance level, and has been chosen so that approximately linear results are obtained for the white points. The condition marked 'tungsten projector, 160 lux', consisted of the chart illuminated in a dark room by means of a tungsten projector the size of whose projected beam exactly covered the chart area and no more, so that the viewing conditions were closely analogous to projecting a transparency. It is seen that although the point labelled 1 for the tungsten projector at 160 lux is higher than that for tungsten room lighting at 24 lux, it is lower than those for Xenon arcs at 1080 lux, north sky at 4300 lux, overcast sky at 10 800 lux, and bright sun at 43 000 lux. Since, in these last four viewing conditions, the chart was seen in an average environment, and not surrounded by a very low luminance as in the case of projected pictures, it is clear that a substantial increase in the luminance of whites can produce a greater increase in brightness, when measured in this manner, than occurs as the result of a dark surround. Thus, according to these results, even prints viewed in a room well-lit by daylight (1000 lux) should appear to have whites of greater brightness than those of projected transparencies at about 160 lux. It is therefore to be

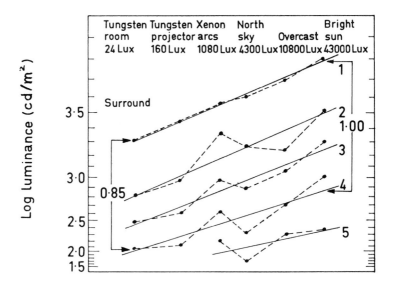

Fig. 13.5. Log luminance (of a $1\tfrac{1}{2}°$ central spot in a 15° surround of 3600 cd/m²) necessary to produce the same brightness as that of a white (1) and four grey (2, 3, 4, 5) samples seen under the conditions shown. The reflectances of the white and the greys were: (1) 85.0, (2) 21.5, (3) 13.3, (4) 5.3, (5) 1.7%. If the visual contrast represented by the spacing between lines 1 and 4 for the bright sun condition is regarded as unity, that for the tungsten room condition is about 0.85 (on this Munsell value type of scale with the surround being set at Munsell value 6).

concluded that prints viewed in good indoor daylight have brighter whites than those of projected transparencies. Thus, although the range of luminances in the prints remains a restriction, their appearance should be greatly improved by viewing them under conditions of high illumination, and this is borne out in practice.

It is interesting to note in Fig. 13.5 that the spacing of the points representing the white and grey squares is roughly similar for most of the six viewing conditions, and, if the Munsell value scale as set is appropriate for these observations, this suggests that the contrast of the grey scale does not alter markedly over this range, although of course the brightness varies. There is, however, some evidence for a slight fall in visual contrast as the luminance level falls, a tendency also reported by others (Breneman, 1962; Stevens, 1961) but it must be remembered that if the surround were allocated a different value on the Munsell scale the spacing of the points would be affected. The figure shows a tendency for the darkest grey (5) to become rather indistinguishable from black as the illumination level drops to 24 lux, and, although this accords with one's general visual impression, the accuracy of haploscopic matching with a surround field of high luminance becomes very low as black is approached, and the position of the points is not very reliable; in fact, the darkest grey (5) could not be measured at all under the two lowest illumination levels. This is probably why the contrast-lowering effect of a dark surround is not shown by the results for the tungsten projector. Brightness-scaling investigations have shown that, as the illumination level falls, dark greys tend to remain fairly constant in appearance, but that blacks actually *increase* in brightness (Stevens, 1961).

It has been shown (Hunt, 1952 and 1965b) that not only does brightness increase with illumination level but substantial increases in colourfulness also take place; this is a further reason why the appearance of colour photographs is improved by viewing them at high levels of illumination (Bartleson, 1965). It thus becomes at least debatable, and no longer a foregone conclusion, whether a projected transparency at 100 lux, or a reflection print at 1000 lux, is to be preferred. Other factors peculiarly deleterious to reflection prints, such as the degradation of dye colour caused by inter-reflections, or the noticeability of stain as such, or the ability to detect quite small departures from optimum in density or colour balance, may give victory to the transparency; but as far as brightness of the whites is concerned the advantage should be with the print, and if the subject matter is carefully handled the restricted luminance range need not have a serious effect. This was illustrated at the 1964/65 World's Fair at New York (Bartleson, Reese, Macbeth, and James, 1964), where a group of reflection prints displayed under very high levels of illumination appeared as attractive as transparencies; however, in this case the print was illuminated at a higher level than the surround so that some 'subtraction of grey' by virtue of simultaneous contrast increased the brightness of the whites above that normally associated with the luminance level used.

REFERENCES

Bartleson, C. J., Reese, W. B., Macbeth, N., and James, J. E., *Illum. Eng.*, **59,** 375 (1964).
Bartleson, C. J., *Phot. Sci. Eng.*, **9,** 174 (1965).
Breneman, E. J., *Phot. Sci. Eng.*, **6,** 172 (1962).
Carnahan, W. H., *Phot. Eng.*, **6,** 237 (1955).
Estes, R. L., *J. Soc. Mot. Pic. Tel. Eng.*, **61,** 257 (1953).
Evans, R. M., Hanson, W. T., and Brewer, W. L., *Principles of Color Photography*, p. 365, Wiley, New York, London (1953).
Hunt, R. W. G., *J. Opt. Soc. Amer.*, **42,** 190 (1952).
Hunt, R. W. G., *J. Phot. Sci.*, **13,** 108 (1965a).
Hunt, R. W. G., *J. Opt. Soc. Amer.*, **55,** 1540 (1965b).
Jones, L. A., and Condit, H. R., *J. Opt. Soc. Amer.*, **31,** 651 (1941).
Land, E. H., *Brit. J. Phot.*, **119,** 858 (1972).
Sony, *Brit. J. Phot.*, **129,** 363 (1982).
Stevens, S. S., *Science*, **133,** 80 (1961).
Williams, F. C., and Clapper, F. R., *J. Opt. Soc. Amer.*, **43,** 595 (1953).

CHAPTER 14

Quantitative Colour Photography

14.1 Introduction

M ANY colour films are successfully exposed without any measurement being made of the illumination levels of the scenes, and even when such measurements are made they are often restricted to the response of an exposure meter to the integrated light reflected by each scene as a whole. Appraisal of the final picture is often by visual inspection without any measurements being made at all.

But even in amateur colour photography measurement plays a very important, even if largely hidden, part in the success or otherwise of the results. Thus, when the final picture is in the form of a reflection print, measurements of the transmittance of the negatives, internegatives, or transparencies from which the prints are made, form a vital part of the printing operation, as will be discussed in Chapter 16; and even when transparencies are produced directly on a reversal film without any printing stage, a very good deal of measurement will have been made by the manufacturer on samples of film cut from the same batch of material as is being used by the customer, and the conditions in which the film is processed will normally be the subject of further extensive measurements.

In professional photography, more elaborate measurements are often made on the scene itself, including, for instance, determining the luminance of an average white (or the maximum and minimum luminances), as well as, or instead of, determining the average luminance of the scene as a whole; and if intermediate stages are used, involving black-and-white separations, internegative or intermediate films, or masking techniques (to be described in Chapter 15), then measurement is usually essential.

If photographic steps are used in conjunction with other media, such as television or half-tone printing, measurement is indispensable in the experimental and setting-up stages, otherwise it becomes virtually impossible to obtain a clear picture of the contribution of each part of the system to the virtues or deficiencies of the final result, and improvements cannot then be systematically sought.

Finally, in some scientific and technical investigations it may be desirable to use colour photography to obtain quantitative results of certain phenomena, and, of course, measurement is then involved.

There are thus many reasons why an understanding of the science of photographic measurement as applied to colour photography should be acquired. The measurement of the sensitivity of photographic materials is called *sensitometry*. Sensitometry is carried out in absolute terms by photographic manufacturers who are naturally greatly concerned with the photographic 'speeds' of their materials, but in many instances of applied colour photography the interest is confined to relative results, and in these cases measurements in absolute terms are not required and simpler techniques can often be adopted. The evaluation of the photographic records obtained in sensitometry is called *densitometry*, because, as the name implies, this usually involves measuring *density*, (defined as $\log_{10}(100/T)$ where T is the percentage transmittance or reflectance of the area being measured).

Sensitometry consists essentially of the three basic steps of all photographic systems, exposure, processing, and evaluation; but each step has to be carried out under closely controlled conditions to obtain consistent results. It is important, however, to see that the controlled conditions are still typical of the actual conditions of use of the material, or the results, though consistent, may not be relevant.

14.2 Sensitometric pictures

Sensitometric exposures are intended to illuminate the photographic material with known amounts of light, the purpose being either to determine the absolute sensitivity of the material, or to calibrate it in relative terms for some particular application. The difficulty of maintaining any real scene constant in luminance at all points means that some artificial 'picture' must be used in practice. This could be a suitable black-and-white or colour photograph, and this type of test object is sometimes used. If the photograph is a transparency it can be printed on to the photographic material under test by placing it in contact and illuminating it with a uniform controlled light source. But, if the size of the image required on the test film is different from that of the original, a projection system must be used and the original can then be either a transparency or a reflection print; transparencies are generally used because of their greater tone range and because of the greater difficulty of keeping a reflection print clean. Instruments for forming images of photographs on to test films for sensitometric purposes are known as *camera sensitometers*; they are mainly used for rather special purposes: thus graininess and sharpness can be evaluated by forming an image greatly reduced in size on to the test film, and then magnifying it again; or if large numbers of identical pictures are required for a test programme it is often convenient to expose them on a camera sensitometer.

14.3 Sensitometric wedges

Using a picture of a scene as a test object, however, has the disadvantage that the identification of individual areas for measurement is rather complicated; and using a projected image for exposing the test-material has the disadvantage that vignetting and flare, caused by the lens, introduce uncertainties into the calibration, unless this is carried out by direct photometry of the image itself, which may be difficult to do with sufficient accuracy. For general purposes, therefore, a very much simplified 'scene' is usually used: this consists of either a stepped or continuous *wedge*. This is a strip of transparent material the transmittance of which varies along its length either in steps, or continuously, in a known manner. The test-material is exposed by contact-printing such a wedge on to it, the instrument in which this is done being called a *sensitometer*; the exposure given at each point is then identified either by the number of the step along the wedge (if it is stepped) or the distance along the wedge (if it is continuous).

For convenience, sensitometric wedges should have, for any one step or position, exactly the same transmittance at all wavelengths of the visible spectrum; in other words, they should be constructed of *non-selective neutral* material. Such material is not easy to find, and some sensitometers, instead of using wedges, use rotating drums that vary the exposure *time*, instead of the illumination, along the strip of test-material. However, changes in time and illumination, although approximately, are not exactly, interchangeable on photographic materials (a phenomenon known as *reciprocity failure*), so that such variable-time exposures are usually only suitable for purely relative work, like checking processing uniformity. For general work, non-selective neutral wedges are required, and these are available commercially either as black-and-white photographic (silver) wedges or as colloidal graphite (carbon) wedges, or as *Inconel* (metal deposit on glass or quartz) wedges. Whenever photographic silver wedges or test-objects are made up, the neutrality of the image should be carefully checked, because preferential scattering at certain wavelengths can cause appreciable coloration.

The colloidal graphite is usually dispersed in gelatin and, if the particle size is small, it has a brownish colour which has to be corrected by incorporating bluish dyes: Wratten neutral density filters are of this form. By making the particle-size larger (M-type carbon) better neutrality is achieved but appreciable scattering occurs which limits the use to non-image-forming situations. The Inconel wedges have very good neutrality, but are expensive and have strong specular reflection which may need to be dissipated (Eastman Kodak Co., 1970).

The size of the steps on a stepped wedge is largely a matter of convenience, but very small steps are difficult to evaluate and are affected by local development-exhaustion effects (*edge-effects*); a step-length of 0.4 inch (10 mm) is widely used, with the width being either the same or extending right across the film. Transmittance increments from one step to the next are convenient if they are in the ratio of the square-root of two, so that the exposure is doubled or halved for every two steps.

14.4 Uniformity of illumination

In camera sensitometers the vignetting of the lens may have to be counteracted by illuminating the test object more strongly at the edges than at the centre, but in contact sensitometers very uniform illumination of the picture or wedge is desirable so that for calibration purposes any residual non-uniformities can be ignored. This means in practice that the light should be uniform to within about ± 1 per cent. If a single source, small enough to be considered a 'point source' is used, it must be about $4\frac{1}{2}$ times the length of the wedge away to provide ± 1 per cent uniformity; thus if the wedge had 21 steps, each 0.4 inch (10 mm) long, the total length of the wedge would be 8.4 inches (21 cm) and the distance of the lamp nearly 40 inches (100 cm). However, it has been pointed out that, if two point-source lamps are used, much better uniformity can be obtained with the lamps

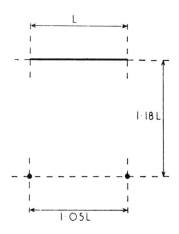

Fig. 14.1. Arrangement of two 'point-source' lights so as to give illumination, along a wedge of length L, uniform to within ± 0.1 per cent.

much closer. Thus a uniformity of ±0.1 per cent is attained with the lamps 1.18 times the length of the wedge (about 11 inches (27½ cm) for an 8.4 inch (21 cm) wedge) away from the plane of the wedge; in this case, as shown in Fig. 14.1, the lamps have to be equidistant from the centre of the wedge and separated by a distance equal to 1.05 times the length of the wedge (Marriage, 1955). This two-lamp arrangement thus enables a much more compact sensitometer to be built with better uniformity: one problem is that unequal ageing of the lamps would cause serious deterioration of the uniformity, but, because the lamps are now so much closer, the illuminance of the test-material is greatly increased, and hence the lamps can be under-run and good stability achieved.

14.5 Exposure time

The time for which the photographic material is exposed should be as near as possible the same as that used in the conditions that the sensitometer is supposed to simulate, unless it is known that the difference in exposure time does not involve reciprocity failure, or that the effects of any reciprocity failure are negligible in the application concerned. If enlarging on to photographic paper is the application, then exposure times of between about 1 and 30 seconds are involved: these can conveniently be given by switching tungsten filament lamps on and off by means of an electronic timing switch; the lamps of course must be run at constant voltage and a constant-voltage transformer is usually adequate for this purpose. With camera films, typical exposure times are usually much shorter and some sort of shutter must be used: for these short exposure times, usually in the range 1/50 to 1/500 second, the illuminance on the film has to be quite high in order to give an adequate exposure. In one convenient form of sensitometer, this is achieved by using a single lamp and a slit which are moved uniformly together along the wedge; narrow slits can then be used to give very short exposure times, and short lamp-to-film distances to give high levels of illumination. Exposure times in the range from 1/1000 to 1/10 000 sec. are usually given in practice by means of electronic flash equipment, and in these cases it is therefore appropriate to use a carefully controlled version of this type of source in the sensitometer, no shutter being necessary.

14.6 Light sources for sensitometry

Except for the very short exposure times just mentioned, tungsten lamps are the most convenient source for sensitometry because of their ease of control and good stability if under-run. When the application that is being simulated also uses tungsten lamps, then they are clearly desirable in the sensitometer because their spectral power distribution is also correct. If the source used in the application is daylight, however, the tungsten lamps must be filtered to simulate the spectral power distribution of daylight, and this can be conveniently done with an appropriate thickness of a blue glass, such as one of the two shown in Fig. 14.2; it may also be desirable to add a heat-absorbing glass in order to reduce the amount of far-red and infra-red light that the tungsten lamps emit relatively copiously and which, as can be seen from Fig. 14.2, the blue glasses do not absorb very fully. Gelatin filters may not be suitable for use in sensitometers because of inadequate permanence under the strong illuminances involved. It should be noted, however, that some glasses change colour appreciably at high temperatures and these should be avoided in sensitometers.

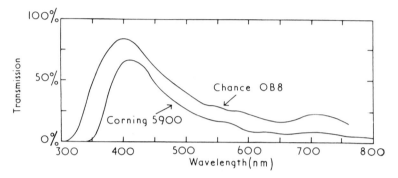

Fig. 14.2. Spectral transmission curves of two blue glasses, useful for simulating daylight with tungsten filament lamps.

14.7 Transmission colour of lenses

Photographic lenses do not transmit all wavelengths equally freely, because of absorptions in the glasses and because of the effects of the anti-reflection coatings on the surfaces of the components; the effect of this on the colour of the light reaching photographic materials in practice is quite large (Williams and Grum, 1960), and may need simulating in sensitometers by the incorporation of suitable filters in the light beam. A British Standard was drawn up to provide tolerances within which the transmission colour of photographic camera lenses should lie (British Standard 3824: 1964), but some lenses of earlier manufacture lie well outside its limits (American Standards Association, PH3.37, 1969).

14.8 Selective exposure of layers

If it is required to expose the layers of a colour material separately, this can usually be achieved by using narrow-cut red, green, and blue filters. Much can sometimes be learned by exposing the layers both separately, to give coloured wedges, and together, in the proportions to give a grey wedge, and then comparing the results; this technique is useful in evaluating *inter-image* effects (see Section 15.5).

14.9 Latent image changes

A variable that needs to be remembered in sensitometric work is the change in the latent image with time. This is not usually a large factor, but it may often occur in the form of a fairly rapid change in the first two or three days, followed by a fairly slow drift. For the highest accuracy it is thus advisable either always to process the material within an hour or so of exposing it, or to allow a few days for ageing at room temperature, followed by cold storage until processing. The former method is more convenient for some purposes, but the latter is useful when large numbers of strips are exposed at one time for subsequent use as process-control checks.

14.10 Controlled processing

Because of the inherent variability of the processing step in photographic systems, special precautions usually have to be taken when processing any material on which measurements are to be made; but care is necessary to ensure that the special precautions do not make the process atypical of those used in practice.

The factors affecting consistency of processing include the chemical compositions of the solutions, the temperatures of the baths, the time spent by the material in each bath, and the nature of the agitation of each solution over the surface of the material; developers usually require much more critical control than the other solutions.

Photographic baths alter with use, or *season*, on account of the chemical interactions between the materials and the solutions. One way of standardizing the chemical composition of the baths is therefore to make them up from fresh ingredients, or to draw fresh supplies from a large bulk mix for each process; but, if this is done, it must be remembered that, although good consistency may be obtained, the process will differ from a normal process unless the formulation of the solutions allows for the seasoning effects that occur in practice. Control of the solutions by chemical analysis may be used, but this is expensive and cannot guard against every possible type of variation: useful results, however, can be obtained by using on every occasion a strip from a *check* or *control* material similar to the test material, to detect any processing variations (Koerner, 1954).

Controlling the times in, and temperatures of, the solutions does not present any very great technical difficulties, but scrupulous care in watching these two factors is well repaid.

The agitation of the solution over the surface of the photographic material greatly affects the rate at which the exhausted products are removed and replaced by fresh supplies, and this in turn affects the rate of the reaction, especially in developers. The difficulty of exactly reproducing practical agitation conditions in special sensitometric processing devices is a strong argument for using practical agitation conditions whenever possible. The following methods of agitation are commonly used: recirculation of the solution through the tank by forcing it out through a pump and back again; gas-burst agitation, by releasing, once every few seconds, a burst of gas at the bottom of the tank and letting it rise to the top (nitrogen is generally used in developers because it has no chemical effect on them); the solutions may be sprayed on to the material from suitable nozzles, or they may be moved across the surface of the material mechanically by means of drums, rollers, vanes, or brushes; and for separate sheets or short lengths of film (but not for continuous lengths of motion-picture film or still films joined together) agitation may be effected by lifting the material out of the solution and letting it drain off at prescribed intervals (the lift-and-drain method); or, if a single sheet is being processed in a dish, a rocking motion may be used provided that regular patterns in the liquid are avoided; finally, processing may be carried out with zero agitation by coating the solutions on the material in viscous layers (Edgecombe and Seeley, 1963).

Sometimes, for instance when the couplers are in three separate developers as in *Kodachrome*, the process is too complicated to duplicate in a rigidly controlled version for sensitometric purposes; in such cases processing variations either have to be averaged out by processing a number of duplicate strips on different occasions, or allowed for by processing with the test material a *control-strip* exposed on a film of known properties. Although these procedures are somewhat cumbersome they do have the merit that the process used for the sensitometric material is the actual practical process itself.

The use of control materials introduces some further possible complications. It is, of course, necessary that such materials be very uniform themselves from one piece to another, and it is therefore essential that all the control material be from the same parent roll of the same batch. The sensitivity of photographic materials varies slowly with time so that a control material cannot be regarded as invariant over long periods; however, the rate of change can usually be greatly reduced by keeping the material at low temperature, and for the most critical work temperatures as low as zero degrees Fahrenheit (minus 18° Celsius) are used (but it is essential to allow the material to reach room temperature before it is unpacked for use, otherwise condensation and local temperature variations will occur). Finally, it is possible for a control material to indicate that a process is on standard when one fault has cancelled out another, for instance a developer having a high temperature and a low concentration of developing agent: under these conditions the test material may not give the same result as in a standard process, although the control material apparently indicates that all is well. To avoid these *film-process interactions*, careful control of the process is therefore still necessary even when control-strips are used, and for very critical work several processings should be used and the results averaged.

It might appear from the above discussion that controlled processing is an impossible undertaking. It is certainly true that to obtain the highest accuracy very great precautions have to be taken, and photographic manufacturers and large scale processing stations generally use both chemical analysis and frequent control-strips to keep their processes on standard; useful, though less accurate

Plate 15

The brilliant colours sometimes seen in petals and leaves may be reproduced partly by transmitted light: in this example the brilliant reddish-magenta colour seen in some of the petals is caused by transmitted light, which produces high luminance and no desaturating surface-reflections. The limited luminance available in the picture causes some interesting changes in the reproduction of the reddish-magenta petals. In the darker areas, the exposure levels in all three layers (or channels) are accommodated, but, as the exposure level rises, first the red, then the blue, and finally the green response reaches its upper limit. Limitation of the red response results in lack of tonal modulation in the petals, limitation of the blue is accompanied by a change of hue from reddish-magenta to magenta (because the red and blue responses are now equal), and limitation in the green results in white (because the red, green, and blue responses are now all equal). Effects of this type occur in photography, television, and printing. (See Section 14.17.) The high luminance of the reddish-magenta petals has caused a cast of this colour on the face, and this should be avoided whenever possible because the effect is always more noticeable in pictures than in real scenes. Reproduced from a 5 × 4 inch *Ektachrome* transparency on a Crosfield *Magnascan* scanner.

Plate 16

Reproduction from the whole 24 × 36 mm area of a *Kodachrome* transparency at a magnification of 1, and of parts of its area at magnifications of 4, 6, 20, 70, 100, 300, and 1000 times. As the magnification is increased, the granular structure of the image becomes more and more apparent (see Sections 18.2 and 18.7), and the sharpness of the image decreases (see Sections 18.9 and 18.15). Photomicrographs by courtesy of G. C. Farnell and Frank Judd, Kodak Research Laboratories, Harrow. Separations made from prints on *Ektacolor* paper on a Crosfield *Magnascan* scanner.

Plate 17

The reproduction of specular reflections of light sources by metallic or other glossy surfaces is difficult in pictures, because of their limited maximum luminance; this is particularly so in reflection prints (see Section 13.10). If the specular reflection is surrounded by darker areas, then simultaneous contrast (see Fig. 5.2.) can provide a useful increase in brightness, as occurs to some extent in this example. Reproduced from a 2¼ inch square *Ektachrome* transparency on a Crosfield *Magnascan* scanner.

results, can still be obtained, however, by less elaborate means by using control-strips with reasonable care and common sense.

When colour photographic materials are being used to record phenomena for subsequent colour measurement, an even better method than the use of separate control-strips is to expose a sensitometric wedge (or other convenient series of controlled patches) on a part of the same piece of film as is being used to record the pictures: differences in processing between the picture and the calibrating exposures are then minimized. On roll films or 35 mm films, part of each film can be used for the wedge and the rest for a series of pictures; on sheet films, part of each sheet must be reserved for the wedge. Alternatively it is possible sometimes to introduce the wedge or a series of patches in the scenes themselves, but, if this is done, due allowance must be made for vignetting in the camera and for any non-uniformity of illumination on the wedge or patches in the scene.

14.11 Visual evaluation

The evaluation of the processed material can be carried out in various ways, some of which are highly sophisticated. The simplest method of all, visual inspection, should not be overlooked, however, because the eye is a very good detector of *differences* in transmittance or reflectance, even though not so precise as an absolute detector. Thus any important differences between a test-material and a control can usually be seen unless they involve changes in blue transmittance seen at low luminances, the eye being rather insensitive even to differences in these conditions. Visual inspection is also very useful in the detection of any streaks or marks on the processed material, such local blemishes being capable of producing most peculiar looking measurements unless they are avoided. Finally visual inspection can usually, but not always, be relied upon to act as a useful check on results obtained by other means.

14.12 Logarithmic scales

It was mentioned in Section 14.3 on sensitometric wedges that the transmittances of successive steps on stepped wedges used for exposing photographic materials usually bore a constant ratio to one another. The reason for the increment being a constant ratio (the transmittances being for example 4, 8, 16, 32, 64, etc.) and not a constant arithmetic difference (4, 19, 34, 49, 64, etc.) are twofold: first, a ratio sequence of transmittances appears much more uniformly spaced to the eye than an arithmetic sequence; and secondly, when alterations are made to the level of the exposure given to a photographic material (by altering the lens aperture, or the scene illuminance, for instance) all the illuminance levels on the photographic material are multiplied by a common factor, and on a ratio sequence this corresponds to the same shift, in terms of number of steps, along the wedge for all parts of the scene. It has therefore become universal practice to evaluate sensitometric results in terms of variations in the logarithm of the exposure (log H), for which a ratio sequence is spaced at equal intervals, instead of in terms of the exposure (H) itself. For similar reasons, instead of using the transmittance or reflectance, T, of photographic materials, a logarithmic function is used, but in this case, although log T could be used, it has long been the practice to use *density*, which is defined as log ($100/T$) when T is expressed as percentage transmittance or reflectance. The logarithms used are in all cases to the base 10, so that a difference in log exposure of one unit, for instance, represents a tenfold change in exposure; and a transmittance of one per cent, for example, is equivalent to a density of 2.0.

It is customary to plot density against log exposure to represent the tone-reproduction characteristics in photography and the D-log H curves thus obtained are known as the *characteristic curves* (see Section 6.3), or *H and D curves* after Hurter and Driffield, who first used such curves. The slope of the characteristic curve is related to the visual *contrast* of the image, and the slope of the straight line tangential to an approximately straight part of a characteristic curve is known as the *gamma* (γ). In a photographic system involving more than one material, the overall gamma of the system is approximately equal to the product of the gammas of the individual parts, but the exact relationships are affected by flare and stray-light considerations. (In television, *contrast* is used to denote the ratio of the luminances of the lightest and darkest parts of the picture. See Section 21.1.) The slope at any given point on the curve is the *point-gamma*, but commonly abbreviated to gamma.

14.13 Densitometers

Much ingenuity has been expended on the design of instruments for measuring density, and many different types have emerged. But modern instruments are usually in one of the three forms shown in Fig. 14.3. The top type represents a simple visual instrument; the middle type, a photoelectric instrument, working on the substitution principle; and the bottom type a direct-reading photoelectric instrument.

In the visual instrument, beams from the transmission and comparison lamps are arranged to illuminate the two halves of a comparison field. (When reflection samples are measured, the two reflection lamps are used instead of the transmission lamp.) The sample is then put in one beam, and the position of a continuously varying neutral wedge, situated in the other beam, is adjusted until the two halves of the field appear to match. The density of the sample is then related to the position of the wedge, which can be read by means of an appropriate scale mounted with it. Any lack of symmetry in the two beams can be allowed for by moving the transmission lamp (or the reflection lamps) so that with a sample of zero density in the instrument a reading of zero is obtained on the wedge, a procedure known as *zeroing*. When a reflection sample is being measured, the zeroing should, in theory, be carried out using a perfect diffuser: a surface of freshly smoked magnesium oxide, or freshly scraped barium sulphate, or magnesium carbonate, powder is a good approximation to a perfect diffuser, but, because of their fragility, specially calibrated white tiles are often used instead.

Visual densitometers are usually simple, fairly inexpensive, and contain little that can go wrong, but they have three disadvantages. First, accurate visual matching is a tiring occupation and fatigue soon sets in if there is much work to

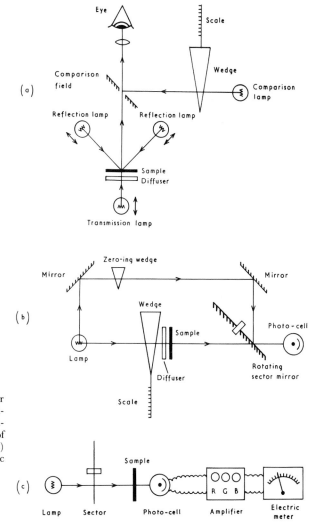

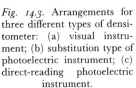

Fig. 14.3. Arrangements for three different types of densitometer: (a) visual instrument; (b) substitution type of photoelectric instrument; (c) direct-reading photoelectric instrument.

do. Secondly, the time taken to make a large number of observations is rather long. Thirdly, the precision obtainable is sometimes less then desirable: in the arrangement of Fig. 14.3(a) it is clear that the higher the density of the sample, the lower will be the luminance of the matching field and hence the worse the precision of matching; the only way to work at constant luminance would be to put the wedge into the same beam as the sample and to adjust it so that the combined density was always equal to some very large fixed density in the other beam; but this is very wasteful of light, with the result that rather low luminance, and hence low precision, usually results for low densities as well as for high densities.

However, when photoelectric cells are used, the sensitivity problems are usually greatly eased, and this method of working becomes possible, as shown in Fig. 14.3(b). One beam of light from the lamp passes through the wedge and the sample on to the photocell; the other beam passes through the zeroing wedge (which is used for zeroing and for providing a density similar to that of the wedge and sample), and then to the photocell by means of an intermittent device such as a rotating sector mirror, which intersects the first beam. If the illuminance on the photocell provided by the two beams is different, as the sector rotates, a variation in current will be produced which can be amplified as an a.c. signal; but if the two beams provide equal illuminances on the photocell no a.c. signal will be produced. The wedge is adjusted in position, either manually or by means of a servo mechanism, until no a.c. signal is produced, and the density of the sample can then be read from the position of the wedge (Hercock and Sheldrick, 1956; Neale, 1956; Harvey, 1956). Substitution photoelectric densitometers of this type are usually easy to use, and give quick and accurate results; they are, however, more complicated and more expensive than visual instruments, and, in common with visual instruments, they depend for their accuracy on the calibration and constancy of their measuring wedges.

The third type of densitometer is a rather simple photoelectric instrument, as shown in Fig. 14.3(c). Light from the lamp passes through the sample and on to a photocell, the output of which is amplified and then measured on an electric meter. Clearly this is a very simple arrangement, but it may be advisable to add a sector, or other means, so that the photocell produces an a.c. signal, which is easier to amplify than a d.c. signal. It is also necessary for either the meter or the amplifier to have a logarithmic output so that a linear scale of density, and not transmittance, results. This third type of densitometer is particularly useful for colour work, because the zeroing adjustment can conveniently consist of a gain control in the amplifier, and this can be triplicated so that three zeros, R, G, B, for reading through red, green, and blue filters can be set up simultaneously, thus enabling the three density readings to be made in succession on each area of the material without having to move it from the measuring position; alternatively zeroing can be carried out by means of three separate aperture adjustments mounted with the red, green, and blue filters. Densitometers of this type depend for their accuracy on the characteristics of the amplifier and meter, but one advantage of this is that all three readings are dealt with similarly (whereas in instruments using wedges any slight lack of neutrality in the wedge has different effects on the three readings). Densitometers of this type include the Eastman Electronic Densitometer Type 31A (MacLeish, 1953), and the MacBeth Quantalog Colour Densitometers. Some of these instruments can measure both transmission and reflection densities, but others are designed specially for one or the other (Watt, 1956).

14.14 Specular and diffuse transmission densities

Black-and-white densities composed of the usual photographic silver deposits not only absorb, but also scatter, the light, and therefore the exact density of a silver image depends on the geometry of the illuminating and viewing optics (Powell, 1956). In Fig. 14.4 are illustrated some practical situations that are affected by this phenomenon. In Fig. 14.4(a) the situation for projection, or enlargement without any diffusion, is shown: most of the light that is scattered by the film will be lost, and the image has the highest possible density, termed *specular density*. In Fig. 14.4(b) a diffuser has been added to the enlarger, and now some of the light that leaves the diffuser at oblique angles, and which would therefore normally miss the objective lens, will be scattered by the image into it, and the density is thus reduced. Since the amount of diffusion tends to increase with density and must clearly be zero with no sample in the beam, the effect of the diffusion is to lower contrast rather than just to reduce density all over; this reduced type of density is termed *diffuse density*. In Fig. 14.4(c) the situation obtaining when a film is viewed on an illuminated opal is shown, and since the geometry is rather similar to that of Fig. 14.4(b) the density is once again diffuse. In Fig. 14.4(d) a rather different situation is shown: here a directional source is used, so one might expect the density to be specular; but the receiver, in this case a piece of paper or a print film being printed by contact, picks up all the light and hence the diffused light is not lost. It can be shown in fact that the densities in cases (c) and (d) are very similar so that once again the density is diffuse. Finally in Fig. 14.4(e) contact printing by means of a diffuse source is used so that both the illuminating and 'viewing' are diffuse, a situation giving densities a little higher than diffuse, but not as high as specular, and termed *doubly-diffuse density*.

Standards have been drawn up defining the geometry of the illuminating and viewing arrangements for specular, diffuse, and doubly-diffuse densitometry (American Standards Association, PH2.19, 1959; International Standards Organization, R.5, 1955), but practical conditions may not exactly duplicate any of the three standards, so that calibration of images in their practical environments is necessary for the highest accuracy with silver images. With the dye images of colour photographic materials the diffusion is generally quite low, so that the effects of illuminating and viewing geometry are fairly small; diffuse density is therefore generally used for colour materials even when they are intended for enlarging or projecting without diffusion. It should be remembered,

245

however, that in these conditions the practical densities may be a little higher than those measured.

In Fig. 14.3(a) and (b) the sample is illuminated diffusely and viewed specularly; in Fig. 14.3(c) the sample is illuminated specularly and viewed

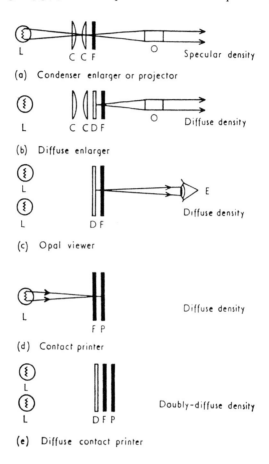

(a) Condenser enlarger or projector

(b) Diffuse enlarger

(c) Opal viewer

(d) Contact printer

(e) Diffuse contact printer

Fig. 14.4. Five different types of optical conditions commonly used in photography, together with the type of density to which each approximates. L = lamp. C = condenser lens. F = film. O = objective lens. D = diffuser. E = eye. P = film or printing paper.

diffusely; thus in both cases diffuse density is read. The specular beams in these instruments are usually confined to within about ±10° from the optical axis; diffuse illumination is usually provided by an opal glass or diffusing plastic, and diffuse collection (as in Fig. 14.3(c)) by placing the photocell close to the sample.

14.15 Printing densities

When the (geometrically appropriate) densities of black-and-white silver images are measured, it is ideal to illuminate the samples with light having a spectral power distribution similar to that used in the practical situation, and to use a detector having a spectral sensitivity similar to that of the detector normally used with the material. However, silver images are usually fairly non-selective, so that it is often possible to use almost *any* light source and detector. The only serious exception to this in practice arises when a yellowish silver image is to be printed on an unsensitized printing material; in this case the visual density, or that recorded photoelectrically with a broad spectral band, gives results of lower density than that 'seen' by the print material, and a detector whose spectral sensitivity approximates that of the print material (having sensitivity only in the blue and ultra-violet parts of the spectrum) has to be used. Densities measured in this way are known as *printing densities*, or sometimes *actinic densities*.

When densities are measured on colour films whose function is to be printed on to other colour films or papers, then once again printing densities have to be measured; but this time, because the three layers of the colour print material have three different spectral sensitivities, three printing densities have to be measured. The simplest way of doing this is to measure the densities through three filters whose spectral transmissions, when combined with the spectral sensitivity of the detector, simulate the effective spectral sensitivities of the three layers. An exact match in sensitivity at all wavelengths is usually very difficult to achieve, and is unnecessary in practice; the main consideration is that the three gammas measured by the densitometer should be similar to those 'seen' by the print material. Densitometers can have calibration adjustments that enable small changes in the gamma of the readings to be made, and these may be used to correct errors in gamma for the dyes of a given dye set; but these adjustments are awkward to make if several dye sets are involved in one set of readings.

For these and other reasons, it has become standard practice in colour densitometers to use an agreed set of red, green, and blue filters in instruments possessing an agreed spectral response. The set of filters used for measuring printing densities of colour films is known as certified MM filters: the components used in this set of filters are given in Table 14.1, and their spectral densities are shown in Fig. 14.5 (a). The agreed spectral response, obtained by multiplying at each wavelength the spectral power distribution of the lamp by the spectral transmittance of the optical components and by the spectral sensitivity of the detector, is shown in Fig. 14.5 (b). The curves in Fig. 14.5 (c) show the spectral response of the instrument through the certified MM filters. The agreed spectral response is conveniently obtained by using a Corning 9780 glass filter of 1.9 mm thickness in the beam. Certified MM filters, used in densitometers having the agreed spectral response, are used for measuring printing densities of both motion-picture and still colour negative films having coloured couplers (Dawson and Voglesong, 1973).

TABLE 14.1

Filters used for densitometry of colour films

Certified filters	Purpose	Components
MM	Measurement of printing densities in densitometers having the agreed spectral response,* for films intended for printing	R Two W_{92} G Two W_{53} B W_{2A}+two W_{48A}
AA	Measurement of integral densities in densitometers having the agreed spectral response,* for films and reflection prints intended for viewing	R W_{29}+W_{25} G W_{74} B Two W_{98}+W_{2E}

Status filters	Purpose	Components
M	Obsolete filters for films intended for printing. Replaced by certified MM filters	R W_{29}+C_{9780} (1.5 mm) +two W_{92} G C_{9782} (1.3 mm)+ three W_{52} B W_{2A}+C_{9782} (1.3 mm) +two 48A
A	Obsolete filters for films intended for viewing. Replaced by certified AA filters	R W_{29}+C_{9780} (2.0 mm) G W_{60}+W_{74}+C_{9782} (1.3 mm) B X_{12740}+C_{9782} (1.3 mm) +three W_{47B}
D	Obsolete filters for reflection prints. Replaced by certified AA filters	R W_{92} G W_{93} B W_{94}
G	Reflection density of printing inks	R W_{25} G W_{58} B W_{47}
V	Measurement of visual density in densitometers having the agreed spectral response*	W_{106}
S	Measurement of sound tracks. (In densitometers having the agreed spectral response which include a 1.9 mm thick C_{9780} filter, this filter must be removed.)	W_{88A}+an interference filter cutting off light beyond 800 nm
Z	For ultra-violet densitometry. (In densitometers having the agreed spectral response which include a 1.9 mm thick C_{9780} filter, this filter must be removed.)	C_{5840} (2.4 mm) (or a W_{18A} can be used for most purposes, but gives densities about 20 per cent greater)

W Indicates Kodak Wratten filter.   X Indicates Kodak experimental filter.
C Indicates Corning glass filter.
* For obtaining the agreed spectral response a Corning 9780 filter of 1.9 mm thickness is convenient.

248

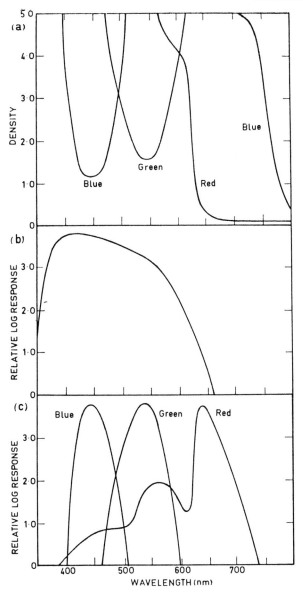

Fig. 14.5. (a) Spectral densities of Certified MM filters. (b) The agreed spectral response for densitometers to be used with Certified MM filters. (c) The agreed spectral response through the Certified MM filters, used for the measurement of printing densities of films intended for printing.

Printing densities were measured in earlier densitometers using Status M filters (Miller and Powers, 1963) each of which contained a Corning glass filter as shown in Table 14.1. The purpose of these glass filters, and of the Corning 9780 filter which is convenient to use to obtain the agreed spectral response, is to absorb light in the far red and infra-red parts of the spectrum. Gelatin filters, and the image dyes used in colour photography, both transmit these wavelengths very freely, and hence even a very small sensitivity of the receptor in this spectral region can completely falsify the results. It is therefore *essential* to absorb all of the far red and infra-red light with glass or interference filters.

Absorption of the far red and infra-red light is also advisable in colour tele-cine equipment (see Section 23.5); in this case, the sensitivity of the photoelectric detectors used may be greater than that of the eye in these regions of the spectrum and differences in the far red and infra-red transmittance of different types of film may then produce gross differences in colour balance and red-image contrast (Kozanowski, 1964).

Since films are usually illuminated with white light when they are being printed, the colour filters in a densitometer should ideally be placed after the white light has passed through the sample; if this is done, and if the spectral power distributions of the white light in the densitometer and in the printer are similar, the correct density readings should be recorded even if the sample fluoresces or scatters to different extents at different wavelengths (assuming that the geometry of the densitometer results in densities of the correct diffusion being read). In practice, however, as far as films are concerned, fluorescence is very unlikely and scattering is low, so that the filters can usually be placed either before or after the sample; and the spectral power distribution of the light source, the spectral transmittance of the optical components, and the spectral sensitivity of the detector, can be lumped together in a single response curve as has been done in Fig. 14.5 (b). (With reflection prints, however, fluorescence may occur.)

Printing densities are nearly always used for colour negative films. In materials incorporating coloured couplers, such as *Kodacolor* negative film and *Eastman Color Negative* film, the printing densities have quite high minimum values because of the absorption of green and blue light by the coloured couplers. Typical sets of curves, showing printing densities plotted against log exposure, for materials incorporating coloured couplers are shown in Fig. 14.6. The films intended for camera use generally have gammas of about 0.65 as shown in Figs. 14.6 (a) and (b); the curves of the two films represented in these figures are similar except that those for films intended for amateur use (Fig. 14.6 (b)) are longer than those intended for professional use (Fig. 14.6 (a)) so as to provide the larger margin for error in exposure level (good *exposure latitude*) desirable in a product intended for amateur snapshots. Fig. 14.6 (c) shows curves representative of an intermediate film (such as *Eastman Color Intermediate* film) with a gamma closely equal to 1.0 so that, when *Eastman Color Negative* film is printed on to it, a positive of gamma 0.65 is obtained, which when printed on to the intermediate film again yields a duplicate negative of gamma 0.65

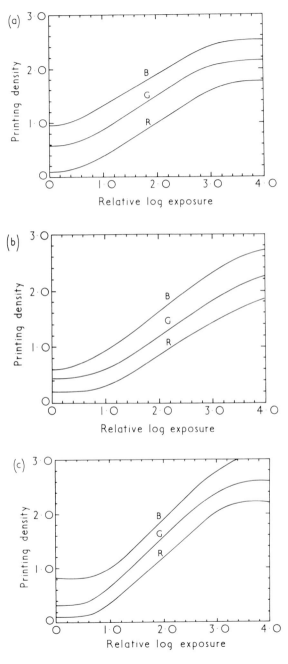

Fig. 14.6. Typical characteristic curves for colour negative films: (a) for professional use (such as *Eastman Color Negative* film); (b) for amateur use (such as *Kodacolor* film); (c) for intermediate use (such as *Eastman Color Intermediate* film). Printing densities are plotted against log exposure for each colour. The high blue and green minimum densities are caused by the colours of the coloured couplers.

251

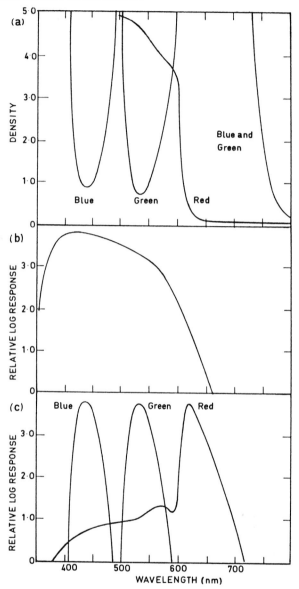

Fig. 14.7. (a) Spectral densities of Certified AA filters. (b) The agreed spectral response (which is the same as in Fig. 14.5(b)) for densitometers to be used with Certified AA filters. (c) The agreed spectral response through the Certified AA filters, used for the measurement of integral densities of films and papers intended for viewing.

(see Plate 11, page 113). (The final gamma obtained when one film is printed on to another is equal to the product of the gammas of the two films used, apart from effects caused by flare and by using curved parts of the characteristic curves.)

14.16 Integral densities

The Certified MM filters used for measuring printing densities are not ideally suited to the dyes commonly employed in films giving positive images, and for general work a set of filters whose components are given in Table 14.1, and whose spectral densities are as shown in Fig. 14.7 (a), is commonly used. These filters are known as Certified AA filters. They are intended to be used with the agreed spectral response mentioned in the previous section and shown again in Fig. 14.7 (b); the curves of Fig. 14.7 (c) show the response through the Certified AA filters.

Earlier densitometers often used Status M or Status A filters (Brewer, Goddard, and Powers, 1955) for films, and Status D filters (Miller and Powers, 1963) for reflection prints; the components used in these filters are also given in Table 14.1.

Arbitrary sets of filters, such as those of Certified AA, do not generate spectral response curves corresponding to any specific application, but they are entirely satisfactory for tests where comparison of the results is restricted to any one set of positive image dyes. For general control and test work on individual positive colour products they are therefore quite suitable. The densities that they read are generally referred to as *integral densities* because each filter measures the total (or integrated) effect of the absorptions of all the dyes having any density in its spectral transmission band.

The colour corresponding to equal values of red, green, and blue integral densities is generally nearly grey for most practical illuminants, and for most colour photographic materials; but exact equality of densities does not necessarily correspond exactly to a grey for any illuminant, and it would be unlikely so to correspond for a particular illuminant used in practice, unless the filters were specially chosen to fulfil this condition for the particular dye set and illuminant involved.

In Fig. 14.8 the characteristic curves of a reversal film measured in terms of integral density are shown. Since, in reversal films, high exposures give low densities, the curves slope downwards, instead of upwards as is the case for negative materials. The curves of Fig. 14.8 are typical of the type of film (such as *Ektachrome Commercial* film) whose primary purpose is to serve as an original for making duplicates; this film therefore has fairly straight characteristic curves and a gamma of about 1.0. Films of the type shown in Fig. 14.8 are usually printed either on to an internegative film having a gamma of about 0.65, with characteristic curves similar to those shown in Fig. 14.6 (a), or on to a reversal film with characteristic curves suitable for producing pictures for projection, as will be discussed in the next paragraphs. (Strictly speaking, this material should have

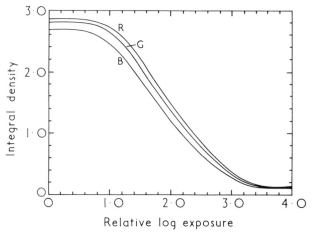

Fig. 14.8. Typical characteristic curves for a reversal film (such as *Ektachrome Commercial* film) designed primarily as an original for making duplicates. Integral densities are plotted against log exposure for each colour.

been measured using printing densities, but integral densities are often used when the image is a positive, even though it is intended for printing.)

In Fig. 14.9 are shown typical sets of characteristic curves of red, green, and blue integral density plotted against log exposure for a reversal (a) and two print (b) and (c) materials, when the colour of the exposing light is such that most exposure levels are reproduced so as to appear grey in the viewing conditions typical of those used in practice for each material. When the viewing conditions consist of projection by tungsten light in a darkened room, the light from the projector appears yellowish (Hunt, 1965), and therefore to obtain results that appear grey the picture has to be slightly bluish (see Section 5.7); this is why the curves of Fig. 14.9(a), which relate to materials intended for tungsten-light projection, are not even approximately coincident, the blue densities being lower than, and the red densities higher than, the green densities, in order to produce the bluish result required. The curves of Fig. 14.9(b) refer to a material intended for projection by arc light, and since this is a whiter source than a tungsten lamp, the curves are now approximately coincident and the greys are reproduced approximately grey on the film. The curves of Fig. 14.9(c) refer to a reflection print material, and again the curves are approximately coincident, and greys are reproduced as near-greys on the paper. (As mentioned in Section 5.6, in some products, greys may be deliberately reproduced with a slight colour bias in order to improve the rendering of skin colours.)

It will be noticed that the curves of Figs. 14.9(a), (b), and (c) are not straight: in addition to the usual 'toe' and 'shoulder' regions at the lowest and highest density levels respectively, the curves exhibit gradually increasing gamma as the densities increase. It is found that these changes in gamma produce more pleasant pictures than are obtained on materials with straight character-istic curves; the high gamma at high density improves the visibility of shadow detail by helping to offset the effect of flare in the camera and in the viewing situation, while the lower gamma at low density prevents 'harshness' in light subject-matter and increases the permissible margin for error in achieving the correct exposure level (see Chapter 6).

For the reversal film of Fig. 14.9(a) intended for projection in a dark room,

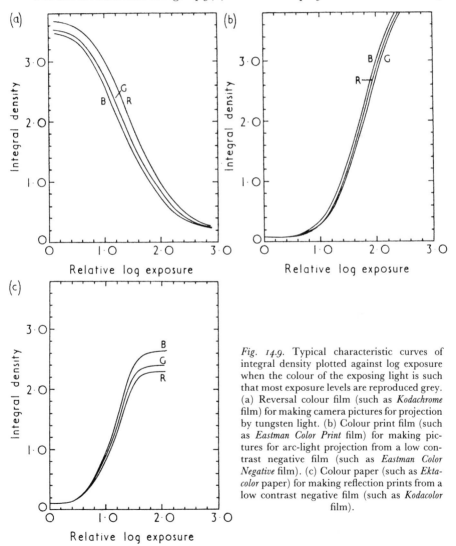

Fig. 14.9. Typical characteristic curves of integral density plotted against log exposure when the colour of the exposing light is such that most exposure levels are reproduced grey. (a) Reversal colour film (such as *Kodachrome* film) for making camera pictures for projection by tungsten light. (b) Colour print film (such as *Eastman Color Print* film) for making pictures for arc-light projection from a low contrast negative film (such as *Eastman Color Negative* film). (c) Colour paper (such as *Ektacolor* paper) for making reflection prints from a low contrast negative film (such as *Kodacolor* film).

the gamma at a density level of 1.0 (corresponding roughly to a medium grey) is about 1.5 and this is found in practice (for black-and-white as well as for colour (Clark, 1953)) to result in pleasing pictures; a gamma of 1.0 is too low because dark areas then appear too light as a result of the lightening effect of the dark surround. (This is discussed in Chapter 6.) The curves of Fig. 14.9(b) refer to a colour print film intended for use with a negative film of gamma about 0.65; the gamma at density 1.0 is therefore much higher, in this case about 2.4, so that the combination yields the required result, which in this case is about 1.6 at density 1.0; this is higher than that for the reversal film shown in Fig. 14.9(a) because this negative-positive system is designed specially for professional motion-picture use, where it is usually possible to keep the lighting-contrast fairly low; and the combination of low lighting-contrast and high photographic-gamma results in a useful gain in colour saturation.

In Fig. 14.9(c) similar curves are shown for a colour paper designed for making reflection prints from a low-contrast negative film. A medium grey in a reflection print is usually reproduced at a density of about 0.5, and the gamma of the paper at this level is about 1.6, which, when combined with the negative gamma of 0.65 yields an overall gamma of about 1.0; this figure is lower than that of transparency systems because the dark areas are not usually lightened by a dark surround (Bartleson and Breneman, 1967).

The changes in gamma with density that occur in films of the type depicted in Fig. 14.9(a) should be corrected if they are to be employed, not for projection, but as originals from which duplicate transparencies or reflection prints are to be made, using materials having the same type of gamma variations; if correction is not made, the gamma changes occur twice in the system and the shadows are too contrasty and the highlights too flat (see Fig. 6.13). Fig. 14.10 shows curves for an internegative film that provides the correction necessary when a negative, made from a transparency with curves as in Fig. 14.9(a), is to be printed on to a paper

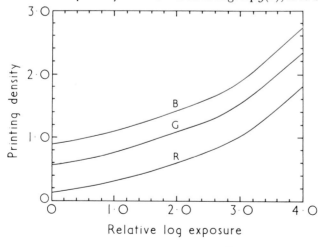

Fig. 14.10. Characteristic curves of an internegative film, with gamma increasing with exposure level so as to counteract the gamma variation in films of the type shown in Fig. 14.9(a).

256

with curves as in Fig. 14.9(c). The gamma varies from about 0.3 in the region where transparency shadows are recorded, up to about 1.0 where the transparency highlights are recorded, so that a negative of gamma about 0.65 can be produced from a transparency with gamma varying from over 2.0 in the shadows to only about 0.65 in the highlights.

14.17 Some effects of curve shape

One of the principal uses of densitometry is to record the shape of the density versus log-exposure curve for the purpose of seeing how the tone reproduction varies with density level. A full evaluation, however, requires the inclusion of proper allowances for camera flare and viewing flare for these are nearly always different, respectively, from sensitometric flare, and stray light in the densitometer; and if the system under study includes any printing stages, whether by contact or by projection, the effect of the printing flare must also be included (see Chapter 6, and Section 13.10).

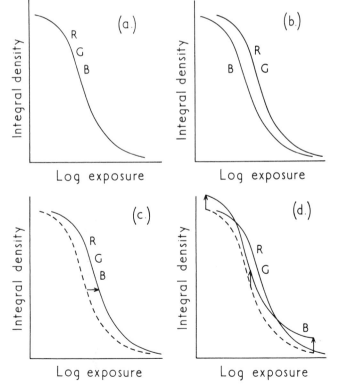

Fig. 14.11. The characteristic curves of a typical reversal film (a) when exposed to light of the correct colour; (b) when exposed to light of a bluer colour; (c) when the blue exposing light is corrected by placing a yellow filter over the camera lens; (d) when attempting unsuccessfully to correct the blue exposing light by placing the yellow filter over the projection lens or over the transparency.

The fact that photographic characteristic curves are never straight at all densities introduces a complication which is illustrated in Fig. 14.11. Suppose for simplicity that, for some reversal colour film and its usual exposing illuminant, the integral densities are such that a scale of greys in its usual viewing illumination is represented by the three curves lying on top of one another, as shown in Fig. 14.11(a). If this film is then exposed in a light of different colour, say a bluer colour, a result similar to that shown in Fig. 14.11(b) might be obtained, the blue curve having shifted along the log exposure axis relative to the other two. This condition can be corrected by placing a yellow filter over the camera lens to give the result in Fig. 14.11(c) in which the blue curve has been moved along the log exposure axis and the three curves are once again superimposed; but it cannot be corrected by placing a yellow filter over the lens of the projector (or over the transparency itself) because this moves the blue curve, not horizontally along the log-exposure axis, but vertically along the density axis as shown in Fig.14.11(d). If the curves were straight, then either vertical or horizontal shifts could be used to superimpose the curves, but, because of the curvature, the vertical shift causes over-correction at low densities and under-correction at higher densities (with over-correction at very high densities, although this is usually of no importance in reversal films); the consequent shift in colour balance, from yellow through grey to blue, results in very unpleasant picture quality. It is for this reason that when a reversal film is used in an illuminant other than that for which it is designed, a correcting filter should be used over the *camera* lens if the best results are to be obtained. Conversely, films with straight characteristic curves should be used if it is known that a wide range of taking-illuminant colour must be accommodated without the possibility of using corrective filters over the camera lens: for with films having straight characteristic curves correction can be achieved by viewing them through correcting filters. (It is interesting also to note that in these cases a considerable measure of correction can also be supplied by the eye adapting to the overall colour of the pictures, and this correction is extremely rapid, as illustrated by the ease with which observers discount the colour of the resultant screen illuminant in the two-colour projections demonstrated by Land (Land, 1959). But, if the characteristic curves of the film are not straight, the colour bias varies with density level and does not have the character of an illuminant change *in the picture*, and therefore cannot be discounted visually to the same extent.)

The limited range of exposure over which photographic materials produce a response can result in important changes in hues in pictures, as exposure level is altered. For instance, a reddish-magenta object normally results in most exposure in the red layer, least in the green, and an intermediate amount in the blue. If the exposure level is gradually increased (by illuminating the object at a higher level or by increasing the lens aperture or exposure time, for instance), the point will come when the red layer reaches its maximum response: this results in a lack of tonal modulation in the colour and a shift in hue towards magenta; as the level of exposure increases further, the response of the blue layer will reach a maximum, and, since the maximum responses possible in the three layers are normally made to be equal, the difference between the red and blue responses will have vanished and the colour becomes magenta; finally, when the response of the green layer reaches its maximum, all three responses will be equal, and the colour becomes white. The colour therefore changes from reddish-magenta, to magenta, and then to white. Only colours for which the responses in two of the three layers are equal (pure red, green, blue, cyan, magenta, or yellow) will be devoid of this type of change in hue before gradually desaturating to white. Similar effects occur in television and in printing, because, in all cases, the three channels, or printing plates, respond over only a limited range. The changes in hue tend to be more noticeable to the eye than the loss of tonal modulation or desaturation towards white. The effects are more likely to occur with saturated colours, because the difference in exposure in the three layers, channels, or plates, is then large; for this reason the effects tend to be more prevalent when high levels of inter-image effect (see Section 15.5) or masking (see Sections 15.6 and 23.13) or matrixing (see Section 19.12) are used. The effect is often most noticeable in reddish-magenta and bluish-magenta colours because they can be of very high colour saturation. Similar effects can also occur as exposure levels are reduced, but, because they then affect mainly the dark parts of the picture, they are usually less noticeable. (See Plate 15, page 240.)

14.18 Colorimetric densities

When comparison of results involving more than one set of image dyes is necessary colorimetric measurements have to be made, and these can be carried out on a densitometer if it is fitted with suitable filters. Thus if the combination of the spectral power distribution of the densitometer lamp (as modified by the optics), the spectral transmittances of the three measuring filters, and the spectral sensitivity of the detector, were such as to duplicate the colour-matching functions

$$E_A(\lambda)\bar{x}(\lambda) \qquad E_A(\lambda)\bar{y}(\lambda) \qquad E_A(\lambda)\bar{z}(\lambda)$$

where $E_A(\lambda)$ represents the spectral power distribution of standard illuminant A, then the three densities measured would be equal to:

$$\log(1/X_A)+k_X \qquad \log(1/Y_A)+k_Y \qquad \log(1/Z_A)+k_Z$$

where k_X, k_Y, and k_Z, are zeroing constants, and X_A, Y_A, and Z_A are the tristimulus values in the CIE XYZ system when the sample is illuminated with standard illuminant A. Readings of these types are known as *colorimetric densities*.

Colorimetric densities enable samples of different dye sets to be compared and when the tristimulus values derived from them are in the same ratios as those for the illuminant being considered, then the sample is exactly grey whatever set of image dyes is being used. This is also true of densitometers that duplicate any linear combination of colour-matching functions, and such instruments can also be used to detect metameric pairs composed of different dye sets; their results can be converted to standard tristimulus values (for the same illuminants) by linear transformations.

14.19 Spectral densities

If integral densities are measured using filters transmitting such a narrow band of wavelengths that the evaluation takes place effectively at one wavelength only for each of the red, green, and blue readings, the results are known as *spectral densities* or *monochromatic densities*. Such densities are important theoretically because for most dyes the transmission densities then become exactly *additive*: that is, the total red density of an image consisting of successive layers of cyan, magenta, and yellow dyes is equal to the sum of the red densities of the three dyes separately; and similar additivity occurs for the green densities, and for the blue densities. For most dyes, spectral transmission densities also obey a *proportionality rule*, in that, for any one dye, a given variation in the amount present alters all spectral densities by the same factor: thus if the red density is increased by 50 per cent, for instance, the green and blue densities will also be increased by 50 per cent.

14.20 Analytical densities

If the interest centres not so much on the total effect of the three dye images together, but on the combination of each separately, then *analytical densities* are used. In the case of transmission densities these can be derived from *spectral densities* by solving three simultaneous equations. This is because, from the proportionality rule, it follows that, if the density of the cyan dye at the chosen red wavelength is C_R, then the densities at the green and blue wavelengths will be $k_1 C_R$ and $k_2 C_R$ where k_1 and k_2 are constants. Similarly if the magenta dye has density M_G at the green wavelength, then the other two densities will be $k_3 M_G$ and $k_4 M_G$; and if the yellow dye has density Y_B at the blue wavelength, the other two densities will be $k_5 Y_B$ and $k_6 Y_B$ where k_3, k_4, k_5, k_6 are constants. Because of the additivity property, the integral transmission spectral densities, I_R, I_G, I_B will therefore be equal to:

$$I_R = C_R + k_3 M_G + k_5 Y_B + r_S$$
$$I_G = k_1 C_R + M_G + k_6 Y_B + g_S$$
$$I_B = k_2 C_R + k_4 M_G + Y_B + b_S$$

where r_S, g_S, b_S are constants to allow for the presence of any constant 'stain' density that does not vary with the concentrations of the three image dyes. The above equations can be solved for the analytical densities C_R, M_G, Y_B, if the constants are known. To find the values of the constants it is necessary to have available a sample of each dye on its own and to measure the ratios of the major absorption to the minor absorptions (Pinney and Voglesong, 1962) and thus find $k_1,, k_2, k_3, k_4, k_5$, and k_6; r_S, g_S, and b_S are found by measuring a suitable area free of image dyes.

Fig. 14.12 shows a transmission spectral-density curve for a neutral grey typical of those in colour photographic films, together with the spectral density curves of the three individual dyes forming the neutral shown. Wavelengths suitable for measuring spectral densities on this material are shown, and it can be seen from the figure how the spectral integral densities are made up from the spectral analytical densities.

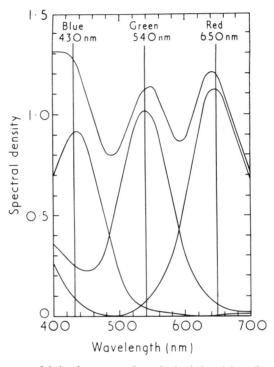

Fig. 14.12. Spectral density curve for a grey that is neutral to 4000 K light, together with the spectral density curves of the three dyes of which it is formed in a typical colour film. The three vertical lines indicate wavelengths that would be suitable to use for measuring spectral densities on such a film.

It is often convenient to multiply the spectral analytical densities, C_R, M_G, Y_B by three factors so that when some chosen grey sample is measured, the modified spectral analytical densities have particular values. For instance, the chosen sample may be such that when viewed by an illuminant of colour temperature 4000 K it is exactly grey and has a visual density of 1.0 (in other words its tristimulus values are all exactly one-tenth of those of the 4000 K illuminant); the multiplying factors are then usually chosen so that the three spectral analytical densities are all equal to 1.0 for this sample. When this is done, equal values of the spectral analytical densities at other density levels also correspond quite closely to exact greys (under the same 4000 K illuminant) for most reasonable dye sets, and the density values then approximate to *equivalent neutral densities*: the equivalent neutral density of any particular amount of dye is defined as the visual density that results when the other two dyes are added in quantities just sufficient to produce a neutral grey (Sant, 1970).

It has been assumed throughout the above discussion that the analytical densities have been calculated from truly monochromatic integral densities. If instead, the integral densities are measured through typical densitometer filters such as those shown in Fig. 14.7, the additivity and proportionality properties of the densities become only approximations, but the errors are not usually serious except for high densities; hence analytical densities are in fact usually obtained from non-spectral integral densities. In cases where significant departures occur from proportionality and additivity of the densities, more elaborate procedures can be used in which measurements are made at additional wavelengths (Baumann, 1980; Muller, 1980).

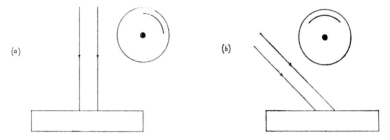

Fig. 14.13. Two alternative arrangements that can be used for reflection densitometry.

14.21 Reflection densities

All surfaces reflect some light from their topmost layers, and in the case of photographic reflection prints this light is unaffected by the image dyes; reflection densitometers are therefore usually designed so as to minimize the amount of this light picked up by the detector. Since most surfaces have some gloss, the detector is placed well away from the position corresponding to the mirror image of the light source (where the surface reflection is at its maximum) and one of the two arrangements shown in Fig. 14.13 is generally used: either the light strikes the sample normally with the detector viewing it from 45°, or the light is incident at 45° with the detector viewing normally. In one instrument an ellipsoidal mirror collects light at 45° in all directions round the normal illuminating beam, thus giving a high efficiency (Watt, 1956). Like transmission densitometry, reflection densitometry is also the subject of standards: in these standards the two beams of light are confined to directions within ±5° of the nominal 45° and perpendicular directions (American Standards Association, PH 2.17, 1958). The insides of reflection densitometers are usually very thoroughly blackened in order to prevent stray light from limiting the maximum densities that can be read.

14.22 Analytical reflection densities

Because of the effects of multiple reflections of the light between the diffusing base and the underside of the topmost layer of reflection print materials (see Section 13.9), the additivity and proportionality properties of dyes do not apply even to monochromatic integral reflection densities. But analytical reflection densities can be found by using a calibration curve to convert from integral reflection to integral transmission density, then calculating the analytical transmission density, and finally using the curve again to convert from transmission analytical to reflection analytical density (Pinney and Voglesong, 1962). At high densities the calibration curve depends on the surface gloss of the sample and the stray-light behaviour of the reflection densitometer used for the measurements, but a typical curve is shown in Fig. 14.14. At low densities, multiple internal reflections cause reflection density to increase rapidly with transmission density; at medium densities, multiple internal reflections become progressively less important as the density increases, so that the rate of change of reflection density with respect to transmission density approaches the value of 2.0 (or 2.13 if one of the beams is incident on the layer at 45° and the other at 90° (Williams and Clapper, 1953)) which would be expected on account of the light having to pass twice through the

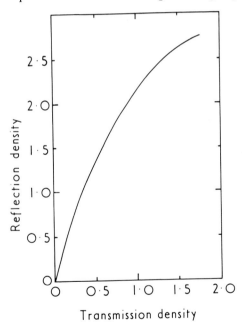

Transmission density

Fig. 14.14. Reflection density plotted against the corresponding transmission density of the image layer for a typical glossy paper and a reflection densitometer capable of reading high densities.

layer in a reflection material. But at high densities the surface-reflection becomes more and more important so that the rate of change drops below 2.0, and finally flattens out to a maximum value that is usually well below twice the maximum transmission value.

For reflection densities between about 0.8 and 1.8 the curve of Fig. 14.14 is often approximately linear, and, for densities within this range, reflection analytical densities can be calculated to a good approximation by the same type of equations as are used for transmission work (described in Section 14.20), the constants in the equations then being evaluated specially for the reflection situation (Onley, 1960).

14.23 Exposure densities

When the performance of camera films is being considered, the log-exposures given to the red-, green-, and blue-sensitive layers by various parts of the scene (sometimes referred to as *actinic exposures*) are of considerable importance.

It is often convenient to consider these log exposures relative to those given by some standard type of object in the scene, such as a perfectly-reflecting, perfectly-diffusing white; differences in log exposure from those given by the standard are called *exposure densities*: thus an object giving a red log exposure of 0.3 less than the standard, for instance, would have a red exposure density of 0.3; an object having a green log exposure greater than the standard by 0.1, say, would have a green exposure density of −0.1. If a reflection densitometer is fitted with filters so that its spectral sensitivities duplicate those of the film in question, then the exposure densities of reflection samples can be measured directly; otherwise, they can be calculated from a knowledge of the spectral characteristics of the light source, sample, and lens transmission colour. In neither case are the effects of atmospheric haze and lens flare automatically allowed for; these may be important in some circumstances, so that it may be necessary to measure exposure densities in actual scenes or in cameras, using a photoelectric photo-meter or tele-photometer incorporating appropriate filters.

14.24 Scales of equal visual increments

One of the reasons for using scales of log exposure and density, instead of exposure and transmittance, is that (as mentioned in Section 14.12) the former scales represent more nearly uniform visual steps. It has been shown, however, that, for reflecting samples, a more uniform scale still is that of Munsell value, which is defined by the empirical formula.

$$R = 1.2219V - 0.23111V^2 + 0.23951V^3 - 0.021009V^4 + 0.0008404V^5$$

where R is the percentage reflectance and V is the Munsell value (Newhall, Nickerson, and Judd, 1943). In Fig. 14.15 Munsell value is plotted against reflection density (log (100/R)). It is clear that, as compared to the Munsell value scale, reflection density over-emphasizes high densities relative to low densities;

and in reflection print work it is found in practice that a given density difference tends to be more important at low than at medium or high densities. It might, therefore, be more useful to plot Munsell value against log exposure for reflection materials; it would not be appropriate to use a Munsell value scale for the *exposure* axis, because original scenes usually include variations in illumination level over their area, and the eye is able to discount such variations to a considerable extent (Evans, 1943), whereas the Munsell value scale applies to conditions of uniform illumination.

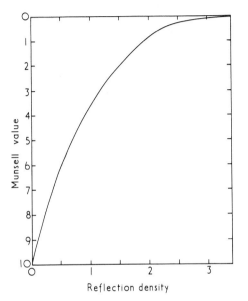

Fig. 14.15. Munsell Value plotted against reflection density. (The reflection density is in this case measured against magnesium oxide as zero, but Munsell Value 10 refers to a perfect diffuser; taking the reflectance of magnesium oxide as 98 per cent, zero density thus corresponds to a Munsell Value of 9.90.)

The definition of Munsell value by the above formula is rather complicated and for most applications the scale of L^*, which is given by $116(Y/Y_n)^{1/3} - 16$ (where Y/Y_n is the reflectance) can be used instead. In the case of samples viewed with a dark surround (see Section 6.5), the resulting drop in apparent contrast might be allowed for by using $L^*_{dark} = 116(Y/Y_n)^{1/4.5} - 16$; a similar allowance for a dim surround might be made by using $L^*_{dim} = 116(Y/Y_n)^{1/3.75} - 16$ (see Section 8.8).

14.25 Tri-linear plots

Colour balance is such an important variable in colour photography that it is often helpful to consider it separately from overall density level. This is often done by plotting density differences on triangular graph paper. An example of a *tri-linear plot* of this type is given in Fig. 16.1 where a green-red printing-density difference is plotted against a blue-red printing-density difference. Tri-linear plots are also used for integral, analytical, colorimetric, and exposure densities.

The visual magnitude of the differences between two colours represented by two points on a tri-linear plot will depend on the spectral absorption or sensitization properties of the film or paper, on the densitometer used, on the absolute density of the colours, and on the viewing conditions. Integral densities are sometimes approximately equally spaced visually for near neutrals of nearly equal absolute density (Staes and Verbrugghe, 1971).

14.26 Stability of dye images

The dyes used for forming the images in colour photographic materials may be affected by heat, humidity, and intense illumination (Hubbell, McKinney, and West, 1967; Adelstein, Graham, and West, 1970; Schafer, 1972; Giles, Forrester, Haslam, and Horn, 1973). The effects vary considerably from one material to another; however, fading of the dyes caused by prolonged periods of intense illumination, and yellowing of unused coupler caused by ultra-violet radiation (often referred to as *print-out*), have perhaps been the features causing most concern in the past (Gale and Williams, 1963). To combat these effects, improved couplers, modified product configurations, and special processing procedures, have been introduced (see Section 17.9) (De Mitri, 1960; Bermaine, 1974; Giles and Haslam, 1974; Tull, 1974; Happé, 1974; Moore, 1974; Rogers, Idelson, Cieciuch, and Bloom, 1974).

The prediction of the long-term stability of image dyes when kept in the dark is usually carried out by means of *accelerated keeping tests* in which the material is subjected to high temperature and humidity. It is found that the time, t_{10}, taken for the dye to fade from a density of 1.0 to a density of 0.9 is related to the temperature in kelvins (T_k) by an equation of the form

$$\log t_{10} = 1/T_k$$

By conducting such tests at two or more high temperatures (such as 60°C and 85°C, for example) it is possible by extrapolation to estimate the time t_{10} for a typical room temperature (such as 24°C). The dyes in some products now have extrapolated values of t_{10} of over 50 years at room temperatures. (Tuite, 1979; Kennel, Sehlin, Reinking, Spakowsky, and Whittier, 1982).

Prediction of the stability of image dyes to fading by light is more difficult because the rate of fading depends not only on the exposure (the product of the illuminance on the material and the time for which it is faded) but also on the illuminance level itself. This 'reciprocity failure' of fading means that results of tests made at high levels of illumination have to be checked for lower levels, thus limiting the extent to which the fading tests can be accelerated. The most critical application of photographic materials for stability to light is the long-term display of colour reflection prints. Some papers now in use for this purpose give pictures with useful lives for typical indoor domestic display estimated at 15 to 20 years.

The perceptibility of changes in the dye images in pictures depends on several factors. If all three dyes change by the same amounts, so that the colour balance is not affected, changes of 10 per cent, or even more, are often not detectable. But, if only one dye fades the change is much more obvious, particularly if it is the magenta dye. A given change is usually more obvious in a reflection print, which can be compared with other objects in the field of view, than in a projected film seen in a dark surround; and changes in negatives can usually be, at least partially, corrected in printing them. A given change of density in the reproductions of well-known objects, such as sky, grass, neutral greys, and particularly Caucasian skin, are usually more noticeable than in less familiar objects or in very colourful objects.

14.27 Photographic speed

One of the most important properties of a film is its photographic *speed*; this enables the photographer to use it at or near the optimum exposure level (exposure being the product of the illuminance on the film and the time for which it is exposed). Methods of determining the speeds of films from their characteristic curves have been the subject of both international and national standards (Zwick, 1979). For negative films, one of the most widely used methods is to determine a speed, S_N, equal to $2^{\frac{1}{2}}/(H_G.H_S)^{\frac{1}{2}}$, where H_G is the 'white light' exposure that produces a green printing density that is 0.15 above the minimum green printing density, and H_S is a similar exposure for whichever is the slowest of the three printing-density characteristic curves; H_G and H_S are measured in lux seconds. For reversal films, a widely used method is to determine a speed, S_R, equal to $10/(H_L.H_S)^{\frac{1}{2}}$ where H_L is the 'white light' exposure that produces a visual density (that is, one based on the CIE $V(\lambda)$ function) that is 0.2 above the minimum density, and H_S is the exposure giving a visual density 2.0 above the minimum density (or such that the line joining the points on the visual-density characteristic curve corresponding to the exposures H_L and H_S is tangential to the curve at the latter point if this is below the density 2.0 above minimum); H_L and H_S are, again, measured in lux seconds. The numbers obtained by these methods are such that the recommended exposure for a typical scene in average bright sunlight is $1/S_N$ or $1/S_R$ seconds at a relative camera-lens aperture of $f/16$. Speeds of this type are often referred to as ASA (American Standards Association) or ISO (International Standards Organization) speeds. Other scales of speed are also used, some of which are logarithmic instead of being arithmetic (as is the case for S_N and S_R); a list of corresponding values for various speed scales is given in Appendix 4.

For some scenes, orthodox methods of estimating the exposure from a knowledge of film speed are not appropriate. An example is the photography of firework displays, for which experience has shown that, for a film speed of 64 ASA, 1/30 second at $f/2.8$ is appropriate for ground displays, and a time exposure, preferably on a tripod, at $f/8$ is appropriate for aerial bursts.

REFERENCES

Adelstein, P. Z., Graham, C. L., and West, L. E., *J. Soc. Mot. Pic. Tel. Eng.*, **76**, 681 (1970).

American Standards Association, A.S.A. Standard PH2.17: Diffuse Reflection Density (1958).

American Standards Association, A.S.A. Standard PH2.19: Diffuse Transmission Density (1959).

American Standards Association, A.S.A. Standard PH3.37: Test method for the selective transmission of a photographic lens (1969).

Bartleson, C. J., and Breneman, E. J., *Phot. Sci. Eng.*, **11**, 254 (1967) and *J. Opt. Soc. Amer.*, **57**, 953 (1967).

Baumann, E., *Phot. Sci. Eng.*, **24**, 11 (1980).

Bermaine, D., *J. Phot. Sci.*, **22**, 84 (1974).

Brewer, W. L., Goddard, M. C., and Powers, S. A., *J. Soc. Mot. Pic. Tel. Eng.*, **64**, 561 (1955).

British Standard 3824: 1964, Specification for Colour Transmission of Photographic Lenses (1964).

Clark, L. D., *J. Soc. Mot. Pic. Tel. Eng.*, **61**, 241 (1953).

Dawson, G. H., and Voglesong, W. F., *Phot. Sci. Eng.*, **17**, 461 (1973).

De Mitri, C., *J. Phot. Sci.*, **8**, 220 (1960).

Eastman Kodak Co., *Kodak Filters*, 1970 Edn., p. 48 (1970).

Edgecombe, L. I., and Seeley, G. M., *J. Soc. Mot. Pic. Tel. Eng.*, **72**, 691 (1963).

Evans, R. M., *J. Opt. Soc. Amer.*, **33**, 579 (1943).

Gale, R. O., and Williams, A. L., *J. Soc. Mot. Pic. Tel. Eng.*, **72**, 804 (1963).

Giles, C. H., Forrester, S. D., Haslam, R., and Horn, R., *J. Phot. Sci.*, **21**, 19 (1973).

Giles, C. H., and Haslam, R., *J. Phot. Sci.*, **22**, 93 (1974).

Happé, L. B., *J. Phot. Sci.*, **22**, 114 (1974).

Harvey, E. A., *J. Phot. Sci.*, **4**, 130 (1956).

Hercock, R. J., and Sheldrick, G. E. A., *J. Phot. Sci.*, **4**, 113 (1956).

Hubbell, D. C., McKinney, R. G., and West, L. E., *Phot. Sci. Eng.*, **11**, 295 (1967).

Hunt, R. W. G., *J. Opt. Soc. Amer.*, **55**, 1540 (1965).

International Standards Organization, Recommendation R5: Diffuse Transmission Density, Photography (1955).

Kennel, G. L., Sehlin, R. C., Reinking, F. R., Spakowsky, S. W., and Whittier, G. L., *J. Soc. Mot. Pic. Tel. Eng.*, **91**, 922 (1982).

Koerner, A. M., *J. Soc. Mot. Pic. Tel. Eng.*, **63**, 225 (1954).

Kozanowski, H. N., *J. Soc. Mot. Pic. Tel. Eng.*, **73**, 939 (1964).

Land, E. H., *Proc. Nat. Acad. Sci.*, **45**, 115 and 636 (1959).

MacLeish, K. G., *J. Soc. Mot. Pic. Tel. Eng.*, **60**, 696 (1953).

Marriage, A., *Science and Applications of Photography*, Royal Photographic Society, London, p. 220 (1955).

Miller, O. E., and Powers, S. A., *J. Soc. Mot. Pic. Tel. Eng.*, **72**, 695 (1963).

Moore, C., *J. Phot. Sci.*, **22**, 117 (1974).

Muller, J., *Phot. Sci. Eng.*, **24**, 17 (1980).

Neale, D. M., *J. Phot. Sci.*, **4**, 126 (1956).

Newhall, S. M., Nickerson, D., and Judd, D. B., *J. Opt. Soc. Amer.*, **33**, 385 (1943).

Onley, J. W., *J. Opt. Soc. Amer.*, **50**, 177 (1960).

Pinney, J. E., and Voglesong, W. F., *Phot. Sci. Eng.*, **6**, 367 (1962).

Powell, P. G., *J. Phot. Sci.*, **4**, 120 (1956).

Rogers, H. G., Idelson, M., Cieciuch, R. F. W., and Bloom, S. M., *J. Phot. Sci.*, **22**, 135 (1974).

Sant, A. J., *Phot. Sci. Eng.*, **14**, 356 (1970).

Schafer, R. K., *Brit. Kinematog. Sound Tel.*, **54**, 286 (1972).

Staes, K., and Verbrugghe, R., *Brit. Kinematog. Sound Tel.*, **53,** 332 (1971).
Tuite, R. J., *J. Appl. Phot. Eng.*, **5,** 200 (1979).
Tull, A. G., *J. Phot. Sci.*, **22,** 107 (1974).
Watt, P. B., *J. Phot. Sci.*, **4,** 116 (1956).
Williams, F. C., and Clapper, F. R., *J. Opt. Soc. Amer.*, **43,** 595 (1953).
Williams, F. C., and Grum, F., *Phot. Sci. Eng.*, **4,** 113 (1960).
Zwick, D. M., *J. Soc. Mot. Pic. Tel. Eng.*, **88,** 533 (1979).

GENERAL REFERENCES

Brewer, W. L., Goddard, M. C., and Powers, S. A., *J. Soc. Mot. Pic. Tel. Eng.*, **64,** 561 (1965).
Duerr, H. H., *J. Soc. Mot. Pic. Tel. Eng.*, **54,** 653 (1950).
Evans, R. M., Hanson, W. T., and Brewer, W. L., *Principles of Color Photography*, Wiley, New York (1953). Chapters XI, XII, and XIII.
Mees, C. E. K., and James, T. H., *The Theory of the Photographic Process*, 4th Edn., Chapters 17 and 18, Macmillan, New York (1977).
Society of Motion Picture and Television Engineers, *Principles of Color Sensitometry*, Society of Motion Picture and Television Engineers, New York (1963).
Syke, G., *J. Phot. Sci.*, **4,** 131 (1956).
Thomas, W., *S.P.S.E. Handbook of Photographic Science and Engineering*, Wiley, New York (1973).
Williams, F. C., *J. Opt. Soc. Amer.*, **40,** 104 (1950).
Williams, F. C., *J. Soc. Mot. Pic. Tel. Eng.*, **56,** 1 (1951).

CHAPTER 15

Masking and Coloured Couplers

1. Introduction – *2*. Contrast masking – *3*. Unsharp masking – *4*. Coloured couplers – *5*. Inter-image effects – *6*. Masking when making separations – *7*. Masking for colorimetric colour reproduction – *8*. Masking for approximate colour reproduction – *9*. Calculation of mask gammas

15.1 Introduction

IN earlier chapters we have seen that all forms of trichromatic colour reproduction introduce errors, and that such pictures do not therefore represent all the colours as they were in the original scene. But we have also seen that the mental standards by which colour in pictures is usually judged are rather imprecise, so that the tolerances are quite large, and, as a result, the errors are often unnoticeable. In certain circumstances, however, the errors can mount up to the point where they are very serious. This is particularly the case where a trichromatic colour reproduction is itself copied by trichromatic means; in particular, in the case of subtractive colour photographs, the unwanted absorptions of the cyan, magenta, and yellow dyes result in dark blues and greens, and in the copy these colours are darkened again, sometimes even to the point where the colour almost vanishes and gives way to black. Furthermore, in reflection prints, as was explained in Chapter 13, there is often a serious problem in accommodating the scene within the limited range of tones normally available on reflecting surfaces such as paper. To provide partial solutions to these problems, recourse is often had to a technique known as *masking*; the principles of masking will now be described, both in general terms and in the form of the use of *coloured couplers* which provide a particularly important method of masking. The analagous technique of *electronic masking* in colour television is discussed in Section 23.13.

15.2 Contrast masking

From 1944 to 1949, before the advent of coloured couplers, the *Kodacolor* system for amateur reflection prints used a negative having dye-image gammas of about

1.0, together with a high contrast paper; the high overall contrast resulted in the system giving colours of high saturation. But the contrast was so high that severe loss of highlight and shadow detail would have been caused, but for the effect of an extra layer in the film which acted as a *mask*. This extra layer was developed in the colour negative as a low contrast positive black-and-white image. The effect of this was to cover light areas of the colour negative with dark deposits of silver, but to leave the dark areas unaltered. The overall contrast of the negative was thus reduced, but without any loss in the saturation of the colours in the negative because the three *colour* layers of the negative still operated at the same high contrasts. At the printing stage a slightly longer exposure was then given, thus enabling burnt-out highlights to be avoided, while the black-and-white mask image present in the negative prevented the shadow areas from becoming blocked up.

The introduction of coloured couplers in the *Kodacolor* system enabled a lower overall contrast to be used, so that the black-and-white mask layer became unnecessary and was abandoned. But the principle involved is of wide interest, particularly when copies of colour photographs have to be made, as when, for instance, a colour print or a colour transparency has to be made from an existing colour transparency.

In Fig. 15.1 the problem is presented in graphical form for the case of making a duplicate transparency from an original transparency, both the original and the duplicate being positives. In Fig. 15.1(a) the characteristic curve (density plotted against log exposure) of a typical reversal colour photographic process is shown (full line). The slope of this curve at the higher densities is usually greater than 45° in order to obtain correct tone reproduction in dark surrounds (see Chapter 6) when projecting with typical levels of ambient lighting and projection-lens flare. But, if the tones of the original transparency are to be reproduced without distortion, a material of gamma 1.0 (characteristic curve at 45°) is required; this is shown by the broken line. Hence, if a material having the characteristic curve shown by the full line were used for making the copy the tones would be distorted (see Fig. 6.13). The object of contrast masking is to avoid this distortion. The method is to make a low contrast negative by contact-printing the original transparency on to a suitable black-and-white film. After processing, this negative, or *mask*, as it is called, is bound up with the original transparency and reduces its contrast, much as did the masking layer in the old *Kodacolor* negative film.

In Fig. 15.1(b) the characteristic curve required by the mask in order to achieve complete tone correction is shown. This curve is constructed by plotting the horizontal distance a of the broken line from the full curve in Fig. 15.1(a), as a density in Fig. 15.1(b) against the log exposure on the broken line. The reason for this construction is that, since the exposure H_A should result in the density D_A (as indicated by the broken line) it is necessary to reduce the exposure to H_A' in order that the full curve should give the density D_A. This is achieved by arranging that all parts of the original which, unmasked, would print at H_A, when masked print

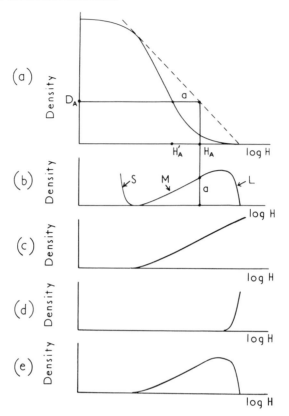

Fig. 15.1. Use of contrast correcting masks in reversal copying. Characteristic curves of (a) the copying material, (b) the ideal mask to correct it, (c) the nearest approximation to the ideal, using a single mask only, (d) the highlight mask required, and (e) the mask made from the original when the latter was bound up with the highlight mask.

at H_A'. Thus all these parts of the original require an increase in density a equal to the difference in exposure between H_A and H_A'. The same argument applies to all values of H_A, and hence the required characteristic curve for the mask is obtained by plotting a against H_A. This curve, shown in Fig. 15.1(b), exhibits first a positive, then a negative, and then another positive characteristic. The reason for this is not far to seek. The central portion of the full curve of Fig. 15.1(a) has a gamma greater than 1.0, and therefore requires negative masking in order to reduce the contrast. The toe and shoulder of the full curve of Fig. 15.1(a), however, have gammas less than 1.0 and therefore require positive masking in order to increase the contrast.

There is no photographic material that has the characteristic shown in

Fig. 15.1(b). But the function of the initial positive part, S, of the curve is to correct the tone rendering of the darkest shadow detail which is often of little importance in the picture and therefore can remain uncorrected without much loss of quality. The negative part, M, of the curve can be obtained fairly easily from a suitable low contrast negative material, as shown in Fig. 15.1(c). A significant improvement usually results from using such a negative mask alone, ignoring both the positive portions S and L of the ideal mask. But such a mask results in a flattening of the highlights in the copy which sometimes robs it of much of the brilliance and sparkle of the original transparency. For certain subjects a marked improvement is therefore gained by making a mask having the positive portion L in addition to the negative portion M. This can only be done by using a rather more complicated procedure, which will be described in a moment, involving the use of a *highlight mask*. But, when a highlight mask is not used, the flattening of the highlights in the copy can be reduced by over-exposing the mask, M, so that the highlights of the original transparency fall in the shoulder region of the characteristic curve of the mask material. In this way the highlights, although reduced in contrast by the low contrast of the copying material, are not appreciably *further* reduced in contrast by the mask, M. This technique of using a mask that *shoulders* is very useful. For full correction of the contrast of the highlights, however, a separate highlight mask may have to be made.

A highlight mask is made by contact printing the transparency on to a very high contrast black-and-white negative material (Fig. 15.1(d)) using an exposure sufficiently short for only the highlight detail to be recorded. This negative mask is then bound up with the transparency when the latter is used to make the negative mask on the material having the characteristic of Fig. 15.1(c). This mask, being a negative, reverses the negative curve of the highlight mask into a positive curve as shown in Fig. 15.1(e). Having made the negative mask in this way, the highlight mask is then discarded, and the mask having the characteristic of Fig. 15.1(e) is bound up in register with the original transparency, before it is printed on the material characterized by the full line of Fig. 15.1(a). This technique is somewhat laborious but in cases where copies of the highest quality are required it is well worth-while.

As mentioned at the beginning of this chapter, in copies, the unwanted absorptions of the cyan, magenta, and yellow dyes occur twice, and blues and greens are often badly darkened. Some reduction in this darkening can be achieved by making the negative mask, M, through a red, orange, or yellow filter; such a mask will have a greater density in areas of red, orange, or yellow, than in areas of green and blue, and hence, in the masked original, blues and greens are lightened relative to reds, oranges and yellows; thus, in the copy, the blues and greens are lighter than they would be if the negative mask had been made without a filter. This technique is often well worth-while adopting. In the old *Kodacolor* process the black-and-white mask layer was situated above the cyan and magenta layers, but below the yellow layer of the negative. The mask layer was sensitive to blue light only, and hence by exposing it with white light through the base the unwanted blue absorptions of the cyan and magenta dyes were printed on to it; the mask was therefore equivalent to a low contrast positive image of the scene made through a yellow filter (Neblette, 1962).

15.3 Unsharp masking

Another useful feature often used in masking was first suggested by Yule (Yule, 1944). Exact registration of the mask when bound up with the original is obviously difficult, and if not perfectly achieved results in halos appearing around any well-defined edges. Yule suggested that the masks should be deliberately made *unsharp* by printing them with a thin spacer between the transparency and the mask materials. This not only helps to obscure slight lack of registration of the masks, but also improves the reproduction of fine detail. A negative mask reduces contrast, but fine detail is seen more clearly if reproduced at high contrast; by having the mask unsharp the fine detail is not resolved by the mask and hence, when it is bound up with the original transparency, it does not reduce the contrast of fine detail, but only of large areas. (See Plate 36, page 545.)

The use of unsharp masks, and negative masks of contrast high enough to provide severe over-correction of tones, together with various other techniques, has enabled Evans (Evans, 1951 and 1954) to obtain with the Kodak *Dye Transfer* system, reproductions that resemble paintings rather than photographs, although photographic techniques are used throughout. These reproductions have been called *Colour Derivations* (see Plate 22, page 352).

15.4 Coloured couplers

The black-and-white masks used in the old *Kodacolor* film enabled the saturation of colours to be increased by raising the contrast of the system without spoiling the tone reproduction, with some correction for the unwanted absorptions of the magenta and cyan dyes. However, a more elegant method of correcting for these unwanted absorptions was introduced in 1949 (1948 in the case of *Ektacolor* film for professional use): the colour-forming couplers were themselves coloured, and in such a way that, as a dye was formed, the transmission of light in the regions of unwanted absorption remained constant. The negatives now become orange in the unexposed area. (See Plate 11, page 113.) The use of these *coloured couplers* in colour negative films is very extensive, and the principles involved will now be described.

A colour negative, consisting as it usually does of three superimposed negative dye-images, is dependent for its success on the ability of the three dyes to make the three negative images easily distinguishable; it is therefore most important that the three dyes absorb only in three well-separated parts of the spectrum. But, if cyan, magenta, and yellow dyes are used, their spectral transmission curves will be similar to those shown in Fig. 4.1, having unwanted absorptions in the green and blue parts of the spectrum. Unfortunately, this type of defect is not confined to cyan, magenta, and yellow dyes; almost all dyes have subsidiary absorptions on the short wavelength side of their main absorption band, so that even if the three dyes were an infra-red absorber, a magenta, and an ultra-violet absorber, for instance, the same difficulty is present. It is the virtue of coloured couplers that they overcome the effects of these unwanted absorptions in a remarkably elegant fashion (Hanson, 1950). The credit for their introduction must be shared by the Research Laboratories of the Eastman Kodak Company and the Ansco Corporation who filed the first patents on the subject on the very same day! In the case of the former company, Dr. W. T. Hanson conceived the idea from first principles, its successful realization coming only after intensive research; in the case of the latter company the fortuitous discovery of a coupler which happened to be coloured in a beneficial way led to the same discovery.

The principle on which coloured couplers work is shown diagrammatically in Fig. 15.2. Suppose that, in the colour negative, the magenta dye, at its maximum concentration, m, has red, green, and blue transmittances of 100 per

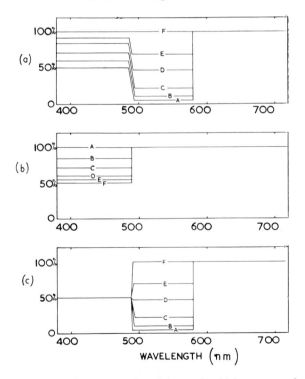

Fig. 15.2. Diagrammatic representation of the way in which a magenta-forming coloured coupler works. (a) Transmission curves of magenta dye at different concentrations. (b) Transmission curves of coloured coupler. (c) Combined transmission curves of the dye and the coupler.

275

cent, 5 per cent, and 50 per cent respectively, as shown in Fig. 15.2(a) by line A. It is thus assumed, for the sake of simplicity, that it is an ideal magenta dye except for a uniform unwanted absorption in the blue. The lines, B, C, D, E, and F, show what the transmittances would be at concentrations $\frac{3}{4}m$, $\frac{1}{2}m$, $\frac{1}{4}m$, $\frac{1}{8}m$, and zero, respectively. It will be supposed that this magenta dye is formed by the colour-development of a suitable coupler in one of the layers of a colour film. Let the concentration of the coupler before development be c. Then the concentrations of the coupler remaining after producing the dye-concentrations A, B, C, D, E, and F will be: zero, $\frac{1}{4}c$, $\frac{1}{2}c$, $\frac{3}{4}c$, $\frac{7}{8}c$, and c respectively.

Suppose, now, that the coupler, instead of being colourless, was *yellow*, having red, green, and blue transmittances (at concentration c) of 100 per cent, 100 per cent, and 50 per cent, respectively. As it is colour-developed to form the magenta dye, its yellow colour in the layer gradually becomes less and less as it is used up, and its transmission curves for the same levels A, B, C, D, E, and F discussed above would be as shown in Fig. 15.2(b). The full transmission curves for the layer are given by combining the appropriate pairs of curves from Fig. 15.2(a) and (b) and these are shown in Fig. 15.2(c). It is seen that the transmittance in the blue region remains constant. When there is no magenta dye, the coupler alone has a transmittance of 50 per cent; when all the coupler has been used, it no longer absorbs at all, but the magenta dye has a transmittance of 50 per cent. At all intermediate stages the blue transmittance of the coupler multiplied by the blue transmittance of the magenta dye is also equal to 50 per cent. (This is because if, at any intermediate stage, the fraction of coupler left is n, the transmittance of the coupler is $(50/100)^n$, and the transmittance of the dye is $(50/100)^{1-n}$, giving a transmittance of the combination of $(50/100)^n \times (50/100)^{1-n}$ which is equal to $50/100$.)

Clearly, with this system, the effect of light on this layer results in variations in the green transmission of that layer, but has no effect on the values of the red and blue transmissions which are fixed at 100 per cent and 50 per cent respectively. The low value of the constant blue transmission can be easily compensated by doubling the blue content of the light used for printing. Thus, the magenta dye and its yellow coupler together form an arrangement by means of which only light in the green part of the spectrum is modulated; hence, from the photographic point of view, the unwanted blue absorption of the magenta dye has been eliminated.

A pink coupler, which forms a cyan dye in another layer, can similarly eliminate the effects of the unwanted green and blue absorptions of that dye. The way in which this takes place is shown in Fig. 15.3. In Fig. 15.3(a), for the sake of simplicity, we have shown the transmission curves of a cyan dye which is ideal except for two uniform unwanted absorptions in the green and blue regions. The line A refers to the dye at maximum concentration, the red, green, and blue transmittances being 5 per cent, 30 per cent, and 40 per cent respectively. The other lines are analogous to those of Fig. 15.2(a). Suppose that the coupler is of a pink colour, having, at maximum concentration, red, green, and blue transmit-

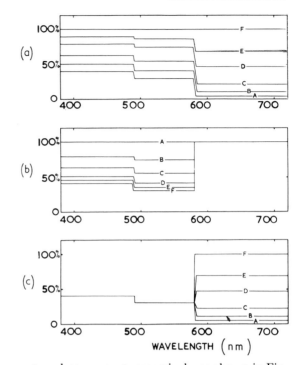

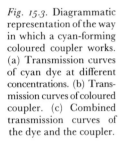

Fig. 15.3. Diagrammatic representation of the way in which a cyan-forming coloured coupler works. (a) Transmission curves of cyan dye at different concentrations. (b) Transmission curves of coloured coupler. (c) Combined transmission curves of the dye and the coupler.

tances of 100 per cent, 30 per cent, and 40 per cent respectively, as shown in Fig. 15.3(b); when this coupler is present with the cyan dye which it forms on colour development, the red-sensitive layer will have the transmission curves shown in Fig. 15.3(c) for the different concentrations. Again it is seen that, where there were varying unwanted absorptions, they are now constant. Hence, by increasing the green content of the printing light by a factor of $3\frac{1}{3}$, and the blue by a factor of $2\frac{1}{2}$, the net result of the effect of light on this layer is merely to modulate the red transmission of the layer.

When actual dyes and coloured couplers are used, the transmissions shown as constant in Figs. 15.2(c) and 15.3(c) are only approximately constant, but this scarcely impairs the degree of improvement resulting. In fact, by allowing these transmissions to rise, by using couplers of deeper colours, the unwanted absorptions of the cyan and magenta dyes used in the print as well as those in the negative can also be, to some extent, compensated. Fig. 15.4 shows curves relating to an actual film; in this case the unwanted blue absorption of a magenta dye has been compensated by using a coloured coupler diluted with an uncoloured coupler. These types of curve can be replotted so as to allow for the increased printing exposure given in the regions of the spectrum where coloured couplers absorb, to obtain *equivalent dyes* for each coloured coupler system (Sant, 1961; Watson, 1966).

277

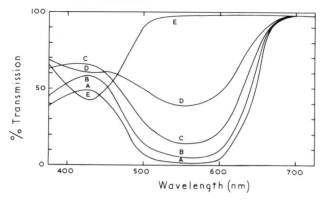

Fig. 15.4. Spectral transmission curves for a yellow-coloured coupler (curve E) and the magenta dye formed from a mixture of this coupler with an uncoloured coupler (curve A). Curves B, C, and D represent intermediate degrees of dye-forming reaction.

The introduction of coloured couplers in colour photography was a major step forward in its technological development, and has resulted in the widespread use of colour negatives, not only for amateur reflection prints, but also for the production of professional motion pictures in colour, for which the positive prints are made by direct printing on to three-layer colour positive stock (see Section 12.11). (See Plate 11, page 113.)

15.5 Inter-image effects

Coloured couplers can, unfortunately, only be used in materials designed to be printed or otherwise duplicated. The presence of the coloured couplers in light areas gives a pronounced orange cast, and the eye is not able to adapt sufficiently to compensate for it. For this reason coloured couplers have no application to normal reversal processes intended for viewing, and the brilliance of some of these is caused, at least in part, by *inter-layer* or *inter-image* effects which have beneficial results not unlike those produced by coloured couplers (Hanson and Horton, 1952; Barr, Thirtle and Vittum, 1969).

There are several ways of demonstrating the presence of inter-image effects, and some of these are illustrated for a reversal material in Fig. 15.5. In Fig. 15.5(a), by plotting the appropriate analytical density, a measure of the amount of cyan image dye present is shown. The curve N shows the amount of cyan dye present in a neutral scale which was produced by giving additive red, green, and blue exposures; the curve R shows the amount of cyan produced when the red exposure only was given. Because these curves are different it is clear that the presence or absence of exposure in the other two layers affects the amount of cyan dye produced: in this case there is less cyan in reds (curve R) than in neutrals

278

(curve N) and hence the effect is to lighten the reds. The effects of the other two layers on the magenta layer, and on the yellow layer, can be shown similarly.

In Fig. 15.5(b) the analytical densities for all three image dyes are shown for the case where the red and blue layers have been given exposure scales and the green layer uniform exposures at different intensities. Any tilt in the curves M_1, M_2, M_3, M_4, representing the amount of magenta dye present, is the result of an inter-image effect, because the green exposure was uniform. (The curves C and Y may be slightly shifted along the log exposure axis by variations in M, but for the sake of simplicity multiple C and Y curves have not been drawn.) Similar sets of curves can be drawn for the cases where the red and the blue layers have the uniform exposures.

In Fig. 15.5(c) results similar to those of Fig. 15.5(b) are shown but this time two layers have uniform exposures and one layer has an exposure scale. Any tilts in the C and M curves indicate inter-image effects. (Again, the curve Y may be shifted along the log exposure axis, but this is not shown.) Similar sets of curves can be drawn for the cases where the green and the red layers have the exposure scales.

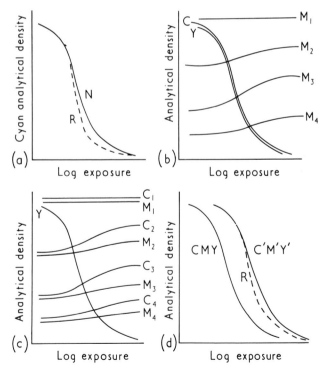

Fig. 15.5. Four different ways of illustrating the nature of inter-image effects which may be occurring.

In Fig. 15.5(d) all three layers have been given exposure scales at two different exposure levels to give two neutral scales (curves C, M, Y, and curves C', M', Y', both sets represented as being superimposed for simplicity). A scale of reds is then exposed having blue and green exposures at the CMY level, but the red exposure at the C'M'Y' level. The resulting curves would be the same as M, Y, and C' if there were no inter-image effects, and inter-image effects are therefore shown up by any differences. Thus the cyan curve in the reds might be like R instead of like C' in which case the difference between the curves R and C' shows the amount of inter-image effect in the cyan image in this scale of reds.

If the results shown in Fig. 15.5(b) and (c) are plotted using *integral* densities instead of analytical densities, any absence of tilt in the approximately horizontal curves indicates either that there is no unwanted dye-absorption operating, and that there is no inter-image effect, or that if there is an unwanted dye-absorption it is being exactly offset by an inter-image effect.

Inter-image effects may be present whenever different development rates occur in adjacent layers. This can happen in several ways.

For instance, as a developer penetrates a multi-layer colour material, it will normally be partially exhausted by the time it reaches the bottom layer; hence, if the development is not carried to completion in all layers, in order to achieve a matched grey scale, it may be necessary to make the bottom layer faster or of higher contrast. But when only the bottom layer is exposed (in the case of a saturated red colour for materials with the conventional layer order), the developer will not be partially exhausted on reaching the bottom layer, because no development will have occurred in the upper two layers. Hence, the speed or contrast of the cyan image will be greater in reds than in greys (an increase in cyan contrast in reds is shown in Figs. 15.5(a) and (d)) and this makes reds lighter and more saturated than they would otherwise be.

Inter-image effects in a multi-layer material can also be caused by the degree of development in one layer being affected by the release from a neighbouring layer of development-inhibiting agents. These agents can be bromide or iodide ions, or special inhibitors released by *development-inhibitor-releasing* (*DIR*) couplers (Barr, Thirtle, and Vittum, 1969; Meissner, 1969; Tull, 1975). (See Section 17.11.)

Inter-image effects that affect colour reproduction adversely can also occur. For instance, if oxidized developer wandered from one layer to another in a coupler-incorporated material, dye of the wrong colour would be formed, and the colour reproduction distorted; this type of contamination is usually minimized by having thin inter-layers between the image-forming layers of multi-layer materials, and these inter-layers may contain chemicals to absorb or immobilize any oxidized developer that reaches them. Another adverse effect occurs if development in an unexposed layer occurs because an adjacent layer is very highly exposed: in Figs. 15.5(b) and (c) this would be shown by the curves M_1 and C_1 dropping off at their right-hand ends; this type of effect has to be avoided by choosing emulsions that are as insensitive as possible to fogging as a result of vigorous development in an adjacent layer. Exposure of a green or red sensitive layer by blue light (sometimes referred to as *punch-through*), such as could occur because of insufficient protection by a yellow filter layer, can also be regarded as an adverse inter-image effect.

15.6 Masking when making separations

In cases where coloured couplers cannot be used, some correction for the unwanted absorptions of cyan, magenta, and yellow dyes can be provided, as we have seen, by using contrast masks exposed through suitable filters. But, in systems where the red, green, and blue records are available on separate negative or positive films, masking can in principle give full correction for unwanted absorptions, and also some correction for various other defects; this separate availability of the three records occurs most commonly when preparing half-tone printing surfaces in graphic-arts applications, but the principles involved are of wider interest and can have other applications.

For clarity, let us start with a simple example. Suppose we wish to correct for the unwanted absorptions of the magenta dye or ink used in a colour reproduction made from a transparency and that we are using a process in which three *separation negatives* are made from the transparency by printing or enlarging it through red, green, and blue filters, on to a suitable black-and-white film. If it were possible, in the final reproduction, to have an image provided by a perfect magenta dye, that had no unwanted absorptions, it would be invisible when viewed through blue and red filters. But images consisting of real magenta dyes, though nearly invisible when viewed through red filters, are visible as low contrast images when viewed through blue filters because of the unwanted blue absorption of magenta dyes. The real magenta dye image may therefore be thought of as consisting, approximately, of a full-contrast perfect magenta image, together with a low contrast unwanted 'yellow' image (giving the blue absorption). In most picture areas that are not white, all three dyes are present to some extent. It is therefore usually possible to reduce the *main* yellow dye image by an amount that is at every point in the picture equal to the low contrast unwanted 'yellow' image, and hence to overcome the effect of the unwanted blue absorption of the magenta dye. The practical procedure of such a scheme, when the original is a colour transparency, is as follows.

The colour transparency is printed on to a low contrast black-and-white film using green light for making the exposure. The mask so obtained is mainly a record of the magenta dye in the transparency, and is therefore also approximately a record of what the distribution of magenta dye will be in the final reproduction. The record, being a black-and-white negative, means that where the magenta dye is going to be heavy the mask is light, and *vice versa*. This mask is then bound up in register with the colour transparency. Its effect is to lighten areas that will be heavy in magenta, relative to those areas that will be light in magenta. Now, although the object of this mask is to overcome the unwanted absorption of the magenta dye, it is not used when the green separation negative is being made: it is only used when making the *blue* separation negative. (The reason for this is that since the unwanted absorption of the magenta dye is a blue absorption, it can only be corrected by reducing the *yellow* dye image appropriately, and hence the mask is used when making the blue separation negative, since it is from this negative that the yellow reproduction-image is produced.) By having the mask over the transparency when the blue separation negative is being made, a low contrast positive image of what the magenta dye will be is added to its exposure. Hence when the positive yellow image is produced from this masked separation negative, there will be less yellow dye at points where the magenta dye will be heavy, but the normal amount of yellow where the magenta dye is absent. The same result can also be achieved by masking the separation negative rather than the transparency, although the mask must now be a positive image.

If the correct contrast is chosen for the mask, the effect of the unwanted absorption can be almost entirely cancelled out in this way. The correct contrast for this very simple case, in which only one unwanted absorption of one dye is being cancelled, is calculated as follows. If the density ratio of the unwanted blue to the wanted green absorption of the magenta dye or ink used in the reproduction is m_B/m, and the gamma of the final image is γ_m, then the gamma of the mask must be such that the main yellow image in the reproduction has superimposed upon it a negative image of gamma $(m_B/m)\gamma_m$. If the gamma of the main yellow image relative to the blue separation negative is γ_y, the gamma of the mask to be used with the blue separation negative must be $(m_B/m)\,(\gamma_m/\gamma_y)$; but if the mask is bound up with the transparency and used when making the blue separation negative, it must have a gamma of $(m_B/m)\,(\gamma_m/\gamma_y)/\gamma_B$, where γ_B is the gamma of the material on which the blue separation is made. The gamma of the film-process combination that should be used for making the mask is obtained by dividing these expressions by γ_t, the gamma of the green filter image in the transparency; the value obtained depends, therefore, on how the mask is used; but it also depends upon how it is made: for if, instead of deriving it from the transparency using a green filter, it is made from the green separation negative, the gamma required is altered. Fig. 15.6 shows the gamma that the mask *material* must have (in its process) in order to produce a mask having the correct gamma for various methods of working.

If the unwanted blue absorption of the magenta dye is the only unwanted absorption in the reproduction system, and the overall gamma of the reproduction is to be the same as that of the transparency, the above formulae simplify because:

$$\gamma_B\gamma_y = 1$$
$$\gamma_G\gamma_m = 1$$
$$\gamma_t = \gamma_m$$

The corresponding values of the gammas required by the mask material are listed in Fig. 15.6 under the column headed 'Simple Case'.

Mask made from	Mask used with	Gamma of mask material	
		General case	Simple case
Transparency using green filter	Blue separation negative	$+\dfrac{(m_B/m)\,(\gamma_m/\gamma_y)}{\gamma_t}$	$+\dfrac{(m_B/m)}{\gamma_y}$
Transparency using green filter	Transparency when making blue separation negative	$-\dfrac{(m_B/m)\,(\gamma_m/\gamma_y)}{\gamma_t\gamma_B}$	$-(m_B/m)$
Green separation negative	Blue separation negative	$-\dfrac{(m_B/m)\,(\gamma_m/\gamma_y)}{\gamma_t\gamma_G}$	$-\dfrac{(m_B/m)}{\gamma_y\gamma_G}$
Green separation negative	Transparency when making blue separation negative	$+\dfrac{(m_B/m)\,(\gamma_m/\gamma_y)}{\gamma_t\gamma_G\gamma_B}$	$+\dfrac{(m_B/m)}{\gamma_G}$

Fig. 15.6. Gammas of mask materials required for correcting the unwanted blue absorption of a magenta dye, when the mask is used in various ways.

m_B/m = ratio of blue and green densities of magenta reproduction dye.
γ_m = gamma of magenta image in reproduction.
γ_y = gamma of yellow image in reproduction relative to that of the blue separation negative.
γ_t = gamma of green-filter image in the transparency.
γ_B = gamma of material on which the blue separation negative is made.
γ_G = gamma of material on which the green separation negative is made.
+ = reversal mask material (or successive negative-positive steps) required.
− = negative mask material required.

15.7 Masking for colorimetric colour reproduction

Yule (Yule, 1938 and 1940) has investigated the possibilities of obtaining colorimetrically correct (see Section 11.5) colour reproduction by means of masking. He concluded that correct duplication of a colour original composed of the *same dyes* as the reproduction was possible if six masks were used, provided that the dyes obeyed certain rules and that tone reproduction and colour balance were properly adjusted.

The rules, which it is necessary, in Yule's theory, for the dyes to obey, concern their properties when recorded on the separation negatives; or, if a densitometer is used which simulates the spectral sensitivities of the three filter-film combinations used for making the separation negatives, the rules can be formulated in terms of the *printing densities* measured on such an instrument. The rules have become known as the *Additivity and Proportionality Rules*, and may be stated as follows (compare Section 14.19):

Additivity Rule. The red printing density of any mixture of the three dyes should be equal to the sum of the red printing densities of the three dyes measured

283

separately; and the same should be true of the green and blue printing densities.
Proportionality Rule. When measured as printing densities, the ratio of the wanted
to the unwanted absorptions for each dye should be independent of the concentration of the dye.

In transparencies these rules are obeyed by most dyes if the red, green, and
blue filters used for exposing the separation negatives transmit light of one
wavelength only; such filters are impracticable because of their very low
transmissions, but the departures from the rules are not too serious if conventional narrow-cut red, green, and blue filters are used. In reflection work the
departures from the rules are greater (but non-linearities of tone reproduction
can reduce the departures usefully).

Assuming, then, that all these conditions have been fulfilled, it is possible to
achieve colorimetric reproduction of an original consisting of a mixture of the
same dyes as the reproduction. The reasons for this are as follows.

Because the original and the reproduction consist of mixtures of the same
dyes, colorimetric reproduction must occur if the printing densities of the
original, O_r, O_g, O_b, and those of the reproduction R_r, R_g, R_b, are the same. The
conditions for colorimetric reproduction are therefore

$$R_r = O_r$$
$$R_g = O_g$$
$$R_b = O_b$$

But the additivity rule enables us to re-write these equations thus:

$$C_r + M_r + Y_r = O_r$$
$$C_g + M_g + Y_g = O_g$$
$$C_b + M_b + Y_b = O_b$$

the reproduction densities having been split up into the contributions from each
of the three dyes, C_r denoting the contribution of the cyan dye towards the total
red printing density, etc. If we re-write these equations thus

$$C_r + k_3 M_g + k_5 Y_b = O_r$$
$$k_1 C_r + M_g + k_6 Y_b = O_g$$
$$k_2 C_r + k_4 M_g + Y_b = O_b$$

where $k_1 = C_g/C_r$, $k_3 = M_r/M_g$ etc., the Proportionality Rule states that the
values of k do not depend on the amounts of the dyes present, and they are
therefore constants. We may therefore solve these equations for C_r, M_g, and Y_b,
and obtain equations of the form:

$$C_r = a_1 O_r + a_2 O_g + a_3 O_b$$
$$M_g = a_4 O_r + a_5 O_g + a_6 O_b$$
$$Y_b = a_7 O_r + a_8 O_g + a_9 O_b$$

If, then, a red separation negative is exposed to have a gamma a_1, and is
combined with a green-light mask of gamma a_2 and a blue-light mask of gamma
a_3, and is then used to produce a cyan dye-image free of tone-distortion, the
correct amount of cyan dye will be produced at each point in the picture. (If the

masks are made from the transparency, they must be exposed on filter-film combinations having the same spectral sensitivities as used for the separation negatives.)

Similarly, a_5 and a_9 represents the gammas of the other two separation negatives, and a_4, a_6, a_7, and a_8 the gammas of the masks necessary to produce the correct amounts of magenta and yellow dye at each point in the picture. Hence, if the photographic steps are free of tone distortion, use of these separation negatives with their masks enables colorimetric reproduction of the original to be achieved.

But what is the position if the original, instead of being composed entirely of mixtures of the dyes used in the reproduction, consists of *any* colours?

Yule has pointed out that, with six masks, colorimetric reproduction is still possible (within the gamut of colours which the reproduction dyes can produce) if the filters used in exposing the separation negatives are such as to modify the spectral sensitivity of the emulsion in such a way as to match a set of *colour-matching functions* (see Section 7.4), provided that the reproduction dyes still obey the Additivity and Proportionality Rules (the dye densities now being measured using spectral sensitivities equivalent to a set of colour-matching functions).

Suppose the separation negatives are exposed with spectral sensitivities that match the CIE colour-matching functions, $\bar{x}(\lambda), \bar{y}(\lambda), \bar{z}(\lambda)$. Then any point in the original will be recorded on the separation negatives as three exposures E_x, E_y, and E_z which are proportional to the tristimulus values X, Y, and Z respectively, of the original. If $O_x = -\log E_x$, $O_y = -\log E_y$, and $O_z = -\log E_z$ we may regard these quantities as the densities of the original to the $\bar{x}(\lambda), \bar{y}(\lambda), \bar{z}(\lambda)$ functions. Similarly if the densities of our reproduction to these functions are denoted by R_x, R_y, and R_z, then the conditions for colorimetric reproduction are simply:

$$R_x = O_x$$
$$R_y = O_y$$
$$R_z = O_z$$

for this would imply that the tristimulus values of the reproduction were the same as those of the original, and hence to the CIE standard observer the original and the reproduction would appear identical (in identical viewing conditions). The Additivity Rule enables us to re-write these equations thus:

$$C_x + M_x + Y_x = O_x$$
$$C_y + M_y + Y_y = O_y$$
$$C_z + M_z + Y_z = O_z$$

the reproduction densities to each of the three functions having been split up into the contributions from each of the three dyes, C_x denoting the contribution of the cyan dye towards the total density to the $\bar{x}(\lambda)$ function, etc. If we re-write these equations thus:

$$C_x + k'_3 M_y + k'_5 Y_z = O_x$$
$$k'_1 C_x + M_y + k'_6 Y_z = O_y$$
$$k'_2 C_x + k'_4 M_y + Y_z = O_z$$

285

where $k'_1 = C_y/C_x$, $k'_2 = C_z/C_x$ etc., the Proportionality Rule states that the values of k' do not depend on the amounts of the dyes present, and they are therefore constants. We may therefore solve these equations for C_x, M_y, and Y_z, and obtain equations of the form:

$$C_x = a'_1 O_x + a'_2 O_y + a'_3 O_z$$
$$M_y = a'_4 O_x + a'_5 O_y + a'_6 O_z$$
$$Y_z = a'_7 O_x + a'_8 O_y + a'_9 O_z$$

C_x, M_y, and Y_z then denote the amounts of cyan, magenta, and yellow dye required at each point in the colour reproduction, in order to obtain a colorimetric match with the original.

The diagonal terms, a'_1, a'_5, and a'_9, of the above equations denote, as before, the gammas of three separation negatives, while the remaining terms denote the gammas of the six masks to be used in conjunction with the separation negatives when the coloured images are being printed. Although, for the sake of simplicity, we have assumed in this discussion that the spectral sensitivity curves used for making the separation negatives were the same as the $\bar{x}(\lambda)$, $\bar{y}(\lambda)$, $\bar{z}(\lambda)$ curves, this is an unnecessary restriction: any set of colour-matching functions, that is to say, any linear combinations of the $\bar{x}(\lambda)$, $\bar{y}(\lambda)$, $\bar{z}(\lambda)$ curves may be used; the only consequence is that the masking equations will call for different gammas for the separation negatives and masks. Similarly, although we have regarded the densities of the dyes as being evaluated using spectral sensitivities equivalent to the $\bar{x}(\lambda)$, $\bar{y}(\lambda)$, $\bar{z}(\lambda)$ functions, the use of any other set (even if different from the set used for making the separation negatives) does not invalidate the theory but only affects the values obtained for the gammas of the separations and masks.

When dye densities are measured using spectral sensitivities equivalent to colour matching functions, the Additivity and Proportionality Rules are not obeyed well. The functions are very much broader than the narrow-cut filters for which these rules hold reasonably well and the rules break down quite considerably. This limits the usefulness of this approach, but departures of characteristic-curve shapes from linearity can be used to counteract the breakdowns of the rules to some extent (Gutteridge, 1972), and the masking can with advantage be based on colours in the original that are metameric matches to the dyes (Clapper, Breneman, and Brownstein, 1977).

15.8 Masking for approximate colour reproduction

Other methods of obtaining approximate colorimetric reproduction in the general case, where the original can consist of any colours, have been investigated theoretically (MacAdam, 1938; Marriage, 1940). Marriage pointed out that masking can either be used in an attempt to correct the spectral sensitivities of the original photographic material, or to correct for the unwanted absorptions of the dyes (or inks), or both. Marriage considered that correcting for the unwanted absorptions was the more important function of masks and applied Yule's theory in this direction (Marriage, 1940). Miller also made this assumption (Miller, 1941) and applied matrix algebra, with great advantage, to the problem of evaluating the gammas of the masks required for this purpose.

Marriage also pointed out that by means of masking it was possible to ensure that at least four colours were always colorimetrically correct (Marriage, 1948). Marriage chose grass-green, skin-pink, white and a grey. Including the latter two colours has the big advantage that all shades of grey from white to black are also reproduced very nearly correctly. This approach was extended by others (Brewer, Hanson, and Horton, 1949) to the case where, instead of seeking colorimetric reproduction of four colours, the masks are chosen so that the errors in a larger number of specially selected colours are kept to a minimum. Using the same criterion of minimum errors, Brewer and Hanson also investigated the relative importance of unwanted dye absorptions and the absence of negative portions in the spectral sensitivity curves (Brewer and Hanson, 1954). They concluded that the unwanted absorptions contributed the greater errors, but that absence of negative portions in the curves was by no means insignificant in comparison.

That it is possible, by suitable choice of masks, gammas, and colour balance, to reproduce any four colours without error can be seen from sets of equations including terms a_{10}, a_{11}, and a_{12}, thus:

$$C_x = a_1 O_x + a_2 O_y + a_3 O_z + a_{10}$$
$$M_y = a_4 O_x + a_5 O_y + a_6 O_z + a_{11}$$
$$Y_z = a_7 O_x + a_8 O_y + a_9 O_z + a_{12}$$

These extra three terms represent the variable of colour balance; they are all zero if colorimetric reproduction of all colours is achieved, but if, as is always the case in practice, there are departures from correct reproduction, use can be made of this variable to reduce the average errors to the minimum. Thus, given four colours that must be reproduced correctly, the three equations for each colour provide 12 equations which may be solved for $a_1, a_2, a_3, \ldots a_{12}$; the values of these terms then give the image gammas, mask gammas, and colour balance required for correct reproduction of the four chosen colours.

Using equations of this basic type, Brewer and Hanson determined the values of twelve coefficients representing image gammas, mask gammas, and colour balances, which resulted in minimal colorimetric errors under various conditions (Brewer and Hanson, 1955). They did not, however, restrict their spectral sensitivity curves to linear combinations of the $\bar{x}(\lambda), \bar{y}(\lambda), \bar{z}(\lambda)$ functions, but in addition tried sets with omitted negative portions, and also a set typical of those commonly used in practice. Sixty different picture test colours were considered in three groups of twenty, one group containing colours covering most of the gamut of the reproduction dyes considered, the other two groups being less saturated and as far as possible typical of average picture-taking experience. Their investigation produced a number of interesting results. First, it was found that, so long as no restriction was placed on the values of the coefficients, that is, so long as six masks of any gammas were allowed, there was very little to choose between one set of sensitivity curves and another. Secondly, the use of a set of sensitivity curves that overlapped one another considerably, such as the $\bar{x}(\lambda), \bar{y}(\lambda)$, $\bar{z}(\lambda)$ functions, led to much higher mask gammas than the use of sensitivity curves that were more separated. Thirdly, omission of required negative portions of sensitivity curves resulted in some increase in mask gammas but was less important than adequate separation of the curves along the wavelength axis. Fourthly, the particular choice of picture test colours affects the values obtained for the coefficients, a group containing more saturated colours leading to higher mask gammas, but the effect is fairly small. Fifthly, the magnitude of the reproduction errors finally obtained was quite small. For typical picture colours, the average error was only about three just noticeable differences (for a two degree field); for more saturated colours the average error was about four times as great.

The practical application of these findings may seem a little remote in that the use of six masks is complicated, photographic processes are often non-linear, and dyes do not generally obey the additivity and proportionality rules. In negative-positive processes, however, coloured couplers provide practicable means of attaining a number of masks, and some products already thus incorporate three. One of the six masks is generally of so low a contrast as to be of little practical consequence, so that the addition of two more masks by means of coloured couplers or inter-image effects could provide all the masking needed. Using the minimum-errors criterion, the Additivity and Proportionality Rules need not be obeyed as far as the image dyes are concerned; but if masking is carried out by means of coloured couplers, it is desirable that the dyes colouring the couplers obey the laws: over the most important density range they will probably do so approximately. It thus seems possible that a colour negative material might have sufficient masks built into it to enable a print to be made from it having remarkably small colorimetric errors for average colours.

It is inevitable, however, that departure of the sensitivity curves from linear combinations of the colour-matching functions renders the whole system vulnerable to failure in the face of any particular colour. And it is interesting to note that the very feature (separation of the sensitivity curves along the wavelength axis) that results in low-gamma masks, also, in general, results in large errors in colours having unusual spectral reflectance curves, such as was mentioned in Sections 5.1 and 9.5 in connection with certain blue flowers. From the practical point of view, however, low-gamma masks are usually desirable in order that the system should not be either very critical to operate or very wasteful of light when coloured couplers are used.

The conditions which call for low-gamma masks are preferable for systems in which no masking is possible, since the errors caused by omitting low-gamma masks are smaller than those caused by omitting high-gamma masks.

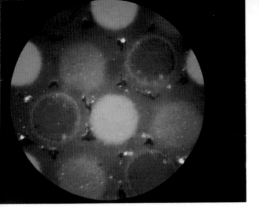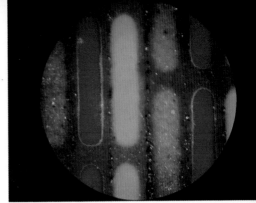

Plate 18

Photomicrographs of the glowing screens of 625-line colour television shadow-mask tubes. *Left:* conventional shadow-mask tube (see Sections 3.3 and 21.4); *right:* Precision In-line tube (see Section 21.6). Total magnification 31 times. At normal viewing distances, the red, green, and blue areas blend into all the intermediate colours needed to form the picture (see Section 3.3). Photomicrographs by courtesy of Thorn Colour Tubes Ltd. Reproduced on a Crosfield *Magnascan* scanner.

Plate 19

A colour television receiver displaying: *top left*, a luminance signal only; *top right*, a luminance signal and an N.T.S.C. I-signal; *bottom left*, a luminance signal and an N.T.S.C. Q-signal: *bottom right*, a complete N.T.S.C. signal consisting of luminance, I, and Q signals, together. The I-signal, giving orange-cyan reproduction, has about a quarter the bandwidth of the luminance signal, but the Q-signal, giving yellow-green-purple reproduction, has only about a tenth (see Sections 22.10 and 22.11). Colour transparencies by courtesy of the British Broadcasting Corporation. Reproduced on a Crosfield *Magnascan* scanner.

Plate 20

The variations in the spectral power distribution of illuminants result in pictures that differ in colour balance. Because pictures usually fill only rather a small part of the total visual field, the eye generally adapts its own colour balance much less to the overall colours of pictures than to those of real scenes (see Fig. 5.4). The series of pictures on these pages show typical variations caused by differences in the colour temperature (see Section 10.11) of the taking illuminant. *Above:* excessive blue shadows caused by blue sky.

Top: night scene with tungsten filament lighting. *Lower left:* excessive blue caused by the blueness of diffuse daylight very late in the evening, together with the orange colour of tungsten-filament flood-lights. *Lower centre:* orange cast caused by low-altitude sunlight. *Lower right:* extreme orange colour of candlelight (see Table 10.4). Reproduced from transparencies exposed on *Ektachrome* Daylight Type film using a Crosfield *Magnascan* scanner. The use of *Type A* or *Type B* films (see Section 5.7), or suitable colour correcting filters over the camera lenses, could have corrected for some of the variations in colour balance: however, such correction is only of limited value in those cases where casts of different colours are present in the same picture; and, because the extent to which the eye adapts to real scenes is limited, partial correction for a colour cast may be more realistic than full correction. In colour television, similar adjustments of colour balance can be made by adjusting the relative gains of the different channels in cameras.

Plate 21

A colour television receiver displaying various P.A.L. signals (see Sections 22.6 and 22.8): *top left*, the luminance signal only; *top right*, the $E_R' - E_Y'$ and $E_B' - E_Y'$ chrominance signals only; *upper left*, the $E_R' - E_Y'$ signal only; *upper right*, the $E_B' - E_Y'$ signal only; *lower left*, the luminance and $E_R' - E_Y'$ signals only; *lower right*, the luminance and $E_B' - E_Y'$ signals only; *bottom*, the complete luminance and chrominance signals together. (When the luminance signal was absent, a uniform luminance was displayed so that the chrominance signals could be seen). The chrominance signals have about a quarter the bandwidth of the luminance signal, but the consequent loss of colour resolution is only noticeable when the luminance signal is absent (see Section 19.7). Colour transparencies by courtesy of the Independent Broadcasting Authority. Reproduced on a Crosfield *Magnascan* scanner.

15.9 Calculation of mask gammas

The calculation of mask gammas is conveniently made by means of matrix algebra[1] (Miller, 1941). If the densities of the unwanted and the wanted absorptions of the reproduction dyes in a neutral density of 1.0 are as follows

	Cyan	Magenta	Yellow
Density to Red light	c	m_R	y_R
Density to Green light	c_G	m	y_G
Density to Blue light	c_B	m_B	y

then, if the overall gamma of the reproduction is to be the same as that of the original, the gammas of the photographic materials for the required masks and separation negatives (when made from the original) are given by:

$$\begin{pmatrix} \gamma_R & \gamma_{GR} & \gamma_{BR} \\ \gamma_{RG} & \gamma_G & \gamma_{BG} \\ \gamma_{RB} & \gamma_{GB} & \gamma_B \end{pmatrix} = \begin{pmatrix} c & m_R & y_R \\ c_G & m & y_G \\ c_B & m_B & y \end{pmatrix}^{-1}$$

where γ_R, γ_G, and γ_B are the gammas for the three separation negatives and γ_{GR} and γ_{BR} are the gammas for masks to be made by green and blue light exposures respectively and used with the red separation negative when producing the cyan image. If the masks are used, not with the separation negative, but with the original, then the gammas have to be divided by that of the red separation negative, γ_R, if this is different from unity and they have to be negatives instead of positives or *vice-versa*. Similarly γ_{RG} and γ_{BG} are the gammas for the masks to be used with the green separation negative when the magenta image is printed, and γ_{RB} and γ_{GB} those for the masks to be used with the blue separation negatives when the yellow image is printed.

Very often several of the six masks called for are of such low contrast that they can be omitted without loss of quality. For instance, the following figures are fairly typical of dyes used in colour photography.

	Cyan	Magenta	Yellow
Density to Red light	0.94	0.05	0.01
Density to Green light	0.10	0.82	0.08
Density to Blue light	0.10	0.25	0.65

The gammas for the required separation negatives and masks are then given by:

$$\begin{pmatrix} \gamma_R & \gamma_{GR} & \gamma_{BR} \\ \gamma_{RG} & \gamma_G & \gamma_{BG} \\ \gamma_{RB} & \gamma_{GB} & \gamma_B \end{pmatrix} = \begin{pmatrix} 0.94 & 0.05 & 0.01 \\ 0.10 & 0.82 & 0.08 \\ 0.10 & 0.25 & 0.65 \end{pmatrix}^{-1} = \begin{pmatrix} 1.07 & -0.06 & -0.01 \\ -0.12 & 1.27 & -0.15 \\ -0.12 & -0.48 & 1.60 \end{pmatrix}$$

Where the signs are positive the masks are negatives, where the signs are negative the masks are positives. The two masks to be used with the red separation negative are of such low gamma (0.06 and 0.01) that they can be neglected; this is because the unwanted red absorptions of the magenta and

---

[1] See Appendix 1.

yellow dyes are small. The most contrasty mask is that made with green light and to be used with the blue separation negative (0.48); this is a consequence of the heavy unwanted blue absorption of the magenta dye.

When the original is a colour transparency, the masks may be used to correct for the unwanted absorptions of the transparency dyes rather than for those of the reproduction dyes; in this case the densities used in the calculation must be those of the transparency dyes. Furthermore, in this type of calculation it is not difficult to correct for the unwanted absorptions of both sets of dyes and hence, at least in principle, to improve on the original transparency.

Some practical systems of masking are discussed in Chapter 28. (See Plate 32, pages 512, 513, for an example of masking.)

REFERENCES

Barr, C. R., Thirtle, J. R., and Vittum, P. W., *Phot. Sci. Eng.*, **13**, 74 and 214 (1969).
Brewer, W. L., and Hanson, W. T., *J. Opt. Soc. Amer.*, **44**, 129 (1954).
Brewer, W. L., and Hanson, W. T., *J. Opt. Soc. Amer.*, **45**, 476 (1955).
Brewer, W. L., Hanson, W. T., and Horton, C. A., *J. Opt. Soc. Amer.*, **39**, 924 (1949).
Clapper, F. R., Breneman, E. J., and Brownstein, S. A., *J. Phot. Sci.*, **25**, 64 (1977).
Evans, R. M., *Penrose Annual*, **45**, 81 (1951)
Evans, R. M., *P.S.A. Journal*, p. 15 (February, 1954).
Gutteridge, C., *Phot. Sci. Eng.*, **16**, 214 (1972).
Hanson, W. T., *J. Opt. Soc. Amer.*, **40**, 166 (1950).
Hanson, W. T., and Horton, C. A., *J. Opt. Soc. Amer.*, **42**, 663 (1952).
MacAdam, D. L., *J. Opt. Soc. Amer.*, **28**, 466 (1938).
Marriage, A., *Phot. J.*, **80**, 364 (1940).
Marriage, A., *Phot. J.*, **88B**, 75 (1948).
Meissner, H. D., *Phot. Sci. Eng.*, **13**, 141 (1969).
Miller, C. W., *J. Opt. Soc. Amer.*, **31**, 477 (1941).
Neblette, C. B., *Photography, Its Materials and Processes*, 6th Edn., p. 463, Von Nostrand, New York (1962).
Sant, A. J., *Phot. Sci. Eng.*, **5**, 181 (1961).
Tull, A. G., *Brit. J. Photog.*, **122**, 955 (1975).
Watson, R. N., *J. Phot. Sci.*, **14**, 304 (1966).
Yule, J. A. C., *J. Opt. Soc. Amer.*, **28**, 419 and 481 (1938).
Yule, J. A. C., *Phot. J.*, **80**, 408 (1940).
Yule, J. A. C., *Phot. J.*, **84**, 321 (1944).

CHAPTER 16

Printing Colour Negatives

16.1 Introduction

BECAUSE of their various advantages (see Section 13.4, for instance), colour negatives are widely used as intermediates for the production of colour photographs. There are, however, many factors that can affect the colour balance and density of a print made from a colour negative; it is therefore usually necessary to adjust the exposure of the print film or paper in both colour and intensity for each negative during the printing operation, in order to obtain correct positive images. But it is extremely difficult to determine, by inspecting it visually, the type and magnitude of the printing adjustments a colour negative requires. Various aids to correct printing are therefore used, their nature depending on the particular application.

16.2 Printing studio negatives

In certain studio work, the lighting can be carefully controlled in both colour and intensity; the transmission colour of the camera lenses can be matched (if necessary by using filters); a single batch of film can usually be used for a considerable period of time; only one film processing location is generally involved; and hence, once the correct printing conditions have been established for one typical negative, if subsequent negatives are all printed with light of the same intensity and colour, the prints will usually be somewhere near optimum in density and colour balance. From such prints, quite reliable visual estimates can usually be made of any changes necessary in the printing conditions to obtain prints of correct density and colour balance at the second attempt; and if fairly high prices can be charged for the final print, the loss in discarding a fair proportion of the first prints (and even some of the second prints) is no great problem. For a variety of reasons, however, it is not always possible to control all the factors as closely as required, and some guidance may be necessary in making the *first* print if it is to be reasonably near optimum.

One method of producing first prints of reasonably good quality is to include a grey card in the scene, and to measure the red, green, and blue printing densities of its image in the processed negative; the correct printing conditions can then be calculated from these densities. If the negative is very large, a grey-card image of measurable size can usually be accommodated outside the area required for the final picture. But with small negatives the presence of the grey card would spoil the picture, and this method can then only be used when (as is often the case in professional work) several negatives are being exposed from similar scenes, as in a series of still pictures of similar subjects; the grey card can then appear only in an extra test-negative of the series. If a grey card was not included in the scene, an area of known colour, such as a skin colour, can be used for measurement instead (see Section 16.11), the printing conditions then being used to give the correct or desired result for the particular chosen colour.

Some use is made of closed-circuit colour-television viewing-devices, in which a positive colour image is derived electronically from the negative and then judged visually to determine the printing adjustments; although such devices are costly, their value can often be justified in professional printing where the total value of the finished prints can be very considerable. (Thomas, Waz, and Dreyfoos, 1970; West, 1971; *British Journal of Photography*, 1972).

The actual adjustment of the colour of the exposure is usually made by inserting pale cyan, magenta, and yellow filters into the beam of an ordinary white-light enlarger, while density is controlled by altering either the exposure time or the lens aperture, or both.

16.3 Printing motion-picture negatives

The problems of printing professional motion-picture negatives are generally similar to those outlined above for studio work, and careful control of lighting, lenses, processing, and film batches is usually exercised to good advantage. Densitometry of images of grey cards or other standard colours, and the use of television grading devices are also very useful. Additional difficulties, however, may be caused by varying lighting conditions on 'location' shots, by artistic requirements calling for prints of unusual density and colour balance in order to give a particular mood, and by the need for many different scenes on the negative film to be printed in quick succession, thus requiring a rapid succession of printing adjustments.

The actual printing operation itself can be carried out either by contact, or by projection (usually referred to in the trade as *optical printing*). Projection must be used, of course, if any change in the size or the shape of the print has to be introduced; but contact printing is simpler, is less affected by dirt or scratches on the negative, and usually gives sharper results. Printers may advance the film continuously or a frame at a time (*step-printing*). The adjustment of density and colour balance, usually referred to as *timing*, may be carried out by inserting pale cyan, magenta, yellow, and neutral filters into the printing beam of a white-light printer; or *additive printers* may be used. In additive printers, separate beams of red, green, and blue light are combined uniformly in the printing gate, the adjustments of colour balance and density then being achieved by altering the relative and absolute intensities of the red, green, and blue beams: although this involves more complicated equipment it has the advantage that the exposures given to the three layers of the print film can be controlled independently, whereas the cyan, magenta, and yellow filters have unwanted absorptions so that each filter affects more than one layer. Additive printers of high efficiency employ dichroic mirrors to separate red, green, and blue beams from a single source, and also to re-combine them.

Because changes in printing conditions may have to be made in rapid succession, it is usual in the professional motion-picture trade for the printing equipment to respond automatically to instructions punched into a paper tape (or other means of storing information) which is fed through the machine in synchronization with the negative film being printed.

16.4 Printing amateurs' negatives

When the colour negative is used for amateur snapshots a wider variety of lighting is encountered and other factors also vary more, so that printing adjustments are even more necessary; but a reasonably low price for the prints is needed for their mass production. It is therefore essential to have a method of printing that is quick, and does not require very highly skilled operators, yet gives a high yield of saleable results at the first printing. These requirements have been the subject of considerable technological efforts, and some remarkably successful solutions have been devised, as will now be described.

16.5 The variables to be corrected

The following factors can affect the colour balance and density of a print made from a colour negative:

(1) intensity of scene illuminant
(2) colour of scene illuminant
(3) scene subject matter
(4) camera lens transmission colour
(5) lens aperture, exposure time, and film speed
(6) film colour balance

(7) film latent-image keeping properties
(8) film processing
(9) printer settings
(10) paper speed and colour balance
(11) paper latent-image keeping properties
(12) paper processing
(13) colour and intensity of print illuminant.

The cumulative effects of factors 1 to 8 on the colour balance and density of a typical sample of amateurs' negatives are shown in Fig. 16.1. It is clear that there is scope here for very wide fluctuations in the prints unless steps are taken to correct for the variations.

16.6 Early printers

In one type of printer (Eastman Kodak No. 1598), which was used in the early 1940s, correction was made for the camera exposure level, for the colour balance of the negative material, and for its processing (items, 1, 5, 6, and 8). Before each film was processed, a small patch of carefully controlled uniform light was printed on to a spare unexposed area at one end of the film. After processing, the red, green, and blue printing densities of this patch were measured, and each negative of the film was then punched with small holes along one edge, the size and positions of the holes indicating the colours and values of correcting filters necessary to adjust the colour of that negative material to a standard. When the negative was printed, the indicated filters were inserted into the printing beam, and then a photocell was used to adjust the position of a neutral density wedge in the beam so that at a fixed printing time the correct density was obtained on the print for negatives of average scene content. The photocell could not of course distinguish between an under-exposed light scene (such as a snow field), and an over-exposed dark scene (such as a coal heap), and would tend to print both as grey. The operator could therefore adjust the exposure above or below that called for by the photocell in order to obtain prints of the correct density, an operation known as *classification*.

A useful measure of success was obtained with this method, but the colour balance of the prints was still rather variable, and it was clear that further control of the variables was desirable.

16.7 Integrating to grey

Modern printers designed for handling amateurs' negatives almost all depend on a principle first described by Evans in the following words: 'A more pleasing effect is often produced in colour prints if they are so made that instead of the colour balance being correct, in which grey is printed as grey, it is so adjusted that the whole picture integrates to grey' (Evans, 1946). At first sight this may sound

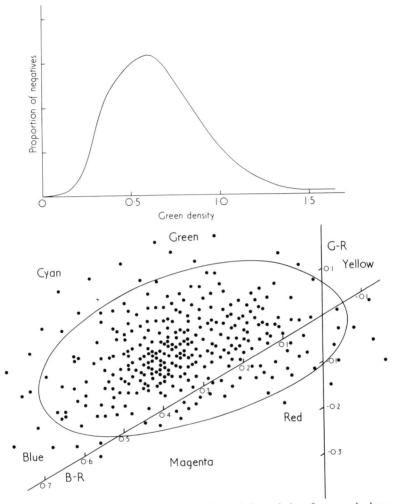

Fig. 16.1. The range of density and colour balance of a typical population of amateurs' colour negatives. The densities plotted are those of whole negatives measured through red, green, and blue filters against an unexposed area of film as zero.

Above: The distribution of green density.

Below: The distribution of colour balance (see Section 14.25).

The colour names indicate the direction in which the lighting or the subject matter would have to be altered in colour (from the average) for the negative to plot in that vicinity. If no corrections were made at the printing stage the prints would have a similar colour bias; the negatives themselves have biases of complementary colours.

The elliptical contour encloses approximately 90 per cent of the points.

an absurd approach to the problem, because if, for instance, it were applied to the case of a portrait of a girl wearing a red dress it would result in the whole picture having a blue-green cast. Evans, however, argued that the presence of the red dress would tend in any case to depress the sensitivity of the eye to red and hence produce a physiological blue-green bias, so that to make the print slightly blue-green would not necessarily be a disadvantage. Whether these physiological effects are large enough in reflection prints to be very important is open to some doubt, although Bartleson has shown that an observer's adaptation certainly is affected to some extent by the colour balance of reflection prints (Bartleson, 1958) and their importance in projected transparencies has been demonstrated by Evans (Evans, 1943) and by Pinney and DeMarsh (Pinney and DeMarsh, 1963). What is beyond dispute, however, is that, as applied in practice, the 'integrating substantially to grey' principle has proved a major factor in the successful operation of amateur colour snapshot systems.

The easiest way of making prints that 'integrate substantially to grey' is to measure the average transmission of each negative through red, green, and blue filters and then expose the red-sensitive layer of the paper to an extent that is inversely proportional to the red transmission of the negative, and the green and blue layers of the paper for extents similarly related to the green and blue transmissions of the negative respectively. In fact, when this is done, it is not true that prints that exactly integrate to grey are obtained. This is because the measured transmissions of the negatives will depend mainly on the dark parts of the scene. Thus the prints will tend to integrate to grey in the shadows more than in the highlights (this effect is reduced, but not eliminated, by the fact that the negatives are of low contrast, the gamma usually being about 0.65).

By taking each negative as the arbiter of the exposure given to the paper, correction is made not only for the exposure level, for the colour balance of the negative material, and for its processing (items 1, 5, 6, and 8), but also for the colour of the scene illuminant, scene subject matter, camera lens transmission colour, and latent-image keeping properties of films (items 2, 3, 4, and 7). This results in a marked improvement in the consistency of the colour balance of the prints made from the majority of amateurs' negatives, but it does of course introduce errors when the subject matter contains large areas of saturated colours, particularly if they are dark (and hence light on the negative). The proportion of amateurs' negatives that suffer from *colour failure* in this way is, however, surprisingly small, and can be looked after by reprinting them with deliberate shifts from the 'integrating substantially to grey' condition: a series of classification adjustments for colour (as well as those used for density) are generally provided for this purpose. The success of the 'integrating substantially to grey' method probably means that the shadow areas of most actual scenes photographed are in fact approximately grey themselves if integrated. Some measurements on outdoor scenes in England showed a marked tendency for whole scenes to integrate substantially to grey (Pitt and Selwyn, 1938).

16.8 The 1599 printer

The first printer to employ the 'integrating to grey' method of printing amateurs' colour negatives was the Eastman Kodak 1599 Printer, which is shown diagrammatically in Fig. 16.2. The negative is illuminated by a diffuser and an array of small lamps whose luminance can be varied; in early models this was done by means of a Variac transformer (as shown in the figure) but later models use a saturable reactor. The diffuser has the dual function of making scratches and other negative defects far less noticeable than would be the case if the beam were specular, and it also spreads some of the light sideways, on to the monitoring photocells. The paper is exposed successively to red, green, and blue light by means of filters situated adjacent to the lens. A bank of photo-voltaic photocells, each one covered by a red, a green, or a blue filter, receive light from the negative and produce three photocurrents which are compared one at a time with that from another photo-voltaic photocell illuminated by a comparison lamp. The difference between the two photocurrents is amplified and used to adjust

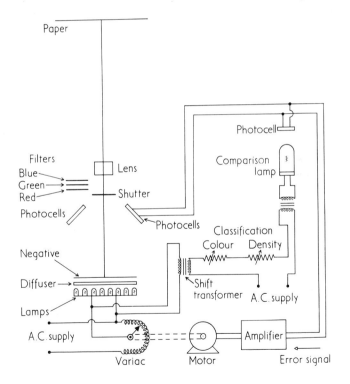

Fig. 16.2. The main features of the Eastman Kodak Printer type 1599.

automatically the voltage of the array of small lamps until the two photocurrents are equal; an exposure for a fixed time is then given for each colour in turn.

The sequence of operations is thus: insert the negative and activate the printer; the red filter moves into place but is covered with a dark shutter that remains in place for 0.3 seconds while the intensity is adjusted correctly for the red exposure; the dark shutter is removed from the beam, the red exposure of about $\frac{1}{2}$ second is given, and the exposure is terminated with the dark shutter; with the dark shutter in place the green filter replaces the red, the lamp voltage is readjusted, and an exposure of about $\frac{1}{2}$ second is given; in the same manner the blue filter replaces the green, the lamp voltage is again readjusted, and an exposure of about $\frac{1}{2}$ second is given. The lamp voltage adjustments are all made during the dark shutter time. Any classification adjustments necessary for colour and density are provided by potentiometers (labelled *colour classification* and *density classification* in the figure) which adjust the voltage of the comparison lamp; this voltage can also be made to depend partly on the voltage of the printing lamps by means of the 'shift transformer': in this way the final intensity of the printing light for each colour can be made to depend partly on the density of the negative, and when suitably adjusted this can increase the yield of good prints appreciably, a technique known as *slope control* (Pieronek, Syverud, and Voglesong, 1956; Hunt, 1960), which will be more fully described later (Section 16.13).

The 1599 printer enabled a very high proportion of amateurs' negatives to be successfully printed at the first attempt and its introduction in 1946 was a major step forward in amateur colour photography. Although the total exposure cycle of the first 1599 printers was about $4\frac{1}{2}$ seconds, a very high rate of printing was achieved by virtue of the fact that each printer comprised three separate channels with three negative gates producing three rows of pictures side by side on a single web of paper. A good operator could in fact reach a rate of printing of around 1000 prints per hour. Later versions of the 1599 printer operated with total exposure cycles of about $2\frac{1}{4}$ seconds.

16.9 Variable time printers

The constant-time variable-intensity principle of exposure used in the 1599 printer means that all negatives are printed at the exposure time that is necessary for the densest negatives likely to be encountered. If, on the other hand, a constant intensity of negative illumination is used and the exposure *time* is varied, then although the densest negative will still require the same exposure time, all other negatives will require shorter exposure times, and hence the average printing time can be substantially reduced. The two principles of 'integrating to grey' and 'variable time' were combined in the Eastman Kodak IVC printer (Pieronek, Syverud, and Voglesong, 1956), but three successive exposures through red, green, and blue filters were still necessary, the times of which were controlled by a photocell charging a capacitor to a predetermined voltage.

16.10 Subtractive printers

For maximum efficiency in a printer it is clearly desirable to avoid wasting either time or light. Time can be saved, not only by adopting the variable-time principle, but also by exposing the three layers of the paper simultaneously instead of sequentially. Unfortunately, however, the use of red, green, and blue filters and of beam combiners tends to waste light.

Greater efficiency is achieved by adopting the subtractive principle. In this case the negative is first printed with white light; then, after the layer that required the least exposure is fully exposed, a filter is inserted to prevent any further exposure of that layer taking place; the exposure of the other two layers then proceeds until one of them is fully exposed, whereupon another filter is inserted to prevent any further exposure of that layer; when the third layer is fully exposed a third filter is inserted to prevent any further exposure of the third layer and the exposing cycle is complete. The three filters used are cyan to terminate the exposure of the red layer, magenta for the green layer, and yellow for the blue layer. The order in which they are inserted is dependent on the requirements of the particular negative, and light is lost only as a result of the unwanted absorptions of the cyan, magenta, and yellow filters during the part of the printing cycle for which they are in the beam. The actual insertion of the filters is carried out automatically as the result of solenoids being actuated by the integration of photocurrents in capacitors.

The Kodak S1 subtractive printer, installed at the Kodak processing station at Hemel Hempstead in 1957 and made available for sale in 1958 (Richardson, 1958; Hunt, 1960) is shown diagrammatically in Fig. 16.3. It is seen that the

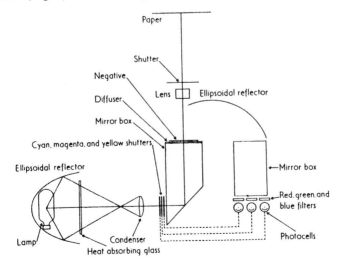

Fig. 16.3. The main features of the Kodak Colour Printer type S1.

cyan, magenta, and yellow shutter filters are placed below the negative instead of above the lens. The reason for this is that, because these filters have unwanted absorptions, their insertion not only terminates the exposure of one layer, but also somewhat reduces the level of exposure of the other two layers. By placing them below the negative, the level of illumination on the photocells is reduced in the same proportion, and as a result the exposure times are appropriately lengthened as required. It is, of course, necessary to place the shutter filters in such a position in the optical system that no shading across the negative occurs as the filters are inserted, for this would result in uneven colour balance across the prints. In the S1 printer a light-integrator (in the form of a box of mirrors with a $45°$ mirror at the bottom) is placed between the filters and the negative thus ensuring complete absence of colour shading.

In the interests of high efficiency the lamp is placed at one focus of a semi-ellipsoidal reflector, a lens collects the light from the other focus and passes it through the filters and into the light-integrator. A diffuser at the top of this integrator ensures that the effects of negative defects are reduced, and also provides light off the optical axis for the photocells. The transfer of the light from the negative to the photocells employs another ellipsoidal reflector, the negative being at one focus and the top of another light-integrating mirror-box at the other. Three photocells are used, one is covered with a red, one with a green, and one with a blue filter, and they are placed immediately beneath the light-integrator; they thus receive substantially scrambled light from the negative. The three photocells charge three separate capacitors. Classification adjustments are provided by altering the predetermined voltages to which the three capacitors have to be charged, thus altering the exposure times for the three layers.

Another printer using the subtractive principle was introduced in the U.S.A. by the Pako Company in 1958 (Blaxland, 1960). In this printer, the cyan, magenta, and yellow filters are placed between the lens and the paper and are controlled by a single photomultiplier tube in front of which rotates a red, green, and blue filter wheel so as to sample the colour of the light reflected from a beam splitter placed between the filters and the paper.

The Eastman Kodak 5S printer, introduced in 1959, utilizes the same basic principles as the S1 printer, but has the cyan, magenta, and yellow filters above the printing lens and the photocells just beneath this lens. The photocells are therefore not affected by the insertion of the cyan, magenta, and yellow filters, and this feature is used to make the printer operate at a lower level of colour correction (Bartleson and Huboi, 1956). In this way the amount of colour failure is reduced, and, although this is an advantage, it is obtained at the expense of some reduction in correction for the other factors, such as lighting, and the colour balance of the negative material and its processing. Because of the lower level of colour correction on this printer (and on certain others), it is necessary to insert filtration into the printing beam so that the cyan, magenta, and yellow shutter filters come in at the same time (sometimes referred to as *dead heat*) for an average negative.

Subsequent developments in printers of this general type have included the Eastman Kodak 2620 printer, in which rates of printing as high as 3000 to 4000 prints per hour can be achieved: these higher printing rates have been made possible by the use of compact tungsten-halogen lamps (see Section 10.2) with very efficient ellipsoidal mirrors, and by handling the negatives in a continuous spliced roll; in the Gretag printer, compensation for the effects of the unwanted absorptions of the shutter filters is provided for in the electronic exposure control circuits and this is also done in the Eastman Kodak 2610 and 3510 printers. Some modern printers can produce prints at rates of 10 000 or more per hour.

16.11 Colour enlargers

The printers referred to above for use with amateurs' negatives mostly produce *en-prints* having widths in the region of $3\frac{1}{2}$ to 4 inches (about 9 to 10 cm); some printers, however, can print sizes up to 5×7 inches ($12\frac{1}{2}\times17\frac{1}{2}$ cm), or even 8×12 inches (20×30 cm). The term 'enlarger', as distinct from 'printer', therefore normally denotes nowadays a piece of equipment that differs not so much in the size of the final print produced as in being less complicated in construction and having greater flexibility in use. Enlargers are widely used in which the colour of the light is altered by simply inserting a pack of uniform filters in the beam, preferably before it passes through the negative so as not to impair the definition of the image. Judgment of the filtration and exposure time necessary to achieve correct colour balance and density in the picture is then made either by trial and error using *test strips* (a technique suitable for home processing), or by measurement of an area on the negative of known subject matter (such as a medium grey or a skin colour); such measurements can be made either on a densitometer, or by means of a special photoelectric device for determining exposure (sometimes known as a *spot monitor*) which can be used at the negative plane or at the paper plane in the enlarger.

16.12 Automatic classification

The adoption of the 'integrating to grey' principle was a major development in the amateur colour snapshot business, and has been incorporated in various types of equipment, including the fairly simple, high output, subtractive printer. This has resulted in colour snapshots becoming widely produced both in manufacturers' processing laboratories and also in independent photo-finishing establishments. High levels of good quality prints are obtained at the first printing, but it is still necessary to use individual classification adjustments to allow for variations in density distribution on the negatives. This is a skilled operation and it would be a considerable advantage to be able to make this step automatic. The need for this is accentuated by the fact that, in many parts of the world, the colour snapshot business has large seasonal fluctuations. It seems unlikely that automatic density classification could ever be correct for all negatives, but even skilled operators make some errors and it would be a useful advantage if a machine could be made that performed no worse than they do. Some printers are now available in which density classification is automatic; in these machines the negative transmittances may be measured in several different areas. Thus, in the Kodak 2620 *(Autoclass)* printer, an upper, a central, and a lower, area are measured; in the Kodak 2610 printer this is elaborated by scanning the central area to measure at 100 different points; in the Kodak 3510 printer this is further elaborated by making all the measurements in all three colours (red, green, and blue); density classification is then based on an appropriate combination of these measurements. Automatic colour classification is then also feasible.

16.13 Factors affecting slope control

Slope control enables adjustments to be made to the colour balance and density of prints made from dense (over-exposed) negatives relative to that of prints made from light (under-exposed) negatives. There are several reasons why such adjustments are necessary.

First, the shapes of the characteristic curves of the negative material play a part. The reason for this is as follows. The photocells will be mainly affected by the darker parts of the scenes, because these parts are the lighter parts of the negatives. In most scenes, however, it is the colour rendering of the medium and light tones that is of the main importance in determining the apparent colour balance of the prints. Hence, unless there is a constant relationship between the colour balance and density of the light and dark parts of the negatives, consistent print quality cannot be obtained. Therefore any departure of the characteristic curves of the negative material from the straight-line condition will affect the results in a way that is dependent on negative density. Some examples will help to demonstrate these effects.

In Fig. 16.4 are shown sets of curves typical of a colour negative material with incorporated masks. The parts of the curves used for a typical scene for various levels of exposure are shown in Fig. 16.4(a) for bluish illuminants and in Fig. 16.4(b) for yellowish illuminants. The separation along the log exposure axis of the points marked 'N' on the green curve is considered to be that used by a normally-exposed scene, and is equal to 1.6 log units, a figure arrived at by taking as an average log luminance range for outdoor scenes a value of 2.2, from which has been subtracted a camera flare factor of 0.6 (James and Higgins, 1960). As the exposure is increased, this range of 1.6 log units will move up the green curve, and the points marked +1, +2, and +3 show the positions of the lightest parts of the scene for exposures of one, two, and three stops more than normal (a stop being a change in exposure level by a factor of 2 or $\frac{1}{2}$). Similarly, the points marked −1, −2, and −3 show the positions of the darkest parts of the scene for exposures of one, two, and three stops less than normal. The sets of points on the red and blue curves show similar data for the red- and blue-sensitive layers of the film; it will be seen that in Fig. 16.4(a) the points have moved one stop up the blue curve and down the red curve as a consequence of the greater blue and lower red content of the bluish illuminant, whereas in Fig. 16.4(b) the points have moved one stop down the blue curve and up the red curve because of the lower blue and greater red content of the yellowish illuminant.

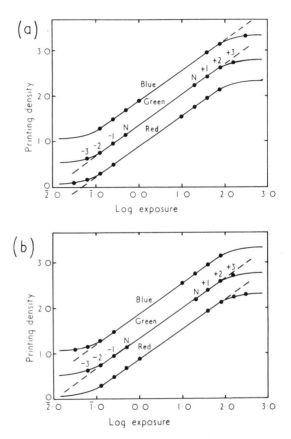

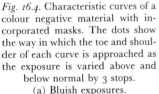

Fig. 16.4. Characteristic curves of a colour negative material with incorporated masks. The dots show the way in which the toe and shoulder of each curve is approached as the exposure is varied above and below normal by 3 stops.
(a) Bluish exposures.
(b) Yellowish exposures.

It is clear from Fig. 16.4(a) that with the bluish illuminant the shadows in under-exposed negatives have higher red densities and slightly higher green densities than if the straight line parts of the curves extended indefinitely (broken lines); this causes the printer to give more red, and slightly more green, exposure, with the result that the lighter areas in the prints are too cyan-blue. On the other hand, with the yellowish illuminant, Fig. 16.4(b), the shadows in under-exposed negatives have higher blue, and slightly higher green, densities than if the curves were all straight; this causes the printer to give more blue and slightly more green exposure, so that the lighter areas in the prints are too yellow-red. At the over-exposure end the high densities will not have much influence on the exposures given by the printer, but, with the bluish illuminant, Fig. 16.4(a), the light parts of the scene have too little blue, and slightly too little green, density so that they will appear too yellow-red in the prints; whereas with the yellowish illuminant, Fig. 16.4(b), the light parts of the scene have too little red, and slightly too little green, density, so that they will appear too cyan-blue on the prints. Thus negative curve-shape can affect print colour balance.

For negatives that are not under- or over-exposed any kinks or bends in a curve (that are not exactly paralleled at the corresponding exposure level in the other two curves), may similarly have important effects on the prints. Since the quality of prints made from severely over- and under-exposed negatives can never be good in any case, slope control is generally aimed mainly at improving those negatives whose important densities lie away from the extreme toe and shoulder portions of the film characteristic: the range covered is usually about three stops (eight times) above and below the normal exposure.

If the negative curves are straight but not parallel, then, as negative density varies, consistency of density and colour balance on the prints will be obtained if the printing system is such as to correct fully for variations in the red, green, and blue transmissions of the negatives; if this full correction is not made then non-parallelism of the negative curves will produce variations in the prints that will be functions of negative density level.

The second factor affecting slope characteristics is the relationship between the spectral sensitivities of the filtered photocells and those of the paper. The spectral density curves of the dyes used in colour negative materials are not 'flat-topped', a typical set being shown in Fig. 16.5. If these dyes each absorbed uniformly throughout three separate parts of the spectrum, in each of which only one layer of the paper had any sensitivity, then, provided each filtered photocell had sensitivity only in one of the three parts, it would not matter how its sensitivity varied within that part. But the dyes have quite marked peaks and hence, if the sensitivity curve of one of the layers of the paper had a peak at a wavelength that coincided with one of the negative dye peaks, but the cor-responding filtered photocell had its maximum sensitivity at a wavelength off the negative dye peak, then as that dye varied in concentration in the negative its effect on the paper could be greater than on the filtered photocell; if this were the case, as the density of negatives increased, the monitoring system would increase the exposures by too little and the prints would be deficient in the corresponding dye. The best way of avoiding this situation is so to filter the photocells that the resulting sensitivity is identical with that of the paper. In practice this is very difficult to achieve because photographic sensitivity generally falls very steeply on the long wavelength side of the sensitivity band, whereas dye absorption curves with steep slopes can only be achieved on the short wavelength side of a transmission band. The best that can be achieved is often similar to that shown in Fig. 16.6.

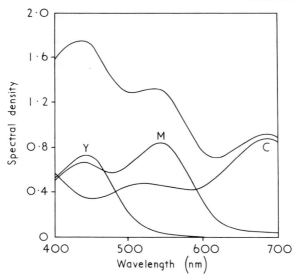

Fig. 16.5. Spectral density curves of the dye images, C, M, and Y, used in a negative material with incorporated masks. The curves shown represent the combined densities of the dyes and the appropriate proportions of unused coloured couplers characteristic of a middle density area. The top curve is equal to the sum of the other three and represents the result when the material is exposed to an approximately neutral subject.

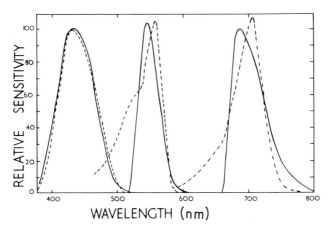

Fig. 16.6. Spectral sensitivity curves of colour printing paper (broken lines) and of photocells filtered with red, green, and blue filters (full lines), typical of those often used in practice.

305

The third factor affecting slope control (in variable-time printers only) is the *reciprocity characteristic* of the paper. In black-and-white materials it is well known that if an exposure time is increased by a factor of ten, for example, to compensate for a reduction in intensity to one-tenth, then the result is not exactly the same. In colour materials slight differences in the reciprocity characteristics of the layers of a paper can result in important variations of colour balance as the exposure time is varied.

16.14 Methods of slope control

In Section 16.8 it was pointed out that the shift transformers of the 1599 printers could make the red, green, and blue exposures partially dependent on negative density; thus, by separate adjustment of the shift transformer in each of the three colour channels, prints of roughly constant density and colour balance could be obtained from negatives of widely different densities.

Another method of slope control, which is widely used on variable time printers, is to make the sensitivity of the electronic integrating circuits vary with time during the exposure cycle (Pieronek, Syverud, and Voglesong, 1956). One way of accomplishing this is as follows. The exposure given by each channel of a variable time printer is not only dependent on the current from the photocell but is also proportional to the voltage through which its integrating capacitor has to be charged before the exposure terminating mechanism is set in motion. If this voltage is made to vary with time, then, since dense negatives require longer exposures than thin negatives, it is possible to alter the exposures in a way that is dependent on negative density, and hence provide slope control.

16.15 Electronic printing

If the information in a picture is made available in the form of electronic signals, those signals can be manipulated with a flexibility that is not normally possible by purely photographic means. Methods of printing have, therefore, been devised in which films are scanned with very small spots of red, green, and blue light, and the light intensities converted into electronic signals by suitable light-sensitive detectors; the electronic signals are appropriately manipulated, and then used to reconstitute light images, which are printed optically on to photographic materials. The manipulations of the signals can be used to improve picture quality in respect of tone and colour reproduction, sharpness, and graininess.

The initial scanning of the film, and the exposure of the final material, can be by means of laser beams deflected by galvanometer mirrors and rotating mirror-polygons, or other suitable devices, as in the *LaserColor* printer (Sealfon, 1979). Alternatively, cathode-ray tubes can be used instead of lasers; or charge-coupled device (CCD) arrays for the initial scanning, and light emitting diode (LED) arrays for the exposing step, can be used. With such systems, fully comprehensive tone-scale adjustment and colour masking can be achieved, so as to produce pictures of very high quality.

It is also possible, by manipulating spatial groups of signals, to reduce graininess and increase sharpness. In one method, the signals are passed through two-dimensional band-pass filters that estimate the average density-gradients in the neighbourhood of a picture element over areas of different sizes and orientations. Graininess is reduced by thresholding, and sharpness is increased by amplifying those signals that carry high-frequency information (Powell and Beyer, 1982; Hunt, 1985). (See Plate 24, page 353.)

Printers that can carry out all the above types of operation are necessarily rather complicated, particularly if high rates of printing are required, but the methods are very powerful, and can be thought of as combining elements of the technologies of both television (see Part Three) and graphic-arts scanners (see Chapter 29).

Image manipulation of the above type can also be carried out when the electronic signals are obtained not from film, but directly from electronic cameras (see Section 13.6).

The facility with which the signals in such electronic equipment can be manipulated makes them useful devices for simulating changes in reproduction systems. Thus, manufacturers of photographic materials can use electronic simulators for investigating the effects on picture quality of changing such parameters as curve shape and inter-image effects (including partial exposure of a layer by light of the wrong colour, sometimes referred to as *punch-through*). The effects of changing the spectral sensitivities of a camera film require that such a film be actually made, but unintended changes in such experimental films (such as changes in curve shape) can be corrected in the simulator. The effects of changing the colours of the dyes of the material used for the final image can be simulated, but only within the gamut of the material actually used at that stage in the simulator. In one such simulator, a flying-spot scanner (see Section 23.9) is used to scan the camera film, and a monochrome cathode-ray tube is used to expose the display material (which can be colour film or paper) successively through red, green, and blue filters; a 798-line raster is used with 1312 picture elements along each line (Giorgianni, 1984).

REFERENCES

Bartleson, C. J., *Phot. Sci. Eng.*, **2,** 32 (1958).
Bartleson, C. J., and Huboi, R. W., *J. Soc. Mot. Pic. Tel. Eng.*, **65,** 205 (1956).
Blaxland, J., *Brit. J. Phot.*, **107,** 638 (1960).
British Journal of Photography, **119,** 166 (1972).
Evans, R. M., *J. Opt. Soc. Amer.*, **33,** 579 (1943).
Evans, R. M., British Patent 660,099; U.S. Patent 2,571,697 (1946).
Giorgianni, E., Video simulation and photography, paper presented at ISCC – GATF Conference on Colour and Imaging, Williamsburg (1984).
Hunt, R. W. G., *J. Phot. Sci.*, **8,** 212 (1960).
Hunt, R. W. G., *Physics in Technology*, **16,** 12 (1985)

James, T. H., and Higgins, G. C., *Fundamentals of Photographic Theory*, 2nd Edn., pp. 237 and 241, Morgan and Morgan, New York (1960).

Pieronek, V. R., Syverud, W. L., and Voglesong, W. F., *Photogr. Sci. Techn. (P.S.A. Technical Quarterly Series II)*, **3,** 145 (1956).

Pinney, J. E., and DeMarsh, L. E., *J. Phot. Sci.*, **11,** 249 (1963).

Pitt, F. H. G., and Selwyn, E. W. H., *Phot. J.*, **78,** 115 (1938).

Powell, P. G., and Beyer, B. E., IEE (London) Conference Publication No. 214, *Electronic Image Processing*, p. 179 (1982).

Richardson, A. W., *Brit. J. Phot.*, **105,** 639 (1958).

Sealfon, P., *New York Times*, p. 41, March 11 (1979).

Thomas, W., Waz, E. M., and Dreyfoos, A. W., *Brit. Kinematog. Sound Tel.*, **52,** 202 (1970).

West, P. A., *Brit. J. Phot.*, **118,** 724 (1971).

CHAPTER 17

The Chemistry of
Colour Photography

1. Colour development – *2.* Developing agents – *3.* Couplers – *4.* Coloured couplers – *5.* The dye-coupling reaction – *6.* The physical form of dye images – *7.* Colour developing solutions – *8.* Silver bleaching – *9.* Processing sequences – *10.* Dye-bleach and dye-removal systems – *11.* Development-inhibitor-releasing (DIR) couplers

17.1 Colour development

IT has been known for many years that, when certain developing agents are used to develop ordinary black-and-white photographic materials, a coloured deposit occurs as well as the silver image. Thus pyrogallol produces a brown stain and, since this is laid down in proportion to the silver image, it can be considered as a dye image. In this case the dye consists of polymerized oxidized developer. This method of forming dye images is sometimes called *primary colour development.*

Another method can be described as *self-coupling development*; in this case molecules of the developing agent, after becoming oxidized by developing the silver image, then react with the unoxidized form to produce the dye image. In this case the developing agent could be a leuco dye. Thus Homolka in 1907 found that indoxyl and thioindoxyl when used as developing agents produced blue and red dye images of indigo and thioindigo, respectively, in addition to the silver image. But none of the above systems of producing dye images has been useful in colour photography, because of the poor properties of either the developing agents that can be used or the dyes that they form or both.

Most modern processes depend on what is sometimes called *secondary colour development* in which the oxidized developer reacts with another substance, normally called a *coupler*, to form the required dye image. The differences between these three types of dye image formation can be illustrated thus:

(*a*) *Primary colour development*

$$\text{Developing agent} + Ag^+Br^- \rightarrow \begin{array}{c}\text{Oxidized developing} \\ \text{agent which is an} \\ \text{insoluble dye.}\end{array} + Ag + H^+ + Br^-$$

(*b*) *Self-coupling development*

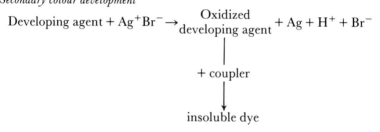

$$\text{Developing agent} + Ag^+Br^- \rightarrow \begin{array}{c}\text{Oxidized} \\ \text{developing agent}\end{array} + Ag + H^+ + Br^-$$

$$+ \text{developing agent}$$

$$\downarrow$$

$$\text{insoluble dye}$$

(*c*) *Secondary colour development*

$$\text{Developing agent} + Ag^+Br^- \rightarrow \begin{array}{c}\text{Oxidized} \\ \text{developing agent}\end{array} + Ag + H^+ + Br^-$$

$$+ \text{coupler}$$

$$\downarrow$$

$$\text{insoluble dye}$$

It will be seen that in all three cases both silver and dye images are formed and hydrogen and bromide ions are added to the solution.

It was Rudolf Fischer who, in 1912, first demonstrated the use of couplers to form image dyes by secondary colour development, and it is interesting that the work he did at this time still forms the basis of most modern colour photographic systems in which the dye molecule is synthesized in development. For not only did he demonstrate the general principle of dye-coupling development, but the art is still mainly confined to one of the two types of developing agent, and to two of the five types of dye, described by him.

The two types of developing agent described by Fischer are para-aminophenol and para-phenylenediamine (or their derivatives).

p-aminophenol   p-phenylenediamine
                       (or p-aminoaniline)

The five types of dye, that Fischer discovered could be formed when silver images are developed by these two developing agents in the presence of suitable couplers, are as follows (where R represents alkyl radicals, CH_3, C_2H_5, etc.):

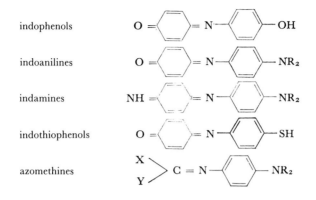

indophenols	
indoanilines	
indamines	
indothiophenols	
azomethines	

These classes of dye were already known when Fischer did his work, but it was he who discovered that the photographic latent image could be used to promote their formation from 'coupler' and 'developing agent'. Of these dyes the indoaniline and azomethine types have been found to possess the necessary properties, but the others usually suffer from some serious defects: thus the indamines cannot be formed quickly enough, and the colour of the indophenols is sensitive to changes in pH (acidity, alkalinity). Since it is only p-phenylene-diamine and its derivatives that can form the indoaniline and azomethine dyes, it has become the most widely used colour developing agent, p-aminophenol being much less useful.

17.2 Developing agents

A good photographic developing agent must react with the latent image at a reasonable rate; it must show good discrimination between the exposed and unexposed silver halide grains so as to develop the image with little or no 'fog'; it must not decompose too rapidly as a result of aerial oxidation or other causes; and for colour work it must of course result in the formation of dyes having good colour and stability. Because of failure in a number of these respects, little application has been found for p-phenylenediamine itself, and its derivatives have been used instead. Combination with the couplers occurs at a free amino (NH_2) group, so that in devising derivatives, whatever other changes are made, a free amino group must be retained.

The only p-phenylenediamine derivatives that have found practical use as colour developing agents are those in which both the hydrogen atoms of one of the amino groups are replaced by alkyl groups thus:

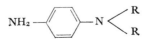

311

Of these N:N-dialkyl-*p*-phenylenediamines the two simplest are:

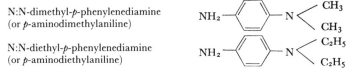

N:N-dimethyl-*p*-phenylenediamine
(or *p*-aminodimethylaniline)

N:N-diethyl-*p*-phenylenediamine
(or *p*-aminodiethylaniline)

The second of these has proved of considerable practical value.

To effect development, the developing agent molecule must donate an electron to a silver ion, and it has been found that the addition of an alkyl group to the benzene ring, at the *ortho* position relative to the coupling amino group, facilitates this donation and thus increases the activity of the developing agent and reduces development times. This leads to structures of the general type:

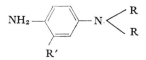

of which a specific example is:

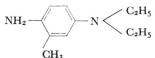

4-amino-3-methyl-N:N-diethylaniline.

A serious hazard with all these developing agents is that on contact with human skin dermatitis is frequently caused; moreover, subjects who are affected in this way do not acquire a resistance to the effect but actually become increasingly sensitive to it. Efforts have therefore been made to produce derivatives that are less prone to cause dermatitis and this often results in one of the groups attached to the substituted amino group being modified, giving the general structure:

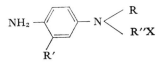

where R and R' are alkyl groups and R'' is an alkylene group bearing the substituent X. Examples of this type of structure are:

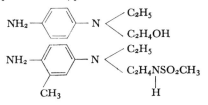

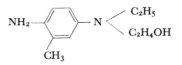

These *p*-phenylenediamine derivatives are easily oxidized by air and they are therefore usually made available as salts or complexes. In this way their stability as dry solids can be made quite satisfactory, and a further advantage is that their solubility in water is much improved, thus facilitating the mixing of developer solutions. The following are examples of developing agent salts:

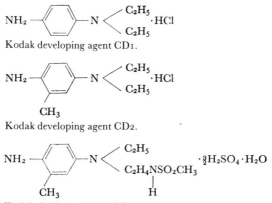

Kodak developing agent CD1.

Kodak developing agent CD2.

Kodak developing agent CD3.

It appears that the $NHSO_2CH_3$ group in the last formula impedes penetration of the developing agent through the skin.

17.3 Couplers

While useful colour developing agents are mainly derivatives of a single basic structure, *p*-phenylenediamine, useful couplers cover a much wider range of compounds. They can, however, be divided into three main groups:

(a) Compounds with an active open-chain methylene ($—CH_2—$) group.

(b) Compounds with an active cyclic methylene ($. .|. .—CH_2—. .|. .$) group.

(c) Phenolic compounds with an active methine ($—CH=$) group.

Yellow couplers are usually in group (a), magenta in groups (a) and (b), and cyan in group (c).

Thus yellow couplers are usually of the form

$$X—CH_2—Y$$

where X = RCO, and Y = R'CO or R'NHCO. The coupling takes place by replacement of the two hydrogen atoms, so that the formation of a yellow dye image can be represented thus:

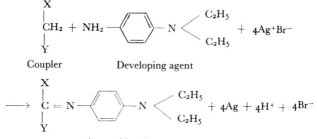

Coupler   Developing agent

Azomethine dye

A most useful class of cyclic methylene magenta couplers are pyrazolones having the form:

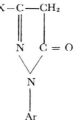

where Ar is an aromatic (containing a benzene ring) group and X is another group. As with the yellow couplers, the coupling occurs by the replacement of the two hydrogen atoms thus:

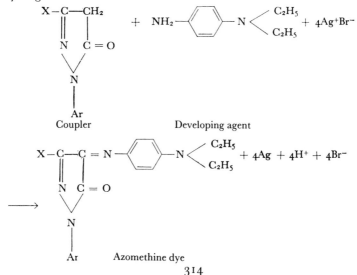

Coupler   Developing agent

Azomethine dye

314

In the phenolic cyan couplers coupling takes place by replacement of the hydrogen atom at the position in the ring opposite (*para*) to a hydroxyl (—OH) group. The simplest compound of this class is phenol:

but couplers can also be based on α-naphthol

and have the general structure

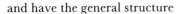

where X and Y are various groups. The coupling reaction can therefore be represented thus:

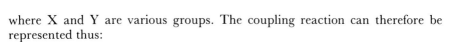

Coupler   Developing agent

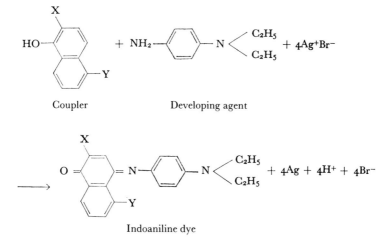

Indoaniline dye

Sometimes the hydrogen atom at the coupling position of the coupler is replaced by a chlorine atom, and in this case the coupling reaction occurs thus:

315

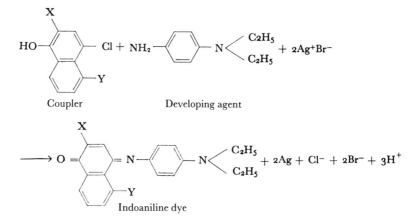

Coupler   Developing agent

Indoaniline dye

It is seen that the same dye is formed as before but each molecule of developing agent now only results in two molecules of silver bromide being reduced instead of four; thus the same amount of dye is produced from only half the amount of latent-image silver.

Thousands of couplers of various structures have been made and tried for colour processes. Amongst the properties of a coupler that are important in deciding upon its usefulness are:

its rate of reaction
its solubility
its proneness to wandering
its colour
the solubility of the dye formed
the stability of the dye formed
the colour of the dye formed.

A high rate of reaction of a coupler with oxidized developer is essential, otherwise the latter will diffuse away from the exposed silver halide grain and produce unsharp images (or even images of the wrong colour if diffusion into another layer occurs). A high rate of reaction is also desirable because very lengthy colour development is generally inconvenient, and it is obviously desirable in most colour processes for the three dye images to be formed at similar rates. The rate of reaction will be affected by the amount of silver halide required to be reduced to form a molecule of dye, but other features of the coupler structure besides the presence or absence of a halogen atom at the point of coupling are important in this respect. If a coupler forms too much dye, there can be mixed with it a *competing* coupler producing with oxidized developer a soluble compound that washes out, thus reducing the final amount of dye in the image.

If the coupler is to be used in the developing solutions, as in the *Kodachrome* type of process, it must clearly be soluble in them; it must also be able to penetrate the gelatin layers and hence the molecule must be fairly small. If, on the other hand, the coupler is to be dispersed in oily droplets in the emulsion layers, as in the *Kodacolor* system, it must have the necessary solubility in the oil used, such as tricresyl phosphate, triphenyl phosphate, *n*-butyl phthalate, or *n*-hexyl benzoate. If the coupler is dispersed directly in the emulsion layers as in the *Agfacolor* system, it must not wander from one layer to another; this feature is generally achieved by making the coupler molecule so large, by the addition of a 'ballast' group, that it cannot pass through the sponge-like network of the gelatin layer. Even when the couplers are dispersed in oily droplets, it is necessary to add a ballast to prevent wandering.

In most instances it is required that a coupler be colourless itself, so that non-image areas are clear; in some colour negatives, however, it can be advantageous for the coupler to be coloured (see Section 17.4) and this again is achieved by modifications to its structure.

As regards the dye formed by the coupler, it must be insoluble (unless the process is one in which the dye is required to transfer from the layer containing

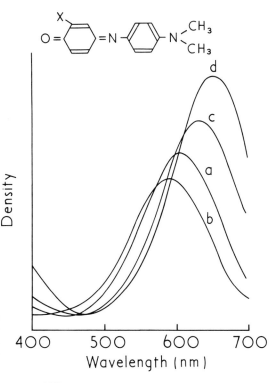

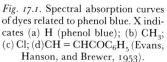

Fig. 17.1. Spectral absorption curves of dyes related to phenol blue. X indicates (a) H (phenol blue); (b) CH_3; (c) Cl; (d) $CH = CHCOC_6H_5$ (Evans, Hanson, and Brewer, 1953).

317

the silver image to a mordanted receiving sheet; in this latter case insolubility after mordanting is what is required). The dye must be stable to light, heat, and humidity, and, while complete stability of any organic dye seems almost impossible to achieve, the couplers now used in many processes give dyes of stability adequate for most purposes.

Finally, of course, the colour of the dye should be as near as possible to the ideal required for the particular process. Some examples of the way in which the colour can be varied in a simple system by structural changes are shown in Fig. 17.1. The colour in the image is also affected by its physical form in the photographic layer, but this will be dealt with in Section 17.6.

In view of all the above considerations, it is not surprising that much effort has gone into modifying coupler structures in order to gain better characteristics. In the case of yellows, the most useful are generally acylacetamides:

An example of this type of coupler is:

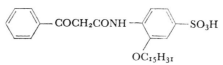

ω-benzoylacet-(2-n-pentadecyloxy)-4-sulphoanilide.

In this example the pentadecyl group ($C_{15}H_{31}$) is the ballast, while the sulphonic acid group (SO_3H) makes the coupler sufficiently hydrophylic and acidic to be compatible with the gelatin layer. More recently, yellow dyes with better light stability have been obtained by using couplers having the general structure

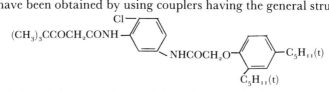

α-pivaloyl-5-[α'-(2,4-ditertiarypentylphenoxy)acetamido]-2-chloroacetanilide.

The most useful magenta couplers found so far are pyrazolones. An example of a pyrazolone derivative yielding a magenta dye is:

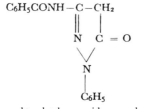

1-phenyl-3-benzamidopyrazol-5-one.

318

To reduce the tendency for unused couplers of this type gradually to become yellowish in the dark, structures of the following type have been introduced:

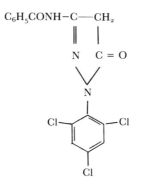

1-(2,4,6-tichlorophenyl)-3-benzamidopyrazol-5-one.

These magenta couplers cited have no large ballast group, and are thus suitable for use in developing solutions.

An example of a cyan ballasted coupler based on naphthol is:

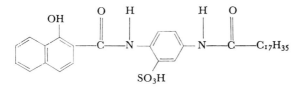

1-stearoylamido-4(1′-hydroxy-2′-naphthoylamido)-benzene-3-sulphonic acid.

In this example, the SO_3H group provides compatibility with the gelatin layer, and the $C_{17}H_{35}$ group provides the ballast to prevent wandering. Improvements in the stability of cyan images when kept in the dark have been obtained by using couplers of the following type:

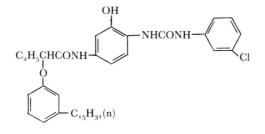

2-m-chlorophenylureido-5-[α-(m-pentadecylphenoxy)butyramido]phenol.

17.4 Coloured couplers

By using couplers that are themselves coloured, as described in Section 15.4, a very convenient method of colour correction by masking can be achieved. A coupler can be made coloured by adding to it a suitable chromophoric substituent, and masking is achieved if this colour is present in proportion to the amount of *unused* coupler in the image (de Ramaix, 1971).

One way of achieving this is to add the chromophore group after development has taken place. For example, if a coupler that forms magenta dye can, in its uncoupled state, be converted to a yellow dye without harming the magenta image, then a yellow positive image can be formed from the unused coupler so as to mask the magenta negative image; an example of this principle was used in the *Icicolor* process in the following way. The magenta-forming coupler, a styryl pyrazolone, was heated with an aromatic aldehyde before coating to form a mixture of a yellow styryl dye and an aldehyde-bis-pyrazolone; in the colour developer, this yellow dye was completely discharged, irrespective of the presence of developable silver halide, and oxidized developer converted the coupler to magenta dye in exposed areas as usual; but subsequent treatment with formaldehyde reformed the yellow dye in the areas where the coupler had not been used and hence masking was achieved. A reddish mask was also formed in the *Icicolor* cyan layer by incorporating the formaldehyde in a stop bath following the colour development step: the simultaneous presence of formaldehyde and colour developing agent (carried over by the film into the stop bath) resulted in the unused cyan coupler being converted to an intermediate compound which was converted to the required reddish colour subsequently in an acid bleach bath (Gehret, 1964; Ganguin and MacDonald, 1966).

Another way of achieving masking by the use of coloured couplers is to add the chromophoric substituent to the coupler before processing and to arrange that the coupling step destroys the colour of the coupler, so that its coloration remains only where it is unused. This can be done by adding the chromophore group to the coupler at the coupling position so that the colour development step consists of replacing it by the oxidized developer fragment. The following is an example of this type of reaction:

$$C_6H_5CONH-\underset{\substack{\| \\ N}}{C}-\underset{\substack{| \\ C=O}}{CH}-N = N-C_6H_5 + NH_2-\langle\quad\rangle-N\begin{smallmatrix}C_2H_5\\C_2H_5\end{smallmatrix} + 2Ag^+Br^-$$

$$\underset{\substack{| \\ C_6H_5}}{N}$$

Coloured coupler   Developing agent

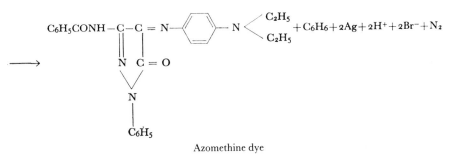

Azomethine dye

It is seen that, as in the case where a chlorine atom was substituted at the coupling position in a cyan coupler, so the substitution of the chromophore at the coupling position of this magenta coupler results in only two silver bromide molecules being required for the formation of each molecule of dye. It is often convenient in practice to work with a mixture of coloured and uncoloured couplers, because by adjusting their relative quantities the amount of coloration in the non-image areas can then be controlled quite accurately.

17.5 The dye-coupling reaction

The equations given earlier for the dye-coupling reactions are only simplifications: in the actual process a number of intermediate compounds are formed, some of which are highly unstable and therefore of very short life. The developing agent, on reacting with the silver ions, is oxidized to a quinone diimine, thus:

$$H_2N \underset{}{\bigcirc} NR_2 + 2Ag^+ \longrightarrow H-N^+ \underset{}{\bigcirc} NR_2 + 2Ag + H^+$$

Two contributing structures to the resonance-stabilized quinone diimine are:

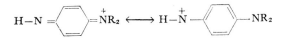

There is good evidence that at the alkalinity used in development (pH 10 to 12) the quinone diimine is the coupling species. A leuco dye is first formed, thus:

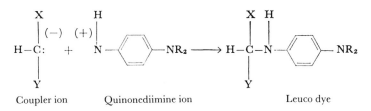

| Coupler ion | Quinonediimine ion | Leuco dye |

321

The leuco dye is then oxidized to the dye:

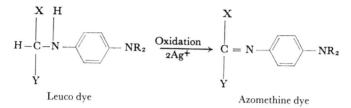

Leuco dye   Azomethine dye

The quinone diimine and its semiquinone precursor are also involved in competing side reactions.

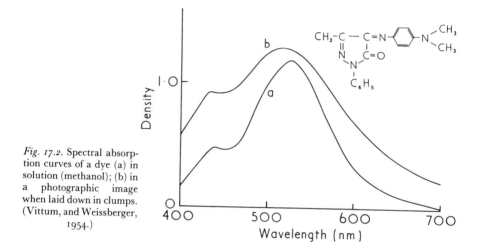

Fig. 17.2. Spectral absorption curves of a dye (a) in solution (methanol); (b) in a photographic image when laid down in clumps. (Vittum, and Weissberger, 1954.)

17.6 The physical form of dye images

If a dye is laid down in large clumps (for example, around large silver grains) with clear interstices, the same degradation of colour will occur as is the case with an ink printed as a half-tone dot pattern (Pollak, 1955; Pollak and Hepher, 1956); the use of the dye at high concentrations over part of the area exaggerates the effects of the unwanted absorptions as compared to the use of the dye at a lower concentration over all the area (Gledhill and Julian, 1963). An example of this is given in Fig. 17.2. If, however, the dye is formed more uniformly and without clear interstices, as would be facilitated if the dye image consisted of several layers of the dye 'grains', then the absorption is more like that in solution. Occasionally, the dye in the image shows a sharper absorption band than in solution, with the absorption peak shifted to shorter wavelengths as shown in Fig. 17.3; this may indicate that the dye in the image is microcrystalline instead of being in the usual amorphous state.

Another physical factor affecting the colour of the dye is the number of times the light passes through it and, as explained in Section 13.9, this has the effect of exaggerating the unwanted absorptions of a dye when it is used in reflection prints.

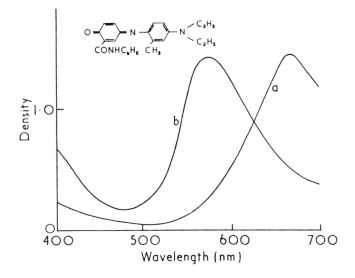

Fig. 17.3. Spectral absorption curves of a dye (a) in solution (n-butyl acetate); (b) in a photographic image when laid down in microcrystalline form. (Vittum and Weissberger, 1954.)

17.7 Colour developing solutions

In addition to the developing agent, a colour developing solution has to have a number of other constituents. First, because most developing agents act as such on photographic latent images only when in alkaline solutions, it is necessary to have suitable alkalis, usually sodium carbonate and caustic soda (sodium hydroxide). Then a buffer is required to maintain the alkalinity at as nearly as possible a constant level, and this may be provided by a sodium carbonate and bicarbonate mixture (the bicarbonate usually being derived from the reaction between the carbonate and the developing agent salt), or a sodium metaborate and borax mixture, or a dibasic and tribasic sodium phosphate mixture, according to the alkalinity (pH) required. A small amount of sulphite acts as a preservative by combining with any developing agent that has become partially oxidized by contact with air (the inert colourless compound formed is far less injurious to the developer than the partially oxidized developing agent which accelerates further oxidation and leads eventually to the formation of brown stains). Then additions are made to reduce the tendency for *un*exposed silver halide grains to develop; these are termed *antifoggants*, and often take the form of bromide ions, added as potassium bromide, or organic antifoggants may be used instead or in addition. Further additions may be made to obtain various effects: competing developing agents to reduce dye formation; anti-stain agents to lower the density in undeveloped areas; and accelerators such as thiocyanate (to act as a silver halide solvent) and benzyl alcohol (which accelerates the dye-forming reaction in certain processes). Constituents may also be added to promote the occurrence of favourable inter-image effects (see Section 15.5).

17.8 Silver bleaching

Although the required dye image is complete as soon as the colour development step has been terminated, unless the dye is at this point transferred to another support, the photographic material must then be given a series of treatments in order that all harmful residues are removed. It is thus necessary to remove the silver image, which, if left behind, would greatly darken the dye image; and also the unused silver halide must be removed, because it darkens on exposure to light and would therefore produce a dark and stained appearance. The metallic silver is oxidized to silver ion, as present in silver halide, by an oxidizing solution, for example ferricyanide; the oxidized silver and the unused silver halide are then removed together in the usual way by *fixing* in *hypo* (sodium thiosulphate). The bleaching reaction can then be represented by:

$$Ag + K_3Fe(CN)_6 + KBr \longrightarrow AgBr + K_4Fe(CN)_6$$

and the fixing step by:

$$AgBr + Na_2S_2O_3 \longrightarrow NaAgS_2O_3 + NaBr$$
$$NaAgS_2O_3 + Na_2S_2O_3 \longrightarrow Na_3[Ag(S_2O_3)_2]$$

the constituent on the right-hand side of the last equation being soluble and therefore washing out. Another common bleaching agent is potassium dichromate; and sometimes ammonium thiosulphate, $(NH_4)_2S_2O_3$, is used for fixing, having the advantage of being more rapid in the case of emulsions containing iodide.

It is sometimes convenient to bleach and fix in the same solution, and such solutions are known as *blixes*. If the blix can be used quickly and then discarded, it can be made by simply combining ferricyanide and hypo. However, this mixture quickly decomposes and therefore for conventional types of processing an iron-sequestrene type of blix is often used. In this type of blix, the ferric iron is protected by a complex with ethylenediaminetetra-acetic acid (E.D.T.A.), and this prevents it from reacting with hypo when it is added to form the blix.

17.9 Processing sequences

It can be seen from the above that the minimum number of solutions required to form a dye image in a photographic emulsion layer is two: a developer and a blix;

in addition, the material requires washing after the blix in order to remove all traces of silver ions, and a wash between the developer and blix is sometimes required to avoid contamination of the latter by the former. In practice, however, it is rarely possible to keep colour processing sequences down to this simple two solution, four step procedure, and they are often elaborated in one or more of the following ways.

Instead of the wash between the developer and the blix an acid *stop* bath is usually preferable since this more effectively isolates the developer from the blix and also provides an abrupt, and therefore well-controlled, end-point to the development process. But because the fixing capacity of blix baths is usually somewhat limited, a stop-fix between the developer and blix is better still. This bath can be based on a mixture of sodium bisulphite to give the acidity, and sodium thiosulphate (hypo) to give the fixing action. The *undeveloped* silver halide is thus removed at this stage, and hence the blix bath has only to bleach and fix the silver that was developed in forming the image. We thus have a three solution, four step process consisting of develop, stop-fix, blix and wash.

These procedures satisfactorily remove the unwanted silver image and the unused silver halide, but in incorporated-coupler materials unused coupler is left behind. Ideally this should also be removed but no simple means have been found for extracting it. The material can, however, be treated with a *stabilizer* designed to reduce the tendency of the unused coupler to react to the detriment of the image. Thus immersion in an alkaline formalin bath sometimes prevents the coupler from reacting with the dye image and reconverting it to the leuco form. The formalin also has a useful hardening action on the gelatin which makes the final result tougher and less liable to scratching.

It is also often found that the stability of the image dyes to fading by exposure to light can be improved by adjusting the final acidity of the image layer to a particular buffered value, and this type of treatment or *conditioning* may also reduce the tendency for the unused coupler to *print out*, that is, become yellow on exposure light.

Better final image stability can therefore often be obtained by adding a stabilizing step or a conditioning step to the three solution process outlined above, giving a four solution process, or a five solution process if both are used. If two extra washes are added, the number of steps may then reach eight: develop, stop-fix, wash, blix, wash, stabilize, wash, condition.

It is possible to combine a fixing stage with the stabilizing step, by adding sodium thiosulphate (hypo) to the alkaline formalin, thus obtaining an alkaline formalin hardening fixing bath. This makes it possible to use a bleach instead of a blix, and sometimes this has certain advantages. Thus some image dyes do not oxidize very readily in the developer from the leuco to the final form, and the bleach, being an oxidizing agent, can be used to complete this reaction. It might be thought that, with fixing at the stabilizing stage, no fixing would be necessary before the bleach so that the stop-fix could be replaced by a simple stop bath. There is sometimes, however, a tendency for developing agent to be adsorbed on to undeveloped silver halide grains in such a way that the stop bath or a wash does not remove it, and the developing agent then oxidizes in the bleach to form undesirable stains. A five solution process using a bleach instead of a blix might therefore be: develop, stop-fix, wash, bleach, wash, alkaline formalin hardening fix, wash, condition.

Finally, by splitting the stop-fix and the stabilize-fix each into their separate components a seven solution process is obtained, which, with four washes, comprises the following eleven steps: develop, stop, fix, wash, bleach, wash, fix, wash, stabilize, wash, condition.

Which particular process is used in any instance will depend on many factors, including the type of emulsion, the couplers, whether optimum image stability is required, the type of processing machine used, and the relative importance of cost, speed, and convenience in getting the final image. Of course, in materials in which the couplers are not incorporated, the colour development stage has to be triplicated so as to produce the cyan, magenta, and yellow images separately, and in reversal processes the colour section has to be preceded by negative (black-and-white) development and re-exposure steps, usually with a stop-bath and a wash in between them. The negative development step may be specially designed to promote the occurrence of favourable inter-image effects (see Section 15.5). Sometimes the re-exposure step is carried out chemically by adding a suitable fogging agent to the colour developer.

Processing times can be shortened by using the solutions at high temperatures, and because colour processes tend to be longer than those for black-and-white materials the black-and-white standard of 68° F (20°C) is often replaced by 75°F (24°C), 85°F (30°C), or even temperatures of 100°F (38°C) or more for special applications. The photographic materials have to be specially hardened to stand up to the higher temperatures and this may be done in manufacture, or the first step in the process may be a special prehardening step (sometimes followed immediately by a neutralizer to prepare the film for the developer). A coupler-incorporated reversal colour film might then have a processing sequence as follows: preharden, neutralize, first development, first stop-bath, wash, colour development (with fogging agent), second stop-bath, wash, bleach, fix, wash, stabilize (Beilfuss, Thomas, and Zuidema, 1966); in this case the fixing and stabilizing steps are not combined, and a conditioning bath is not included.

If a film is to be developed with a sound track it is usually necessary for this to be a silver image, because all dye images tend to transmit in the infra-red where the photocells commonly used in sound projectors are quite sensitive (so that the effective modulation of the light by the dye is spoiled). Special processing of the edge of a colour film where a sound track has been recorded or printed is therefore normally required, and this can be achieved by applying special solutions to the sound track area only. In the processing sequence just described for a coupler-incorporated reversal film, a sound track fixing step can be added before the colour developer (so that no. reversal image can be formed) and a sound-track developer, giving a silver image, can be added after the bleach (so that the negative image is not fixed out in the fixer). In this way the sound track area has a negative silver image, and no positive dye image is formed.

Increasing the temperature of developers used for colour materials may necessitate changes both in the materials themselves and in the developing solutions, because the extra hardening necessary may decrease the penetrability of the solutions. For this reason it is not normally possible to make colour materials that can produce the same results at appreciably different developer temperatures (Fritz, 1971).

In order to obtain images that are as sharp as possible, films usually incorporate an *anti-halation layer* to absorb light reflected from the bottom surface of the film base. Thus the back of the film base may be coated with a thin gelatin *backing-layer* containing dyes that dissolve or bleach during processing; or a layer of silver between the bottom emulsion layer and the base may be used, in which case it is removed by the usual bleaching and fixing steps. But another form of anti-halation layer that is widely used consists of colloidal carbon in a resin, coated on the bottom surface of the base; this backing layer is made so that it hydrolyses in alkali, and it therefore softens in the developer and can be removed. However, it is usually necessary to rub the surface in order to ensure that the backing is completely removed, and this is more conveniently done in a pre-bath than in the developer; processes intended for films having this type of backing may therefore incorporate an alkaline pre-bath and buffing stage. The buffing step is really only convenient when the film to be processed consists of long lengths, and the resin type of backing is therefore usually confined to motion-picture films or still-picture films that can be joined together in long lengths for processing.

17.10 Dye-bleach and dye-removal systems

In the colour development systems based on Rudolf Fischer's work the image dyes are formed in the layers of the material during photographic development of the silver image. But successful results have also been obtained with systems in which the dyes are formed by ordinary chemical means first and then incorporated in the layers; variation in concentration of the dye from point to point in the layer is then achieved by bleaching or removing it as a function of the silver photographic image.

In one such method the silver of the developed photographic image is used to bleach azo dyes in the presence of halogen acids. The reaction can be represented thus:

$$RN = NR' + 4H^+Cl^- + 4Ag \xrightarrow[\text{Catalyst}]{\text{Organic}} RNH_2 + R'NH_2 + 4Ag^+Cl^-$$

Azo dye Bleached dye

In this case the dye is bleached most where the material has been exposed most, and hence a positive image is obtained directly. The material requires fixing after dye-bleaching, in order to remove the silver halide formed at this step. This type of process is sometimes known as *Gasparcolor* because Gaspar patented a number of its features. It was used commercially by Ilford for the material on which they made positive reflection prints from positive transparencies from 1953 to 1962. The processing steps comprised: develop, fix, dye-bleach, fix (Collins, 1960; *British Journal of Photography*, 1962). The dye-bleach bath usually contains a silver-halide complexing agent, such as thiourea, $(NH_2)CS(NH_2)$, which, by forming a complex with the silver chloride, keeps the silver-ion concentration low and hence maintains the reaction in the right direction. The function of the organic catalyst is to become reduced by the silver and then to become re-oxidized by reducing, and thus bleaching, the dye; this is important because both the dye and the silver are insoluble. A typical catalyst is 2,3 diaminophenazine. In the *Cibachrome* version of the process, colour saturation is increased by inter-image effects caused by adding the catalyst to the developer instead of to the bleach (Meyer, 1965 and 1974).

An alternative to bleaching the incorporated dyes is to alter their solubility as a function of the silver photographic image. In the *Polacolor* system (Crawley, 1963) use is made of the fact that the solubility of catechol or hydroquinone varies as it is used to develop silver halide. Thus hydroquinone, for example, being a weak acid, is soluble in aqueous alkali as an anion (negative ion) which is very active as a developing agent. As a result of developing the silver halide, however, it is oxidized to quinone which is of low solubility. These reactions can be represented thus:

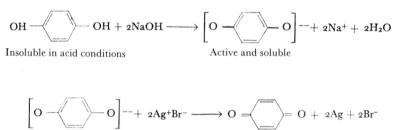

If now a dye molecule is suitably attached to a hydroquinone molecule, the solubility of the combination can be made to alter in the above way, so that the dye is insolubilized around the developed silver. This process can be represented as shown on opposite page.

After the formation in this way of an insoluble dye image, the rest of the dye can be washed out in an alkaline solution, and the silver bleached and fixed in the usual way. Most insoluble dye will be formed where most silver has been

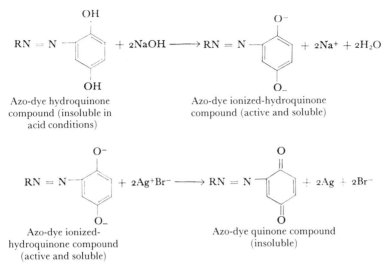

Azo-dye hydroquinone compound (insoluble in acid conditions)

Azo-dye ionized-hydroquinone compound (active and soluble)

Azo-dye ionized-hydroquinone compound (active and soluble)

Azo-dye quinone compound (insoluble)

developed and therefore the dye image is distributed in the same sense as the silver image, thus normally giving a negative.

It is possible, however, to use this system to give direct positive images by using, not the insoluble dye, but the soluble dye. In this case, after soaking in alkali to activate and solubilize the hydroquinone, the material is held in intimate contact with a suitable receiving sheet containing a mordant. Any dye that is still soluble then transfers to the receiving sheet where it is insolubilized by combining with the mordant. In this case most dye is obtained on the receiving sheet where least silver has been developed and hence a direct positive image is obtained. This arrangement has the advantage that the unwanted silver image and the unused silver halide are left behind in the original material which is discarded, and the processing is therefore reduced to the very simple sequence: develop, and transfer; this enables the entire process to be completed in about one minute.

The way in which the *Polacolor* system operates is depicted in Fig. 17.4. The negative material consists of three silver-halide layers sensitized in the conventional manner and coated in the usual order, red next to the base, green in the middle and blue at the outside. On the base side of each layer, however, is coated a layer of dye-developer, cyan next to the red layer, magenta next to the green layer, and yellow next to the blue layer, as shown in Fig. 17.4. The yellow dye-developer layer acts as the usual yellow filter layer preventing blue light from reaching the red and green sensitive layers. The negative is processed in the camera, one picture at a time, by drawing it, together with the receiving sheet, through a pair of pressure rollers; the receiving sheet has attached to it small pods containing a viscous solution of alkali, and passage through the rollers ruptures the pods and enables the alkali to activate the dye-developers. The dye-developers then diffuse in all directions but preferentially towards the receiving sheet because of the falling concentration gradient in that direction. Wherever they encounter developable silver halide grains the dye-developer molecules will be insolubilized and will not transfer, as required to give a positive image. In order to obtain a correctly coloured image, however, it is also necessary that the process of insolubilization of each of the three types of dye-developers be specific to its own layer; thus the cyan dye-developer must be insolubilized only by exposed silver halide grains in the red sensitive layer. To achieve this, the *Polacolor* system depends on the fact that before the cyan dye-developer can reach the green layer it must pass through the red layer, a spacer layer, and the magenta dye-developer layer; it will therefore take much longer to reach the green layer than the magenta dye-developer which is immediately next to it. Hence, by the time the cyan dye-developer has reached the green layer, the latter has been more or less fully developed by the magenta dye-developer and can therefore have little or no effect on the migrating cyan dye-developer molecules. Similar diffusion-time factors are used to prevent the development of the blue sensitive layer by either the cyan or the magenta dye-developers; equally, any development of the

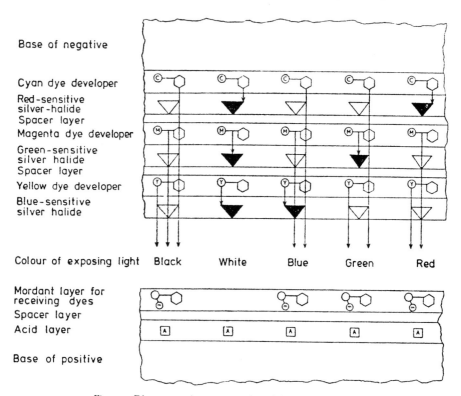

Fig. 17.4. Diagrammatic representation of the *Polacolor* process.

green or red layers by back diffusion of the yellow or magenta dye-developers is also avoided by diffusion-time considerations. The structure of the receiving sheet is as shown at the bottom of Fig. 17.4. Its top layer contains the mordant to insolubilize the dye-developer molecules emerging from the negative material. If the receiving sheet contained nothing else, the prints would be strongly alkaline on removal from the negative, because of the transfer of alkali, as well as solubilized dye. Strongly alkaline prints would be very unpleasant to handle, and the alkaline condition would adversely affect the permanence of the dye images. It is therefore necessary to neutralize the alkali and this is the function of the bottom layer of the receiving sheet. This contains acid molecules, made non-wandering by having long-chain ballast groups attached to them. The spacer layer between the mordant layer and the acid layer slows down the rate at which the alkali reaches the acid layer, thus enabling the negative and mordant layer to remain strongly alkaline as required during the development, but towards the end of the development time sufficient alkali reaches the acid layer to react with it and form enough water to swell the spacer layer and assist in the neutralization reaction.

In 1972 a new version of the *Polacolor* process (SX 70) was introduced in which the negative and receiving materials are combined into a single 16-layer integral sandwich, and hence the need to peel the receiving sheet apart is avoided. Because the image is viewed from the same side of this material as that on which the exposing light is incident, the camera (which is a folding one) contains a mirror to give correct-reading pictures. After the exposure is made, the camera automatically extrudes the picture through rollers, that break a pod of processing solution (attached to each sheet of material) which diffuses into an air gap between two of the layers of the material, and thence into the various layers themselves, producing a finished colour print a few minutes later. The general structure of the material is similar to that shown in Fig. 17.4 but with the receiving sheet integral with the negative (Land, 1972).

Since this material is processed outside the camera it is necessary to protect it from the light while processing is taking place: this is accomplished by having an opaque layer between the negative base and the cyan dye-developer layer, and by having in the pod a dye that is black at the high pH occurring when processing starts, but which becomes colourless at the low pH occurring when processing is finished; this black dye diffuses in and above all the light-sensitive layers and, together with the opaque layer, protects them from light during processing. It is also necessary to provide a white diffusing layer against which the final image can be seen: the pod therefore also contains a polymer that solidifies during processing and forms a white (titanium dioxide) layer between the opaque layer and the transferred image dyes. The base of the receiving layer is transparent so that the image can be viewed through it. Some compensation for the effect of ambient temperature on diffusion rates is obtained by incorporating, in the spacing layers, polymers whose temperature coefficient is such as to slow down diffusion as the temperature rises (*British Journal of Photography*, 1972).

In the Kodak system of *Instant* colour photography, marketed from 1976 to 1986 (Hanson, 1976) the final image is viewed from the opposite side to that on which the exposure was incident, so that the image is right-reading and it is unnecessary to incorporate a mirror in the camera (although some cameras using this system incorporate a pair of mirrors in order to achieve a compact shape). The film structure is as shown in Fig. 17.5.

The light from the camera lens first passes through a transparent cover sheet, then through a thin air gap (into which the viscous processing fluid is subsequently injected), and then, after an ultraviolet absorbing layer, the blue, green, and red light-sensitive layers are reached, each having adjacent to it a uniform layer of an immobile dye releaser from which the yellow, magenta, and cyan images, respectively, are derived. Oxidized-developer scavenger layers are situated between the pairs of light-sensitive and dye-releaser layers, and all these layers are coated on an opaque black layer which is coated on a reflecting white layer and an image-receiving layer, the latter being next to the transparent film-base support.

The light-sensitive silver halide grains contain internal sensitization sites and this results in exposure producing latent images that are *internal* to the grains:

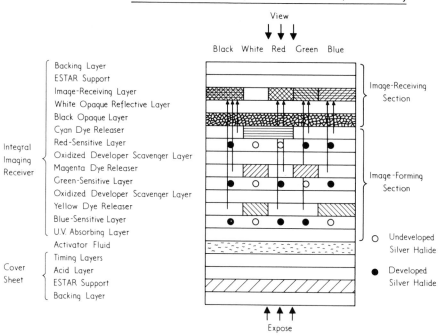

Fig. 17.5. Diagrammatic representation of the Kodak System of *Instant* colour photography.

this makes it possible for subsequent development to produce *positive* silver images (unlike the negative silver images developed in the *Polacolor* systems).

After exposure, the processing fluid is injected into the thin air gap by passing the picture unit through a pair of pressure rollers that rupture a pod containing a suitable viscous mixture. This mixture includes enough carbon to protect the light-sensitive layers from any further exposure through the transparent cover sheet so that the picture unit can be ejected from the camera at this stage, protection against exposure from the support side being provided by the opaque black layer. The processing fluid also contains an *electron transfer agent* (ETA) that, together with a nucleating agent, develops the unexposed, but not the exposed, silver halide grains, thus giving a positive silver image. However, unlike an ordinary developer, the ETA is not used up in the development process but is continuously regenerated. This is achieved by the reduction of the oxidized form of the ETA (produced as a result of development) to its original unoxidized form when it reacts with one of the dye releasers, and this reaction causes release of diffusible dyes from the dye releasers, hence producing the required positive dye images. The released dyes migrate from their layers through the opaque black and the reflecting white layers to the image-receiving layer where a mordant renders them immobile. All the layers in the picture unit are sufficiently thin for the final image to appear adequately sharp at normal viewing distances. The oxidized-developer scavenger layers prevent ETA oxidized by development in one light-sensitive layer from reaching the dye releaser associated with a different light-sensitive layer. The acid layer contained in the cover sheet is used to stop the development process and neutralize the alkaline processing fluid so as to stabilize the final print, and the timing layer in the cover sheet ensures that these reactions occur at the appropriate time. The processing fluid also causes the cover sheet to adhere to the rest of the unit in the finished picture. The image first begins to appear in the receiving layer in less than a minute, but the build up of the full density requires several minutes, as in the case of the integral *Polacolor* system.

A modified version of the Kodak *Instant* system has been made available in which the dye images are transferred to a separate receiving sheet. Known as *Ektaflex*, it is available as a negative or reversal material, and provides a means of making prints from negatives or positives with minimum processing requirements. After exposure, the Ektaflex film is fed into a simple processor in which it is immersed in a single solution and then sandwiched on to the receiving paper; after waiting several minutes, the materials are peeled apart to produce the print.

A feature of all systems in which the dyes are present in the photographic material at the time of exposure is that some light is lost by absorption. These light losses are minimized if the structure is as shown in Figs. 17.4 and 17.5: the dyes are beneath, rather than mixed with, their associated light-sensitive layers; and the light passes through the dyes in the order yellow, magenta, cyan, so that the only unwanted absorptions involved are the green and red absorptions of the yellow dye and the red absorption of the magenta dye, which are usually quite small.

A similar transfer process can also be devised for systems using conventional colour development. Of course, the dyes can only transfer if they are highly mobile, but the couplers from which they are formed must be non-wandering, or highly immobile, since they must be incorporated in the original emulsion layers. One solution to the problem is to attach a large ballast group to the couplers at the coupling position; the oxidized developer can then replace the ballast group on coupling, thus forming a mobile dye from an immobile coupler. A reaction of this type in the case of a yellow coupler can be represented by the following:

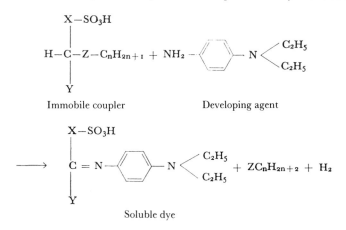

If the value of n in the C_nH_{2n+1} group is fairly large ($n = 15$ or more), then the coupler molecule will be large enough to be fairly immobile; the dye molecule, however, has no ballast, and is therefore quite small, and is soluble on account of the sulphonic acid (SO_3H) group. After development it can therefore transfer to a receiving sheet, if the latter is held in moist contact with it; a suitable mordant in the receiving sheet can then immobilize the dye once again to give an insoluble final image. Thus, again, the processing sequence is simply: develop, and transfer.

17.11 Development-inhibitor-releasing (DIR) couplers

In Section 17.4 it was pointed out that, by attaching a chromophore group to a coupler at the coupling position, a very convenient method of masking could be achieved. If, instead of a chromophore group, a compound that inhibits development is attached at the coupling position, masking can also be achieved, together with some other advantages. On coupling, such *development-inhibitor-releasing (DIR) couplers* produce local concentrations of compounds that retard development. It is possible to arrange for the compounds to travel from one layer to another so that the amount of retardation in one layer is proportional to the amount of development in another layer: thus, if a DIR magenta-forming coupler

334

retards the formation of yellow dye in an adjacent layer, it is possible to balance the magnitudes of the effects so that the amount of unwanted blue absorption of the magenta dye is always off-set by the reduction in the amount of yellow dye formed.

It is also possible to use DIR couplers to reduce graininess and improve sharpness. Thus, the release of development-inhibitor, as coupling takes place around a grain, results in dye clouds of smaller size, and, as well as improving sharpness, this necessitates developing more grains for a given density and hence reduces graininess. Edge effects can also be enhanced by using DIR couplers, since the concentration of development-inhibitor tends to fall near the edge of an area being developed, and hence increases in density differences are obtained at edges (Barr, Thirtle, and Vittum, 1969). (See Sections 18.8 and 18.16.)

REFERENCES

Barr, C. R., Thirtle, J. R., and Vittum, P. W., *Phot. Sci. Eng.*, **13,** 74 and 214 (1969).
Beilfuss, H. R., Thomas, D. S., and Zuidema, J. W., *J. Soc. Mot. Pic. Tel. Eng.*, **75,** 344 (1966).
British Journal of Photography, **109,** 904 (1962).
British Journal of Photography, **119,** 872 (1972).
Collins, R. B., *Phot. J.*, **100,** 173 (1960).
Crawley, G., *Brit. J. Phot.*, **140,** 76 (1963).
De Ramaix, M., *Phot. Sci. Eng.*, **15,** 262 (1971).
Evans, R. M., Hanson, W. T., and Brewer, W. L., *Principles of Color Photography*, Wiley, New York, p. 260 (1953).
Fritz, N. L., *Proc. Am. Soc. Photogram.*, Workshop on Color Aerial Photography and Plant Sciences and Related Areas (March 1971).
Ganguin, K. O., and MacDonald, E., *J. Phot. Sci.*, **14,** 260 (1966).
Gledhill, R. J., and Julian, D. B., *J. Opt. Soc. Amer.*, **53,** 239 (1963).
Gehret, E. C., *Brit. J. Phot.*, **111,** 818 (1964).
Hanson, W. T., *Phot. Sci. Eng.*, **20,** 155 (1976).
Land, E. H., *Brit. J. Phot.*, **119,** 858 (1972), and *Phot. Sci. Eng.*, **16,** 247 (1972).
Meyer, A., *J. Phot. Sci.*, **13,** 90 (1965).
Meyer, A., *Phot. Sci. Eng.*, **18,** 530 (1974).
Pollak, F., *J. Phot. Sci.*, **3,** 112 (1955).
Pollak, F., and Hepher, M., *Penrose Annual*, **50,** 106 (1956).
Vittum, P. W., and Weissberger, A., *J. Phot. Sci.*, **2,** 81 (1954).

GENERAL REFERENCES

Bent, R. L., Brown, G. H., Glesmann, M. C., Harnish, D. P., Tremmel, C. G., and Weissberger, A., *Phot. Sci. Eng.*, **8,** 125 (1964).
Evans, R. M., Hanson, W. T., and Brewer, W. L., *Principles of Color Photography*, pp. 257–266, Wiley, New York (1953).
Glafkides, P., *Photographic Chemistry* (translated by K. M. Hornsby), **2,** 593–615, Fountain Press, London (1960).
Hornsby, K. M., *Basic Photographic Chemistry*, Fountain Press, London (1956).
James, T. H., and Higgins, G. C., *Fundamentals of Photographic Theory*, 2nd Edn., pp. 119–122, Morgan and Morgan, New York (1960).
Mees, C. E. K., and James, T. H., *The Theory of the Photographic Process*, 4th Edn., pp. 335–372, Macmillan, New York (1977).

Neblette, C. B., *Photography: Its Materials and Processes*, 6th Edn., pp. 240–248, Van Nostrand, New York (1962).

Thirtle, J. R., and Zwick, D. M., in *Encyclopedia of Chemical Technology*, 3rd Edn., **6,** 617, Wiley, New York (1979).

Vittum, P. W., *J. Soc. Mot. Pic. Tel. Eng.*, **71,** 937 (1962).

Vittum, P. W., and Weissberger, A., *J. Phot. Sci.*, **6,** 157 (1958).

CHAPTER 18

Image Structure in Colour Photography

18.1 Introduction

IF a colour photograph is examined by eye without any magnification, the only structure visible may be that of the subject matter of the picture. But, if increasing degrees of magnification are introduced, it will be seen that, just as in the case of black-and-white photography, the image is composed of a granular structure.

18.2 Magnifications

Let us consider a 35 mm slide as an example. If held in the hand, and viewed at about 10 inches, or 250 mm, distance, no granular structure will be visible; in this case, assuming a visual resolution of 20 cycles per degree (objects of $1\frac{1}{2}$ minutes of arc in diameter visible), the eye cannot see detail finer than about 5 cycles per mm, so that the smallest object visible would be about 1/10 mm in diameter. If, however, the slide is projected using a projection lens of 100 mm focal length, the magnification introduced will be 250/100, that is $2\frac{1}{2}$ times, if the screen is viewed from near the projector; or twice this amount, that is 5 times, if the observer sits half-way between the projector and the screen: the limits of resolution on the slide then become $12\frac{1}{2}$ or 25 cycles/mm respectively (see Table 18.1). At these magnifications the picture is usually still largely free of granular appearance, but

TABLE 18.1

Visual resolutions at various magnifications

Viewing situation	Magnification	Visual resolution				Structure visible
		High-contrast object		Low-contrast object		
		c/mm	μm	c/mm	μm	
Naked eye Projected with 100 mm lens	1	5	100	1	500	None
At projector	2½	12½	40	2½	200	None
Half-way from screen	5	25	20	5	100	Little or none
One-tenth from screen	25	125	4	25	20	Clumps of grains
Microscope	250	1250	0.4	250	2	Individual grains
Electron microscope	2500	12500	0.04	2500	0.2	Dye globules (if any)

if the magnification is increased much further some structure usually becomes apparent. If the screen is viewed from a distance equal to one-tenth of the projector-to-screen distance, so that the magnification is 25 times, then areas that appeared uniform before will now appear definitely granular, much less sharp, and lacking in fine detail: we are now seeing down to about 125 cycles/mm (so that objects of about 4 micro-metres diameter will be visible). If now, by using a microscope, we increase the magnification by a further ten times, to 250 times, so that we can see down to 1250 cycles/mm (0.4 micro-metre objects), we will see that the image is composed of small blobs of different colours. A further increase of magnification to about 2500 times, so that we can see down to 12 500 cycles/mm (0.04 micro-metre objects), might reveal that the blobs themselves consisted of clouds of small droplets of dye. (See Plate 16, pages 240, 241.)

The basic reason for the granular structure of most colour photographic images is that they are derived from silver-halide photographic emulsions which are themselves composed of discrete crystals, or *grains* as they are usually called. The average size of these grains varies from about 0.5 micro-metre (μm) in diameter for slow emulsions such as are used in print films, up to about 1.5 μm in diameter for fast emulsions such as are used in X-ray films. In any one emulsion the grains usually cover quite a range of sizes. Since the larger the grain the more light it can absorb in a given time, the range of sizes can be useful in providing grains having a range of sensitivities: this can result in an emulsion that can accommodate a wide range of exposures, that is, one which possesses good *exposure latitude*.

The mean grain-size in emulsions used for camera films in colour photography can be regarded as about 1 μm in diameter; hence the 25 times magnification involved in the close inspection of the 35 mm projected slide, with its 4 μm limit of resolution, does not enable us to see the grains. Why, then, does the picture look granular at this magnification? The reason is that the grains are

338

not present in a regular array, but are distributed more or less randomly, and this results in clumps of grains and grain-free areas of much more than 4 μm diameter being present: it is these areas, perhaps sometimes extending up to about 40 μm in diameter, that makes the picture appear granular.

In colour images the developed silver grains and the undeveloped silver-halide grains are usually all removed in the processing sequences, so that the blobs in the image are small volumes of cyan, magenta, or yellow dyes produced by the colour development step. If the couplers forming these dyes are uniformly dispersed in the emulsion layers, or are provided from colour developing solutions, then the small volumes of dye formed round each developed silver grain have only molecular structure; but if the coupler is dispersed in small oil globules then the volumes of dye have a globular sub-structure. The volumes of dye, which we may call *colour grains*, may be similar in size to, or slightly larger than, those of the silver-halide grains: thus the separate volumes may be about 1 μm in diameter (although they occur frequently merged together into larger volumes); but the individual oil globules, when present, are usually about 0.1 to 0.2 μm in diameter. It is thus clear why magnifications of 250 times (giving resolution down to 0.4 μm) are necessary to see the colour grains, and 2500 times (giving resolution down to 0.04 μm) are necessary to see the oil globules. These figures are summarized in Table 18.1 where limits for visual resolutions are given not only for high contrast detail, but also for low contrast detail for which the resolution is much reduced.

In motion-picture colour photography, higher magnifications are usually involved than in still photography, and the granular structure of the image is usually more apparent. Since the granular structure is random, for a given area of the scene it is different on successive frames of a motion-picture film. The visual result is that the granular pattern appears to be in rapid random motion, a phenomenon sometimes called *boiling*, because of its similarity to the random movements on the surface of a boiling liquid. The tendency of the eye to be attracted to motion in the field of view can make the grain in motion pictures more evident than in still pictures viewed at the same magnification. Focal lengths of typical motion-picture projection lenses (Happé, 1971) together with the magnifications involved for viewing at the projection position, and at distances equal to four-fifths, three-fifths, a half, and one-sixth, of the screen-to-projector distance, are given in Table 18.2.

It is usually considered good practice for the nearest seat in a cinema to be not closer than 0.87 times the screen width, and for the most distant to be not more than 6.0 times the screen width, away (Wheeler, 1969). The width of the picture on the film is typically about a fifth of the focal length of the projector lens (see Table 18.2); hence, as can be seen from the geometry of Fig. 18.1, the closest seat will be about 5/0.87, or nearly six times as close to the screen as the projector (ignoring any anamorphic factor). The last line of figures in Table 18.2 therefore represent the maximum magnifications for viewers in the front row. The magnifications involved for viewing at four-fifths, three-fifths, and midway from the screen are also given in Table 18.2. Three-fifths of the screen-to-projector distance is roughly equal to three screen-widths and this is sometimes regarded as a typical average viewing situation. Remembering that at magnifications greater than about 5 times, the granular structure of a typical film begins to become noticeable, the figures given in Table 18.2 suggest that film graininess is only likely to be apparent in 35 mm films for viewers closer to the screen than to the projector, but for 16 mm films it is likely to be apparent for viewers in an average viewing position, while for Super 8 and 8 mm films it is likely to be apparent from all viewing positions. These are broadly the results obtained in practice, although, of course, the grain tends to be worse for higher speed films (which usually have larger grains in order to obtain more speed) than for lower speed films; and grain may also be aggravated by duplication, particularly if internegative or intermediate films are used.

TABLE 18.2

Typical magnifications

Film Size	35 mm	16 mm	8 mm and Super 8
Typical projection-lens focal length	100 mm	50 mm	25 mm
Picture width on films (projection-aperture)	21 mm	10 mm	4.4 or 5.3 mm
Magnifications for various viewing positions			
At projector	$2\frac{1}{2}$	5	10
At four-fifths from screen (typical television viewer)	3	6	12
At three-fifths from screen (typical cinema seat)	4	8	17
Mid-way from screen	5	10	20
At a sixth from screen (closest cinema seat)	15	30	60

When film is used in television, the magnifications obtained are the same as those given in Table 18.2 for projectors, if the projector-to-screen distance is regarded as five times the television screen width (see Fig. 18.1); this is because the ratio of the typical projection-lens focal-length to film-width is five to one. An average television viewing distance is usually regarded as being equal to about five or six picture heights, so that, allowing for the 4 to 3 aspect ratio of television pictures, this is equivalent to about four picture widths, or four-fifths of the equivalent screen-to-projector distance. It is thus clear from Table 18.2 that grain should not be apparent when 35 mm film is viewed on television (a magnification of 3) but is likely to show when 8 mm is used (a magnification of 12); with 16 mm film, with a magnification of 6, the situation is a borderline one, and this suggests that grain may sometimes be apparent. This is broadly the experience in practice.

340

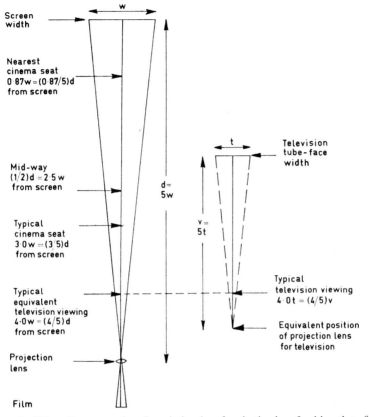

Fig. 18.1. When films are projected, typical ratios of projection-lens focal-length to film-width are 5 to 1, and this results in typical screen-to-projector distances, *d*, being equal to five times the screen width, *w*. For television viewing, the distance, *v*, of the equivalent projector is therefore 5*t*, where *t* is the width of the television tube face. Various typical viewing positions are shown.

18.3 Graininess and granularity

The extent to which the granular structure of a photographic image is apparent to an observer is termed the *graininess*; the physical property of the photographic materials that cause graininess is termed the *granularity*. Graininess is thus a subjective term; granularity is an objective term. At a given magnification, a particular piece of film illuminated in a given manner has an invariant granularity; but its graininess at that magnification will vary according to the conditions of viewing: for instance, if surrounded by an area of much higher luminance, the graininess may become imperceptible, but if surrounded by an area of much lower luminance, the graininess may be very apparent. Graininess

341

is usually objectionable in pictures because it results in uniform areas being reproduced with spurious texture (or motion, in cinematography), and because it results in fine detail being fragmented.

18.4 Granularity of silver images

Black-and-white photographic emulsions have a granular structure because they consist of the discrete silver-halide crystals, referred to as grains. On development, filaments of silver grow from the grains (initially from specks of latent-image), until a fully developed grain consists of a tangled mass of silver filaments; this would be much larger than the original grain were it not for the presence of the gelatin (in which the grains are imbedded when the film is coated) which confines the developed silver to the neighbourhood of the site of the original grain. The individual silver filaments are so fine that only an electron microscope can resolve them, and groups of many filaments tend to act as a single unit as far as affecting light is concerned. Since the refractive index of developed silver grains is very different from that of its gelatin matrix, silver grains not only absorb light but also scatter it considerably. This is illustrated by the fact that the *Callier Coefficient* (the ratio of specular to diffuse density; see Section 14.14) can be as high as about 2 for silver images; but for the dye images used in colour photography it is usually not higher than 1.1, because the refractive indices of the dye particles and gelatin are usually nearly equal in the dry film. With black-and-white silver images, therefore, the granularity may vary with the geometry of the illuminating and viewing beams of light, and hence any measuring system should duplicate the appropriate practical conditions in this respect; with colour images this is less necessary, but is still desirable because the surface of the film sometimes exhibits distortions (see Section 18.16).

The granularity of silver images has been extensively studied both theoretically (Selwyn, 1935) and practically (Selwyn, 1939; Selwyn, 1943; Higgins and Stultz 1959). Granularity is commonly evaluated in practice by measuring the density of the sample through an aperture of suitably small size (obtained with a microscope), while the sample (or the aperture) travels along a circular path whose radius is large compared to the size of the aperture (see Fig. 18.2). The fluctuations in density, as the path is followed, then provide a measure of the granularity. Experiments have shown that granularity measured in this way correlates well with graininess determinations, in that, for a given set of viewing conditions (including constant sample and surround luminances), granularity measurements usually rank samples in the correct order for graininess. According to Selwyn's theory, the granularity should be measured as the square root of the average of the squares (*root mean square* or *r.m.s.*) of the density fluctuations; this amounts, in fact, to measuring the standard deviation of the density fluctuations, σ_D. In granularity machines it is therefore desirable for the light transmitted by the sample through the aperture to be logarithmically amplified, to produce signals proportional to density, and then for these to be measured on

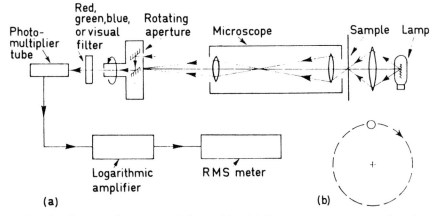

Fig. 18.2. One type of r.m.s. granularity machine. (a) General arrangement: an enlarged image of the sample is focused by a microscope on a screen with a rotating aperture, and the r.m.s. meter measures the fluctuation in an electrical signal that is proportional to the logarithm of the light transmitted by the aperture. (b) Path followed by the aperture (Rotthaler, 1974).

an r.m.s. electrical meter (as indicated in Fig. 18.2). Selwyn's theory predicted that the granularity measured in this way would be inversely proportional to the square root of the area, a, of the scanning aperture, so that

$$\sigma_D (2a)^{1/2} = K$$

where K is a constant known as *Selwyn granularity*, and represents, for any given image, its basic density-fluctuation independent of scanning-aperture area. For silver images, this relationship has now been confirmed by experiment. (When the area of the scanning-aperture becomes large, slow drifts in density over the sample area, and the presence of dirt, tend to make the experimental results too high unless extreme precautions are taken; a scanning aperture of about 24 μm diameter has been found to give more reproducible results than larger or smaller apertures (Mees and James, 1966)). Since the square root of the scanning-aperture area is proportional to its linear size, σ_D is proportional to the linear magnification of the sample, for black-and-white silver images.

A widely-used method of expressing granularity measurements is to multiply the standard deviation of the density fluctuations, σ_D, by 1000 to obtain

R.M.S. Granularity = 1000 σ_D

using a circular scanning aperture of 48 μm diameter, on samples of average density 1.0, in a diffuse density measuring mode. If r.m.s. granularity = 10 (a typical value for a negative film) then $\sigma_D = 0.01$; this means that 30 per cent of the density fluctuations (assuming they have a statistically normal distribution) will exceed ± 0.01, 5 per cent will exceed ± 0.02 and 0.3 per cent will exceed ± 0.03.

343

From Table 18.1, it might be thought that the scanning aperture of 48 μm should correspond to a magnification of about $2\frac{1}{2}$ times, but it has been estimated (Stultz and Zweig, 1959) that it represents a magnification of about 12 times. This difference arises because the better figures for visual resolution given in Table 18.1 refer to detail of high contrast, whereas graininess near the threshold of its visibility is usually of quite low contrast; it is therefore more appropriate to use the poorer figures for visual resolution given in Table 18.1 for low contrast detail, and if this is done it is seen that a 48 μm aperture does correspond to a magnification of about 12 times. It can be seen from Table 18.2 that a magnification of 12 times corresponds to the viewing of projected 8 mm and Super 8 film at a distance of about four-fifths from the screen.

Selwyn's theory also predicted that granularity would be proportional to the square root of the density. For black-and-white silver images of limited grain-size spread in which all grains are fully developed, this square-root relationship holds; but with many combinations of emulsion and developer, the spread of grain-size, and partial development of some grains (especially at high densities), and other effects, often make the granularity increase more slowly with density at medium densities than the square-root relationship would suggest.

When the granularity of negative materials is being considered, it must be remembered that these materials are usually used by printing them on to high-contrast print-films or papers; the granularity of the print will then depend on the optical sharpness of the printing step and on the granularity and gamma of the print-material as well as on the granularity of the negative. Thus a somewhat unsharp printing step will tend to reduce the effects of negative granularity, but a high gamma in the print material will tend to increase them.

18.5 Wiener spectra

Consideration of these factors is facilitated by considering the nature of the density fluctuations more fully. The output from the micro-densitometer in a granularity machine can be analysed by passing the electrical signals through narrow-band frequency filters: if this is done, the power of the variation in the electrical signals can be measured for different frequencies (Doerner, 1962; Wall and Steel, 1964; De Belder and De Kerf, 1967; Verbruggher, De Belder and Langner, 1967). Since the electrical frequencies correspond to spatial frequencies on the film, the amount of density fluctuation at different spatial frequencies can be determined: examples of such measurements, which are known as *Wiener spectra*, are given in Fig. 18.3; in this figure P is plotted against f, where f is the spatial frequency in cycles/mm, and

$$P = \sigma^2_{Df}$$

σ_{Df} being the standard deviation of the density fluctuations in a narrow band of spatial frequencies centred on f, for the particular area of scanning aperture chosen for use in the micro-densitometer. It is customary to plot the square of σ_{Df},

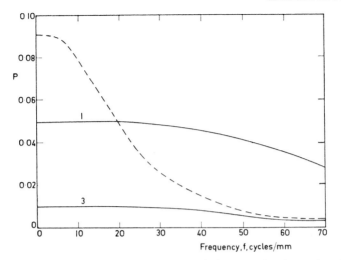

Fig. 18.3. Wiener spectra for a negative film (curve 1), for a print film (curve 3), and for a print made from the negative film on the print film (broken curve). The square, P, of the standard deviation of the density fluctuations, σ_{Df}, in a narrow band of frequencies centred on the frequency, f, is plotted against f.

because, when this is done, the total density-fluctuation, σ_D, for all the frequencies, is given by:

$$\sigma_D = A^{\frac{1}{2}}$$

where A is the area under the Wiener spectrum curve. Thus, for a given area of scanning aperture, the r.m.s. granularity is proportional to the square root of the area under the Wiener-spectrum curve (Rotthaler, 1974).

In Fig. 18.3, curve 1 shows the Wiener spectrum for a negative film, and curve 3 that for a print film. It is clear that both curves are fairly flat apart from curving downwards at the high-frequency end: this reduction at high frequencies is caused by the finite size of the scanning aperture on the granularity machine, and, without this effect, the curves would probably be completely flat (over the range of frequencies normally considered), a result sometimes referred to as 'white noise'. The fact that curve 1 is wholly above curve 3 shows that the negative film has higher density-fluctuations than the print film at all spatial frequencies resolved by the scanning aperture.

The range of frequencies over which P is gradually reduced to zero depends on the size of the scanning aperture: the smaller the aperture, the more high-frequency response will be retained. Hence, if the square root of the area under the Wiener-spectrum curve is to be correlated with graininess, it is necessary to choose the area of the scanning aperture so as to make the high-frequency losses occur at the same frequencies as in the eye, with due allowance being made for

345

any magnifications involved. As can be seen from Fig. 18.13, the eye also has reductions in response at low frequencies, and it may be desirable to allow for this by including an electrical filter providing appropriate low-frequency attenuation in the output from the granularity machine. Alternatively, by using no low-frequency filter, and by using an extremely small scanning aperture, a Wiener spectrum having maximum values of P can be obtained: to correlate such spectra with graininess, it is then necessary to allow for the effects of the eye by suitably weighting the values of P at different frequencies (using the square of a function that might be similar to that shown in Fig. 18.13) before integrating the area under the curve.

The broken curve in Fig. 18.3 shows the Wiener spectrum that might be obtained from a print made using the films having the Wiener spectra shown by curves 1 and 3.

If the printing step were perfectly 'sharp', so that it resulted in no loss of information, and if the print film had a gamma of 1.0, then the values of P for the print would be simply the sum of those for the two films at each frequency. But it is clear from Fig. 18.3 that, at low frequencies, the value of P for the print is greater than the sum of the individual values for the two films: this is because the gamma of the print film, being greater than 1.0, results in an increase in the contribution of the negative density-fluctuations to the total. It is also clear from the figure that at high frequencies the value of P for the print approaches that of the print film alone: this is because the lack of sharpness in the printing step, and in the print film, results in the negative density fluctuations being blurred away.

These effects can be expressed by the equation (Doerner, 1962):

$$P_T = P_3 + P_1 \gamma_3^2 M_2^2 M_3^2$$

where P_T is the value of P for the print, P_3 that for the print material, and P_1 that for the negative film: γ_3 is the gamma of the print material; M_2 is the modulation transfer function (to be described in Sections 18.12 and 18.13) of the optics of the printing step, and M_3 is the photographic modulation transfer function of the print material. The factors M are functions of frequency, and they are normally about 1.0 at low frequencies, and gradually decrease to zero as the frequencies become high: they thus allow for the fact that both the printing step and the print material will tend to blur out the negative density-fluctuations to progressively greater degrees as the spatial frequency increases.

If it is required to correlate P_T with the visual effects of the density fluctuations, and the precaution has not been taken of restricting the scanning-aperture size appropriately and using a suitable low-pass filter, then it is necessary to multiply P_T by W_E^2, where W_E is the suitable weighting function for the eye, due allowance being made for the magnifications involved. If the area under the curve of $P_T W_E^2$ is A_E then an eye-weighted density fluctuation, σ_{DE}, is given by

$$\sigma_{DE} = A_E^{\frac{1}{2}}$$

For black-and-white silver images, σ_{DE} is proportional to linear magnification, m, but for colour images this is not generally true.

Because print granularity is proportional to $A^{\frac{1}{2}}$ or $A_E^{\frac{1}{2}}$, it is clear that it will increase as γ_3, the print-material gamma, increases. The gamma of most print materials becomes progressively lower as the density level on the print decreases, and this tends to make print granularity decrease with print density.

18.6 Graininess in prints

The graininess that is seen in prints depends not only on the print granularity, but is also markedly affected by the density of the area concerned in the print: for a given print granularity, as the print density increases, the graininess generally becomes less and less noticeable, because of the difficulty of seeing into the shadows (Lythgoe, 1932). Thus, print granularity depends on negative granularity and on print-material granularity; it decreases with decreasing sharpness in the printing; and it decreases with decreasing print-material gamma (and hence usually with decreasing print density); but, for a given print granularity, graininess decreases as print density increases. As a result of these conflicting effects, in black-and-white silver negative-positive reflection-print systems, print graininess is usually greatest at a print density of about 0.6, and, as this usually corresponds to a negative density of about 0.8, granularity of silver negative images is often measured at about this density level.

18.7 Granularity of colour images

Since colour images are normally derived from black-and-white silver images, it might be expected that colour granularity would be similar to black-and-white granularity. There are similarities, but there are also a number of important differences.

The individual silver grains of black-and-white images are opaque, and they scatter light as well as absorbing it. The dye clouds of colour images are partially transparent, and are usually non-scattering. The dye clouds in the fully processed and dried photographic layer are usually very far from spherical. Although the clouds may be roughly spherical when formed, this takes place in the swollen wet condition of the layer during processing; on drying, the layer shrinks to only about a fifth of its wet thickness and so the dye clouds become very flattened as shown in Fig. 18.4.

The total visual graininess is made up of superimposed cyan, magenta, and yellow grain patterns, and it might therefore be thought that the graininess would appear to consist of fluctuations in both brightness and colour. However, for most applications, films are used near enough the threshold of perceptible graininess for the fluctuations in colour to be largely unnoticeable. To see colour differences

Thickness of layers
(a) during development
(b) after drying

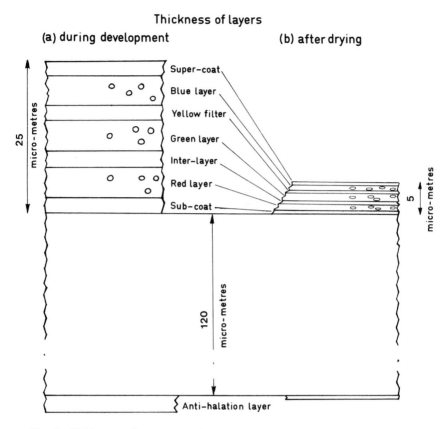

Fig. 18.4. Thicknesses of components of a typical multi-layer colour film (a) when swollen during aqueous development, (b) after drying. The shrinking that takes place during drying results in dye-clouds that were spherical during development being squashed down to about one-fifth of their thickness after drying.

it is necessary for each of the three types of retinal cone to be represented in each elemental area; to see brightness differences it is only necessary for one of the three types of cone to be represented (or, more precisely, any one of two of the three types, since the short-wavelength cones contribute negligibly to brightness). Bearing in mind the random distribution of the cones in the retina, this makes the magnification at which brightness differences are visible about 4 times less than that at which colour differences are visible (Hunt, 1967). It has therefore been found that the graininess of colour films correlates well with just luminance fluctuations. These can be measured on the type of equipment shown in Fig. 18.2 with a detector-filter combination whose spectral response duplicates the spec-

348

tral luminance function of the eye (visual filter in Fig. 18.2); or a composite granularity can be derived by adding together weighted proportions of the cyan, magenta, and yellow granularities as determined separately through red, green, and blue filters, which are also indicated in Fig. 18.2 (Ooue, 1960; Zwick, 1963). The proportions combined are usually about 30 per cent for the cyan, 60 per cent for the magenta, and 10 per cent for the yellow; these are very similar to the proportions of red, green, and blue signal in the luminance signal in colour television (see Section 22.3). Although these proportions are appropriate for films viewed by transmission, when the final image viewed is a reflection print, the proportions then become about 40 per cent for cyan, 45 per cent for magenta, and 15 per cent for yellow, for daylight illuminants, or about 50 per cent for cyan, 40 per cent for magenta, and 10 per cent for yellow for tungsten light (Sawyer, 1980).

For black-and-white films (as mentioned in Section 18.4) it has been established, both by theory and by experiment, that r.m.s. granularity, σ_D, is proportional to magnification: for colour films this is not always so. This difference is demonstrated in Fig. 18.5 (Zwick, 1963), where a dye image has higher values of σ_D than a silver image at magnifications below 20, but lower values above 20. Furthermore colour films vary from one to another in their σ_D-magnification relationships. Hence, if two colour films are compared for r.m.s.

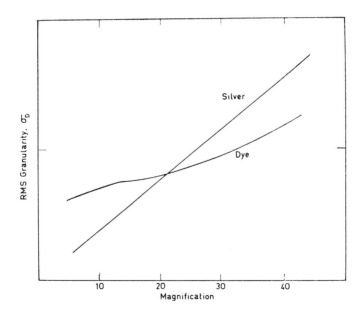

Fig. 18.5. For silver images, the standard deviation of the density fluctuations, σ_D, is proportional to magnification. But for the dye images used in colour film this may not be so, and different colour films may have σ_D-magnification relationships that differ from one another.

349

granularity with one scanning aperture and found to be the same, it cannot be assumed that they will also have equal r.m.s. granularities when a scanning aperture of different size is used. Graininess determinations confirm that two colour films may have similar graininess at one magnification, but different graininess at another magnification. This phenomenon is caused by the nature of the dye clouds. For instance, if one film had fairly small but dense dye clouds, its r.m.s. granularity would fall very quickly at low magnifications; but in another film, in which the dye clouds tended to be more diffuse and of lower density, the r.m.s. granularity could be smaller than that of the first film at high magnifications, but larger at low magnifications. A consequence of this is that granularity measurements on colour films should ideally be made with scanning apertures appropriate to the magnifications likely to be used with each film.

For black-and-white films, granularity increases with density. For colour films this is not always so. At high densities the dye clouds may merge together so that granularity can decrease. In a coupler-incorporated film, at very high

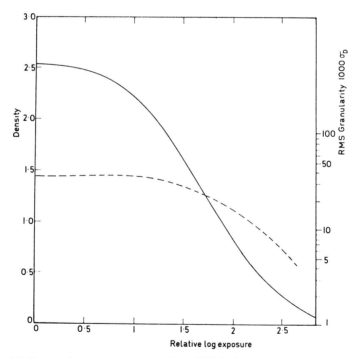

Fig. 18.6. Density, log-exposure characteristic (full line), and r.m.s. granularity, log-exposure characteristic (broken line), for a reversal colour film. Granularity reaches a maximum at a density of about 2.0, but maximum graininess usually occurs at a density of about 1.0, because of the visual difficulty of seeing the effects of granularity in dark areas of the picture.

densities, all the coupler may be converted into dye, in which case the nonuniformity will be only that of the oil globules or molecular dispersion (apart from the gaps at the sites of bleached silver-halide grains, but the effect of these is usually negligibly small). It is therefore important that the granularity of colour films be measured at the density at which graininess is most apparent: for reversal films this is usually between 0.9 and 1.0, and measurements on reversal films are therefore often made at a density of 1.0 (Zwick, 1972); for negative films the situation is more complicated, as will be discussed below.

In Fig. 18.6 r.m.s. granularity measurements are shown for a colour reversal film (Davies, 1970). It is seen that granularity rises with density up to a maximum at about 2.0 and then remains fairly constant; but, as already explained, the higher granularity at the higher densities is reduced in visibility by the relative darkness of the areas of high density as normally viewed (Lythgoe, 1932), and graininess is usually at a maximum at densities of about 1.0. In this figure the r.m.s. granularity is plotted on a log scale, and this has the advantage of being more nearly visually uniform than a linear scale.

For colour negative films, the graininess produced on prints made from them will depend, as in black-and-white systems, on the granularity of the negative film, on the sharpness of the printing step, on the granularity, sharpness, and gamma of the print material, and on the density level at which the negative is printed on the print material. It has been shown that for colour images, as for silver images, if, as is usual, the granularity of the print material is much less than that of the negative film, the granularity of the print increases approximately proportionally with the gamma of the print material (Zwick, 1965). It has also been shown that an 8 per cent change in gamma causes a 6 per cent change in r.m.s. granularity: this change in granularity corresponds to about one just noticeable difference (j.n.d.) in graininess for critical densities for uniform fields: for typical pictures one j.n.d. corresponds to a change of about 15 per cent in r.m.s. granularity (Zwick and Brothers, 1975). Since, unlike silver negatives, colour negatives do not have granularities that are simply related to negative-density, it is desirable to measure granularity at various density levels on negative films; and graininess on print materials should be determined from negatives of various densities, such as arise from different levels of camera exposure. The same arguments apply whether the print material is a film or a paper, although different magnifications may be involved in the two cases.

In Fig. 18.7 r.m.s. granularity measurements are shown for a colour negative film (and for a black-and-white film for comparison), and it is seen that for this particular film the granularity first rises with density, reaches a maximum, and then falls, after which it rises again to reach a second but lower maximum and then falls again (Morris and Wait, 1971). This type of variation of granularity with density is peculiar to colour films: one way in which it can arise is as a result of a layer of a colour film being coated in two parts. It has already been mentioned that one way of obtaining a wide exposure latitude in films is to use emulsions with a wide range of grain sizes. If, however, it could be arranged that all the large fast grains were coated at the top of a layer and all the small slow grains at the bottom, then an increase of speed would be obtained, because the fast grains would receive the light without any of it being absorbed or scattered by grains above them; furthermore, an increase in exposure latitude would be obtained because the slow grains would be reduced in speed because of absorption and scattering above them. One way of achieving this type of result is to coat the layer in two component layers: first a slow, finer-grained emulsion, and then a fast, coarser-grained emulsion on top. (Kennel, Sehlin, Reinking, Spakowsky, and Whittier, 1982; Vervoort and Stappaerts, 1980). If this were done with a black-and-white film, granularity would increase with density rapidly at first, as the fast emulsion was exposed, and then more slowly as the slow emulsion was exposed. But with colour emulsions, if the granularity of the fast emulsion were to decrease at high densities because of dye clouds merging together, then it would be possible for the granularity to occur as shown in Fig. 18.7. When films of this type are used, it is important to avoid under-exposing if minimum graininess is required; this is contrary to black-and-white practice where, because granularity increases with negative density, graininess is minimized by giving the minimum satisfactory negative exposure. In some fast colour negative films, the fast red-

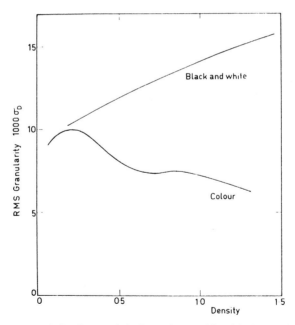

Fig. 18.7. R.m.s. granularity characteristics for a colour, and for a black-and-white, negative film. For the colour film granularity rises rapidly at first, and then drops down again, as exposure is increased: with this type of film it is important to avoid under-exposure if minimum graininess is required.

352

Plate 22

These two pictures are *Colour Derivations* (see Section 15.3). Using only photographic methods, such as masking, flashing, and reversal of tones and colours, a wide variety of results can be obtained, similar to those often achieved with the aid of computers. Originals by courtesy of Richard Tucker, F.R.P.S. Separations made on a Crosfield *Magnascan* scanner.

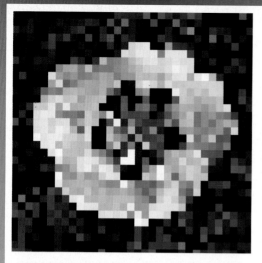 

Plate 23

Graphic arts scanner systems can be used to provide extensive manipulation of images (see Section 29.9). In these examples, the original picture of the peony (shown at 1) has been altered in the following ways:

2: background leaves removed, size reduced, and extra images added with white petals changed to cyan, magenta, and yellow.

3: white petals changed to green with yellow segments.

4: white petals changed to yellow in medium shadows and to red in dark shadows; background lightened and changed to red in lightest areas.

5: the colour allotted to each small square is the average for that square; when viewed from a distance, this picture looks like the unaltered original.

6: white petals changed to cyan.

7: white petals changed to yellow, and sawtooth waveform applied to horizontal lines of the picture.

8: all parts of the flower changed to colours depending on their densities.

9: same as for 8, but with the background given various colours depending on the density and on the distance from the bottom of the picture.

10: white petals changed to cyan, magenta, or yellow, according to angular position.

Image manipulations by David MacKenzie using Crosfield *Studio* electronic page make-up equipment. Reproduced from a 36 × 24 mm *Ektachrome* transparency on a Crosfield *Magnascan* scanner system.

5	6
7	8
9	10

1	2
3	4

Plate 24
Upper. Reproduction of an enlargement made from part of a 12 × 16 mm negative, which was exposed under adverse lighting conditions. *Lower*. The same picture after image enhancement: intermediate electronic signals were adjusted to improve tone and colour reproduction, to increase sharpness, and to reduce graininess (see Section 16.15). Separations made on a Crosfield *Magnascan* scanner. Pictures by courtesy of the Eastman Kodak Research Laboratories.

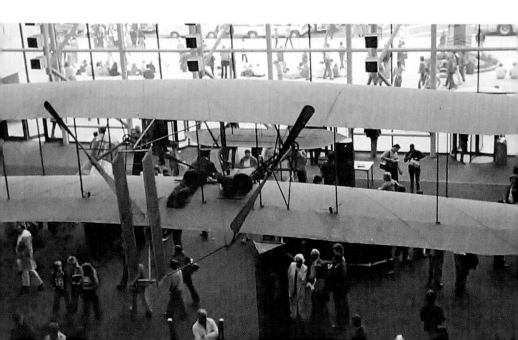

sensitive emulsion is coated above the slow green-sensitive emulsion in order to increase the effective speed of the former (Meyers and Dalton, 1979; Maude, 1980; Vervoort and Stappaerts, 1980). (See Plate 10, pages 112, 113.)

18.8 Reducing granularity of colour systems

Methods of reducing the granularity of colour films have included improvements in the basic speed-grain relationship of the silver-halide emulsions, and also improvements in the granularity of the dye cloud relative to that of the basic silver halide. Colour development tends to give images of higher density and gamma than the corresponding silver images, and hence to obtain the required tone reproduction the amount of silver halide coated has to be reduced. This, however, tends to increase granularity because a given density is now produced by developing fewer centres in a given area. Granularity can be reduced, therefore, if the dye-forming step is made less efficient so that more centres have to be used. This can be done by reducing the amount of coupler available so that not all the oxidized developer can form dye, or by adding a competing coupler, either to the film or to the developer, which couples with some of the oxidized developer and forms a soluble dye which is subsequently removed. Citrazinic acid, for example, can be used for this purpose as a competing coupler in developers (Thirtle and Zwick, 1964). Similar reductions in efficiency can also be made by using developer-inhibitor-releasing (DIR) couplers (see Section 17.11) (Barr, Thirtle, and Vittum, 1969). By these and other methods, the size of the dye clouds can be reduced, and the number of them per unit area for a given density can therefore be increased, and hence the granularity reduced. It has been shown that for colour images, as for silver images, r.m.s. granularity, for a given size of scanning aperture, is inversely proportional to the square root of the number of image centres (Zwick, 1965).

If the surface to volume ratio of silver halide grains is increased, it is possible to increase the amount of sensitizing dye adsorbed to the surface of the grains for a given weight of silver halide. The surface to volume ratio can be increased by decreasing grain size, but this reduces speed; however, the surface to volume ratio can also be increased significantly by changing grain shape. One way of doing this is to make the grains of tabular shape. As a result of the increased amount of sensitizing dye present, the speeds of these *T-grains* are increased; and they may have sufficiently enhanced green and red speed relative to their natural blue speed to make it unnecessary to use a yellow filter layer, and omission of this filter layer can further increase speed (see Plate 10, pages 112, 113.)

When a negative film and a print film are used it is often important to know to what extent each is contributing to the final graininess seen on the screen. With motion-picture films, a useful technique to adopt is to make, in addition to a normal print, a print in which all the frames on the print film are made from one stationary frame of the negative film: the negative then contributes a stationary pattern of grain, but all the moving grain, or boiling, must come from the print film. The procedure can be further elaborated by having only one of the three layers of the negative film printed as a stationary frame, and in this way the contributions of the cyan, magenta, and yellow layers of the negative film to graininess can be separately assessed (Zwick, 1963).

18.9 Sharpness

Because light is diffused in photographic layers, it is clear that, as the magnification is increased, some noticeable blurring of edges in pictures must occur, and eventually fine detail will be completely obscured. The extent to which this is visible in pictures is usually referred to as *sharpness* (or unsharpness, to be precise). The granular structure of the image can also affect sharpness, but is usually only a minor factor; there are other factors that can be important, however.

Thus, if the optical image falling on the photographic material is blurred, the picture will, of course, be unsharp no matter how little the light is diffused in the layer: this can arise because of incorrect focusing of the camera or printer, or because of the optical aberrations in the lens, or because of movement of the image during the exposure, or because of the inability of the camera lens to bring objects at different distances all in focus at the same time. Another possible cause of unsharpness is diffusion of chemical constituents during processing causing the photographic image to be only diffusely related to the optical image. The sharpness of pictures is an important feature affecting their quality, and must therefore be considered in addition to graininess when assessing image structure. (See Plate 16, pages 240, 241.)

18.10 Focusing

If sharp pictures are to be obtained, it is clearly essential that the image on the photographic material should be in optical focus. Very simple cameras usually have their lenses fixed in position relative to the film plane, and this severely limits the range of distances from the camera for which objects will be in focus (as will be discussed further in the next Section). More elaborate cameras provide facilities for adjusting the distance of the lens from the film, and the adjustment may be set from one of a small number of scene types (such as views, groups, and portraits), or by guessing the distance in metres or feet, or by the use of an optical rangefinder, or by inspection of the image as in single-lens reflex cameras. Automatic focusing devices are also obtainable, based on several different principles. The Honeywell *Visitronic* system, by means of two lenses a few centimetres apart, forms two images on a sensor comprising an array of charge-coupled devices (CCDs); a mirror is used to move one of the images until the two sets of signals produced by the two images on the array have maximum correlation: the position of the mirror is then correlated with the distance of the object being photographed and is used to set the focus automatically (British

Journal of Photography, 1977). The Polaroid *Sonar Autofocus* system relies on the time taken for a burst of ultrasonic waves to reach the object and be reflected back to a detector on the camera (Mannheim, 1978; Crawley, 1979). Other devices emit a flash of light or infra-red radiation and depend on the amount reflected back from the scene, the amount being correlated with distance for scenes of average reflectance. All these devices are useful, but can sometimes give false readings from unusual scenes or situations, such as photography through a glass window; and none of them solve the commonly-met problems of having to produce sharp images of objects at different distances from the camera simultaneously.

18.11 Depth of field

When we look at the objects in a scene, there is a strong tendency for all of them to appear to be in focus at the same time even though they may be at a wide range of distances from us. This is partly because the diameter of the part of the lens of the eye normally used is only a few millimetres; partly because the eye looks at a scene by scanning from one object to another, changing its focus rapidly in the process; and partly because of the emphasis on object-recognition in vision rather than any preoccupation with optical phenomena. (Of course, the reduced ability to focus with increasing age is noticeable, often because of difficulty with reading.) An apparently faithful reproduction of a scene, therefore, often requires that objects at a whole range of different distances from the camera should be sharp in the picture. (See Plate 14, page 177.) To achieve this *depth of field* places certain constraints on the characteristics of camera lenses.

For calculating the range of distances over which objects in a scene look acceptably sharp in pictures, it is commonly assumed that point-objects should produce images having visual angles not exceeding $1/1000$ radian (that is, 3.6 minutes of arc, or a diameter of 0.25 mm viewed at 250 mm). Hence, if objects in a plane at a distance, l_1, from the camera lens are sharply focused, those in planes at distances, l_0 and l_2, from the lens will be just acceptably sharp if points in these planes form images that subtend $1/1000$ radian at the camera lens, assuming that the picture is viewed at correct perspective; it can be shown that this image-spread will occur when the angles subtended by the lens at these planes differ from that at the plane distant l_1, by $\pm 1/1000$ radian. Therefore, if the diameter of the lens is d, it must follow that:

$$d/l_0 - 1/1000 = d/l_1 = d/l_2 + 1/1000$$

which can be rewritten as

$$l_0 = 1000d/(n+1) \quad l_2 = 1000d/(n-1)$$

where $n = 1000d/l_1$. The quantity $1000d$ is known as the *hyperfocal distance* and is the value taken by l_0 when l_1 is infinity. The equations for l_0, and l_2, given above, show that, if the lens is focused for a distance equal to $1/n$ of the hyperfocal

355

distance, then the depth of field is from $1/(n+1)$ to $1/(n-1)$ of the hyperfocal distance. If more stringent requirements for sharpness are necessary, then the hyperfocal distance must be increased from $1000d$ to a higher figure, such as $2000d$.

The depth of field thus depends only on n, that is, on d/l_1; hence, for any given value of l_1 (the distance of sharp focus), the depth of field depends only on d, the diameter of the camera lens. The depth of field does not depend on the focal length of the lens at all. But the illuminance produced on the film by a lens of a given diameter is inversely proportional to the square of the focal length, so that, for a given depth of field and film speed, short focal-length lenses can operate at lower lighting levels than long focal-length lenses.

Correct perspective is obtained in pictures when the viewing distance is equal to the focal length of the camera lens multiplied by the total magnification at which the film image is viewed. In popular photography, this may occur when an enlargement of about 150 × 200 mm size (250 mm diagonal) is viewed at 250 mm; this is because the focal lengths of typical camera lenses are usually roughly equal to the diagonal of the film frame, and hence with this size of enlargement the viewing distance is equal to the focal length of the lens multiplied by the magnification. Popular size 'en-prints' of about 75 × 100 mm (125 mm diagonal) are thus normally viewed at about twice the distance for the correct perspective and this tends to increase their depth of field; correct perspective for this print size requires a focal length of about twice the diagonal of the film frame, which is usually regarded as in the 'telephoto' range.

18.12 Modulation transfer functions

It might be thought that a simple way of measuring sharpness would be to measure the finest pattern of dots or lines that could be resolved. Such measurements of *resolving power* are useful in some applications, especially if the contrast of the test object is chosen appropriately, but they do not always correlate well with apparent picture sharpness; a picture with higher resolving power may appear less sharp than another picture with lower resolving power.

A very useful concept when dealing with the sharpness of imaging systems is the *modulation transfer factor*. This provides a measure of the degree to which a system reduces the contrast of detail of a certain fineness. The modulation transfer factor consists of the ratio of a measure of the contrast in the image to the same measure of contrast in the object, at the fineness of detail considered. The factor usually has a value of about 1.0 (or 100 per cent) for coarse detail, and zero for detail that is so fine as to be completely undetectable in the image: it is the value of the factor at all the intermediate degrees of fineness that provides the useful measure of sharpness.

To calculate modulation transfer factors, it is necessary to have a means of measuring the contrast in the image relative to that in the object, for various degrees of fineness of detail. This is usually achieved by using *sine-wave test-objects*:

356

in these, as shown in Fig. 18.8, the luminance, L, varies sinusoidally along its length (that is, variations in L are proportional to sin x where x is the distance along the test object); the frequency of the sinusoidal variation is usually altered in steps along the object, so as to provide areas having patterns with various fineness of detail (Lamberts, 1963).

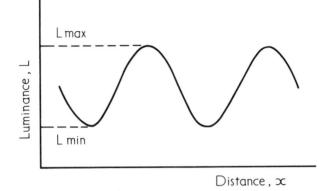

Fig. 18.8. Distribution of luminance, L, along the length, x, of a sinusoidal test object of amplitude $L_{MAX}-L_{MIN}$. The variations in L are proportional to sin x.

The measure of contrast, C, used for both the test object and for the image is usually:

$$C = \frac{L_{MAX}-L_{MIN}}{L_{MAX}+L_{MIN}}$$

where L_{MAX} and L_{MIN} are the maximum and minimum values of L in the object or image. If, at some frequency on the sine-wave test-object, the values of C are C_o for the object, and C_i for the image, then the modulation transfer factor is given by the ration C_i/C_o. The way in which the modulation transfer factor varies with the frequency of the sinusoidal pattern is the *modulation transfer function* (MTF), and this is usually plotted as a graph, as shown in Fig. 18.9, with frequency as abscissa and modulation transfer factor as ordinate. It is seen that in this case, which refers to the on-axis performance of a lens, contrast is reduced to 50 per cent at 50 cycles/mm and to 20 per cent at 100 cycles/mm.

There are three reasons why test objects are used that have a sinusoidal distribution of luminance variation along their length. First, it has been shown that a sinusoidal distribution of light remains sinusoidal (even though the amplitude may change) after imagery, regardless of the characteristics of the image-forming system. (This is only true in *linear* systems, which are systems in which the output signal is proportional to the input signal; if a non-linear stage is involved, as can occur if the light is transduced into electrical signals, as in television, or into silver dye deposits, as in photography, then distortions from the sinusoidal distribution occur: the effects of this will be discussed in Section

357

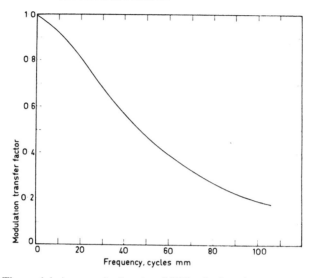

Fig. 18.9. The modulation transfer function (MTF) of a lens (measured on its axis). Contrast is reduced to 50 per cent at 50 cycles/mm and to 20 per cent at 100 cycles/mm.

18.13.) Secondly, any other distribution of light can always be made up of a collection of sinusoidal distributions of appropriate frequencies and amplitudes. An example of the way in which a square-wave distribution can be synthesized in this way is shown in Fig. 18.10 (Selwyn, 1959). Thirdly, when a sinusoidal image passes through several image-forming systems, the overall modulation transfer factor at any frequency is equal to the product of the individual modulation transfer factors of the various stages of the system: hence the overall modulation transfer function can be obtained by taking individual functions and calculating

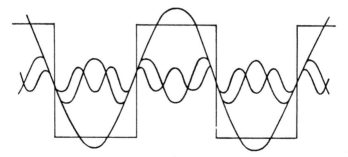

Fig. 18.10 A square-wave distribution of light can be approximated by a series of superimposed sine waves whose amplitudes are added together. Only the fundamental frequency and the first two harmonics are shown; perfect re-constitution of the square wave requires an infinite series of harmonics.

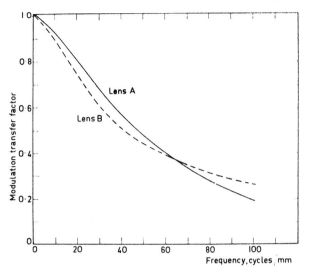

Fig. 18.11. Modulation transfer functions of two lenses (measured on their axis). Lens B has the higher on-axis resolving power, but lens A would give sharper on-axis pictures in systems in which all frequencies above 60 cycles/mm were ineffective.

the products at every frequency (but, again, this is only true if all the components of the system are linear).

Fig. 18.11 shows the on-axis modulation transfer functions for two lenses, and illustrates how resolving power and sharpness might fail to correlate. Lens B has the greater resolving power, in that it shows higher resolution at all frequencies above about 60 cycles/mm. But lens A, although having a lower MTF at high frequencies, has a higher MTF at frequencies up to about 60 cycles/mm. Hence, in an application where other factors (such as diffusion in photographic layers or the limit of visual acuity) result in all information above 60 cycles/mm being totally lost, lens B would show the poorer sharpness. Of course, a full evaluation of two lenses would require MTF's to be compared at various positions in their field as well as on-axis.

18.13 Photographic modulation transfer functions

As already mentioned, when the light is transduced into electrical signals as in television, or into silver dye deposits as in photography, non-linear processes can occur; in this case, a sine-wave input no longer results in a pure sine-wave output of the same frequency: harmonic frequencies are also generated.

The scattering of light within a photographic emulsion layer is a linear process, and hence the MTF of the diffusing properties of such a layer can be determined if evaluations are made of the ratio

$$C_i = \frac{H_{MAX} - H_{MIN}}{H_{MAX} + H_{MIN}}$$

for different frequencies, where H refers to the exposure within the film. The values of H have to be determined by developing the film, measuring the densities produced, and using a $D - \log H$ curve of the film to derive H. This is a strictly valid procedure only if the $D - \log H$ curve used is appropriate for all the frequencies considered: this is only the case if the photographic processing acts equally on all exposures irrespective of their spatial frequency (there being no *adjacency effects* or *edge effects* see Section 18.16). In such cases the MTF of the diffusing properties of a photographic material can be determined and the results are independent of level of exposure and degree of development. The exposure is normally made using a lens, and hence the photographic record results from the combined MTFs of the lens and the photographic diffusion: the MTF of the photographic diffusion alone can be obtained by dividing the combined result, at each frequency, by the MTF of the lens alone.

If, however, the degree or manner of development is affected by the spatial frequency of the exposure, then, to determine H, different $D - \log H$ curves should be used for each frequency: but this complication is not usually included, a single $D - \log H$ curve being commonly used. In this case, we obtain what we may call the *photographic modulation transfer factor*, C_e/C_o, where C_e, the effective photographic contrast, is given by:

$$C_e = \frac{H'_{MAX} - H'_{MIN}}{H'_{MAX} + H'_{MIN}}$$

where H' is the effective exposure obtained by using the single $D - \log H$ curve. The use of a $D - \log H$ curve which is different from the true one, for any frequency, generally introduces non-linearity into the system because a change in curve shape or gamma usually occurs. The calculated values of H' do not then correspond to a pure sine wave. But curves of photographic modulation transfer factor, which we may call *photographic modulation transfer functions*, or *photographic MTF*, based as they are on the effective exposure, H', include the effects of photographic processing, and this is useful in assessing the performance of a photographic material as a recording medium, particularly when adjacency effects occur.

In Fig. 18.12 several photographic MTF curves are shown. It is clear that one of these curves has values of photographic MTF that are greater than 1.0 (100 per cent) at low spatial frequencies: this is caused by adjacency effects in the processing. Such adjacency effects result in spatially-varying images being recorded with higher contrasts than is the case for the relatively large uniform areas used in measuring the $D - \log H$ curve. Adjacency effects can occur at high, as well as at low, spatial frequencies: they are certainly not confined to frequencies for which the photographic MTF is greater than 1.0, and they may in fact represent a larger proportion of the response at higher frequencies (but this is not revealed by the photographic MTF curves).

When photographic MTF curves have values significantly less than 1.0 at very low spatial frequencies (as can be seen, for example, at the left-hand end of curve P in Fig. 18.12), this is likely to be caused by light reflected from the base of the photographic material over relatively large distances, a phenomenon known as *halation*: it is the function of anti-halation layers (see Sections 17.9 and 18.16) to reduce halation as much as possible.

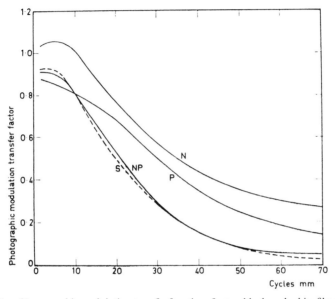

Fig. 18.12. Photographic modulation transfer functions for two black-and-white films: N, for a negative film; P, for a print film; NP, for the cascaded combination of N and P (that is, the products of the individual values of curves N and P at each frequency). Curve S shows the print-through photographic MTF for the system comprising film N contact-printed on to film P. Curve S was obtained by measuring the density fluctuations in the print film, P, for patches printed on it from recordings of sinusoidal patches of various frequencies exposed on the negative film, N; a $D - \log H$ curve representative of the system for large uniform areas was used to derive the effective exposure levels on the negative corresponding to the density fluctuations in the print film; the ratio of the contrast of these effective exposure levels to the contrast of the pattern of light falling on the negative film during exposure, is plotted as the photographic modulation transfer function, S, of the system.

When one film is printed on to another, it is useful to know how the photographic MTFs of the individual films combine into a single photographic MTF of the system as a whole. If the photographic materials behaved linearly, then their MTFs could be combined by multiplying together at each frequency the factors for each film and printing step involved. But we have already seen that adjacency effects introduce non-linearities into photographic MTF curves as normally measured, and considerable non-linearity may also arise at the printing stage: this is because the input to the second film is based on the transmittance of the first film, the gamma of which is frequently very different from 1.0, especially in the case of negatives. (Even if a negative has a gamma of 1.0, it is actually *minus* 1.0, so that its output is proportional, not to the input, but to the reciprocal of the input.)

It has been found in practice, however, that, in spite of these non-linearities, if the photographic MTFs of the individual films are cascaded, using simple multiplication of the individual factors at each frequency, then the results can agree reasonably well with a composite photographic MTF determined experimentally. The composite MTF is obtained by measuring, at each frequency, the density fluctuations of the final photographic image, and, using a composite $D - \log H$ curve for the whole photographic system (sometimes referred to as a *print-through* curve), obtaining corresponding effective exposures, H', and thence C_e, for comparison with C_0. The agreement between such photographic print-through MTFs and cascaded individual photographic MTFs is best if the MTF of the printing stage is not appreciably inferior to that of the combined system preceding this stage; and if the gamma of the negative stage is low and that of the positive is high (which is usually the case in practice). In Fig. 18.12, curve N represents the photographic MTF of a negative film, curve P that of a print film, curve NP that of curves N and P cascaded together and curve S shows the print-through photographic MTF for the system comprising film N contact-printed on to film P (using a print-through $D - \log H$ curve). The agreement between curves NP and S is seen to be reasonably good (Lamberts, 1961): this suggests that the MTF of the contact printing step was nearly 1.0 at all the frequencies involved and that the effects of non-linearities were small.

Photographic print-through MTFs have been used to compare different combinations of colour photographic films, including colour negative, reversal, internegative, intermediate, reversal intermediate, and print films, in various combinations (Norris, 1971). When, instead of using contact-printing, the print is made by employing a lens to form an image of the negative on the print material, a technique known as *optical* printing, then the MTF of the printing lens must be included in calculating cascaded photographic MTFs.

18.14 Acutance

Overall assessments of sharpness are sometimes made in terms of the areas under photographic MTF curves of the type shown in Fig. 18.12. One such measure is *system modulation transfer acutance*, or *SMT acutance* (Crane, 1964). In making such assessments, allowance has to be made for the fact that the human eye, as shown in Fig. 18.13, has a maximum response at about 3 cycles/degree (corresponding to about 10 cycles/mm on the retina or about 0.75 cycles/mm at 250 mm viewing distance), with lower responses at both higher and lower frequencies (Schade, 1956; DePalma, and Lowry, 1962). The empirical formula for SMT acutance derived by Crane is as follows:

$$120{-}25 \log{(C_1 + C_2 + C_3 + \dots\dots\dots\dots C_n)}$$

where

$$C_1 = (200\ m_1/a_1)^2,\ C_2 = (200\ m_2/a_2)^2,\ \text{etc.,}$$

and a_1 is the area (in units of mm^{-1}) under the MTF curve for the first element of the system, a_2 for the second element, etc., and m_1, m_2, etc., are the magnifications of each element, calculated as the ratios of the image width on the observer's retina to the image width in the element concerned. The constants in the formula were chosen so that one unit in SMT acutance corresponds to a just-perceptible difference in sharpness, and values over 90 represent 'good' sharpness, and values over 80 'fair' sharpness. A complete system can consist of many elements such as: camera, negative film, first printer, first intermediate film, second printer, second intermediate film, third printer, print film, projector, screen and observer: eleven stages in all, each with its own value of C to be included in the formula.

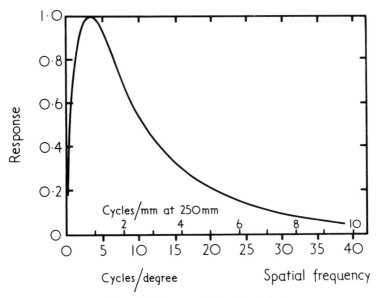

Fig. 18.13. The modulation transfer function of the human eye.

Although SMT acutance gives useful results, better correlations with subjective judgments of sharpness can be obtained, particularly when systems include films having considerable chemical adjacency effects, by using the area under a single photographic MTF curve representing the whole system: such single curves are obtained either by direct measurement, using a $D - \log H$ print-through curve for the whole system, or by cascading the photographic MTF curves of all the stages by evaluating the products at each frequency (Gendron, 1973). This *cascaded modulation transfer acutance*, or *CMT acutance*, is then given by:

$$111 - 21 \log (200/a_s)^2$$

where a_s is the area under the cascaded photographic MTF curve for the system. Care has to be taken to cascade the curves using a frequency scale for each stage that correctly allows for changes in magnification. Other methods of evaluating MTF curves in such a way as to correlate with visual sharpness have also been explored (Granger and Cupery, 1972).

18.15 Sharpness of colour images

As has already been mentioned, sharpness can be lost by light being diffused in photographic emulsion layers. In the visible part of the spectrum, light is not heavily absorbed by the silver halide grains, and therefore the image is not usually concentrated at the top of an emulsion layer. (Even when absorbing dyes are added, as in some print films, which will be discussed in Section 18.16, the exposure tends to be fairly uniform throughout the layer, because it is typically only about five grains thick). In fine-grain emulsions light is diffused by diffraction and by Rayleigh scattering (the particles being smaller than the wavelength of light and scattering blue light much more than red light); in medium-grain, and in coarse-grain, emulsions the light is diffused by Mie scattering (the particles being of size similar to the wavelength of light) and by reflection and refraction by the grains. The effects of greater scattering of blue light in a layer can be offset if it is absorbed more heavily than red light. In multi-layer colour materials there is an inevitable tendency for the image in the top layer to be the sharpest, and that in the bottom layer to be the least sharp, because of the diffusing effects of the upper layers on the light as it passes through them; the upper layers of such materials are therefore made with as low a turbidity as possible. In spite of this, in a material of conventional layer order (yellow at the top, cyan at the bottom) there is a tendency for the yellow image to be sharper than the magenta or cyan images, so that a narrow white line on a dark background may tend to be reproduced bluish.

When photographic MTFs are measured for colour films it is found that the results show the consequences of these diffusing processes; this can be seen from Fig. 18.14, where the top, yellow, layer has the highest photographic MTF, and the bottom, cyan, layer has the lowest (except at low frequencies).

In Fig. 18.14, log scales are used for both photographic MTF and for spatial frequency. The advantage of using a log scale for photographic MTF is that it is more nearly visually uniform than a linear scale. The advantage of using a log scale for frequency is that it tends to emphasize the low frequencies at which the eye has maximum response, at the expense of the high frequencies at which the eye has minimum response; however, it does not allow for the fact that the eye also has a low response at very low frequencies (see Fig. 18.13).

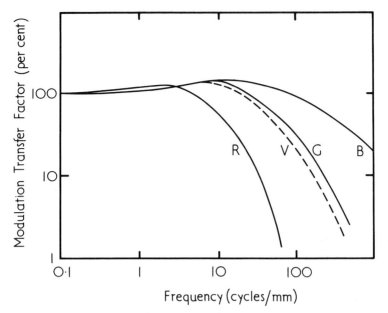

Fig. 18.14. Photographic modulation transfer functions (expressed as a percentage and plotted on a log scale) for a multi-layer colour film as measured with a red (R), a green (G), a blue (B), and a visual (V) filter, the latter designed to give the micro-densitometer a spectral response equal to that of visual luminance. It is seen that the blue-filter photographic MTF, affected mainly by the top, yellow-image, layer, is the sharpest, and that the red-filter photographic MTF, affected mainly by the bottom, cyan-image, layer, is the least sharp; the green-filter photographic MTF, affected mainly by the middle, magenta-image, layer, has intermediate sharpness, and its photographic MTF is similar to that of the photographic MTF obtained with the filter (V) simulating the visual luminance response. Note that the frequency is also plotted on a log scale.

18.16 Increasing sharpness of colour films

Because the sharpnesses of the images in the different layers of colour films are unequal, it is sometimes possible to improve the overall sharpness by using an unconventional layer order. It is for this reason that in some colour print films (where use can be made of the slower chloride or chloro-bromide emulsions with their natural sensitivities shifted towards the ultra-violet, instead of being in the blue part of the spectrum) the magenta layer is coated on top and the yellow layer at the bottom (as mentioned in Section 12.11). Since the contributions of the photographic MTF of the individual layers to the overall photographic MTF is usually greatest from the magenta image, and least from the yellow, it is clearly preferable to have the magenta image sharpest and the yellow least sharp. Films of camera speed can also sometimes be made with the magenta layer on top and the yellow layer at the bottom, by using a very fast blue layer that can be exposed

365

through a yellow filter layer which is then situated above all the light-sensitive layers (Moser and Fritz, 1975).

Several other methods have been used for increasing the photographic modulation transfer factors in colour films. First, there has been a tendency for the individual layers in colour films to become much thinner. The total coated dry thickness of a colour film (which may have as many as 10 or more layers) may now be only about 5 to 10 μm (Meyer, 1965; Engel, 1968), as indicated in Fig. 18.4, whereas in the 1940s the total coated dry thickness was typically about five times as great (Thirtle and Zwick, 1979). Clearly the thinner the layers, the sharper the pictures are likely to be, especially in the lower layers. To coat such extremely thin layers has required the development of special techniques: in one method, the liquid emulsions are extruded from slits in hoppers to form a multilayer wet sandwich, which slides on to the moving film-base web, a meniscus being maintained at the point of contact with the aid of air suction (Hanson, 1977 and 1981).

Secondly, in print films and papers, where photographic speed is not likely to be very important, each layer may have a dye incorporated in it (or in an adjacent interlayer) that absorbs light of the colour to which the layer is sensitive: thus a green-sensitive layer would have a magenta dye present, a red-sensitive layer, a cyan dye, and a blue-sensitive layer, a yellow dye. The presence of these absorbing dyes reduces the intensity of light travelling sideways in the layers and hence improves sharpness (Hanson and Kisner, 1953; Bello, Groet, Hanson, Osborne, and Zwick, 1957; Davies, 1970).

Thirdly, unsharpness can be caused during colour development by the sideways migration of oxidized developer or long-lived intermediates of the development reaction. This has to be minimized by chemical means. If chemical migration takes place vertically from one layer to another, colour contamination (formation of dye of the wrong colour) can occur; this is often minimized by using suitable compounds in inter-layers between the image forming layers (Thirtle and Zwick, 1964).

The promotion of *adjacency effects* or *edge effects* is a fourth and very important way of improving sharpness and enables some small detail to be reproduced at contrasts that are actually higher than that of coarse detail; in this way photographic modulation transfer factors greater than 100 per cent can be achieved, and instances of this can be seen in Fig. 18.14. Edge effects can occur as a result of development products being able to escape sideways when the area being developed is near an area not being developed, that is, near an edge. The result is that the density can rise near an edge because the development products cause less restraint on the development. Conversely the density can fall on the lower-development side of the edge, as a result of the products of development that have migrated into it causing less development than normal. Fig. 18.15 shows examples of micro-densitometric traces across images of edges in developed film samples for lines of various widths. It is seen that lobes or 'ears' of increased density at the edges occur with the broad lines, and for narrow lines the

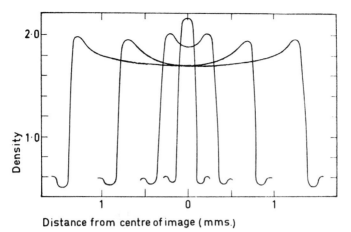

Distance from centre of image (mms.)

Fig. 18.15. Micro-densitometer traces (density plotted against distance along the image) for a series of lines of different width. For thick lines, lobes of extra, and reduced, density appear near the edges; for thin lines the lobes of extra density merge together to give lines of increased density.

lobes merge to form a line of higher density. Development products can also migrate from one layer to another, producing further concentration gradients at image boundaries, thus increasing the magnitude of edge effects and improving sharpness further.

Developer inhibitor releasing (DIR) couplers (see Section 17.11) provide a particularly efficient mechanism for producing edge effects. Since the inhibitor is released from these compounds when they dye-couple during colour development processes, controlled amounts of development inhibitors diffuse away from development sites to inhibit development adjacent to these sites. Thus a concentration gradient of inhibitor is available at a development edge to allow increased development at the edge of a strongly developing area, relative to its centre, and to restrict development on the less strongly developing side of the edge, relative to development farther from that boundary (Barr, Thirtle, and Vittum, 1969).

A fifth way in which sharpness is sometimes improved is for a colour film to be physically distorted during processing, perhaps by differential tanning of the gelatin in image and non-image areas, so as to produce a relief image on its surface. This can result in improved sharpness when the film is viewed by means of specular light, as in projection. For this reason, the evaluation of the sharpness of films in which this effect occurs should always be undertaken with an optical system having similar geometry to that which is used with the film in normal practice, even though the dye images themselves may be non-scattering. Sometimes the surface relief effect is caused in the bleaching step in a process, especially when the bleaching agents used are dichromates; ferricyanide bleaches usually give little or no relief image. If the relief image is too large, unpleasant halo effects can result (Bello and Zwick, 1959), and if tanning around individual grains takes place the granularity may be increased (Zwick, 1962).

If a film has no relief image and has a very smooth surface, *Newton's rings* (circular patterns of interference colours) may be formed if the film is printed by contact on to another smooth film. This can be avoided by coating one of the films with a roughening agent, or by printing with a thin film of a liquid (having the same refractive index as gelatin) between the two films (Zwick, 1962); this latter technique is known as *liquid-gate printing*, and it also has other advantages, such as reduction in the visibility of dust or dirt in the printing gate (Stott, Cummins, and Breton, 1957; Turner, Grant, and Breton, 1957).

A sixth method of increasing sharpness, which is also used in black-and-white films, is the incorporation of an *anti-halation layer* to absorb light reflected from the bottom surface of the film base. This can either be a thin *backing layer* (so called because it is coated on the back of the film) containing dyes or a pigment such as carbon, or a layer of silver coated between the bottom image layer and the base; in either case the absorption provided has to be removed during processing, or the film will appear unpleasantly dark (see Sections 17.9 and 18.13). (See Plate 10, pages 112, 113.)

A final factor that may affect sharpness, if the picture is being projected, is the screen. A smooth matte-white painted screen is not likely to result in any loss in sharpness, but beaded, grained, or lenticular screens, if viewed too closely, could result in sharpness being lost.

See Plate 13 (pages 176, 177) for an example of the remarkable sharpness achieved in a modern colour film.

The combined influence of sharpness and graininess on the quality of colour prints has been studied by Bartleson (Bartleson, 1981 and 1985).

18.17 Mottle on papers

When colour papers are used, the roughness of the surface of the paper support may cause the image to be slightly mottled. Since a mottling of the yellow image is much less visible than of the cyan or magenta images, it can be advantageous in colour papers to coat the yellow layer next to the support instead of the cyan layer (as is customary with films). (See Plate 10, pages 112, 113.)

18.18 Image structure in transfer systems

Colour photographic systems in which the image is at some stage transferred from one layer to another (as, for instance, in the Kodak *Dye Transfer*, the Kodak *Instant*, and the *Polacolor* processes) may result in images of very low or zero granularity, as a result of sideways spreading of the dye clouds during the transfer step. Unfortunately, however, the sideways diffusion also reduces sharpness, and such systems may be restricted to situations where no magnification is involved, such as normal viewing of reflection prints, unless special precautions are taken.

REFERENCES

Barr, C. R., Thirtle, J. R., and Vittum, P. W., *Phot. Sci. Eng.*, **13,** 74 and 214 (1969).
Bartleson, C. J., *J. Phot. Sci.*, **30,** 33 (1981), and **33,** 117 (1985).
Bello, H. J., and Zwick, D. M., *Phot. Sci. Eng.*, **3,** 221 (1959).
Bello, H. J., Groet, N. H., Hanson, W. T., Osborne, C. E., and Zwick, D. M., *J. Soc. Mot. Pic. Tel. Eng.*, **66,** 205 (1957).
British Journal of Photography, **124,** 950 (1977).
Crane, E. M., *J. Soc. Mot. Pic. Tel. Eng.*, **73,** 643 (1964).
Crawley, G., *Brit. J. Phot.*, **126,** 340 (1979).
Davies, B. J., *Brit. Kinematog. Sound Tel.*, **52,** 118 (1970).
De Belder, M., and De Kerf, J., *Phot. Sci. Eng.*, **11,** 371 (1967).
De Palma, J. J., and Lowry, E. M., *J. Opt. Soc. Amer.*, **52,** 328 (1962).
Doerner, E. C., *J. Opt. Soc. Amer.*, **52,** 669 (1962).
Engel, C. E., *Photography for the Scientist*, p.58, Academic Press, London (1968).
Gendron, R. G., *J. Soc. Mot. Pic. Tel. Eng.*, **82,** 1009 (1973).
Granger, E. M., and Cupery, K. N., *Phot. Sci. Eng.*, **16,** 221 (1972).
Hanson, W. T., and Kisner, W. I., *J. Soc. Mot. Pic. Tel. Eng.*, **61,** 681 (1953).
Hanson, W. T., *J. Soc. Mot. Pic. Tel. Eng.*, **90,** 791 (1981).
Hanson, W. T., *Phot. Sci. Eng.*, **21,** 293 (1977).
Happé, L. B., *Basic Motion Picture Technology*, p. 309, Focal Press, London (1971).
Higgins, G. C., and Stultz, K. F., *J. Opt. Soc. Amer.*, **49,** 925 (1959).
Hunt, R. W. G., *J. Roy. Television Soc.*, **11,** 220 (1967).
Kennel, G. L., Sehlin, R. C., Reinking, F. R., Spakowsky, S. W., and Whittier, G. L., *J. Soc. Mot. Pic. Tel. Eng.*, **91,** 922 (1982).
Lamberts, R. L., *J. Opt. Soc. Amer.*, **51,** 982 (1961).
Lamberts, R. L., *Applied Optics*, **2,** 273 (1963).
Lythgoe, R. J., *Med. Res. Council Spec. Rep. Ser.*, No. 177 (1932).
Maude, N., *British Journal of Photography Annual*, p. 143 (1980).
Mannheim, L. A., *Brit. J. Phot.*, **125,** 430 (1978).
Mees, C. E. K., and James, T. H., *The Theory of the Photographic Process*, 3rd Edn., pp. 523–530, Macmillan, New York (1966).
Meyer, A., *J. Phot. Sci.*, **13,** 90 (1965): See Fig. 1.
Meyers, M. B., and Dalton, P. J., *Brit. J. Phot.*, **126,** 832 (1979).
Morris, R. A., and Wait, D. H., *J. Soc. Mot. Pic. Tel. Eng.*, **80,** 819 (1971).
Moser, J. S., and Fritz, N. L., *Phot. Sci. Eng.*, **19,** 243 (1975).
Norris, J. C., *J. Soc. Mot. Pic. Tel. Eng.*, **80,** 30 (1971).
Ooue, S., *J. Appl. Phys. (Japan)*, **29,** 685 (1960).
Rotthaler, M., *Brit. Kinematog. Sound Tel.*, **56,** 51 (1974).
Sawyer, J. F., *Photographic Image Quality*, Oxford Symposium, Royal Photographic Society, p. 222 (1980).
Schade, O. H., *J. Opt. Soc. Amer.*, **46,** 739 (1956).
Selwyn, E. W. H., *Phot. J.*, **75,** 571 (1935).
Selwyn, E. W. H., *Phot. J.*, **79,** 513 (1939).
Selwyn, E. W. H., *Phot. J.*, **83,** 227 (1943).
Selwyn, E. W. H., *J. Phot. Sci.*, **7,** 138 (1959).
Stott, J. G., Cummins, G. E., and Breton, H. E., *J. Soc. Mot. Pic. Tel. Eng.*, **66,** 607 (1957).
Stultz, K. F., and Zweig, H. J., *J. Opt. Soc. Amer.*, **49,** 693 (1959).
Thirtle, J. R., and Zwick, D. M., *Encyclopedia of Chemical Technology*, **5,** 840 (1964).

Thirtle, J. R., and Zwick, D. M., *Encyclopedia of Chemical Technology*, 3rd Edn., **6,** 617, Wiley, New York (1979).
Turner, J. R., Grant, D. E., and Breton, H. E., *J. Soc. Mot. Pic. Tel. Eng.*, **66,** 612 (1957).
Verbrugghe, R., De Belder, M., and Langner, G., *Phot. Sci. Eng.*, **11,** 379 (1967).
Vervoort, A., and Stappaerts, H., *Brit. Kinematog. Sound Tel.*, **72,** 148 (1980).
Wall, F. J. B., and Steel, B. G., *J. Phot. Sci.*, **12,** 34 (1964).
Wheeler, L. J., *Principles of Cinematography*, 4th Edn., p. 308, Fountain Press, London, (1969).
Zwick, D. M., *J. Soc. Mot. Pic. Tel. Eng.*, **71,** 15 (1962).
Zwick, D. M., *J. Phot. Sci.*, **11,** 269 (1963).
Zwick, D. M., *Phot. Sci. Eng.*, **9,** 145 (1965).
Zwick, D. M., *Phot. Sci. Eng.*, **16,** 345 (1972).
Zwick, D. M., and Brothers, D. L., *Phot. Sci. Eng.*, **19,** 235 (1975).

GENERAL REFERENCES

Dainty, J. C., and Shaw, R., *Image Science*, Academic Press, London (1974).
Kowaliski, P., *Applied Photographic Theory*, Wiley, New York (1972).
Lamberts, R. L., *J. Opt. Soc. Amer.*, **49,** 425 (1959).
Lamberts, R. L., *J. Soc. Mot. Pic. Tel. Eng.*, **9,** 635 (1962).
Mees, C. E. K., and James T. H., *The Theory of the Photographic Process*, 4th Edn., Macmillan, New York (1977).
Thirtle, J. R., and Zwick, D. M. *Encyclopedia of Chemical Technology*, 3rd Edn., **6,** 617, Wiley, New York (1979).

PART THREE

COLOUR TELEVISION

CHAPTER 19

The Transmission of Colour Television Signals

1. Historical introduction – *2*. Bandwidth – *3*. Interlacing – *4*. Single side-band transmission – *5*. The field sequential system – *6*. Blue saving – *7*. Band saving – *8*. Colour-difference signals – *9*. Band sharing – *10*. The effect of band sharing on monochrome receivers – *11*. Carrier sharing – *12*. The effects of signal processing on colour reproduction – *13*. Gamma correction – *14*. Noise reduction – *15*. Higher definition television – *16*. Videoconferencing

19.1 Historical introduction

FROM the earlier capability of transmitting Morse code, radio techniques had been developed by the early 1920s to the point where the broadcasting of sound programmes, and the transmission of still pictures at slow speed, were achieved. But the transmission of pictures with movement had to await further developments. The methods ultimately used were described in a remarkably prophetic paper written by A. A. Campbell-Swinton in 1908 (Campbell-Swinton, 1908); but, before his prophecies were fulfilled, another approach was to be tried (Hawker, 1983).

The first demonstration of a working television system was given by J. L Baird at Frith Street, London, to forty members of the Royal Institution and other guests, in 1926 (Shiers, 1976). In the system used, the scene was imaged on to a rotating disc; a series of holes in the disc (see Fig. 20.1) resulted in the light from the scene being sampled at successive points along a series of slightly curved lines. This light was imaged on a photoelectric cell to produce a sequence of electrical signals, representing the amount of light by their amplitudes and the position of each picture point by the elapsed time since the start of each complete picture scan. Transmission of the signals was by means of the then standard speech-carrying telephone cables and sound-broadcasting carrier waves. A reconstituted display of the picture was achieved by modulating the intensity of a neon lamp in the same time sequence, and viewing it through another disc, having holes like the first, presented in synchronization. The number of lines in the picture was only 30, so that the definition was extremely limited, and the number of pictures produced per second was only $12\frac{1}{2}$, so that the displayed picture flickered badly. By 1928 Baird was able to show colour pictures, using the successive frame method (see Section 3.2) in which red, green, and blue, pictures are produced in succession. Also in 1928, an alternative system for studio use had been developed in which the *flying-spot* principle was used: the scene itself was illuminated by a spot of light that moved in a series of lines, and suitably placed photoelectric cells then picked up the light reflected by the scene. Experimental broadcasting using the system began in London in 1929, simultaneous sound being added in 1930, outdoor televising in 1931, and the televising of films in 1932. Responsibility for running these broadcasts was taken over by the British Broadcasting Corporation (BBC) in 1932, and they continued until 1935 (Bridgewater, 1977).

The low definition and pronounced flicker of the 30-line, $12\frac{1}{2}$ pictures per second, system were serious disadvantages, and Baird devised a 240-line, 25 pictures per second, system, but it could only be used in the studio; for large studio shots, a film camera was used to produce a photographic record which was processed in 64 seconds, and this was scanned instead of the scene itself; for small studio shots, the flying-spot method was used.

Meanwhile work had been in progress elsewhere, notably at the Electric and Musical Industries (EMI) research laboratories at Hayes in England, to implement systems of television that did not require rotating discs, or other mechanical scanning devices (such as rotating mirror-drums, which were also used in some of Baird's equipment). To avoid the need for mechanical scanning it was necessary to do two things. The first was to derive a television signal by scanning a light-sensitive surface with a beam of electrons: television camera tubes operating on this principle were first devised by V. K. Zworykin at the Westinghouse research laboratories in the U.S.A., and by the team led by J. D. McGee at EMI (McGee, 1976). The second necessity was to display a television picture by scanning a light-emitting surface with another beam of electrons, and a device for doing this, even though at that time somewhat crudely, already existed in the form of the cathode-ray tube. Using these electron scanning devices, the EMI team, under the overall direction of Isaac Shoenberg, developed a system having 405 lines and 25 pictures per second.

The stage was now set for trials to see whether the Baird or the EMI system was the better for broadcast television. Accordingly, in November 1936, the BBC in London started transmissions using the two systems on alternate days. It became quickly apparent that the future of television lay with the EMI system. Not only did it have more lines, and therefore more detail in its pictures, but its cameras were light and mobile and could be moved around in studios and used for outside broadcasts. The Baird 240-line system, on the other hand, used either a camera and flying-spot illumination system which were immobile, or a film

374

camera with in-built processing and scanner which was too large to move around; and neither alternative could be used for outside broadcasts. As a result, use of the Baird system was discontinued in February 1937 (Birkinshaw, 1977).

The EMI system was then used for increasingly ambitious studio productions, and for outside broadcasts that included, in 1937, the coronation of King George VI and the Wimbledon tennis championships. By 1939 there were about 20 000 sets in use, but the service was discontinued during the period of the second world war; it was restarted in 1946, and was the first regular public service of high-definition television, although a system using 180 lines was started in Berlin in 1935. (Sidey, Longman, Glencross, and Pilgrim, 1981.)

We must now examine some of the basic requirements of systems of high-definition broadcast colour television.

19.2 Bandwidth

It will be appreciated that, just as two powerful sound radio stations cannot be received without interference unless their frequencies (or wavelengths) are adequately separated, in a similar way each television station must have its own adequate frequency space, or *bandwidth* as it is usually termed. However, the radio spectrum is a limited one, and television has to be fitted in to the existing demands made upon it by sound radio, radio telephony and telegraphy, police and military radio communications, shipping and aircraft signals, radar, etc. Television, by its very nature, requires far more bandwidth than is required for transmitting sound, and the problem of finding adequate room for each station is much more acute. In *colour* television it is necessary to transmit, not one picture, but three. If this were done in such a way as to take three times as much bandwidth as is used for monochrome television the problems of fitting all the stations in without interference would be tremendous. It is therefore important to save as much bandwidth as possible when transmitting colour television pictures and much effort has been applied to this end.

The reason why television requires so much more bandwidth can be seen in the following way. In sound radio it is a common experience that when a receiving set is slightly detuned from a station the low notes fade out first, and the higher notes last, so that in its detuned position the reception becomes squeaky and high-pitched. The reason for this is that the high notes are of higher frequency than the low notes, and hence modify the carrier-wave frequency more. Thus, consider a medium-wave station operating at a carrier frequency[1] of, say, 1000 kHz (which corresponds to a wavelength of about 300 metres[2]), and transmitting sound frequencies of about 50 Hz (low notes) and about 5000 Hz

[1] The unit used for frequency is the hertz (Hz) which is equal to 1 cycle per second. In radio and television the kilo-hertz (kHz), 1000 cycles per second, and the mega-hertz (MHz), 1 000 000 cycles per second, are widely used.
[2] The frequency multiplied by the wavelength is always equal to the speed of propagation of the wave, in this case the speed of light which is 3×10^8 metres per second.

(high notes). Modulation of a 1 000 000 Hz carrier wave at 50 Hz will produce some energy at frequencies of 1 000 050 and 999 950 Hz (corresponding to wavelengths of 299.985 and 300.015 metres) which are very little different from the basic 1 000 000 Hz. But modulation at 5000 Hz results in some energy at frequencies of 1 005 000 and 995 000 Hz (corresponding to wavelengths of 298.5 and 301.5 metres), a much more significant change.

In television, the number of modulations required per second is very much greater than the maximum required in sound radio. Let us take, for example, a system employing 525 lines and 30 complete pictures per second. The number of lines limits the fineness of the detail that can be resolved in the vertical direction to a grid of 525 black and white horizontal stripes, that is $\frac{1}{2} \times 525$ pairs of black and white stripes. If the system is to have the same resolving power in the horizontal direction as in the vertical, it must be possible to resolve $\frac{1}{2} \times 525 \times 4/3$ pairs of stripes, the factor of $4/3$ being introduced to allow for the fact that television pictures are not square, but have an aspect ratio of $4/3$. Thus each line of the system may receive $\frac{1}{2} \times 525 \times 4/3$ complete modulations, from black through white to black again, per scan. But there are 525 lines, and 30 pictures per second, so that the number of modulations possible per second is given by:

$$\tfrac{1}{2} \times 525 \times (4/3) \times 525 \times 30 = 5\ 512\ 500$$

or approximately 5.5 MHz. In practice, however, owing to the necessity of transmitting synchronising information in addition to the picture information, not all 525 lines, and not all of the time allocated to each line, are used in the picture. The simple calculation given above is also complicated by the fact that breaking the picture up into lines results in a reduction in the resolution from half the number of lines to about three-quarters of that number (that is three-eighths of the total number of lines); this factor of about three-quarters is known as the *Kell factor*. Hence the maximum frequency actually used in this system is generally only about 4 MHz (Jesty, 1957).

It is clear, from the above, that television systems are required to transmit very much higher frequencies than ordinary sound radio systems; and, since the modulating frequency can be as high as about 5 MHz, the carrier frequencies have to be not less than about 50 MHz, corresponding to a wavelength of about 6 metres. (Incidentally, it is for this reason that, whereas sound-radio signals can be transmitted round the earth's surface, television signals are limited to rather less than a 100-mile radius at ground level, because signals of frequencies in the 50 MHz range are not appreciably diffracted or reflected by the ionosphere in the upper atmosphere, and therefore can only be received satisfactorily within the transmitting station's horizon, or via a satellite relay station.) Typical carrier frequencies are given in Table 19.1.

A carrier wave of 50 MHz modulated at frequencies up to 5 MHz will produce some energy over the frequency range 45 to 55 MHz (corresponding to a wavelength range of 6.7 to 5.5 metres). The more detail required in a television picture, the greater will be the bandwidth of the signal and the greater will be the

376

TABLE 19.1

Carrier frequencies used in broadcast television

VHF (Very High Frequencies)	30 to 300 MHz
UHF (Ultra High Frequencies)	300 to 3000 MHz
Band I	41 to 68 MHz
Band II	87.5 to 100 MHz
Band III	162 to 230 MHz
Band IV	470 to 558 MHz
Band V	582 to 860 MHz

amount of frequency or wavelength space required to accommodate it. In fact the calculation shows that the maximum modulating frequency increases in proportion to the square of the number of lines in the picture; it is also proportional to the number of pictures per second.

The 525 line, 30 pictures per second, system is that which has been adopted in the U.S.A. In Great Britain a 405 line, 25 pictures per second, system using about 3 MHz maximum modulating frequency was employed exclusively until 1964, when a 625 line, 25 pictures per second, system using about $5\frac{1}{2}$ MHz maximum modulating frequency was introduced in addition; this 625 line system has been adopted as standard for Europe, and is the system used in Europe for colour. All countries now use either the 525 or the 625 line system for colour.

19.3 Interlacing

The range of modulating frequencies required in these systems would in fact be higher still, but for the use of a technique known as *interlacing*. When the number of individual pictures per second composing the display is too low, an unpleasant flickering is apparent. Although about 25 pictures per second are used in photographic motion pictures, it is usually arranged for the light to be interrupted by a shutter once (or sometimes twice) during the projection of each picture as well as between successive pictures, so that the light flickers at about 50 times per second, and this is not very noticeable. In television, however, interruption of the electron beam during the scanning of a picture would result only in the disappearance of the part of the picture being scanned at the time; and interruption of the light emitted by the fluorescent screen of the tube would produce variations in luminance over the area of the picture because the interruption would occur at different points in the time cycle of the afterglow of the phosphor at different parts of the picture.

But 25 to 30 pictures per second is too low a frequency for flicker to be avoided and hence some means of increasing the frequency is required. To transmit twice the number of pictures per second would require twice as much bandwidth, but, by using interlacing, the apparent flicker frequency is doubled without any increase in bandwidth being necessary. In an interlaced picture the electron beam first produces all the odd lines of the picture, the first, the third, the fifth etc., and then adds the even lines in between them. The parts of the picture composed by all the even or all the odd lines are called *fields*. Thus, by means of interlacing, a 25 pictures per second system involves 50 fields per second; and, although each field contains only half the total number of picture lines, 50 fields per second are almost as good as 50 pictures per second, as far as absence of flicker is concerned. Flicker becomes more noticeable as the luminance of the picture is increased, and one advantage of the 30 pictures per second (60 fields per second) systems over the 25 pictures per second (50 fields per second) systems is that higher picture luminances can be used without flicker being unpleasantly noticeable. Interlacing is already used in monochrome television so that its use in colour television does not give a further bandwidth advantage.

19.4 Single side-band transmission

Another method of reducing the amount of bandwidth required is to transmit only those frequencies that are equal to or *higher* than that of the carrier wave; the frequencies below the carrier frequency are exactly similar to those above, and to transmit both these *side-bands* of frequency is therefore unnecessary, since either side-band carries all the information. Consequently most television stations filter out most of the frequencies below (or above) the carrier frequency, and transmit only one of the side bands, a technique known as *single* (or *asymmetric* or *vestigial*) *side-band transmission*. Hence the bandwidths required are approximately equal to the maximum modulating frequency, as indicated in Section 19.2, and not to twice that value.

Single side-band transmission is already used in monochrome television so that its use in colour television does not give a further bandwidth advantage.

19.5 The field sequential system

The simplest colour television system is the field sequential system, in which red, green, and blue filters rotate in front of a single television camera, and similar filters rotate in synchronism over the viewing tube of the receiving set, as shown in Fig. 19.1. Unfortunately, however, there are three reasons why such a system cannot easily be adopted for wide scale broadcasting. In the first place, the picture frequency necessary to avoid flicker and colour break-up in such a system is about three times that necessary in black-and-white systems, so that existing black-and-white television receiving sets would not be able to receive the colour broadcasts in monochrome without modifications being necessary; such systems are called *incompatible*. In the second place the rotating filter wheel is an undesirable feature in the receiver. (It has been suggested that a stationary filter be used, consisting of narrow red, green, and blue strips, with a lenticular screen oscillating so that the viewer sees the picture through the strips of only one colour at a time, the colours being changed in rapid succession; but rapidly moving parts are still required.) In the third place, the transmission of three times as many pictures per second requires the use of three times as much bandwidth; for a red, a green, and a blue picture would have to be transmitted for every black-and-white picture, entailing, on a 25 pictures per second system, for instance, 75 colour pictures per second (these colour pictures are generally referred to as *frames*). With interlacing this becomes 25 pictures, 75 frames, and 150 fields per second. This field frequency is sufficient to overcome all flicker from the colour filter wheel, except for very rapidly moving objects which can show some fringing or 'break-up', particularly if they are highly coloured.

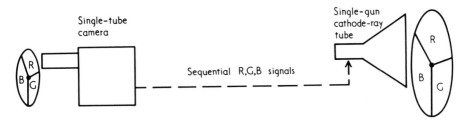

Fig. *19.1*. Diagrammatic representation of field sequential system.

Interlacing does not alter the maximum modulating frequency of a system if the speed at which the electron beam moves along each line is unchanged. The modulating frequencies in a field sequential system of 150 interlaced fields per second are therefore the same as those of a 75 (non-interlaced) colour-pictures per second system and are therefore three times those of a black-and-white system of 25 pictures (or 50 interlaced fields) per second. Maximum modulating frequencies of up to about 15 MHz thus becoming necessary, and this means that the number of colour television stations that could operate within a given band of wavelengths would be reduced to one-third of the number of black-and-white stations. For this reason, and because of the incompatibility of high picture-frequency systems with black-and-white receivers, the simple field sequential system of colour television is not well suited to public broadcasting. However, it can give good quality with fairly simple equipment and techniques, and for this reason it finds applications in closed circuit colour television; it has been used, for instance, to display positive pictures from colour negative film as an aid to photographic printing (see Sections 16.2 and 16.3). Its application to public broadcasting has been confined to the short period from 1950 to 1951 when the version proposed by C.B.S. (the Columbia Broadcasting System) was adopted by the Federal Communications Commission (F.C.C.) in America: 144 fields per second were used with 405 lines and only 2 MHz for each of the three pictures (which were therefore of rather poor definition), making a total of 6 MHz bandwidth. Defence requirements in 1951 prevented further use of this system and subsequently an entirely different system was adopted in the U.S.A.

379

19.6 Blue saving

Various methods of reducing the bandwidth required by colour television have been suggested, one of which we may call 'blue saving'. The human eye is able to distinguish fine detail illuminated by red or green light much better than when blue light is used. But, in a colour television system, it is only the fine detail that results in the higher modulating frequencies being used. Thus a picture of one thick black tree trunk seen against a white sky, will result in modulations every time the scanning beam crosses the tree trunk, which, for each colour, will be once per line per picture, or, in a 525 line 30 picture per second system, a mere 15 750 times per second.[1] But a picture of a forest of 200 tree trunks will naturally result in 200 times the number of modulations of the scanning beam, and hence a frequency of 3 150 000 times per second. Since, however, at normal viewing distances the eye is incapable of seeing the 200 tree trunks in the blue picture there is no point in transmitting them. Thus an electronic filter could be fitted somewhere in the blue channel which effectively eliminated all signals of frequencies higher than those that result in detail that is just perceptible in the blue picture at normal viewing distances. In this way the bandwidth required for the blue picture can be reduced from 5 MHz to about 1 MHz.

Blue saving does not help in the field sequential system because, even if the blue picture were transmitted with reduced bandwidth, the red and green pictures would still each have to be scanned in a third of the time.

19.7 Band saving

The ability of the eye to see fine detail depends for the most part on differences in *luminance* in the pattern and only to a much smaller extent on *colour* contrast. Thus the visibility of white letters on a dark grey background, when viewed from a distance, is not improved very much by adding colour contrast to the existing luminance contrast: yellow letters on a dark blue background, or orange letters on a dark green background, for instance, having the same luminance differences, are scarcely any clearer.

This suggests that if the information in a colour television picture could be divided into its luminance content and its colour content, then only the luminance information need be transmitted at high definition, and bandwidth could be saved by transmitting the colour information at reduced definition. This is in fact what is done, with remarkably good effect.

From the television camera, three electrical signals, E_R, E_G, E_B (usually expressed as voltages), are obtained that are proportional at each point of the picture to its red, green, and blue contents, R, G, B, as analysed by the spectral sensitivities of the three channels of the camera. (Proportionality between the electrical and corresponding optical signals at both the camera and the receiver is

---

[1] Higher harmonics will also occur, but they do not vitiate the argument.

380

assumed for the moment for the sake of simplicity; the effects of the non-linearities that occur in real systems will be considered later.) The luminance, L, at any point in the picture will be given by:

$$L = L_R R + L_G G + L_B B$$

where L_R, L_G, L_B are the luminances of the units in which the red, green, and blue contents are measured. It is therefore possible to produce an electrical signal E_L, that is proportional to the luminance, L, by adding together the same proportions of the signals E_R, E_G, E_B; thus, assuming that the constants of proportionality between the electrical and optical signals are the same for all three channels:

$$E_L = L_R E_R + L_G E_G + L_B E_B$$

If now the three signals transmitted were not E_R, E_G, E_B, but E_L, and two of the other signals, say E_R and E_B, then the signal E_L could be transmitted with broad bandwidth, and the signals E_R and E_B with narrow bandwidth. The receiver would then have to recover the E_G signal necessary to produce the final display by performing the operation

$$E_G = \frac{1}{L_G} E_L - \frac{L_R}{L_G} E_R - \frac{L_B}{L_G} E_B$$

When this is done, if two colours form a pattern of such fineness of detail that the chromaticity difference is not transmitted by the E_R and E_B signals, but the luminance difference is successfully transmitted by the E_L signal, then the receiver will display a chromaticity equal to the average of that of the two colours, upon which will be superimposed the luminance difference produced by the E_L signal. That this procedure results in the display of the correct luminance can be seen as follows. Suppose that the true E_R and E_B signals for some areas are altered to $E_R + A$ and $E_B + B$ as a result of the chromaticity averaging. The resulting signals used for the display will then be modified to:

$$E_{RM} = E_R + A$$

$$E_{GM} = \frac{1}{L_G} E_L - \frac{L_R}{L_G} (E_R + A) - \frac{L_B}{L_G} (E_B + B)$$

$$E_{BM} = E_B + B$$

The luminance displayed will then be proportional to:

$$L_R E_{RM} + L_G E_{GM} + L_B E_{BM}$$

$$= L_R (E_R + A) + E_L - L_R (E_R + A) - L_B (E_B + B) + L_B (E_B + B)$$

$$= E_L$$

It is thus clear that the luminance displayed is the same as that corresponding to the unmodified signals E_R and E_B, and hence the luminance is correctly reproduced in spite of errors in chromaticity. This is an important result and is known as the *constant luminance principle*.

It is found that by transmitting a separate high definition luminance signal very considerable savings in bandwidth can be achieved. If the system is such that the luminance signal has a bandwidth b, then the two other signals only require about $\frac{1}{4}b$ each, making a total of $1\frac{1}{2}b$ (Hunt, 1967), instead of $3b$ required by the field sequential system. (See Plate 21, page 289.) Moreover, further reductions in bandwidth can be made when two of the signals require much less bandwidth than the third, as will be discussed in Sections 19.9 and 19.11. Systems employing a luminance signal and two colour signals thus have a decisive bandwidth advantage over those employing three colour signals.

There are two further advantages arising from the use of a separate luminance signal. First, the modifying signals, A and B, introduced above, can arise from any source, and hence if the E_R and E_B signals suffer from interference, for instance, they will not affect the luminance displayed. This has a beneficial effect, because the eye is more sensitive to luminance changes than to chromaticity changes (it is on this principle that the flicker photometer depends); it is found that about $2\frac{1}{2}$ times as much 'noise' can be tolerated in the colour signals as in the luminance signal. In general, the E_L signal will suffer from interference as well, but the effects of this will not be made worse by the presence of interference in the E_R and E_B signals.

The second additional advantage of one of the three transmitted signals being a luminance signal is that it can be used very effectively for the production of monochrome pictures on black-and-white receivers: it is then only necessary for these receivers to ignore the colour signals in order to produce monochrome versions of colour transmissions. The use of a luminance signal therefore greatly facilitates *compatibility*.

Although the algebra showed that the correct value of E_L was always obtained, even when E_R and E_B were in error, this will not be true if negative values of E_{GM} are obtained: the receiver cannot produce negative amounts of green light. The modified green signal, E_{GM}, is given by:

$$E_{GM} = \frac{1}{L_G}E_L - \frac{L_R}{L_G}(E_R + A) - \frac{L_B}{L_G}(E_B + B)$$

If A and B are sufficiently large and positive the value of E_{GM} can become negative. This will happen for saturated purple colours, for instance. Saturated purples are matched by mixtures of red and blue only, so that E_G will be zero. Hence, for these colours

$$E_G = O = \frac{1}{L_G}E_L - \frac{L_R}{L_G}E_R - \frac{L_B}{L_G}E_B$$

and hence

$$E_{GM} = -\frac{L_R}{L_G}A - \frac{L_B}{L_G}B$$

Suppose we have a fine pattern of light and dark saturated purple of the same chromaticity. The E_R and E_B signals will be modified to average values, so that for the light areas E_{RM} and E_{BM} will be smaller than E_R and E_B. A and B are therefore both negative and hence E_{GM} is positive and will result in the area being lightened to the correct luminance. But for the dark areas, E_{RM} and E_{BM} will be larger than E_R and E_B, so that A and B will be positive, making E_{GM} negative. But the receiver cannot produce the 'negative' amount of green light necessary to reduce the luminance, and so the luminance of the dark part of the pattern is too high. This type of error, however, is only likely to occur in patterns such that the green signal is small (that is, fairly saturated purples) and the luminance difference large.

19.8 Colour-difference signals

There are several advantages if, instead of transmitting the signals E_L, E_R, E_B, the luminance signal is accompanied by two *colour-difference* or *chrominance* signals, such as $E_R - E_L$, and $E_B - E_L$. The receiver then recovers a signal $E_G - E_L$ by performing the operation:

$$E_G - E_L = \frac{1 - L_R - L_G - L_B}{L_G}E_L - \frac{L_R}{L_G}(E_R - E_L) - \frac{L_B}{L_G}(E_B - E_L)$$

The advantages of using difference signals are only fully realized, however, if the luminance signal is compounded from the E_R, E_G, E_B signals so that it is equal to

$$lE_R + mE_G + nE_B$$

where
$$l = L_R/(L_R + L_G + L_B)$$
$$m = L_G/(L_R + L_G + L_B)$$
$$n = L_B/(L_R + L_G + L_B)$$

so that
$$l + m + n = 1.0$$

We shall call this new luminance signal E_Y (the suffix Y indicating, not yellow, but the Y of the CIE X, Y, Z system) to distinguish it from E_L; E_Y is still a true measure of luminance but is now expressed in units $L_R + L_G + L_B$ times as large as those used for E_L.

The colour difference signals now become $E_R - E_Y$ and $E_B - E_Y$, and the receiver can recover a signal $E_G - E_Y$ by performing the operation

$$E_G - E_Y = -\frac{l}{m}(E_R - E_Y) - \frac{n}{m}(E_B - E_Y)$$

(to which the expression for $E_G - E_L$ reduces when E_Y, l, m, and n are substituted for E_L, L_R, L_G, and L_B). This is a simpler operation than that necessary to recover the signal $E_G - E_L$, and has the important advantage that, since only the two low definition signals $E_R - E_Y$ and $E_B - E_Y$ are involved, a mixing circuit of low frequency response can be used for this operation with consequent savings in the cost and complexity of the receiver. The three signals transmitted then become the sharp luminance signal, E_Y, and the two unsharp colour-difference signals, $E_R - E_Y$ and $E_B - E_Y$, as shown in Fig. 19.2.

At the receiver the high definition signal E_Y can now be added to all three signals, $E_R - E_Y$, $E_G - E_Y$, $E_B - E_Y$, to obtain the signals E_R, E_G, E_B necessary

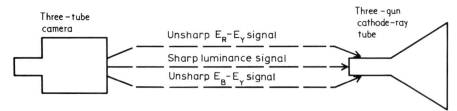

Fig. 19.2. Diagrammatic representation of sharp luminance, unsharp chrominance, system.

for the display. One convenient way of doing this is to apply the three low definition colour-difference signals $E_R - E_Y$, $E_G - E_Y$, $E_B - E_Y$, to the grids of three electron guns, and the same high definition luminance signal E_Y to their three cathodes (see Section 21.1).

The use of colour-difference signals of this type has further advantages, which, however, are only fully realized if another arbitrary condition is applied. It is further arranged that the relative sensitivities of the three channels of the camera are such that for whites, greys, and blacks $E_R = E_G = E_B$. Because $l+m+n = 1$, and $E_Y = lE_R+mE_G+nE_B$, it follows that for whites, greys, and blacks $E_R = E_G = E_B = E_Y$; hence for these colours, the colour-difference signals $E_R - E_Y$ and $E_B - E_Y$ are both zero. This has two further advantanges.

First, variations in the relative strengths of the three signals E_Y, $E_R - E_Y$, $E_B - E_Y$, do not affect the colour balance of the grey scale in the reproduction.

Secondly, because most scenes consist mainly of colours of fairly low colour saturation, the need for transmitting information additional to that contained by the E_Y signal is reduced, and hence *cross-talk* (that is, interference between the luminance and colour-difference signals) in band sharing systems (to be considered later) is minimized.

The constant luminance principle still applies when colour-difference signals are used: this can be shown, as above, by calculating the displayed luminance when spurious signals A and B are added to the colour difference signals.

19.9 Band sharing

The reduction of the bandwidth required from $3b$ for a full-definition system to only $1\frac{1}{2}b$ for a luminance-signal system is obviously a most important saving. It would be ideal, however, if colour television signals could be sent out with the use of no more bandwidth than for monochrome signals. This may at first sight seem impossible, for clearly there is more information in a colour picture than in a black-and-white one. But if the monochrome signal was not using its bandwidth to the greatest efficiency, then the colour information might be added to it, without the bandwidth having to be increased at all. This is the principle of *band sharing*.

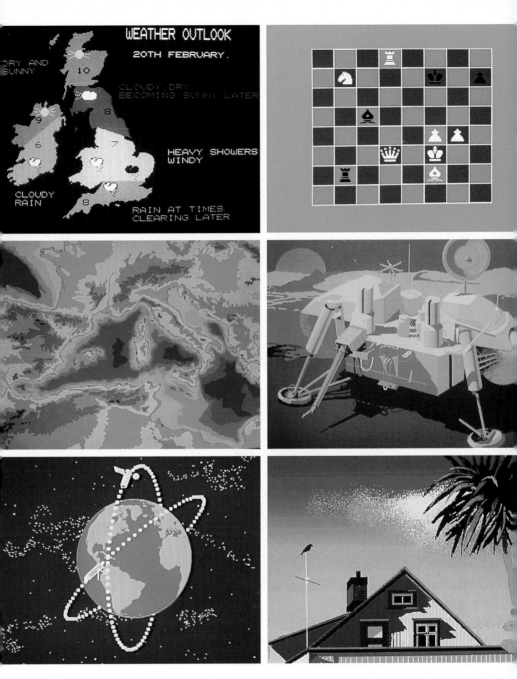

Plate 25

All these television pictures were generated by computers. *Top:* pictures transmitted in the field-blanking time (Teletext, level 4; see Section 25.4). *Centre* and *bottom*: video graphic pictures for normal transmission (see Section 25.6). Pictures by courtesy of the Independent Broadcasting Authority (top and bottom) and the British Broadcasting Corporation (centre). Separations made on a Crosfield *Magnascan* scanner.

Plate 26
Lower left: pilot seated in trainer cockpit, with simulated runway displayed in his field of view. *Other pictures:* examples of the realistic simulations possible by computer-driven raster displays. (See Section 25.6.) Pictures by courtesy of Rediffusion Simulation Limited. Separations made on a Crosfield *Magnascan* scanner.

Plate 27

Both these pictures were generated entirely on computer-controlled equipment (see Section 25.6). *Top:* an example of the type of result obtainable on the Quantel DPB 7000 series Digital Paint Box. Manipulation of an electronically-sensed 'brush' over a tablet produces results on a colour monitor similar to those that would be produced by an artist; different types of brush stroke can be simulated. Picture by courtesy of Quantel Limited. *Bottom:* similar type of result obtained on a Crosfield computer-controlled scanner, by Peter McAllister. Picture by courtesy of Crosfield Electronics Limited. Separations made on a Crosfield *Magnascan* scanner.

It was pointed out earlier in this chapter that the higher modulating frequencies are only produced by fine detail. Since a television picture is scanned by a series of horizontal lines, some modulating frequencies will be more commonly produced than others (Mertz and Gray, 1934). Thus in a 525 line 30 pictures per second system, a single vertical bar would be traversed by the scanning spot 15 750 times per second, and thus would give a 15 750 Hz modulation. If we call this *line frequency f*, it is clear that h vertical bars, equally spaced in the picture, will give fundamental frequencies of hf Hz. The only way in which fundamental frequencies of $(h+\frac{1}{2})f$, where h is a whole number, can be produced are for the bars to be displaced by the width of half a bar on each successive line of a field.

In Fig. 19.3 the situation is depicted diagrammatically for the simple case of a 9-line field with an aspect ratio of 4 to 3. Patterns of bars resulting in fundamental frequencies of f, $2f$, $3f$, and $6f$ are shown in the first column, while the second column shows patterns giving fundamental frequencies of $1\frac{1}{2}f$, $2\frac{1}{2}f$, $3\frac{1}{2}f$, and $6\frac{1}{2}f$. The third, fourth, and fifth columns show patterns of sloping bars that

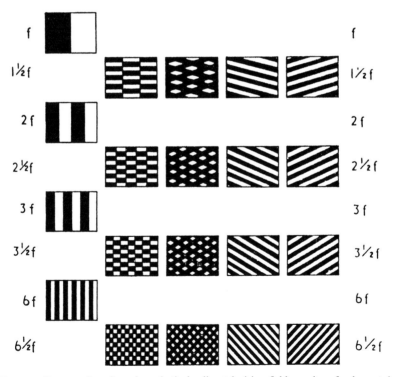

Fig. 19.3. Patterns that, for a hypothetical 9-line television field, produce fundamental frequencies f, $1\frac{1}{2}f$, $2f$, $2\frac{1}{2}f$, $3f$, $3\frac{1}{2}f$, $6f$, and $6\frac{1}{2}f$, where f is the line frequency.

approximate to the pattern of the second column, and would therefore produce substantially the same fundamental frequencies $1\frac{1}{2}f$, $2\frac{1}{2}f$, $3\frac{1}{2}f$, and $6\frac{1}{2}f$. It is seen that the angle of the pattern giving a fundamental frequency of $6\frac{1}{2}f$ is approximately $45°$. With a field of $\frac{1}{2}\times525$ lines, instead of 9 lines, the fundamental frequency corresponding to a pattern of lines at $45°$ is given by: $(\frac{1}{2}\times525\times4/3+\frac{1}{2})f$ which is almost identical with the maximum frequency of the system: $(\frac{1}{2}\times525\times4/3)f$. The $45°$ pattern therefore represents the end of the series of patterns giving the intermediate fundamental frequencies $(h+\frac{1}{2})f$.

Now in the vast majority of television scenes, fragments of vertical bar patterns of the first column of Fig. 19.3 are far more common than fragments of the chequered or sloping bar patterns of the other columns of Fig. 19.3; hence in the average television transmission there are peaks of energy at frequencies which are multiples of f, and there are troughs in between; an example of this is shown in Fig. 19.4.

Furthermore, in most television scenes there is less fine detail than coarse detail, so that the higher frequencies carry less energy than the lower. It has therefore become customary to transmit the colour-difference signals on sub-carriers having frequencies high in the luminance band, and carefully chosen to be suitable odd multiples of half (or sometimes quarter) the line frequency, f, in

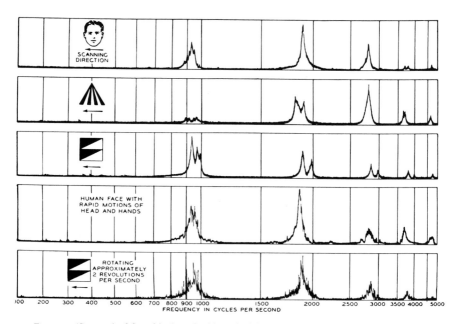

Fig. 19.4. 'Spectra' of four black-and-white television transmissions. The energy is concentrated at frequencies that are multiples of the line frequency (in this case 940 Hz). (Mertz and Gray, 1934.)

order to coincide with the energy troughs of the luminance signal. In Fig. 19.5, a hypothetical system in which the two colour-difference signals are transmitted on two separate sub-carriers high in the luminance band, is shown diagrammatically.

In these diagrams, which are frequently used in discussions on television transmission, the energy transmitted at each frequency is plotted against the frequency. The diagram is not a true plot, however, but only a diagrammatic representation in which the maximum energy permissible is shown at all frequencies. It will be noted that the carrier frequency is not in the centre of the band, as one would expect on the grounds that its modulation by the signal would produce a symmetrical frequency pattern, because vestigial sideband transmission is being used (see Section 19.4); but it is not usually possible to cut out *all* of the unwanted side-band and it is usually present with a much restricted bandwidth as shown in Fig. 19.5. In band sharing systems, double-side-band transmission is generally used for the colour-difference signals, however, in order to reduce the intensity of *cross-talk*, that is, interference between them, or between them and the luminance signal (see Fig. 22.11).

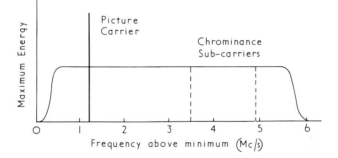

Fig. 19.5. Hypothetical band sharing method of transmitting colour television signals. The two chrominance (or colour-difference) subcarriers have frequencies that are odd multiples of half the line frequency, and are high in the luminance band.

The colour-difference signals will result in the sub-carrier having energy not only at its own frequency, but also at various frequencies above and below it; but, again, because of the relative rarity of chequered or sloping bar patterns the energy will be concentrated at frequencies that differ from the sub-carrier frequency by multiples of the line frequency f. Hence, if the sub-carrier frequency is $(h+\frac{1}{2})f$, the energy of the sub-carrier signal will be concentrated at frequencies $(h+\frac{1}{2})f+jf$, where j is another whole number. (This corresponds to odd multiples of half the line frequency.) The frequencies at which the sub-carrier usually has energy therefore correspond to those at which the main carrier usually does not have energy. The two energy-frequency distributions are therefore mainly interleaved. The receiving set has to distinguish between the luminance and colour-difference signals and this is accomplished by means of electronic devices that divide the incoming signals into those of frequencies that approximate either to multiples of the line frequency f, or to odd multiples of half the line frequency. The former are treated as luminance signals, and the latter as colour-difference signals, and for the reasons already stated this is substantially a correct interpretation. Sometimes, however, as for instance with some of the patterns shown in Fig. 19.3, a luminance signal will be interpreted and displayed by the receiver as a colour-difference signal, and vice versa. This can only happen in fine detail, however, and then only with certain types of picture pattern, such as fine diagonal stripes on a jacket which are sometimes reproduced with spurious random colours.

19.10 The effect of band sharing on monochrome receivers

When a band sharing signal is picked up by an unmodified monochrome receiver, the energy is not divided into two parts, corresponding to multiples of the line frequency and to odd multiples of half the line frequency, as in a colour receiver. The result is that the colour-difference signals are displayed as luminance signals, but are broken up into a chequer-board pattern by the oscillation imposed by the sub-carrier, successive lines of each *field* displaying the pattern shifted horizontally by one pattern unit. This does have a disturbing effect on monochrome reception of colour transmissions, but since the colour-difference signals are zero for whites, greys, and blacks, the effect is absent for these colours and is small for the prevalent pale colours: in saturated colours, however, it can be noticeable, although the sub-carrier oscillation causes luminance to be added in the light parts of the pattern and subtracted in the dark parts, so that the average luminance when viewed from a distance sufficient for the pattern not to be resolved will be unaltered; (actually, because of the non-linearity of the light output of the receiver relative to its electrical input, a net gain in luminance occurs from the pattern, but this partially offsets another error that causes the luminance of saturated colours displayed on monochrome receivers to be too low, as will be discussed in Section 19.13 on Gamma Correction.)

19.11 Carrier sharing

Although the choice of odd multiples of half the line frequency high in the frequency range is helpful in avoiding interference between the colour-difference and luminance signals there is really insufficient space for two sub-carriers (each with side-bands of about $\pm\frac{1}{4}b$) in the luminance band. The concept of *carrier sharing* has therefore been introduced (Fig. 19.6). In this scheme the two colour-difference signals are transmitted at the same sub-carrier frequency, but using either signals a quarter of a cycle out of phase with one another, or transmitting each colour-difference signal only on alternate lines of each field of the picture; the receiving sets then have to be fitted with phase-sensitive detectors or line-delay arrangements so that the two signals can be distinguished. This is discussed further in Chapter 22.

The systems widely used for broadcast television make use of band saving and the constant luminance principle (by using a luminance type of signal for one of the three signals), colour-difference signals, band sharing, and carrier sharing. They are described in detail in Chapter 22. We must now consider two other important topics: the effects of signal processing on colour reproduction, and the effects of non-linearities in the processing of the signals.

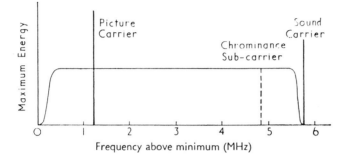

Fig. 19.6. Carrier sharing system. The two chrominance (or colour-difference) signals both have the same carrier frequency, which is an odd multiple of half the line frequency and is high in the luminance band, but the two signals have a 90°-phase difference, or are transmitted on alternate lines of each field.

19.12 The effects of signal processing on colour reproduction

As explained in Chapter 7, if three reproduction stimuli R, G, B have been chosen for a colour television display device, then the camera should have spectral sensitivity curves corresponding to the three colour-matching functions, $\bar{r}(\lambda), \bar{g}(\lambda), \bar{b}(\lambda)$, which define the amounts of these three stimuli needed to match each wavelength of the spectrum. A typical set of functions of this type was shown in Fig. 7.4 for monochromatic stimuli, and in Fig. 19.7 sets are shown for phosphors (N.T.S.C.) considered to be representative in the early days of colour television (broken lines), and for phosphors (B.R.E.M.A.) considered to be representative of present practice (full lines); the chromaticities of these two sets of phosphors are shown in Fig. 21.7. These colour-matching functions for red, green, and blue primaries have both positive and negative regions, and, although this presents certain difficulties in adopting them in television cameras, their use could ensure colorimetrically correct reproduction of all colours within the triangle formed in the chromaticity chart by the three points representing the stimuli R, G, B. This arrangement is shown as the first alternative in Fig. 19.8.

The incorporation of the negative regions of the spectral sensitivity curves of the camera is very awkward, but the same result can be achieved by using an all

389

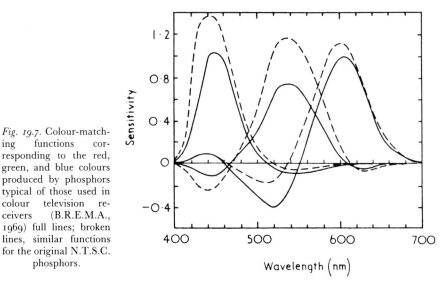

Fig. 19.7. Colour-matching functions corresponding to the red, green, and blue colours produced by phosphors typical of those used in colour television receivers (B.R.E.M.A., 1969) full lines; broken lines, similar functions for the original N.T.S.C. phosphors.

	Camera sensitivity curves	Extra manipulation of signals	Receiver repro-duction colours	Final result displayed
1	Colour-matching functions with negative portions	None	Colours corresponding to camera colour-matching functions	Colorimetrically correct reproduction of all colours within receiver gamut
2	All positive colour-matching functions	Matrixing at camera (or at receiver)	Colours correspond-ing to matrixed camera colour-matching functions	
3		None	Colours of the same hue as those cor-responding to camera colour-matching functions	All colours slightly desaturated
4	Positive parts of colour-matching func-tions which have some negative portions	None	Colours correspond-ing to complete camera colour-matching functions	Some errors in most colours
5		Matrixing at camera (or at receiver)		Small errors in most colours

Fig. 19.8. The effects of alternative camera sensitivity curves on colour fidelity.

390

positive set of colour-matching functions such as CIE $\bar{x}(\lambda), \bar{y}(\lambda), \bar{z}(\lambda)$ functions and then obtaining the required tristimulus values R, G, B, from X, Y, Z, preferably *before* transmission, so as to avoid having to carry out this step at the receiver. The manipulations required are the same as those described in Section 8.4 for *colour transformations*, and are usually referred to in colour television as *matrixing*. The operations required are of the form

$$E_R = a_1 E_X + a_2 E_Y + a_3 E_Z$$
$$E_G = a_4 E_X + a_5 E_Y + a_6 E_Z$$
$$E_B = a_7 E_X + a_8 E_Y + a_9 E_Z$$

where a_1 to a_9 are constants. This arrangement is shown as the second alternative of Fig. 19.8. (Herman, 1975).

It is not necessary with this arrangement for the all-positive camera sensitivity curves to be the $\bar{x}(\lambda)$, $\bar{y}(\lambda)$, $\bar{z}(\lambda)$ curves: any set of colour-matching functions could in principle be used. But if one of the three curves is the $\bar{y}(\lambda)$ curve, then the corresponding tristimulus value is equal to Y, which enables the E_Y signal to be obtained directly. However, this apparently advantageous arrangement involves complications because of the effects of non-linearities in practical systems (see Section 20.7), and it tends to be wasteful of light in the camera.

Matrixing became practicable with the introduction of the Plumbicon tube into camera design, because of its excellent signal-to-noise ratio and its linear characteristic (see Fig. 20.6).

The third alternative arrangement shown in Fig. 19.8 is to adopt the Ives-Abney-Yule compromise described in Section 7.9. The all positive sensitivity curves are now chosen so as to be equal to a set of colour-matching functions corresponding to stimuli P_1, P_2, P_3, having the same dominant wavelength as the reproduction stimuli R, G, B. This arrangement has the advantages that it confines errors to slight losses in colour saturation which are not very noticeable; it avoids the abrupt change in colour fidelity experienced in the first two arrangements as the edges of the R, G, B triangle are crossed; and it does not require a matrixing operation for the signals, a step which inevitably tends to reduce signal-to-noise ratio (because, although signal voltages can be added or subtracted at will, uncorrelated 'noise' voltages always add).

The fourth alternative arrangement shown in Fig. 19.8 is widely used and consists of using the positive parts only of the colour-matching functions corresponding to the reproduction stimuli. Since one or more of such colour-matching functions are usually negative at any wavelength, this alternative tends to produce some errors in most colours. It is found, in practice, however, that the errors involved are not prohibitively large. Although the arrangement is almost certainly not the best from the point of view of colour fidelity, it does have the advantage of being simple, and experience with colour photography has shown that the exact shape of the three spectral sensitivity curves is not very critical for most colours (see Fig. 4.3 for instance).

Another advantage of using the positive parts only of the colour-matching functions is that the resulting red and green curves usually overlap less than is the case for typical all-positive curves; this can be seen from Fig. 19.9 where the narrowest set of all-positive colour-matching functions (Yule, 1973) are compared to a set of sensitivity curves typical of those used in cameras and approximating the positive parts of colour-matching functions of the type shown in Fig. 19.7. This greater wavelength separation of the positive parts means that they can be produced more efficiently in cameras using dichroic beam-splitting mirrors (see Chapter 20). This advantage is gained at the price of departing irrevocably from correct colorimetric analysis of the scene, but Sproson has shown that if positive-part sensitivity curves are used with matrixing the colorimetric errors likely to occur in practice can be reduced to quite small values,

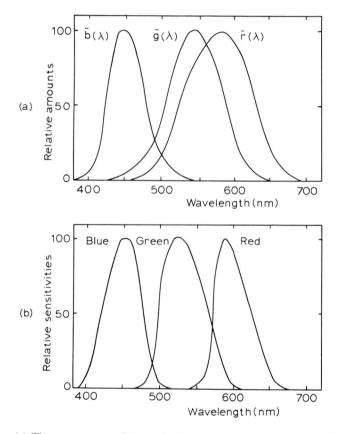

Fig. 19.9 (a) The narrowest possible set of colour-matching functions. (b) Sensitivities typical of those used in television cameras and approximating the positive parts of the colour-matching functions of Fig. 19.7.

if the constants in the matrix are worked out empirically to minimize the errors for typical scene colours (Sproson, 1966). This is shown as the fifth alternative in Fig. 19.8; it is also widely used (see Section 20.10).

The neglect of the negative portions of the matching functions results, in the fourth alternative of Fig. 19.8, in the luminance signal being based, in effect, on a three-humped spectral sensitivity curve instead of on the single-humped $\bar{y}(\lambda)$ curve which is required to give true luminance; the result is that the luminances displayed on both monochrome and colour receivers suffer from errors, and these can be quite large, notably in the case of blues which are lightened appreciably. The matrixing used in the fifth alternative of Fig. 19.8 can also result in these luminance errors being greatly reduced.

The fifth alternative can be operated by carrying out the matrixing either at the camera or at the receiver. European practice is to do it at the camera, and with cameras using Plumbicon tubes, with their nearly linear response, it is theoretically correct to matrix the actual signals produced by the tubes. In the U.S.A., if matrixing is carried out, it is often done at the receiver (or monitor); since the signals at this stage are no longer linearly related to the optical image (for reasons to be described in the next section), it is not theoretically correct to matrix them at that stage. However, calculations and practical tests have shown that the resulting errors are not too large. An advantage of matrixing at the receiver is that, as new phosphors are brought out, the matrixing can be adjusted accordingly (DeMarsh, 1974).

19.13 Gamma correction

So far, a linear relation has been assumed between the electrical and corresponding optical signals at both the camera and the receiver. But the light output from receiver tubes is not linear: it is approximately proportional to the cube of the applied voltage. Thus if the logarithm of the resulting tube luminance is plotted against the logarithm of the applied voltage the slope of the line obtained (that is, the gamma) is about 3 (2.8 ± 0.3 is the accepted index for colour receivers).

From the point of view of signal-to-noise ratio, a high gamma is desirable because the darker portions of the picture, where noise is most obvious, tend to be reproduced nearly black. But while a monochrome picture which has a gamma of about 3 might be tolerable, a colour picture will exhibit severe colour distortion.

Suppose E_R, E_G, E_B are intended to produce a colour $R(R)+G(G)+B(B)$ on a linear display. If, instead, they are applied to a cube law display, the resulting colour is $R^3(R)+G^3(G)+B^3(B)$. For example, if $R=1$, $G=\frac{1}{2}$, $B=\frac{1}{2}$ and unit quantities of (R), (G), and (B) result in a white (W), then the intended colour is equivalent to $\frac{1}{2}(W)+\frac{1}{2}(R)$. But the displayed colour will be $R=1$, $G=\frac{1}{8}$, $B=\frac{1}{8}$, or $\frac{1}{8}(W)+\frac{7}{8}(R)$. Hence the luminance has decreased, and the saturation has increased. A simple means of correcting for this is to pre-distort the signals E_R, E_G, E_B, at the transmitter to $E_R^{1/\gamma}$, $E_G^{1/\gamma}$, $E_B^{1/\gamma}$; the luminance signal is then transmitted as $E_Y' = lE_R^{1/\gamma} + mE_G^{1/\gamma} + nE_B^{1/\gamma}$, and the colour-difference signals as $E_R^{1/\gamma} - E_Y'$ and $E_B^{1/\gamma} - E_Y'$.

At the receiver the corresponding green colour-difference signal $E_G^{1/\gamma} - E_Y'$ is obtained thus:

$$E_G^{1/\gamma} - E_Y' = -\frac{l}{m}(E_R^{1/\gamma} - E_Y') - \frac{n}{m}(E_B^{1/\gamma} - E_Y')$$

Then by adding E_Y' to all three colour-difference signals the voltages $E_R^{1/\gamma}$, $E_G^{1/\gamma}$, $E_B^{1/\gamma}$, are recovered, and can be applied to the appropriate tubes (with power law γ) to give the correct R, G, B.

This method gives distortionless large-area reproduction. But the luminance carried by E_Y' (which would be displayed by a monochrome receiver all over the picture, and by a colour receiver in fine detail) is $(E_Y')^\gamma$. The ratio of luminance carried by E_Y' to the true luminance is:

$$\frac{(E_Y')^\gamma}{E_Y} = \frac{(lE_R^{1/\gamma} + mE_G^{1/\gamma} + nE_B^{1/\gamma})^\gamma}{lE_R + mE_G + nE_B}$$

For the worst case (saturated blue) this ratio is:

$$\frac{n^\gamma E_B}{nE_B} = n^{\gamma-1} = 0.11^{1.8} = 0.019$$

assuming values of 0.11 for n and 2.8 for γ which are fairly typical.

So in this case the luminance signal E_Y carries only 2% of the true luminance; hence small-area saturated colours are reproduced too dark. Also compatibility suffers, as a monochrome receiver will display too little luminance; however, in practice, the effect of the non-linearity of the cathode-ray tube characteristic on the dots produced by the chrominance signals increases the 2% quoted to about 5% (see Section 19.10). Thus monochrome errors are not too bad, and large area colour is correct. But as E_Y' does not carry all the luminance, the remainder must be carried by the sub-carrier modulation, and hence the constant luminance principle is not obeyed, with the result that the subjective effect of noise and interference is increased.

A further point is that, as the sub-carrier modulation is severely limited in bandwidth, definition will suffer because the luminance content of the sub-carrier will also be limited in bandwidth. But for white, $E_R = E_G = E_B$, and $\frac{(E_Y')^\gamma}{E_Y} = 1$, and for the more prevalent neutral shades the ratio will not be very much less than unity. Hence the above shortcomings are evident only for the higher saturations (see Fig. 22.4).

There are several alternative methods for gamma correction, but in general these involve additional complications at the receiver. For instance, if the luminance signal was composed *before* E_R, E_G, and E_B were predistorted and the signals transmitted were $E_R^{1/\gamma} - E_Y^{1/\gamma}$, $E_B^{1/\gamma} - E_Y^{1/\gamma}$, and $E_Y^{1/\gamma}$ then the above difficulties would not arise. But the recovery of the green signal is then much more complicated, requiring the signals first to be raised to the power γ, then mixed to obtain E_G, then re-distorted to the power $1/\gamma$, before finally applying them to the tube.

So far, it has been assumed that the signals are pre-distorted by the full factor of $1/\gamma$. In practice, however, the gamma correction applied to the signals only amounts to raising them to the power of $1/2.2$, whereas colour receivers usually have gammas of about 2.8 (± 0.3). The result of this is that in the final display the gamma of the picture is increased over that of the original scene by a factor of $2.8/2.2$, that is 1.273 times. This increase in displayed gamma is necessary (see Section 6.5) in order to overcome the reduction in apparent contrast caused by the dim surround conditions in which television is normally viewed (Bartleson and Breneman, 1967; Pitt and Winter, 1974); but, as can be seen from Fig. 19.10, increases in purity and shifts in dominant wavelength occur. The increases in purity can be beneficial in compensating for any losses of saturation caused by the

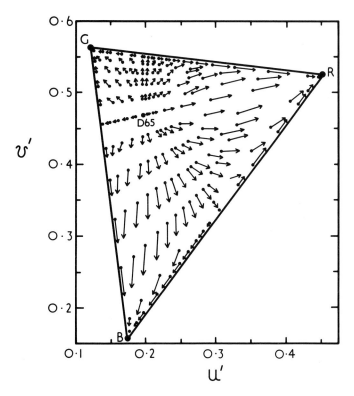

Fig. 19.10. Shifts in reproduced chromaticities resulting from altering the overall gamma of a television system from 1.0 to 1.273; a gamma of about 1.25 is required in order to offset the contrast lowering effect of the dim surround typical of normal television viewing conditions. (After Brown, 1971. Fig. 3.)

effects of the dim surround or by other factors; but the shifts in dominant wavelength can cause errors in hue. It has also been shown that, when the ambient lighting is increased to give a surround of luminance similar to the average picture-luminance, then the added flare light reduces the gamma from 1.25 to about 1.0 as required (Novick, 1969).

It has also been shown that to achieve a gamma of 1.25 in ambient lighting giving a dim surround whose luminance is about one-tenth (a typical value) of picture peak-white, the picture must be such that in the absence of ambient light it has a gamma of 1.5 (DeMarsh, 1972). This can be achieved by making adjustments to the black level of the camera so that its gamma is about 0.54 instead of 0.45 (1/2.2), or by making similar adjustments to the receiver so that its gamma is about 3.3 instead of 2.8, or by making partial adjustments to both.

When the transfer characteristic required from a gamma-reducing circuit is plotted as a voltage obtained against the voltage applied, the power-law functions involved call for ever-increasing amplification factors as the applied voltage is decreased to approach zero. Because these very high amplification factors are not realizable, the power functions are usually replaced by linear functions for low applied voltages, and the output voltages are then lower than they should be. This can cause undesirable increases in colour saturation, especially for red, orange, and yellow colours, which often have very low blue signals. The form of the linear approximation is often different from one camera to another, and this can cause undesirable changes in colour reproduction as cuts are made between cameras in a programme.

19.14 Noise reduction

Spurious signals in television give rise to random moving specks, generally called *noise*. Their harmful effects on picture quality can be reduced for stationary objects by averaging the signals over more than one frame, and this technique can result in marked improvements to the appearance of the pictures, even when some of the subject matter is moving (Sanders, 1980).

19.15 Higher definition television

The 525 and 625 line systems used for broadcast colour television were developed at a time when compatibility with monochrome receivers and economy of bandwidth were extremely important considerations. The advent of different forms of transmission, including satellite broadcasting, and both conventional and fibre-optic cable systems, means that signals requiring much broader bandwidth can be considered.

The Japan Broadcasting Corporation (NHK, Nippon Hoso Kyokai) and the BBC have proposed 1125 and 1501 line systems (Fink, 1980), as shown in Table 19.2. Neither of these systems would be compatible with existing systems, but proposals have been made for a 1249 line system (as also shown in Table

19.2), which would be compatible with existing 625 line systems. Compatibility could be achieved by transmitting the signal in two parts, one which would produce a 625 line picture on existing receivers, and the second which, together with the first, would produce a 1249 line picture on higher definition receivers; various ways have been proposed for combining the signals, one of which would add extra picture areas on either side of the compatible picture area to give an aspect ratio of about 8 to 3. As indicated in Table 19.2, a wider aspect ratio is usually considered a desirable feature of higher definition systems. An improved separation of the luminance and chrominance signals is also usually planned, either by using separate bandwidths, or by using more sophisticated methods of band sharing, or by using time sharing over line-scan periods by means of time compression devices. Displayed pictures of larger size are usually proposed, either by means of larger shadow-mask type tubes, or by projection devices, or by flat panel displays when technology permits. Broadcast transmission would have to be in the gigahertz (GHz) range of frequencies (usually 10 to 90 GHz, that is, 10 000 to 90 000 MHz).

TABLE 19.2

Parameters for proposed higher definition television

System	1125 line	1501 line	625 Compatible
Aspect ratio	5:3 or 2:1	8:3	4:3 or 8:3
Lines per picture	1125	1501	1249
Field frequency	60 Hz	60 Hz	50 Hz
Interlace ratio	2:1	2:1	2:1
Luminance bandwidth	20 MHz	50 MHz	10 MHz
Chrominance bandwidths	7.0 MHz	12.5 MHz	4 MHz
	5.5 MHz	12.5 MHz	4 MHz

When space satellites are used for broadcasting, the form of the signals has to be revised to minimize power consumption, and proposals have been made to avoid luminance-chrominance band sharing by using time compression techniques over line-scan periods. For receivers to take full advantage of this, they would have to have the option of operating in this alternative mode.

19.16 Videoconferencing

A great deal of time and money is spent in attending business conferences, and the use of television is attractive as a more adequate substitute than is offered by telephone calls. The use of standard broadcast signals in this application is difficult because of their wide bandwidth, and systems using suitable compression of the information have been developed. In one such system, digital signals of only 2 megabits per second are used; but higher definition is obtained than would normally be expected from this rate, by storing the signals frame by frame and transmitting only changes in the picture. For typical conference activity, the restricted transmission of movement provided by this system can be quite adequate. Two cameras can be used to cover the wide angle presented by a group of about six people sitting at a table, and the two pictures can then be displayed one above the other on the same monitor. This 'face to face' mode of operation can be supplemented by a 'graphics' mode, in which a third camera is used to give signals of high definition from a static display, the complete transmission of which can take up to about 2 seconds. (Nicol and Duffy, 1983.)

REFERENCES

Bartleson, C. J., and Brenemen, E. J., *Phot. Sci. Eng.*, **11**, 254 (1967).
Birkinshaw, D. C., *J. Roy. Television Soc.*, **16**, 32 (1977).
B.R.E.M.A., *Radio and Electronic Engineer*, **38**, 201 (1969).
Bridgewater, T. H., *J. Roy. Television Soc.*, **16**, 25 (1977).
Brown, G., B.B.C. Research Report 1971/4 (1971).
Campbell-Swinton, A. A., *Nature*, **78**, 151 (18 June 1908).
De Marsh, L. E., *J. Soc. Mot. Pic. Tel. Eng.*, **81**, 784 (1972).
De Marsh, L. E., *J. Soc. Mot. Pic. Tel. Eng.*, **83**, 1 (1974).
Fink, D. G., *J. Soc. Mot. Pic. Tel. Eng.*, **89**, 89 and 153 (1980).
Hawker, P., *J. Roy. Television Soc.*, **20**, 273 (1983).
Herman, S., *J. Soc. Mot. Pic. Tel. Eng.*, **84**, 267 (1975).
Hunt, R. W. G., *J. Roy. Television Soc.* **11**, 220 (1967).
Jesty, L. C., *Wireless World*, **63**, 304 (1957).
McGee, J. D., *Brit. J. Phot.*, **123**, 81 (1976).
Mertz, P., and Gray, F., *Bell. Syst. Tech. J.*, **13**, 464 (1934).
Nicol, R. C., and Duffy, T. S., *Professional Video*, **10**, 36 (November, 1983).
Novick, S. B., *Brit. Kinematog. Sound Tel.*, **51**, 342 (1969).
Pitt, I. T., and Winter, L. M., *J. Opt. Soc. Amer.*, **64**, 1328 (1974).
Sanders, J. R., *J. Roy. Television Soc.*, **18**, 29 (1980).
Shiers, G., *J. Roy. Television Soc.*, **16**, 6 (1976).
Sidey, P., Longman, R., Glencross, D., and Pilgrim, A., *J. Soc. Mot. Pic. Tel. Eng.*, **90**, 1165 (1981).
Sproson, W. N., private communication (1966).
Yule, J. A. C., *J. Color and Appearance*, **2**, 30 (1973).

GENERAL REFERENCES

Carnt, P. S., and Townsend, G. B., *Colour Television: N.T.S.C. Principles and Practice*, Iliffe, London, (1961), and *Colour Television, Vol. 2, P.A.L., S.E.C.A.M. and Other Systems*, Iliffe, London (1969).
Fink, D. G., *Color Television Standards*, McGraw-Hill, New York (1955).
Gouriet, G. G., *An Introduction to Colour Television*, The Television Society, London (1955).
Jesty, L. C., *J. Television Soc.*, **7**, 488 (1954).
Kaufman, M., and Thomas, H., *Introduction to Color T.V.*, John F. Rider, New York (1954).
Loughren, A. V., *J. Soc. Mot. Pic. Tel. Eng.*, **60**, 321 and 596 (1953).
Pearson, D. E., *Transmission and Display of Pictorial Information*, Pentech Press, London (1975).
Proc. Inst. Radio Engnrs., **39**, 1124–1331 (1951).
Proc. Inst. Radio Engnrs., **41**, 838–858 (1953).
Proc. Inst. Radio Engnrs., **42**, 5–344 (1954).
Proc. Inst. Radio Engnrs., **43**, 742–748 (1955).

Sproson, W. N., *Colour Science in Television and Display Systems*, Adam Hilger, Bristol (1983).

The Television Society, London, *A Bibliography of Colour Television* (1954).

The Television Society, London, *A Bibliography of Colour Television Supplement* (1955).

Zworykin, V. K., and Morton, G. A., *The Electronics of Image Transmission in Colour and Monochrome*, Chapman & Hall, London (1954).

CHAPTER 20

Electronic Cameras

20.1 Introduction

TELEVISION systems operate by converting the information from a two dimensional image of a scene into a one-dimensional signal; after transmission this signal is re-formed once more into a two-dimensional reproduction. This conversion of information from two dimensions to one is accomplished at the camera by the process of *scanning* the original image line by line in time-sequence. The two-dimensional reproduction is reformed at the receiver by the inverse process of building up the picture line by line from the transmitted signal which is varying with time.

20.2 Early camera tubes

The cameras used in the earliest experimental forms of television broke the picture into a series of lines by optical or mechanical means. Thus if a disc of the type shown in Fig. 20.1 were rotated in front of an image, an analysis into a series of slightly curved lines would result. These optical and mechanical scanning devices, however, were only capable of producing pictures of rather low definition, sometimes having fewer than 100 lines per picture. The advent of high definition television, involving 400 or more lines per picture, was made possible by electronic scanning devices, of which the earliest was the *iconoscope* invented by Zworykin in 1928. In this device an image of the outside scene is made to fall on a mosaic of activated silver specks on a layer of mica; each speck is separated from the others and acts as a small photocell. When light falls on a cell, electrons are emitted as a function of the amount of incident light and hence an image of

Picture area

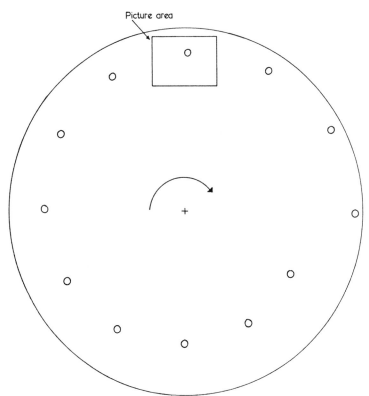

Fig. 20.1. Rotating disc used for mechanical scanning in early television systems.

electric charge is formed. This *photocathode* surface is then scanned by a beam of high velocity electrons (accelerated through 1000 volts) which causes secondary emission of electrons, and this results in a net loss of electrons from the mosaic to a collecting plate so that current flows from the collector; but areas that have been exposed to light have already lost some electrons by photoemission, and they therefore lose fewer electrons by secondary emission and hence result in less current flowing when they are scanned by the electron beam. The current is therefore related to the intensity of the light in the image; the change in current is in fact roughly proportional to the square root of the change in intensity of the light, and, although this may at first sight seem an undesirable relationship, television receivers usually have display devices in which the light output is approximately proportional to the square or cube of the applied signal, so that the two devices together give a linear or 1.5 power-law result (the latter being required for viewing with a dark surround, as discussed in Chapter 6). As is the case for all television camera tubes (or *pick-up* tubes, as they are sometimes

401

called), the various components of the iconoscope are all housed in an evacuated glass envelope.

In the *image iconoscope*, greater sensitivity is achieved by forming the optical image on a semi-transparent photocathode, and then accelerating the photoelectrons produced so as to make them strike a storage plate with an image-wise distribution; this results in secondary emission of electrons from the storage plate, so that when this is scanned by a beam of high velocity electrons, the current from a collecting plate is a function of the light intensity of the scene, as in the iconoscope.

The iconoscope and image iconoscope tubes are not suitable for colour television because the uncontrolled secondary electrons distort the tone reproduction as a function of the luminance of neighbouring areas; since the distribution of light in the three colour images is different, the distortions would also be different, and hence spurious colours would be produced.

The first public television service having high definition, which was broadcast by the BBC in London in 1936, depended on a different type of pick-up tube developed independently by the team led by J. D. McGee of E.M.I and known as the *Emitron*. An improved version known as the *C.P.S.* (cathode-potential-stabilized) *Emitron* or *Orthicon* was introduced about ten years later.

Fig. 20.2 illustrates the main features of the C.P.S. Emitron or Orthicon tube. The image is formed on a thin mica sheet which carries a transparent conducting layer on the front and a photosensitive layer on the back. The photoelectrons emitted from the back surface are collected at an anode leaving an image of electric charge. This electrical image is then scanned by a beam of low velocity electrons, whose energy is just insufficient to reach the photosensitive layer in black areas of the scene; but in light areas the existence of a positive charge, caused by the loss of photoelectrons, enables the scanning beam to reach the photosensitive layer and to replace the lost electrons, so that current flows from the electron gun to the photosensitive layer, as a function of the luminance of the image, as the electron beam scans the picture: the flow of this current causes changes in the voltage on the transparent conducting layer, from which the image signal is derived. This type of tube does not suffer from local tone-distortions

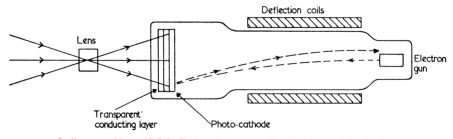

20.2. Ordinary orthicon (C.P.S. Emitron) camera tube. Full lines: light; broken lines: electrons.

which occur in iconoscopes, but its sensitivity is rather too low for effective use in colour cameras. It has an almost linear relationship between signal and luminance, which means that the signal must be adjusted (*gamma-corrected*) before being supplied to a receiver. (Benson, 1981.)

20.3 Tubes suitable for colour

In Fig. 20.3 the main features of the *image orthicon* tube are illustrated. This tube combines the low-energy scanning beam of the orthicon, with the electron-image stage of the image iconoscope, and adds an electron-multiplier stage for increased sensitivity. The image is formed on a photosensitive layer, and the photoelectrons produced from this photocathode are focussed in an imagewise way on the 'target'. The target consists of an extremely thin glass membrane with a fine metal mesh, located very closely parallel to it, on the side on which the photoelectrons are incident. The photoelectrons cause secondary electrons to be emitted from the glass membrane and these are collected on the wire mesh; the glass is so thin that the resulting charge pattern on the glass appears on both sides of it. When the glass is scanned by a low-energy electron beam, electrons therefore reach the target only as a function of the luminance of the image of the scene, as in the ordinary orthicon. Changes in image luminance therefore cause changes in the electron beam current, which, on returning from the target, enters the first stage of a five stage electron multiplier tube.

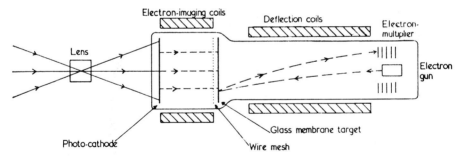

Fig. 20.3. Image orthicon camera tube. Full lines: light; broken lines: electrons.

The sensitivity of image orthicon tubes is very high, enabling good black-and-white pictures to be obtained at scene-illumination levels down to about 20 lux. Like the orthicon, the image orthicon has a linear response; but, as the luminance level is raised, a point is reached where the wire mesh is unable to collect many more secondary electrons and the response then flattens out (Neuhauser, 1956), and the secondary electrons not collected by the mesh fall back on to the target and produce areas of negative charge which result in black halos appearing around very bright objects such as light sources. For colour work these distortions being, in general, different in the three pictures, would produce intolerable spurious colours, so the tubes must be operated below the bend, or 'knee', of their characteristics. The range of tones over which a colour image-orthicon tube operates linearly is about 90 to 1, expressed as the ratio of peak-signal to R.M.S.-noise; in log-units this amounts to 1.95, or in decibels of signal amplitude 39 dB.[1] In practice, however, the need for gamma correction and *aperture correction* (increasing the amplitudes of the high frequencies in the signal so as to offset fine-detail contrast-reduction caused by the finite size or 'aperture' of the electron spots, see Section 20.11) usually reduce the peak signal-to-noise ratio below this level.

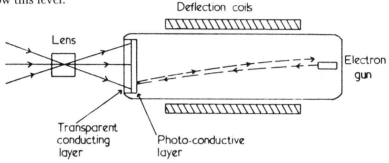

Fig. 20.4 Vidicon tube. Full lines: light; broken lines: electrons.

In Fig. 20.4 the main features of the *vidicon* tube are shown. The image of the outside scene is formed on a thin layer of photoconductive material which has in contact with it (on the incident-light side) a transparent conducting layer which is kept at a voltage positive with respect to that of a low-energy electron gun on the other side of the layer. The result is that, in the absence of light, the photoconductive layer acts as an insulator and prevents current flowing from the gun to the transparent conducting layer; but when light falls on the device, the photoconductive layer becomes conducting, so that electrons move towards the transparent conducting layer, and hence the potential of the surface of the photoconductive layer rises as a function of the light intensity; then the electron beam, as it scans the image area, deposits electrons so as to reduce the charge once more, and hence a current flows to the transparent conducting layer, in accordance with the luminance of the image, to provide the required signal.

It is clear that the vidicon is a somewhat simpler device than the image orthicon but it is less sensitive, requiring scene illumination levels of around 500 lux for good black-and-white pictures. The response characteristics of the vidicon

[1] In electrical engineering, power is proportional to the *square* of the voltage or current so that a change of one $\log_{10}$ unit in voltage or current is equal to twenty (not ten) decibels. In television, if the transducer-output voltage or current is considered as *directly* proportional to the amount of light, a given change in signal voltage or current expressed in decibels must be divided by twenty to obtain the corresponding change in $\log_{10}$ of the amount of light; conversely, changes in $\log_{10}$ of the amount of light must be multiplied by twenty to obtain the corresponding change in voltage or current in decibels.

are like the iconoscope's, the signal being roughly proportional to the square root of the incident light, so little gamma correction is needed; but, unlike the iconoscope's, the vidicon's response is not affected by local luminance variations. Unlike the orthicon, the tone range handled without distortion is very adequate, signal to noise ratios of 300 to 1 (50 dB) being quoted, although aperture correction often causes a reduction to more like 100 to 1 (40 dB). In Fig. 20.5 response curves for a typical vidicon (Neuhauser, 1956) are shown on a log-log plot, so that straight lines of gamma (slope) equal to about 0.5 are obtained; the gammas of typical vidicon tubes usually fall in the range 0.5 to 0.6. The different lines in Fig. 20.5 were obtained simply by altering the voltage difference between the transparent conducting layer and the electron gun, and this provides a very convenient sensitivity adjustment (in a colour camera it can be used to equalize the three signals given by white). The vidicon has a very low and stable dark current and this is important in colour cameras, otherwise shadows and blacks may be reproduced coloured; in fact all its tonal transfer characteristics are very stable and linear (on a log-log plot), being free from any 'knees' such as occur with the image orthicons. But, in addition to limited sensitivity, the vidicon does possess one further disadvantage: it has a time lag which can cause slight smearing of the picture; however, the lag is dependent on the light level, and with scene illumination levels above 1000 lux it is fairly negligible. The vidicon, with its wide stable range of tone reproduction, is well-suited to deriving television signals from photographic film images. Vidicon tubes are smaller than image orthicons and therefore enable smaller colour cameras to be constructed.

The *Plumbicon* tube (De Haan and Van Doorn, 1964) operates on exactly the same principles as the vidicon tube but the photoconductor, instead of being antimony sulphide (Sb_2S_3) or Selenium (Se) is lead monoxide (PbO).

The sensitivity of the Plumbicon is intermediate between that of the image-orthicon and the vidicon, good black-and-white images being obtained down to scene-illumination levels of about 100 lux. The gamma of the response curve, as

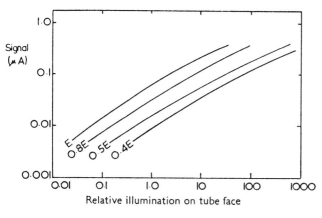

Fig. 20.5. Response characteristics for vidicon tube at various voltage differences (E, $0.8E$, $0.5E$, and $0.4E$, where E is the maximum possible) between the transparent conducting layer and the electron gun.

shown in Fig. 20.6, is about 1.0 (on a log-log plot) so that gamma correction is required. The range of tones reproduced without distortion is about 80 to 1 or 38 dB at scene-illumination levels of 1500 lux, and about 65 to 1 or 36 dB at 250 lux. The image area used is very small, 12×16 mm or less, so that lenses of large relative aperture can be used without serious loss of depth of field (see Section 18.11), and very compact colour cameras can be made. The dark current is very low and the tone characteristics are very reproducible and linear, and are independent of light level, local luminance differences in the image, operating voltage, and ambient temperature. The Plumbicon does not suffer from any appreciable time lag effects (except at very low light levels). Plumbicon tubes are very widely used in colour television cameras.

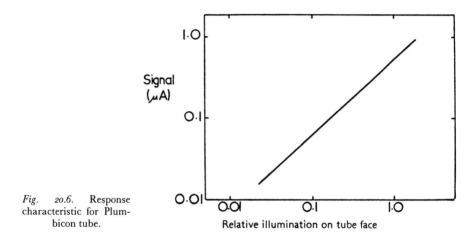

Fig. 20.6. Response characteristic for Plumbicon tube.

Tubes using photoconductors of selenium, arsenic, and tellurium, *Saticon* tubes, have also been introduced; their properties are similar to those of Plumbicon tubes, but have less time lag at low light levels (Benson, 1981).

20.4 Spectral sensitivities of television camera tubes

It is obviously necessary that camera tubes used for colour should have sensitivity throughout the visible spectrum, but the exact form of the spectral sensitivity curve is relatively unimportant because the desired curves (see Section 19.12) can usually be closely enough approximated by means of suitable filtration. In Fig. 20.7 the spectral sensitivities of typical image orthicon, vidicon, and Plumbicon tubes are shown for comparison. It is clear that all of the tubes have relatively more sensitivity in the blue than in the red regions of the spectrum; but blue filters are usually less efficient than red, and conditions of dim illumination often occur with tungsten lighting which is particularly poor in blue content. Hence it is usually found that in colour cameras the red and blue channels have similar effective sensitivities, with the green somewhat more sensitive. The sensitivity of the ordinary Plumbicon tube is too restricted in the red to give proper rendering of colours that have important changes in spectral reflectance beyond 640 nm; however, by matrixing, this can be corrected to a useful degree for the colours that are most likely to occur in practice (Monteath, 1966). Plumbicon tubes with increased far-red sensitivity have also been made (De Haan and Van Doorn, 1965).

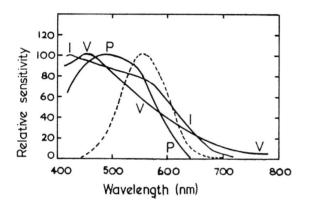

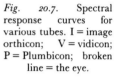

Fig. 20.7. Spectral response curves for various tubes. I = image orthicon; V = vidicon; P = Plumbicon; broken line = the eye.

20.5 Camera arrangements

For closed circuit television, where the signals travel entirely by cable, the simple field sequential system of colour television can be used: in this case a red, green, and blue filter wheel rotating in front of what is essentially a black-and-white camera (but with the frame frequency suitably increased) is all that is needed. But for broadcast colour television the need to transmit mixtures of the red,

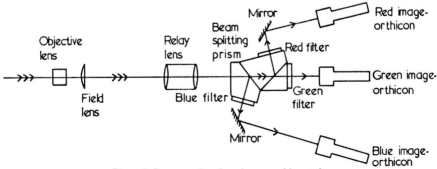

Fig. 20.8. Camera using three image-orthicon tubes.

407

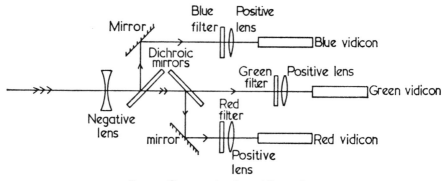

Fig. 20.9. Camera using three vidicon tubes.

green, and blue signals requires that they be present simultaneously, and hence different arrangements are necessary for both the cameras and the receivers. Three different types of camera are shown diagrammatically in Figs. 20.8, 20.9, and 20.10.

In Fig. 20.8 a camera employing three image orthicon tubes is shown (Bertero, 1963). The objective lens forms an image in the plane of its associated field lens and a relay lens then forms secondary images on the three camera tubes. The light is split by a dichroic (interference) prism assembly into red, green, and blue components before being imaged on to the three image orthicon tubes. Colour filters are added to improve the approximation to the required spectral sensitivity curves, and neutral density filters may also be added to equalize the signals produced by white.

Fig. 20.9 shows the main features of a colour camera that employs three vidicon tubes (James, 1959). In this camera a relay lens is avoided by using a telephoto lens system and inserting the dichroic beam splitting mirrors between

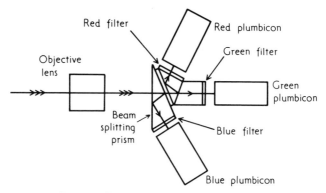

Fig. 20.10. Camera using three Plumbicon tubes.

408

the front (negative) part of the lens and the rear (positive) part; the rear lens, being situated after the beam is split into three, has to be triplicated, of course. By suitable choice of focal lengths, the angles of the beams of light between the front and rear lenses is kept down to $\pm 4\frac{1}{2}°$ which makes the optics easier to design.

Fig. 20.10 depicts a Plumbicon camera arrangement. A very compact dichroic prism arrangement, situated between the lens and the tube faces, is used to split the beam into three, and this, combined with the small size of the Plumbicon tubes, results in a very neat camera arrangement (Van Doorn, De Lang, and Bouwis, 1966). Cameras of this general type are very widely used.

20.6 Image equality in colour cameras

In a colour camera it is necessary for the three images to be identical in tone reproduction, and, to achieve this, care has to be taken at several stages. First, the flare characteristics of the three optical paths should be closely similar. Secondly, each of the three tubes should have uniform sensitivity and tone reproduction characteristics all over its image area, or colour shading over the picture will occur (although some correction for non-uniformity over the picture can be provided subsequently electronically). Finally, the tone reproduction characteristics of the three tubes should be closely similar to one another, or a grey scale will be reproduced coloured at some densities.

The camera tubes used in colour television are normally capable of providing all the definition required by the television system being used, but unless the images formed on the three tubes are geometrically identical and in exact registration with respect to the electron scanning, poor definition, and colour fringing, occur. Very precise optical and electronic components are therefore required and the optics have to be mounted very rigidly. (In the image orthicon cameras it was also necessary for the electron imaging sections of the three tubes to be identical.)

20.7 R-Y-B cameras

The problems of definition are somewhat alleviated in the case of the blue tube because the blue contribution to the high definition luminance signal usually amounts to only about 10 per cent of the total, and the colour-difference signals, where the blue contribution is much more important, are of low definition. From the point of view of definition alone, the ideal camera would be one in which the three signals produced were not red, green, and blue, but red, luminance, and blue. Optically this is not difficult to arrange, requiring only that the green filter be broadened to admit a little red and blue light so as to result in that channel having a spectral sensitivity similar to the $\bar{y}(\lambda)$ function. The registration problems would then be alleviated because all the high definition would be confined to one of the three images, and lack of registration with the other two would only show up if it was so large as to be resolved by the low definition colour-difference signals. The advantage, however, is not found in practice to be as large as might be expected, because the eye seems to be very sensitive to colour fringes around edges in a picture: the mis-registration tolerances in an R-Y-B camera are therefore only about twice as great as for an R-G-B camera, instead of four times as might be expected from bandwidth considerations. Splitting the light into luminance, red, and blue signals, is also less efficient optically than splitting it into green, red, and blue signals (see Section 20.10).

The production of a separate luminance signal in the camera leads to a rather complicated situation regarding gamma correction. The luminance signal, E_Y, would have to be gamma-corrected to $E_Y^{1/\gamma}$ before transmission, and would therefore be related to the red, green, and blue signals E_R, E_G, and E_B, thus:

$$E_Y^{1/\gamma} = (lE_R + mE_G + nE_B)^{1/\gamma}$$

where l, m, and n are the usual luminance factors. The other two signals available at the receiver would be $E_R^{1/\gamma} - E_Y^{1/\gamma}$ and $E_B^{1/\gamma} - E_Y^{1/\gamma}$, and although $E_R^{1/\gamma}$ and $E_B^{1/\gamma}$ are easily recovered (by adding $E_Y^{1/\gamma}$ to each), the recovery of $E_G^{1/\gamma}$ is a very complicated business, requiring the signals first to be raised to the power γ, then added and subtracted, and then raised to the power $1/\gamma$. The added complexity that this would call for in the receiver (involving extra cost and loss of signal-to-noise ratio) has prevented R-Y-B types of camera from being used in this way.

Various schemes have been suggested for recovering an approximate $E_G^{1/\gamma}$ signal by less complicated procedures at the receiver: for instance, if the circuits used were the same as those for use with the normal luminance signal $(E_Y' = lE_R^{1/\gamma} + mE_G^{1/\gamma} + nE_B^{1/\gamma})$ then the actual green signal recovered, $E_g^{1/\gamma}$ would be given by (see Section 19.13)

$$E_g^{1/\gamma} = E_Y^{1/\gamma} - \frac{l}{m}(E_R^{1/\gamma} - E_Y^{1/\gamma}) - \frac{n}{m}(E_B^{1/\gamma} - E_Y^{1/\gamma})$$

Remembering that $l+m+n = 1$, this becomes

$$E_g^{1/\gamma} = \frac{1}{m}E_Y^{1/\gamma} - \frac{l}{m}E_R^{1/\gamma} - \frac{n}{m}E_B^{1/\gamma}$$

Since $-\frac{l}{m}E_R^{1/\gamma} - \frac{n}{m}E_B^{1/\gamma} = E_G^{1/\gamma} - \frac{1}{m}E_Y'$ we obtain

$$E_g^{1/\gamma} = E_G^{1/\gamma} + \frac{1}{m}(E_Y^{1/\gamma} - E_Y').$$

For greys $E_Y^{1/\gamma} - E_Y'$ is zero, but for colours it is always positive. Hence the green signals are distorted by being increased as saturation increases: thus green colours would be increased in saturation, but red and blue colours would be desaturated and altered in hue. These errors can be reduced by subtracting from $E_g^{1/\gamma}$ a signal that is a suitable function of colour saturation, and this could be obtained by rectifying the chrominance signal (James and Karowski, 1962), but only at the cost of complicating the receiver.

Because of the above receiver complications it is preferable with an R-Y-B camera for a low definition E_G signal to be recovered from the E_R, E_B, and E_Y signals at the transmitter; in the next section it is shown that by suitable processing of these signals before transmission the correct signals can be recovered without any extra complications at the receiver, and without losing the advantages of having a separate high-definition luminance signal produced by the camera.

20.8 Four-tube cameras

The difficulties over recovering the green signal properly, when an R-Y-B camera is used, appear in a somewhat different form if a separate luminance signal is derived in addition to all three colour signals by using four camera tubes. In such an R-G-B-Y camera, the two colour-difference signals are derived from the red, green, and blue tubes in the usual way, by first forming the E_Y' signal

$$E_Y' = lE_R^{1/\gamma} + mE_G^{1/\gamma} + nE_B^{1/\gamma}$$

and then subtracting this signal from $E_R^{1/\gamma}$ and $E_B^{1/\gamma}$ to obtain

$$E_R^{1/\gamma} - E_Y'$$

and

$$E_B^{1/\gamma} - E_Y'$$

The fourth tube is then used to provide a true gamma-corrected luminance signal $E_Y^{1/\gamma}$. As usual, only the luminance signal is of high definition, so that the red, green, and blue tubes which contribute only to the colour-difference signals can be of lower definition and require less precise registration. The receiver then attempts to recover the required colour signals $E_R^{1/\gamma}$, $E_G^{1/\gamma}$, and $E_B^{1/\gamma}$, but, because in general E_Y' and $E_Y^{1/\gamma}$ are not equal, all three colour signals are now only recovered in modified form, $E_{MR}^{1/\gamma}$, $E_{MG}^{1/\gamma}$, and $E_{MB}^{1/\gamma}$, thus:

$$E_{MR}^{1/\gamma} = E_R^{1/\gamma} - E_Y' + E_Y^{1/\gamma} = E_R^{1/\gamma} + (E_Y^{1/\gamma} - E_Y')$$

$$E_{MB}^{1/\gamma} = E_B^{1/\gamma} - E_Y' + E_Y^{1/\gamma} = E_B^{1/\gamma} + (E_Y^{1/\gamma} - E_Y')$$

$$E_{MG}^{1/\gamma} = E_Y^{1/\gamma} - \frac{l}{m}(E_R^{1/\gamma} - E_Y') - \frac{n}{m}(E_B^{1/\gamma} - E_Y')$$

Remembering that $E_Y' = lE_R^{1/\gamma} + mE_G^{1/\gamma} + nE_B^{1/\gamma}$ and that $l+m+n = 1$, the last expression reduces to:

$$E_{MG}^{1/\gamma} = E_G^{1/\gamma} + (E_Y^{1/\gamma} - E_Y')$$

Thus it is seen that, whereas in the R-Y-B camera only the green signal was distorted, in the four-tube camera all three colour signals are distorted, and this might at first sight seem to be a worse situation. But the distortion now consists of an equal addition to all three signals and this is equivalent to a small addition of white light, which is usually a fairly unimportant defect. Hue errors are thus largely avoided, and it is found in practice that the errors caused by a four-tube camera are appreciably less noticeable than those produced by an R-Y-B camera (Abrahams, 1963). It must be remembered also that, even with a basic R-G-B camera, colours of high colour saturation are reproduced too dark in small areas because of gamma-correction effects, so that the extra luminance resulting from the use of the fourth camera tube will operate in the direction of correcting this defect.

However, it is possible to process the signals from a four-tube camera so as to produce truly correct signals for transmission, at least for large areas. The procedure is to form an additional low-definition true luminance signal from the three colour tubes, thus:

$$E_{YC}{}^{1/\gamma} = (lE_R + mE_G + nE_B)^{1/\gamma}$$

This then has subtracted from it the low-definition E_Y' signal to obtain a low-definition luminance correcting signal

$$E_{YC}{}^{1/\gamma} - E_Y'.$$

This low-definition correcting signal is then subtracted from the high definition true-luminance signal $E_Y{}^{1/\gamma}$ to give a corrected luminance signal for transmission:

$$E_Y{}^{1/\gamma} - (E_{YC}{}^{1/\gamma} - E_Y').$$

When the receiver adds this signal to its colour difference signals, in large areas, where $E_Y{}^{1/\gamma} = E_{YC}{}^{1/\gamma}$, the correct colour signals are obtained thus:

$$E_R{}^{1/\gamma} - E_Y' + E_Y{}^{1/\gamma} - (E_{YC}{}^{1/\gamma} - E_Y') = E_R{}^{1/\gamma}$$
$$E_G{}^{1/\gamma} - E_Y' + E_Y{}^{1/\gamma} - (E_{YC}{}^{1/\gamma} - E_Y') = E_G{}^{1/\gamma}$$
$$E_B{}^{1/\gamma} - E_Y' + E_Y{}^{1/\gamma} - (E_{YC}{}^{1/\gamma} - E_Y') = E_B{}^{1/\gamma}$$

Circuits performing these operations are practicable and since all the extra complication is at the transmitter and none at the receiver, the use of luminance correcting signals of this type is the preferred procedure with four-tube-cameras.

A similar form of correction can also be applied to three-tube R-Y-B cameras. However, in this case it is necessary first to derive a low definition E_G signal from the E_R and E_B signals and from a modified E_Y signal whose bandwidth has been suitably restricted; this derivation must be carried out with linear signals (voltages proportional to amounts of light) in order to obtain the correct result, and with cameras using Plumbicon tubes this is facilitated by the fact that these tubes have gammas of about 1.0 as shown in Fig. 20.6.

The four-tube camera makes it possible to run the colour signals at a higher gamma than the luminance signals, and in this way colour saturation can be increased to overcome system deficiencies that tend to lower saturation (such as neglect of the negative lobes of the theoretical sensitivity curves; or the addition of the unwanted white signal described earlier, if this is not corrected by a luminance correcting signal).

A four-tube camera employing an image orthicon for the luminance signal and three vidicons for the colour signals has been described (James, 1963); but most four-tube cameras employ four vidicons or four Plumbicons. Fig. 20.11 illustrates the arrangement in a four-tube vidicon camera (Abrahams, 1963); this

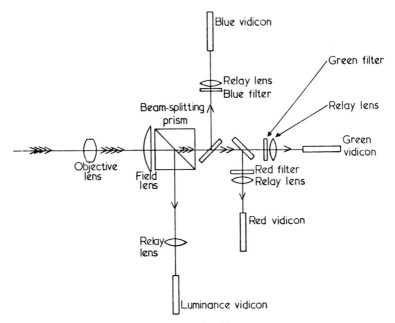

Fig. 20.11. Four-tube vidicon camera.

type of camera is well suited to deriving colour television signals from colour film (see Chapter 23). For ordinary broadcast work, four-tube cameras employing four Plumbicons are generally used (Underhill, 1967; Parker-Smith, 1967; James, Perkins, Pyke, Taylor, Kent, and Fairbarn, 1970). In the latter type of

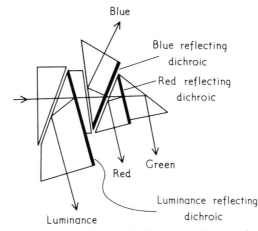

Fig. 20.12. Dichroic beam-splitting prism block of the type used in some four-tube cameras.

camera a compact, four-way, beam-splitting, dichroic prism block is used of the type shown in Fig. 20.12.

The advent of automatic registration devices (see the next Section) has to some extent removed the need for the four-tube approach to camera design (Parker-Smith, 1970).

20.9 Automatic registration

One way of improving the registration of the images in television cameras is to provide servo devices that automatically correct for any errors that happen to be present at any given time. These automatic registration devices can consist of a built-in test pattern, which can be injected into the camera optically, and from which error signals can be derived for adjusting the positions of the three electron beams; or error signals can be detected from ordinary subject matter on the basis of the assumption that a difference signal (for example, the signal from the green tube subtracted from that of another tube) will be minimal when the camera is correctly registered (Wood, 1970). Some cameras combine both these features: a specially injected test signal for a very comprehensive series of adjustments, including tilt and twist of the images; and a dynamic error-detection and servo system for continual adjustment of the camera in some features while it is actually in use (Underhill, 1971).

20.10 Spectral sensitivities used in cameras

A set of three colour-matching functions may be regarded as the theoretically correct basis for the spectral analysis of scenes by colour television cameras (see Sections 7.4 and 19.12); any departure from this condition may result in colours that look alike in the scene being reproduced differently, or in colours that look different in the scene being reproduced alike.

It has been shown that the spectral sensitivity curves of camera tubes, when modified by the spectral transmittances of their associated dichroic filters and colour filters, can be made to approximate a set of colour-matching functions quite well (Jones, 1968). It is not necessary to use the colour-matching functions corresponding to the receiver phosphors because the signals from the tubes can be matrixed (see Section 19.12) and thus transformed, in effect, from one set of primaries to another. It is thus possible to use a set of all-positive colour-matching functions for the tubes and filters, the negative lobes of the functions for the receiver phosphors being provided in the matrix transformation. The desire to avoid any unnecessary loss of light in the dichroic beam splitter, so as to obtain the best signal-to-noise ratio from the camera, makes the narrowest colour-matching functions (see Fig. 19.9) preferable to others. The values required in the matrix are dependent on the chromaticities of the phosphors, and hence standardization of the phosphor chromaticities is very desirable. A standard has been drawn up for use in Europe (B.R.E.M.A., 1969) and in Fig. 20.13 the broken lines show the corresponding colour matching functions; the full lines in Fig. 20.13 show the extent to which a typical camera, using matrixing, duplicates these curves.

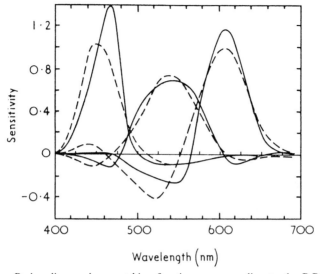

Fig. 20.13. Broken lines: colour-matching functions corresponding to the B.R.E.M.A. phosphors typical of those used in domestic receivers (B.R.E.M.A., 1969); full lines: typical camera sensitivity curves obtained with matrixing.

The practical choice of spectral sensitivities is usually based on a compromise aimed at achieving a balance between several conflicting requirements. Thus, if the coefficients of the matrix are too high, the signal-to-noise ratio may be adversely affected. It is also desirable to equalize the light levels on the three tubes to avoid coloured lag-effects. Loss of light, and high matrix coefficients, make the choice of the $\bar{y}(\lambda)$ curve unsatisfactory for one of the three functions (although attractive as a basis for the luminance signal required for transmission). A compromise used in one type of camera is to use a narrowed $\bar{y}(\lambda)$ curve for one function, to provide high-definition approximate-luminance information in fine detail, with the low-definition red, green, and blue signals being derived from it and from the other two (red and blue) tubes by matrixing (Underhill and Crowley, 1970). Other compromises lead to somewhat different results, but the curves used in practice are usually narrower than the narrowest set of colour-matching functions, as shown in Fig. 19.9. But, although use of a set of colour-matching functions, together with matrixing appropriate to the receiver phosphors, makes possible colorimetrically correct reproduction of all colours within the gamut of colours matchable by the phosphors, operation of the system with an overall gamma greater than 1.0 introduces errors as shown in Fig. 19.10; thus the advantages of using colour-matching functions are less clear-cut than in a system of gamma 1.0, and optimum matrixing coefficients usually have to be established empirically (Sproson, 1978 and 1983).

Matching the spectral sensitivities of groups of cameras used in the same studio is very important, and, because a perfect match is difficult to obtain, final adjustments of colour balance may better be made using for test purposes a skin colour rather than a white or grey (Knight, 1972).

20.11 Aperture correction

In camera pick-up tubes, even if the optical and electric-charge images are perfectly sharp, the picture signal produced by the tube corresponds to an image whose sharpness is reduced by being scanned by an electron spot of finite size or *aperture*. This results in a basic limitation in the number of points per picture width that can be resolved, and also in a reduction in contrast across vertical edges; the contrast across vertical edges can, however, be increased again, a technique known as *horizontal aperture correction*. One way of achieving this type of correction is to delay the signal for a time comparable with that taken to scan a distance similar to the effective aperture of the electron spot, and then to subtract from it a correcting signal that is a function of the average of the undelayed signal and a signal delayed twice as long; in this way a sort of unsharp masking effect can be achieved (see Section 15.3). The correcting signal is less sharp than the main signal, and is made less contrasty; the result of subtracting it from the main signal is therefore to reduce contrast except in fine detail. Hence, upon increasing the overall contrast of the composite signal, the contrast of the coarse detail is restored, but that of the fine detail is enhanced.

In a similar way it is possible to introduce *vertical aperture correction* to correct for any tendency for the aperture of the electron spot in the camera pick-up tube to be greater than the spacing between neighbouring picture-lines, and to increase the contrast across horizontal edges. In this case the main signal has to be delayed by a time equal to that taken to scan one *line*, and the correcting signal is then a function of the average of the undelayed signal and a signal delayed twice as long. In this case the unsharpness of the 'mask' is equivalent to about 3 field-lines or 6 picture-lines.

In three- and four-tube cameras, the advantages of aperture correction are only fully realized if the registration of the images is very good, because of the high sensitivity of the eye to colour fringing. It is, however, possible to derive a common 'masking' signal from *one* of the camera tubes and to use it as the correcting signal for *all* the main signals. When this is done in three-tube cameras, some of the advantages of four-tube cameras are obtained because the enhancement of contrast across edges, being derived from a single camera-tube, cannot be affected by camera mis-registration and therefore cannot contribute to colour fringing. This is known as the *contours out of green* technique, because the green camera-tube (having the spectral sensitivity most similar to that cor-

Plate 28
The use of computer-generated pictures to depict what a new building will look like from different viewing points is a great aid to design. This example shows three views of a commercial bank building (see Section 25.6). Animator: Michael Collery. Pictures by courtesy of Cranston/Csuri Productions, Inc., Columbus, Ohio. Separations made on a Crosfield *Magnascan* scanner.

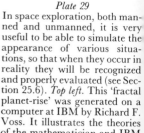

Plate 29

In space exploration, both manned and unmanned, it is very useful to be able to simulate the appearance of various situations, so that when they occur in reality they will be recognized and properly evaluated (see Section 25.6). *Top left.* This 'fractal planet-rise' was generated on a computer at IBM by Richard F. Voss. It illustrates the theories of the mathematician and IBM Fellow Benoit B. Mandelbrot, and is an excerpt from his book *The Fractal Geometry of Nature* published by W. H. Freeman. Picture by courtesy of Benoit B. Mandelbrot (copyright, 1982). *Bottom left.* View of Uranus as seen from Voyager 2. *Top right.* View of Jupiter from Voyager 2. *Bottom right.* View of Saturn from its moon Mimas. These last three pictures are also computer-generated and are all by courtesy of James F. Blinn of the Jet Propulsion Laboratory, California Institute of Technology. Separations made on a Crosfield *Magnascan* scanner.

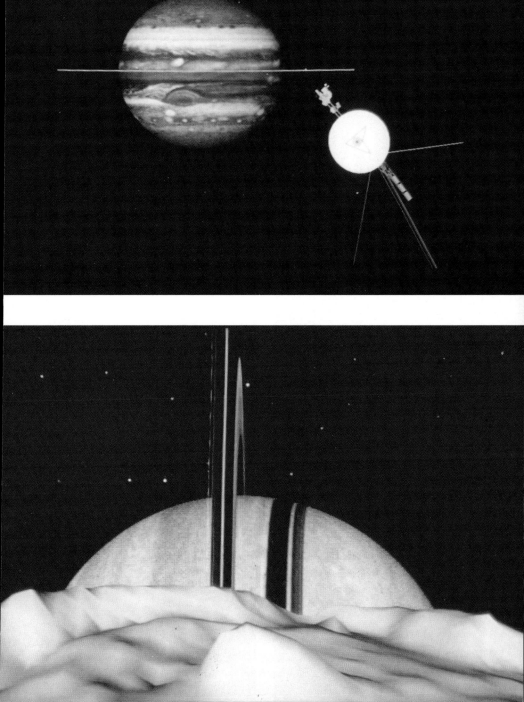

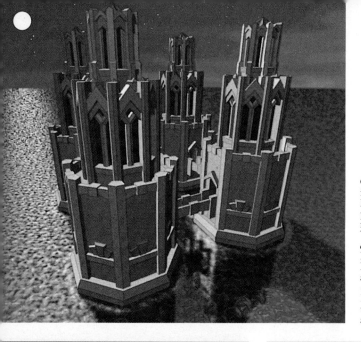

Plate 30
Computer-generated pictures incorporating shadows, catchlights, reflections, and transparency (see Section 25.6). Animators: *top*, Ned Greene (New York Institute of Technology); *centre left*, Hsuen-Chung Ho and Michael Collery, *bottom right*, Hsuen-Chung Ho, *centre right* and *bottom left*, Michael Collery (Cranston/Csuri Productions Inc., Columbus, Ohio). Separations made on a Crosfield *Magnascan* scanner.

responding to a true luminance signal) is generally the preferred source for the correcting signal (De Vrijer, Tan, and Van Doorn, 1966).

20.12 Simple cameras

For broadcast television, it is usually justifiable to employ complicated, and normally quite expensive, cameras, in order to obtain high performance; but, for many closed circuit television applications, colour pictures are only economic if less expensive cameras can be used. Some cameras using only two tubes have been designed (Takemura, Sato, and Tajiri, 1973; Flory, 1973), but the simplicity of cameras using only a single tube makes them more suitable for these applications, even if, as is usually the case, some sacrifice is made in the sharpness of the pictures obtainable. These cameras usually have a pattern of stripes of differently coloured filters bonded on to the photosensitive element of the pick-up tube (normally of the vidicon or Plumbicon type). Some designs have incorporated differently coloured stripes at different angles (Attew, 1972), but vertical stripes are more usual (Weiner, Grey, Borgan, Ochs, and Thompson, 1955).

In the Ferguson *Videostar* camera, a tube is used having vertical stripes of green, cyan, and white filters. The three colours together provide an approximate luminance signal; a blue signal is obtained by subtracting the green from the cyan, and a red signal by subtracting the cyan from the white. The red, green, and blue signals are then used to generate the chrominance signals (and also, if necessary, a luminance correcting signal of the type described in Section 20.8).

In the Sony *Trinicon* camera, vertical red, green, and blue stripes are used, together with transparent stripes of indexing electrodes. The voltage applied to the indexing electrodes causes an offset in the signals and this is reversed at the end of each television line by reversal of the voltage; by subtracting successive line outputs (using a delay line), a triad-based square-wave is obtained which facilitates the formation of an accurately phased chrominance signal.

Another approach to the problem of making a simple camera is to use a solid-state electronic sensor, instead of camera tubes operating in evacuated glass envelopes. The solid state sensor usually consists of an array of charge coupled devices (CCDs) covered with a chequer-board pattern of red, green, and blue filters, one type of which is shown in Fig. 20.14 (Dillon, Lewis, and Kaspar, 1978). In the pattern shown in this figure, there are twice as many green areas as red or blue, and this is because the green signal contributes more to the luminance signal than the red or blue signals (see Section 22.3). The light-sensitive element behind each filter produces an electric charge whose magnitude is related to the light transmitted by the filter. The charges produced by all the elements in a row of the array are extracted from the device by moving them all along from one element to the next at regular time intervals; this produces a time dependent signal at the end of the row for one line of the picture. The same procedure is carried out in sequence for all the rows, to produce time dependent

417

signals for the entire picture. The rate at which the signals are clocked out must of course be appropriate for the particular television system being used, and would normally include interlacing.

R	G	R	G	R	G	R	G
G	B	G	B	G	B	G	B
R	G	R	G	R	G	R	G
G	B	G	B	G	B	G	B
R	G	R	G	R	G	R	G
G	B	G	B	G	B	G	B
R	G	R	G	R	G	R	G
G	B	G	B	G	B	G	B

Fig. 20.14. A small part of the pattern of a red, green, and blue filter array that can be used with a CCD area sensor.

The factors affecting the choice of spectral sensitivities for the three channels of an array sensor differ from those of a camera employing a dichroic beam-splitting device, in that the use of broad sensitivities, instead of wasting light, actually uses it more efficiently (as is the case for the cone mosaic in the retina of the eye). Overlapping sensitivity curves, such as those corresponding to all-positive sets of colour-matching functions, may, therefore, be used without sacrificing sensitivity. However, as the sensitivity curves are broadened, the required matrix coefficients become larger, and this worsens signal-to-noise ratios; so, once again, a compromise has to be made with respect to conflicting factors.

The use of an array of the type just described leads to a very simple design of camera, it only being necessary to provide a suitable lens for imaging the scenes on to the compact sensor. But the array itself is difficult to construct with enough elements to give pictures of high definition. For a system having 600 active lines and a 4:3 aspect ratio, the number of elements required for a monochrome picture is about half a million ($600 \times 600 \times 4/3$), and hence would be about a million for colour, if the filter pattern of Fig. 20.14 were used.

Compared to normal broadcast type cameras, simple cameras are usually smaller and lighter and are, therefore, more suited to location work, especially in remote or difficult areas. Their convenience for these applications can be further enhanced by incorporating a compact cassette video tape recorder as an inherent part of the camera. This can also be done with the more conventional three-tube type of camera, if of the compact type. Examples of cameras having incorporated tape recorders are the Sony *Betacam*, the RCA *Hawkeye*, and the Kodak *Kodavision*. Equipment of this type is suitable for making amateur video recordings.

418

The advent of portable camera and recording equipment has made possible the technique of *electronic news gathering* (ENG). Using this type of equipment news events can be recorded in the field and relayed back to a broadcasting station for transmission with very little delay.

In the *Mavica* system developed by Sony as an all electronic still camera for amateur use (Crawley, 1981), a CCD array is used having about 280 000 elements, arranged in a 570 (horizontal) by 490 (vertical) pattern. The information corresponding to the image is recorded on discs coated with a magnetic recording layer, and the discs are reusable. The image can be played back on a domestic colour television receiver, or used to produce a reflection print on paper by using a *Mavigraph* printer (see Section 13.6). The photographic speed of the Mavica system is 200 ASA (see Section 14.27); its resolution is limited by the geometry of its sensor and is less than that of broadcast television or photographic cameras for still photography.

REFERENCES

Abrahams, I. C., *J. Soc. Mot. Pic. Tel. Eng.*, **72**, 594 (1963).
Attew, J. E., *J. Roy. Television Soc.*, **14**, 123 (1972).
Benson, K. B., *J. Soc. Mot. Pic. Tel. Eng.*, **90**, 708 (1981).
Bertero, E. P., *J. Soc. Mot. Pic. Tel. Eng.*, **72**, 602 (1963).
B.R.E.M.A., *Radio and Electronic Engineer*, **38**, 201 (1969).
Crawley, G., *Brit. J. Phot.*, **128**, 1112 (1981).
De Haan, E. F., and Van Doorn, A. G., *J. Soc. Mot. Pic. Tel. Eng.*, **73**, 473 (1964).
De Haan, E. F., and Van Doorn, A. G., *J. Soc. Mot. Pic. Tel. Eng.*, **74**, 922 (1965).
De Vrijer, F. W., Tan, A. L., and Van Doorn, A. G., *J. Soc. Mot. Pic. Tel. Eng.*, **75**, 1080 (1966).
Dillon, P. L. P., Lewis, D. M., and Kaspar, F. G., *I.E.E.E. Trans. Electron Devices*, **25**, 97 and 102 (1978).
Flory, R. E., *J. Soc. Mot. Pic. Tel. Eng.*, **82**, 29 (1973).
James, I. J. P., *J. Brit. Inst. Radio Engrs.*, **19**, 165 (1959).
James, I. J. P., private communication (1963).
James, I. J. P., Perkins, D. G., Pyke, P. J., Taylor, E. W., Kent, D. E., and Fairbairn, I. A., *Radio and Electronic Engineer*, **39**, 249 (1970).
James, I. J. P., and Karowski, W. A., *J. Brit. Inst. Radio Engrs.*, **23**, 297 (1962).
Jones, A. H., *J. Soc. Mot. Pic. Tel. Eng.*, **77**, 108 (1968).
Knight, R. E., *Brit. Kinematog. Sound Tel.*, **54**, 272 (1972).
Monteath, G. D., *J. Television Soc.*, **11**, 109 (1966).
Neuhauser, R. G., *J. Soc. Mot. Pic. Tel. Eng.*, **65**, 636 (1956).
Parker-Smith, N. N., *Brit. Kinematog. Sound Tel.*, **49**, 100 (1967).
Parker-Smith, N. N., *I.E.E. Conference Publications*, No. 69, p. 147 (1970).
Sproson, W. N., *Proc. I.E.E.*, **125**, 603 (1978).
Sproson, W. N., *Colour Science in Television and Display Systems*, p. 27, Adam Hilger, Bristol (1983)
Takemura, Y., Sato, I., and Tajiri, A., *J. Soc. Mot. Pic Tel. Eng.*, **82**, 12 (1973).
Underhill, W. T., *J. Roy. Television Soc.*, **11**, 167 (1967).
Underhill, W. T., *J. Roy. Television Soc.*, **13**, 191 (1971).
Underhill, W. T., and Crowly, V. D., *I.E.E. Conference Publications*, No. 69, p. 138 (1970).
Van Doorn, A. G., De Lang, H., and Bouwis, G., *J. Soc. Mot. Pic. Tel. Eng.*, **75**, 1002 (1966).
Weiner, P. K., Grey, S., Borgan, H., Ochs, S. A., and Thompson, H. C., *Proc. Inst. Radio Engrs.*, **43**, 370 (1955).
Wood, C. B. B., *I.E.E. Conference Publications*, No. 69, p. 50 (1970).

CHAPTER 21

Display Devices for Colour Television

1. Introduction – *2*. The trinoscope – *3*. Triple projection – *4*. The shadow-mask tube – *5*. The Trinitron – *6*. Self-converging tubes – *7*. The Eidophor – *8*. Laser displays – *9*. Beam-penetration tubes – *10*. Tubes with liquid crystals – *11*. Phosphors for additive receivers – *12*. The chromaticity of reproduced white – *13*. The luminance of reproduced white

21.1 Introduction

T HE earliest television systems, which depended on a mechanical or optical device for breaking the picture up into lines, such as the rotating disc shown in Fig. 20.1, used similar non-electronic devices for displaying the picture at the receiver. The advent of high definition television, with the invention of the iconoscope, required much more rapid means of 'writing' the picture, and the required means was found in the cathode-ray tube, the principle of which is to produce light by exciting a fluorescent layer with a beam of electrons moved rapidly over its surface.

The main features of a cathode-ray tube of the type used in television are shown diagrammatically in Fig. 21.1(a). A *cathode* is coated with a suitable electron-emitting material, and hence, when its temperature is raised by a heater, a supply of electrons is provided. The cathode is housed in an evacuated glass envelope, so that the electrons can be accelerated towards an *anode* which is kept at a highly positive potential relative to the cathode; the accelerated electrons strike a *screen*, consisting of a layer of phosphor coated on the inside of the glass envelope, and thus light is produced. The amount of light produced is regulated by altering the voltage difference between the cathode and the *modulator grid* (sometimes called the *control grid* or often just the *grid*). The position at which the light is produced on the screen is regulated by *focusing* devices and by *deflection* devices. The focusing devices, which produce suitable magnetic or electrostatic

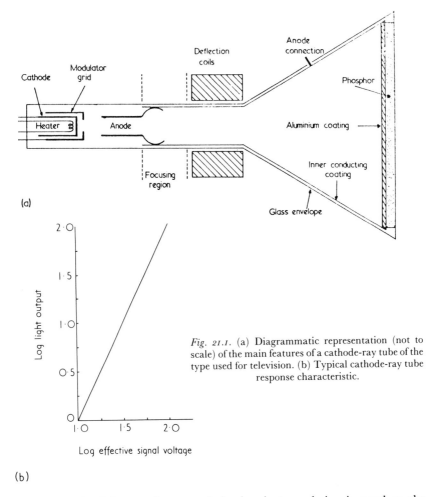

(a)

(b)

Log light output

2·0

1·5

1·0

0·5

0

Log effective signal voltage

1·0   1·5   2·0

Fig. 21.1. (a) Diagrammatic representation (not to scale) of the main features of a cathode-ray tube of the type used for television. (b) Typical cathode-ray tube response characteristic.

fields in the neck of the envelope, result in the electrons being imaged on the phosphor as a small spot, instead of as a large patch. The deflection devices usually consist of coils producing magnetic fields (electrostatic deflection is generally used only in oscilloscopes): these magnetic fields cause the electron spot to be moved very rapidly across the face of the tube so as to produce the picture by scanning the phosphor line by line in the required manner. Accurate scanning is facilitated by coating the inside of the conical part of the tube with a conducting layer which is held at the same voltage as the anode thus providing for the moving electron beam a volume virtually free of any electrostatic fields. The phosphor usually has coated on its inner side a very thin layer of aluminium which is also held at anode voltage and this not only assists in eliminating electrostatic fields, but also prevents secondary emission of electrons from the phosphor, and also avoids waste of light by reflecting forwards in a useful direction the light emitted by the phosphor backwards (towards the cathode) which would otherwise be lost.

Typical operating values for cathode-ray tubes used for television, are as follows. Anode voltages range from about 16 kV (kilovolts) for black-and-white tubes to 25 kV for colour tubes and 100 kV for projection tubes. The difference between the modulator grid and the cathode voltages may range from some tens of volts negative (modulator grid voltage *below* cathode voltage) when all the electrons are prevented from reaching the phosphor and black is produced, to some tens of volts positive when maximum luminance (*peak white*) is being produced. The voltage difference that produces black is referred to as the *black-out bias* or *black-sit* (adjustment of which on a domestic receiver is sometimes labelled '*brightness*'). It is convenient to express the modulator-grid to cathode voltage difference as a difference from the black-out bias, and when this is done it may be referred to as the *effective signal voltage* or the *drive* (adjustment of the amplification of which on a domestic receiver is sometimes labelled '*contrast*'). The amount of current flowing in the electron beam (the *beam current*) usually varies from zero when black is being reproduced to some hundreds of microamps when peak white is being reproduced. The luminance at which peak white is produced is usually about 100 cd/m² for ordinary viewing, but may be several times this figure for projection tubes operated at anode voltages as high as 100 kV. These high luminances would be unsatisfactory for direct viewing because the 50 or 60 fields per second flicker in television pictures is only unobtrusive at the lower luminance levels.

The actual tubes used in practice are more elaborate than indicated in Fig. 21.1(a), and extra grids and electrodes may be included: thus a *screen grid* at a few hundred volts above the cathode voltage, and focusing electrodes at a few thousand volts above the cathode voltage, may be used, in which case the electrons are accelerated through a series of voltage steps between the cathode and the anode.

When the effective signal voltage, E, is zero, the beam current, by definition, is also zero. As E is increased, the beam current, I, increases, and the effect is analogous to opening the iris diaphragm on a lens. If the effective diameter of the electronic iris is roughly proportional to E, the area of the iris will be proportional to E^2, and hence the beam current, I, will be roughly proportional to E^2. The power in the beam is equal to the current multiplied by the accelerating voltage, V, and hence is equal to VI. The accelerating voltage, V, is equal to the anode voltage relative to the cathode voltage, and although the latter may vary as E is varied, its maximum variation is usually only about 100 volts and this is negligible compared to the anode voltage of tens of thousands of volts. The accelerating voltage, V, may therefore be regarded as constant, so that the power is proportional to the beam current, I. If the phosphor is such that the amount of light produced is proportional to the power, and hence proportional to the beam current, I (such phosphors are often referred to as being *linear*), then, since I is proportional to E^2, the amount of light produced by this hypothetical tube will also be proportional to E^2.

The tone reproduction characteristics of real cathode-ray tubes are such that the light output is approximately proportional to a function intermediate between the square and the cube of the effective signal voltage. When the logarithm of the light output is plotted against the logarithm of the effective signal voltage, a straight line is therefore obtained, as shown in Fig. 21.1(b). A slope (or *gamma*) of 2.8 ± 0.3 for this line has been agreed upon as standard for colour television (B.R.E.M.A., 1969). The range of tones that a cathode-ray tube can display, when no ambient light is allowed to fall on the face of the tube, depends on the amount of 'spill-over' light in the tube itself, and this may be low enough to enable the range to reach 100 to 1 (corresponding to 2.0 log units); but in practice the ambient lighting in the viewing situation usually lightens the blacks to the point where they have a luminance of at least 2 per cent of the maximum luminance, so that the range of tones is not more than 50 to 1 or 1.7 log units; if this is seen in dim-surround viewing conditions it will be equivalent to 1.7/1.25 or about 1.4 log units of scene luminance, which is much less than the figure of 2.2 log units typical of outdoor scenes (Jones and Condit, 1941; Hunt, 1965). If the ambient and spill-over light amounts to about 5 per cent, then the range becomes 20 to 1 and this is equivalent to a log range of only 1.3, which is similar to that of typical reflection prints (see Section 13.10). Practical figures are generally regarded to be within the 2 to 5 per cent range (Wentworth, 1955).

The application of the cathode-ray tube principle (fluorescence caused by electron bombardment) to the specific problems of colour television display devices has been the subject of intense technological effort and enormous ingenuity has been expended in trying to arrive at inexpensive reliable receivers for domestic use; some of these display devices will now be described, together with others for more special applications. (Sproson, 1983.)

21.2 The trinoscope

The principle of the *trinoscope*, perhaps the least sophisticated colour television display device, is illustrated in Fig. 21.2. The three colour images are displayed on three separate cathode-ray tubes, and images of two of them are then combined with the third by means of semi-reflecting dichroic mirrors. To obtain maximum efficiency, special red-emitting, green-emitting, and blue-emitting phosphors can be used in the three tubes; colour trimming filters can be used in addition if necessary. The main problem in the trinoscope is to get exact geometric registration, and matching tone reproduction characteristics, of the three images all over the picture area, and it is evident that in this respect the device has problems similar to those of colour cameras. It is therefore necessary to use very high quality cathode-ray tubes and ancillary gear, and very precise mirrors rigidly mounted in exact position relative to the tubes. This makes it a

423

costly device, but it has the virtue of giving images of higher luminance than most other display devices. Its main use is as a device from which telerecordings on film can be made (Venis, 1969); it is far too expensive, as well as being too bulky, to be considered for use as a domestic receiver.

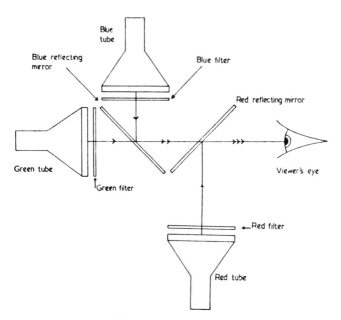

Fig. 21.2. The trinoscope.

21.3 Triple projection

When projection television devices are being used, it is possible to use the triple projection principle (see Section 3.1) by having three projection television tubes arranged so as to throw red, green, and blue images on to a single reflecting screen (Federman and Pomicter, 1977). The usual problems of registration have to be overcome, but when the final display is wanted in projected form, the method is usually the best to adopt (unless field sequential displays can be used). Triple projection is used for the display of colour television pictures to large audiences, and sometimes also for displaying the terrain in simulators used for training the crew of aircraft and ships (Marconi, 1969).

21.4 The shadow-mask tube

The undesirable bulk and registration problems of the trinoscope and triple projector are avoided in the *shadow-mask tube*, by incorporating all three electron

424

guns in the same cathode-ray tube. Electronic registration is still required, of course, but, by having the three guns in the same tube, the same magnetic fields can be used for moving the three electron beams throughout the scanning sequence for each field of the picture. This is a considerable help, but there are many residual problems caused by the fact that, because the three electron beams do not originate from the same place, they do not in fact scan the picture identically. Thus when the beams are scanning the corners of the picture they have further to travel to the screen than when they are scanning the centre; hence if their convergence is correct for the centre it will be too great for the corners, and so the magnetic fields have to be altered during the scanning by means of special current wave-forms applied to the electro-magnets. (Law, 1977).

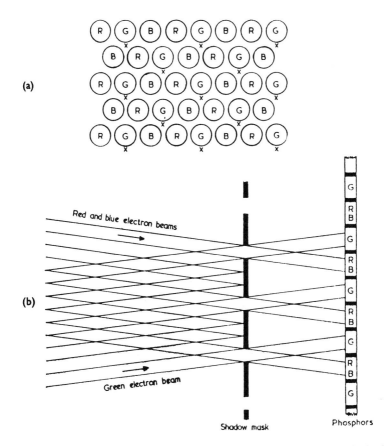

Fig. 21.3. The shadow-mask tube: (a) arrangement of dots, with positions of holes in the shadow mask marked thus: X. (b) Electron beam paths in a vertical plane: the red and blue electron beams and phosphors are separated horizontally.

425

Before reaching the screen the three beams meet a metal plate with about 400 000 holes in it, situated about 18 mm from the phosphors. The three phosphors are laid down as dots (see Figs. 3.1 and 21.3), and the geometry of the electron beam directions, the positions of the holes, and the positions of the dots, is such that all the red-phosphor dots are irradiated only by the gun to which the red signal is applied, the green-phosphor dots by the green-signal gun, and the blue-phosphor dots by the blue-signal gun. The rows of dots do not have to be aligned with the lines of the picture, but *moiré patterns* caused by beats between the line structure and the dot pattern arise at certain angles: as these are worst at $\pm 30°$ and negligible at $0°$ it is arranged for the lines of the picture and the lines of the dots to be more or less parallel. For 525-line displays, the shadow-mask usually has about 357 000 holes, which provide about 520 lines of holes with about 690 holes in each line; hence the maximum definition of the tube amounts to about 345 picture-point pairs along a line and 260 picture-point pairs vertically. If the three electron beams were small enough to irradiate, on the average, not more than one line of holes and its associated triad of phosphor dots then the tube would not restrict the definition much in a 525-line system, but this would be a rather critical condition in which to operate and each electron beam normally irradiates two or three lines of holes and their associated triads of dots; there is therefore some theoretical loss of definition, but other factors, such as interlacing, may make the loss unimportant in practice (Jesty, 1958). For 625-line displays the tubes usually have about 440 000 holes providing about 575 lines of holes, with about 770 holes in each line.

The shadow-mask tube is capable of giving pictures of very good colour quality and is widely used both for high quality monitors and for domestic receivers. Very great care has to be taken in manufacture to see that the pattern of phosphor dots exactly coincides with the pattern of the holes in the metal plate, otherwise colour contamination occurs and produces serious errors of hue. The holes in the metal plate are therefore etched by photo-engraving using a master negative to ensure absolutely correct geometry. The inside of a tube is then coated with a photoresist containing the red-phosphor, and then its metal plate is mounted in position and a light source placed in the position from which the electron beam carrying the red signal will finally appear to emanate. The light passing through the holes in the metal plate then causes the photoresist to form hardened dots of phosphor; the metal plate is then removed and the unhardened phosphor washed away leaving the required pattern of dots. The same metal plate is then replaced in exactly the same position and the green-phosphor dots formed in a similar way, and finally the blue-phosphor dots are formed similarly. In a 56 cm tube the distance between adjacent dots is only about 0.4 mm so that it can be seen that very great accuracy is required in carrying out all these operations (Wright, 1971). (See Plate 18, page 288.)

Shadow-mask tubes are often used in colour *video display units* (*VDUs*) for viewing data generated by computers. In this case, the viewing distance is usually only about a half to one metre, instead of about two to three metres typical for viewing normal pictorial television; it is therefore necessary to use tubes having finer dot structures in VDUs. The size of the dot structure is usually quoted as the *triad pitch*: by this is meant the distance, p, between adjacent holes in the mask; adjacent rows of phosphor dots are then separated by $\frac{1}{2}p$ (see Fig. 21.3(a)), and the distance between adjacent phosphor dots is $p/\sqrt{3}$. Thus in the case where (for typical pictorial television) adjacent phosphor dots are separated by 0.4 mm, the triad pitch, p, is given by $0.4 \times \sqrt{3}$, which is equal to about 0.7 mm. For VDUs, triad pitches of about 0.3 mm (or sometimes about 0.2 mm) are usually used. In each vertical triad pitch there are two lines of holes (see Fig. 21.3).

The three beams of a colour VDU must be very accurately registered all over the display area, because mis-registration is very noticeable when small symbols are displayed, especially, as is often the case, against a black background. For this reason, special registration adjustments are usually provided in VDUs and registration to within a half, or a third, of a television line width is usually desirable. These displays are also often not interlaced.

The effective spot size in shadow-mask tubes may be regarded as corresponding to the diameter where the luminance is half the maximum, when all three guns are firing. In VDUs this spot size is usually about twice the triad pitch; it cannot be smaller than this, because smaller spot sizes result in the spot having variable colour when writing small symbols and make it difficult for the eye to locate the centre of a spot or a line. Thus in the case of a triad pitch of 0.3 mm, the spot size would be about 0.6 mm. When using VDUs it is not normally necessary to use the line standards adopted for broadcast television. For a display height of 280 mm (typical of tubes having a diagonal of 19 inches or 48 cm), a spot size of 0.6 mm corresponds to $280/0.6 = 467$ lines. However, the use of more lines than this is common, and as many as 1000 are sometimes used; the excess lines are useful in reducing the incidence of spurious patterns (aliasing) and of jagged edges to lines that should be straight.

For pictorial television, spot sizes may be similar in diameter to the triad pitch, because small symbols are not often displayed; thus, for a display height of 325 mm (typical of tubes having a diagonal of 22 inches or 56 cm), a triad pitch and spot size of 0.7 mm corresponds to $325/0.7 = 464$ lines; the use of more lines than this (525 or 625) in practice, again reduces the incidence of aliasing and jagged edges.

Although the shadow-mask tube is the dominant display device for colour television, much effort has been, and is still being, devoted to finding alternative devices to give improved performance, better convenience, or reduced costs. Some of the alternatives that have been suggested will now, therefore, be reviewed.

21.5 The Trinitron

A three-gun tube in which the phosphor is laid down in stripes is the *Trinitron*. In this tube the three electron beams lie in the same horizontal plane, and a metal plate with vertical slots in it is positioned so that the electrons from one beam can only reach vertical stripes of red phosphor, those from another beam only stripes of green, and those from the third only stripes of blue (as already described in Section 3.4).

This tube has certain advantages over the shadow-mask tube. First, deflection of the three electron beams is easier because the gun construction enables the neck of the tube to be smaller. Secondly, the displayed picture emits twice as much light per unit area: this is because, for the same spot size, the beam current can be increased by a factor of 1.5 times; and because the stripes of phosphor cover 1.33 times as much area of the tube face-plate. Thirdly, vertical resolution is not affected by the screen structure so that there is no moiré pattern or loss of vertical resolution by the screen. Fourthly, adjusting the convergence to obtain registration of the three images is easier because the three beams are in a single plane. The triads of phosphor stripes may be up to about half a millimetre wide, giving about 600 triads in a tube of 300 mm width. For equal horizontal and vertical definition the luminance signal should be able to resolve about 350 cycles per line (for example $525 \times \frac{1}{2} \times (4/3)$ black-white pairs in a system having 525 actual picture lines); the number of triads of vertical lines required is therefore ideally not less than about 700, but, as in the shadow-mask tube, smaller numbers can be used without too much apparent loss of definition because the actual visual appearance is complicated by interlacing and various other factors (Jesty, 1958). The Trinitron tends to be used for smaller displays than the shadow-mask tube.

21.6 Self-converging tubes

In Trinitron and conventional shadow-mask tubes, it is necessary to provide *dynamic convergence correction*. This is required because (as mentioned in Section 21.4) stronger magnetic fields are needed to bring the three electron beams into coincidence around the centre of the picture, than those required for the corners, which are further away and therefore have longer electron paths. As the three electron beams scan the picture, the amount of convergence is therefore adjusted dynamically according to their position in the scan.

In the *Precision In-line tube* (Neate, 1973), the three electron guns are arranged parallel to one another in the same horizontal plane, as in the Trinitron tube; but, instead of providing dynamic convergence correction, a special deflection coil is accurately cemented to the neck of the tube. This coil is designed to converge the three electron beams on to the shadow-mask at all positions in the picture. But such a coil can only be made to do this for horizontal or for vertical fans of electron beams: in this case the horizontal fans are converged and the vertical fans converge before the mask is reached and then spread out into short vertical lines. However, by making the shadow-mask have vertical slots, instead of holes, the efficiency with which it allows the electrons through is about 16 per cent, which is similar to that of a conventional shadow-mask tube (although less than that of a Trinitron tube, which is about 20 per cent).

After passing through the slots, the electrons then land on the red, green, and blue phosphors, which are laid down in stripes, as in the Trinitron tube. By making the slots in the shadow-mask discontinuous, the mask is sufficiently rigid to enable it to be made with a spherical profile, as in the conventional shadow-mask tube, rather than cylindrical as in the Triniton tube. The stripes of phosphor are 0.0108 in (0.27 mm) wide, so that each colour is repeated every 0.0324 in (0.81 mm). The geometry of the phosphor stripes, the slots, and the electron guns, is arranged to result in the electrons from each gun landing on phosphor of only one colour. The electron guns are mounted 0.200 in (5.08 mm) apart from one another. The Precision In-line tube is particularly suitable for small and medium picture-sizes (see Plate 18, page 288); it can also be made using dots rather than short vertical lines.

The shadow-mask, Trinitron, and self-converging types of tube, described in the last three Sections, are widely used in monitors and in domestic receivers.

21.7 The Eidophor

The Eidophor, invented by Fischer of Zurich, was developed primarily as a solution to the problem of showing television pictures to a large audience (Baumann, 1952; Sponable, 1953). This problem calls for a projection system, and the Eidophor is particularly interesting in that the image is formed by the light from a conventional carbon or xenon arc source, and not by fluorescence caused by electron bombardment.

The principles of the Eidophor are illustrated in Fig. 21.4. Light is picked up by a large aperture concave mirror and reflected horizontally so as to produce an enlarged aerial image of the source; a lens then further magnifies this image: after being reflected downwards from a plane mirror, on to another concave mirror, this second concave mirror sends the light back up through some slots in the plane mirror, to form an image of the source in a projection lens. The picture is produced on the surface of the second concave mirror and the projection lens projects it on to a second plane mirror, and so horizontally to the screen. The first plane mirror is situated so that its centre is at the centre of curvature of the second concave mirror. Its slots are arranged so that, in the absence of any image on the second concave mirror, all the light is returned back to the source: this is achieved because the image of each slot formed by the second concave mirror falls exactly on another slot, and each reflecting strip (between the slots) on another reflecting strip. The optical system is therefore a 'Schlieren' system in which all the direct light is just blocked and only when some disturbance is introduced does any light reach the screen. The disturbance in this case is provided by the second concave mirror being covered with a very thin film of oil on which ripples are produced: the effect of a ripple is to deflect some of the light off the reflecting strips and through the slots, and so on to the screen. The ripples are produced by scanning the surface of the oil with an electron beam whose intensity is modulated according to the image to be reproduced. As the electron beam scans the surface, electrons are deposited and an image of electrical charge is built up. Where this charge is present, electrostatic forces deform the surface, and as the deformation increases with the charge, so the light on the screen is modulated in accordance with the intensity of the electron beam; hence an image is produced.

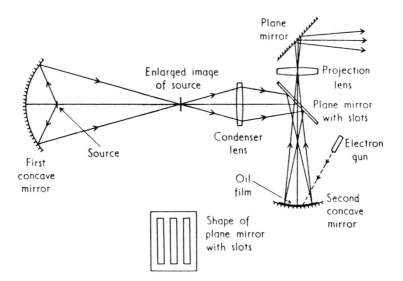

Fig. 21.4. The Eidophor.

The conductivity and viscosity of the oil are chosen so that the deformation at each spot does not fully decay until the electron beam is just about to traverse the spot again: in this way each part of the image is emitting light on to the screen for most of the time, giving a very considerable gain over the cathode-ray tube, each point of which, even with its afterglow, is emitting appreciably for only a small percentage of the time. One of the disadvantages of using an optical system depending on Schlieren principles is the very critical nature of the optics, but the ripples on the oil are made small enough, relative to their distance from the slots in the plane mirror, for their refractive effect to be accompanied by diffraction; the slots are then made smaller than the reflecting strips and the alignment is rendered less critical.

The Eidophor principle has been used for colour television in several different ways. If a field sequential system can be used, then a rotating filter array in front of the projector lens is all that is required. For simultaneous systems, the triple-projection method can be used, either by using three completely separate Eidophor systems, or by using a single light source, split by means of dichroic mirrors into three beams, which then pass through three separate oil-films and projection lenses. Single light-source, single electron-beam, systems have also been described, in which the three signals are carried by the electron beam in the form of high-frequency spot-wobblings in different directions: these result in the oil film having formed on it diffraction gratings whose depth and direction control the magnitude of the red, green, and blue components of the beam of light (Glenn, 1970).

21.8 Laser displays

For special applications, especially when very high light levels are required, displays have been described in which red, green, and blue laser beams are combined, modulated by suitable electro-optical cells, and then made to scan the picture by means of rotating or vibrating mirrors (Stone, Schlafer, and Fowler, 1969). The high light levels that result from the use of lasers make this type of display especially useful for telerecording on to photographic colour film (see Section 23.11). In one laser display system, an argon laser provides the blue and green primaries, and a dye laser, pumped by excess power from the argon laser, provides the red primary (Lobb, 1983). Laser displays have the advantage of high resolution and high luminance, but the disadvantages of less reliable beam deflection technologies and low efficiencies (for example, 10 kilowatts to produce 300 lumens).

21.9 Beam-penetration tubes

By forming layers of different phosphors on top of one another on the inside of the face plate of cathode-ray tubes, it is possible to produce a different colour in each layer. To do this it is necessary to use electron beams of different velocities, only the higher velocity beams reaching the lower layers. The difficulties of making electron beams of different velocities scan the picture in register with one another usually confine this type of tube to the use of two layers only, the colours produced ranging from that of the top layer to that of an additive mixture of the light emitted by both layers. Such tubes are useful for the display of data or graphics.

In one type of beam-penetration tube, the *Penetron*, a single electron beam is used. At low accelerating voltage the beam excites only a red-emitting phosphor; at a high voltage it penetrates through the layer of red phosphor to excite a green-emitting phosphor situated below the red phosphor. To achieve this, green phosphor particles are covered with barrier layers, on top of which layers of red phosphor are deposited; these 'onion layer' particles are then coated uniformly to form the screen. The green phosphor is much more efficient than the red, so that, at the high accelerating voltage, the colour produced is almost the pure green of the green phosphor, the red light from the red phosphor being small in comparison. Hence, by varying the accelerating voltage, a range of colours can be obtained from the red of the red phosphor to almost the green of the green

431

phosphor; intermediate orange and yellow colours are produced by using intermediate voltages, thus providing four different colours in all. The beam current is reduced as the accelerating voltage is increased so as to produce the four colours at similar luminances. Because the colours are produced with a single beam of electrons, the images are considerably sharper than those obtainable with typical shadow-mask tubes. Because of the time taken to change the accelerating voltage, these tubes normally operate, not by scanning with a raster of lines, but by writing symbols or shading areas directly (called *stroke* operation, see Section 25.6), and this is done one colour at a time.

21.10 Tubes with liquid crystals

By viewing a monochrome cathode-ray tube display through a liquid crystal that can be switched electronically to transmit different colours, colour displays having high spatial resolution can be achieved. In one such system, a phosphor that emits both red and green light is used, and the liquid crystal is switched to transmit only the red light or only the green light on alternate fields. Such a two-colour system is only suitable for the display of data or graphics.

21.11 Phosphors for additive receivers

In the early days of colour television the phosphors used were cadmium borate for the red, zinc silicate for the green, and calcium magnesium silicate for the blue; but in 1951 new red and blue phosphors were discovered, so that the following set became established:

Red: zinc phosphate $x = 0.674, y = 0.326; u' = 0.485, v' = 0.528$
Green: zinc silicate $x = 0.218, y = 0.712; u' = 0.078, v' = 0.576$
Blue: zinc sulphide $x = 0.154, y = 0.068; u' = 0.175, v' = 0.174$

The N.T.S.C. standards (see Chapter 22) were drawn up assuming the receiver primaries:

Red:	$x = 0.67$	$y = 0.33$	$u' = 0.477$	$v' = 0.528$
Green:	$x = 0.21$	$y = 0.71$	$u' = 0.076$	$v' = 0.576$
Blue	$x = 0.14$	$y = 0.08$	$u' = 0.152$	$v' = 0.195$

The N.T.S.C. primaries are thus similar in chromaticity to the set of phosphors available in 1951. The luminance of the pictures obtained with these phosphors was rather limited, however, and in 1961 the green and the red phosphors were changed and the following set of 'sulphide phosphors' became established:

Red: zinc cadmium sulphide	$x = 0.663, y = 0.337; u' = 0.464, v' = 0.531$
Green: zinc cadmium sulphide	$x = 0.285, y = 0.595; u' = 0.119, v' = 0.560$
Blue: zinc sulphide	$x = 0.154, y = 0.068; u' = 0.175, v' = 0.174$

The chromaticities of these phosphors are shown by the points marked S in Fig. 21.5; those of the N.T.S.C. receiver primaries, marked N, are also shown for comparison. It is seen that the red sulphide phosphor is more orange than the red N.T.S.C. primary, the green is more yellow and of lower purity, and the blue is less green and of similar purity. Although the loss of colour gamut for blue-green and green colours is appreciable with the sulphide phosphors, the luminance of the picture is doubled, the efficiency with which a white is produced in a shadow-mask tube being increased from about 1 lumen per watt to about 2 lumens per watt, and experiments have shown that the increased colourfulness caused by the gain in luminance more than offsets the effects of the lower purity (see Section 11.9 (Matthews, 1963)). If the receiver white corresponds in chromaticity to that of a full radiator at a colour temperature of 9300 K, instead of to that of Standard Illuminant C (colour temperature approximately 6500 K), as specified in the N.T.S.C. standards, the efficiency rises to about 2.6 lumens per watt, when using the sulphide phosphors.

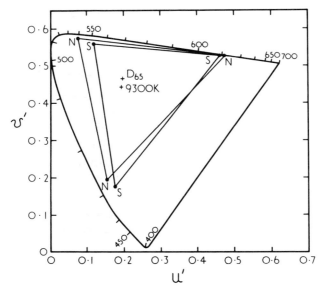

Fig. 21.5. Chromaticities of the sulphide phosphors, S, and of the N.T.S.C. receiver primaries, N, together with those of standard illuminant D_{65} and of a white of colour temperature 9300 K.

In 1964, it was discovered that the rare earth, yttrium vanadate, activated with europium, provided a red phosphor superior in both chromaticity and efficiency to the red sulphide phosphor, and this enabled a 9300 K white to be produced at an efficiency of 2.9 lumens per watt. The usual phosphor set thus became:

Red: europium yttrium
vanadate $\qquad x = 0.675, y = 0.325; u' = 0.486, v' = 0.527$
Green zinc cadmium
sulphide $\qquad x = 0.285, y = 0.595; u' = 0.119, v' = 0.559$
Blue: zinc sulphide $\qquad x = 0.154, y = 0.068; u' = 0.175, v' = 0.174$

The chromaticity of the new red phosphor is very close to that of the N.T.S.C. red receiver primary, so that the gamut of reproducible colours is represented in Fig. 21.5 by the triangle formed by the green and blue S-points and the red N-point.

In Fig. 21.6 the relative spectral power curves of the sulphide phosphors are shown by the full lines, and that of the europium yttrium vanadate phosphor by the broken line.

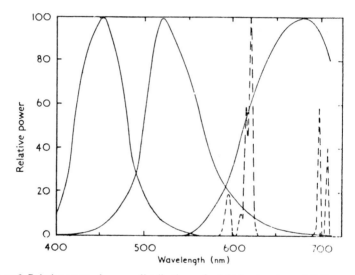

Fig. 21.6. Relative spectral power distributions of sulphide phosphors, full lines, and of europium yttrium vanadate phosphor, broken line.

In 1969 a set of chromaticities representative of phosphors used in European domestic receivers was agreed as (B.R.E.M.A., 1969; Sproson, 1978):

Red:	$x = 0.64$	$y = 0.33$	$u' = 0.451$	$v' = 0.523$
Green:	$x = 0.29$	$y = 0.60$	$u' = 0.121$	$v' = 0.561$
Blue	$x = 0.15$	$y = 0.06$	$u' = 0.175$	$v' = 0.158$

These chromaticities (and their tolerances) are shown in Fig. 21.7, together with those of the original N.T.S.C. primaries. It is interesting to note the consequent changes in the corresponding colour-matching functions, given in Fig. 19.7: the movement of the 1969 green phosphor in towards the red has resulted in the red matching function having a greater negative lobe in the green part of the spectrum.

434

Changing the chromaticities of the phosphors from those assumed for the N.T.S.C. system affects the theoretical values that should be adopted for the factors l, m, and n, used in forming the luminance signal (see Section 22.3), but the extent of the changes is usually considered to be too small to be of practical importance.

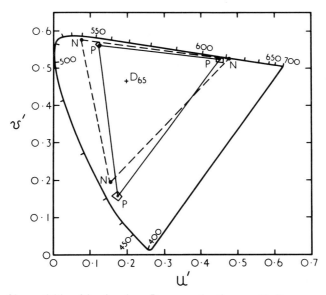

Fig. 21.7. Chromaticities of the phosphors, P, representing those used in European receivers in 1969 (B.R.E.M.A., 1969) together with those of the original N.T.S.C. primaries, N. The rectangles round the points, P, show the tolerances for the European phosphors.

21.12 The chromaticity of reproduced white

As mentioned in Sections 5.7 and 19.8, it is usually arranged that television cameras produce equal red, green, and blue signals for a standard white surface, whatever the colour of the taking illuminant. This arrangement has the advantage that it imitates the adapting effects of the eye, which largely compensate for changes in illuminant colour so as to maintain the appearance of whites approximately constant; it also has the advantage of keeping the magnitudes of the chrominance signals to a minimum by making them zero for whites, greys, and blacks, whatever the illuminant colour.

But what effect does the above practice have on the best chromaticity to choose for the display of white on the receiver? In the N.T.S.C. system the luminance signal is compounded on the assumption that equal red, green, and blue signals produce the chromaticity of Standard Illuminant C on the display. Therefore the equalization of the red, green, and blue signals for the standard

435

white in the taking illuminant may be thought of as roughly equivalent to filtering the light of the taking illuminant so that it always has approximately the same chromaticity as Standard Illuminant C.

In monochrome displays the relative luminances of objects will therefore always be similar to those in Standard Illuminant C (they would be identical if the filtering effect of equalizing the red, green, and blue signals was equivalent to using the appropriate spectral-power converting filter; see Section 10.3). The relative increase in the luminances of reddish objects in tungsten light, for instance, is therefore lost; but this type of effect is probably not very important (it has been shown that in black-and-white photography a spectral sensitivity broader than that required for producing correct luminances is preferable; Mouchel, 1963).

In colour displays, if (as assumed by the N.T.S.C. system) the receiver reproduces equal-signal white at a chromaticity equal to that of Standard Illuminant C, then the reproduced chromaticities in the picture will approximate to those which the scene would have had if it had been illuminated by light of Illuminant C quality. The *appearance* of these chromaticities will, however, depend on the conditions under which the display is viewed: in tungsten light, the picture will appear bluish, in clear north-sky light it will appear yellowish. It might be argued that, since television pictures are viewed more in the evenings than in the day time, the equal-signal white should be reproduced at a chromaticity typical of tungsten light, say 2856 K, but, if this is done, the picture becomes intolerably yellow in ambient daylight illumination. On the other hand, if the equal-signal white is reproduced at an ambient daylight chromaticity (6500 K, for instance) the picture does not appear intolerably bluish in ambient tungsten light illumination because the latter is often, or can be made to be, of sufficiently low intensity for the receivers to dominate the adaptation of the eye. This is why black-and-white receivers are usually made so as to have correlated colour temperatures of not less than 6500 K.

The higher the luminance of the television picture the greater is the extent to which it can dominate the adaptation, and since the blue-emitting phosphors usually have higher efficiencies than those emitting light of longer wavelengths, the luminance of television pictures generally increases as the correlated colour temperature of the reproduced white is increased. In black-and-white television, correlated colour temperatures as high as 9300 K are therefore widely used, and this chromaticity has sometimes been used for equal-signal white in colour displays. However, with modern phosphors the luminance obtainable at 6500 K can be close to that obtainable at 9300 K (Hirsch, 1968; Lamont, 1968).

The best chromaticity at which to reproduce equal-signal white is thus affected by several factors, but in colour displays it is found that, for most ambient conditions met with in domestic situations, correlated colour temperatures in the range 6500 to 9300 K are acceptable, but the lower figure is preferable in appearance (Zwick, 1973). In Europe Illuminant D_{65} (whose correlated colour temperature is about 6500 K) has been chosen as standard (B.R.E.M.A., 1969).

The use of a chromaticity different from that of Standard Illuminant C should really involve a change in the proportions l, m, and n, of the red, green, and blue signals in the luminance signal, but the extent of the changes is usually considered to be too small to be of practical importance.

It should be borne in mind that there is a case for compensating somewhat less than completely for changes in illuminant colour at the camera, because this is also true of the eye (see Section 8.10). Thus, the camera should produce signals such that the standard white is reproduced slightly yellowish when the illuminant is tungsten light, for example. Even further departures from the conditions for which whites are reproduced as whites may be desirable if special 'moods' are required for artistic reasons.

21.13 The luminance of reproduced white

The introduction of the rare-earth red phosphor enabled whites of luminance about 50 cd/m² to be attained. If the screen has the same luminance at all angles of viewing, this corresponds to the emission of about 50π, or 160, lumens/m². For a screen area 0.15 square metres, the total emission is therefore about 24 lumens. At 3 lumens per watt this would require about 8 watts, or a beam current of about 0.3 milliamps at 25 kV anode voltage.

Subsequent improvements to the phosphors, such as the use of europium activated yttrium oxy-sulphide for the red, and copper activated zinc cadmium sulphide for the green, made whites of about 85 cd/m² attainable, and slight modifications to the chromaticities used, and improved screening techniques, have further increased the luminance to about 120 cd/m² (Wright, 1971). By filling the interstices between phosphor dots with a black absorbing material, it is possible to increase the transmission of the glass face-plate of the tube (which is normally grey to reduce the effects of ambient illumination) and this further increases the luminance.

High luminance in the display is desirable both because colourfulness increases with luminance, and because the higher the picture luminance the less harmful will be a given level of ambient illumination. Flicker caused by the field frequency, however, becomes more noticeable as the luminance rises.

REFERENCES

Baumann, E., *J. Brit. Inst. Radio Engnrs.*, **12**, 69 (1952).
B.R.E.M.A., *Radio and Electronic Engineer*, **38**, 201 (1969).
Federman, F., and Pomicter, D., *J. Roy. Television Soc.*, **16**, vii (May–June, 1977).
Glenn, W. E., *J. Soc. Mot. Pic. Tel. Eng.*, **79**, 788 (1970).
Hirsch, C. J., *J. Soc. Mot. Pic. Tel. Eng.*, **77**, 202 (1968).
Hunt, R. W. G., *J. Phot. Sci.*, **13**, 108 (1965).
Jesty, L. C., *Proc. Inst. Elec. Engnrs.*, **105B**, 425 (1958).
Jones, L. A., and Condit, H. R., *J. Opt. Soc. Amer.*, **31**, 651 (1941).
Lamont, H. R. L., *J. Roy. Television Soc.*, **12**, 62 (1968).
Law, H. B., *J. Soc. Mot. Pic. Tel. Eng.*, **86**, 214 (1977).

Lobb, D. R., *S.P.I.E. Proc.*, **18,** Advances in laser scanning and recording, Geneva (1983).
Marconi, *Marconi Companies and their People*, **19,** 8 (February 1969).
Matthews, J. A., Private communication (1963).
Mouchel, P., *J. Phot. Sci.*, **11,** 291 (1963).
Neate, J., *Television*, **23,** 344 (1973).
Sponable, E. I., *J. Soc. Mot. Pic. Tel. Eng.*, **60,** 337 (1953).
Sproson, W. N., *Proc. I.E.E.*, **125,** 603 (1978).
Sproson, W. N., *Colour Science in Television and Display Systems*, p. 100, Adam Hilger, Bristol (1983).
Stone, S. M., Schlafer, J., and Fowler, V. J., *J. Information Display*, p. 41 (January 1969).
Venis, R. J., *Brit. Kinematog. Sound Tel.*, **51,** 379 (1969).
Wentworth, J. W., *Color Television Engineering*, McGraw Hill, New York, p. 153 (1955).
Wright, W. W., *J. Roy. Television Soc.*, **13,** 221 (1971).
Zwick, D. M., *J. Soc. Mot. Pic. Tel. Eng.*, **82,** 284 (1973).

CHAPTER 22

The N.T.S.C. and Similar Systems of Colour Television

22.1 Introduction

THE National Television Systems Committee (N.T.S.C.) of the U.S.A. recommended a system of transmitting colour television signals that involves both band sharing and carrier sharing (Loughren, 1953), and in 1953 this system was adopted by the U.S.A. Federal Communications Commission (F.C.C.) for general use in that country. In this Chapter we shall examine the N.T.S.C. system to show how its features are related to its basic premises; the colorimetric calculations involved are therefore given in sufficient detail to enable the interrelation of each feature with the others to be fully appreciated. Modifications of the N.T.S.C. system, such as the P.A.L. and S.E.C.A.M. systems, will also be considered.

22.2 N.T.S.C. chromaticities

The N.T.S.C. system is intended to be used with receiver colours having the following chromaticity co-ordinates (the original specifications were given in the

XYZ system but the corresponding values in the approximately uniform U'V'W' system are also given for convenience):

	x	y	z	u'	v'
Red (R)	0.67	0.33	0.00	0.477	0.528
Green (G)	0.21	0.71	0.08	0.076	0.576
Blue (B)	0.14	0.08	0.78	0.152	0.195
White (S_C)	0.310	0.316	0.374	0.201	0.461

The positions of these stimuli in the u', v' diagram are shown in Fig. 22.1.

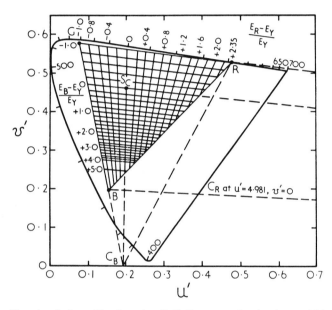

Fig. 22.1. The triangle formed by the points R, G, B representing the chromaticities of the red, green, and blue receiver stimuli in the N.T.S.C. system. S_C represents standard illuminant C, the stimulus matched by equal amounts of R, G, and B. C_R, S_C, and C_B represent the stimuli corresponding to the variables $R-L$, L, and $B-L$, where $L = 0.299R+0.587G+0.114B$. The lines drawn from C_B represent $(R-L)/L$ constant, those from C_R represent $(B-L)/L$ constant. Since E_R, E_G, E_B, E_Y are proportional to R, G, B, L, respectively, these are also lines of constant $(E_R-E_Y)/E_Y$ and $(E_B-E_Y)/E_Y$, the values of which are given in the figure.

As was explained in Chapter 7, the choice of the chromaticities of the reproduction primaries implies that the camera should have spectral sensitivity curves corresponding to the three colour-matching functions of those primaries. In Section 19.12, various means of approximating these functions were described. In this chapter we shall assume that a satisfactory approximation has been used.

22.3 The luminance signal

A luminance signal, E_Y, and two chrominance signals, $E_R - E_Y$ and $E_B - E_Y$, are used,[1] and it is arranged that the two chrominance signals are both zero for Standard Illuminant C (S_C). This has the advantage that for whites, greys, and blacks illuminated by daylight, the bandwidth restricted chrominance signals are little used. The choice of the chromaticity for which the chrominance signals are zero, together with that of the chromaticities of the red, green, and blue reproduction stimuli, fixes the values of l, m, and n in the expression

$$E_Y = lE_R + mE_G + nE_B$$

where $l+m+n = 1$ (see Section 19.8).

These values can be calculated by using the laws of colour mixture in any convenient chromaticity diagram: we will use the x, y diagram, but the same answer would be obtained in the u', v' or any other diagram. We proceed as follows.

The luminances L_X and L_Z of unit quantities of X and Z are both zero and so the luminance L_Y of unit quantity of Y may be set arbitrarily at unity. With the amounts of R, G, and B measured in luminance units, a fact we indicate by using the symbols R_L, G_L, B_L we may therefore write:

$$0.33(R_L) \equiv 0.67(X) + 0.33(Y) + 0.00(Z)$$
$$0.71(G_L) \equiv 0.21(X) + 0.71(Y) + 0.08(Z)$$
$$0.08(B_L) \equiv 0.14(X) + 0.08(Y) + 0.78(Z)$$

But, for S_C, $E_R = E_G = E_B$, and hence the corresponding amounts of R, G, and B light will also be equal for S_C; hence we may write:

$$(S_C) \propto 0.333(R) + 0.333(G) + 0.333(B)$$

But when the amounts of R, G, and B are measured in luminance units (with S_{CL} indicating that the amount of S_C is also measured in luminance units) this becomes:

$$0.333l(R_L) + 0.333m(G_L) + 0.333n(B_L) \equiv 0.333(l+m+n)\,(S_{CL})$$

because l, m, n are proportional to the luminances, L_R, L_G, L_B of the units used for measuring the amounts of R, G, and B (see Section 19.8). But $l+m+n = 1$. Therefore the equation for (S_{CL}) reduces to:

$$1.0(S_{CL}) \equiv l(R_L) + m(G_L) + n(B_L)$$

Hence by substitution:

$$
\begin{aligned}
1.0(S_{CL}) \equiv\ & (0.67l/0.33)\,(X) + l(Y) + (0.00l/0.33)\,(Z) \\
& + (0.21m/0.71)\,(X) + m(Y) + (0.08m/0.71)\,(Z) \\
& + (0.14n/0.08)\,(X) + n(Y) + (0.78n/0.08)\,(Z)
\end{aligned}
$$

But we also have, with the amount of S_C written in luminance units:

---

[1] The need for gamma correction is ignored for the moment; the complications it introduces will be considered later.

$$0.316(S_{CL}) \equiv 0.310(X)+0.316(Y)+0.374(Z)$$
$$\therefore \quad 1.0(S_{CL}) \equiv (0.310/0.316)(X)+1.0(Y)+(0.374/0.316)(Z)$$

By comparing the two equations for 1.0 (S_{CL}) we obtain

$$0.67l/0.33+0.21m/0.71+0.14n/0.08 = 0.310/0.316$$
$$l \quad + \quad m \quad + \quad n \quad = 1.0$$
$$0.00l/0.33+0.08m/0.71+0.78n/0.08=0.374/0.316$$

These three simultaneous equations may be solved for l, m, and n in the normal way to obtain the result:

$$l = 0.299 \qquad m = 0.587 \qquad n = 0.114$$

Hence the luminance signal E_Y in the N.T.S.C. system is made up thus:

$$E_Y = 0.299E_R+0.587E_G+0.114E_B$$

These values of l, m, and n, are universally used, even though camera sensitivity curves corresponding to different sets of phosphors and white points may be used for scene analysis.

22.4 (R) (G) (B) to (X) (Y) (Z) transformation equations

The evaluation of l, m, and n also enables the transformation equations between the (R), (G), (B) and (X), (Y), (Z) systems to be obtained. We have:

$$0.299(R_L) \equiv 1.0(R)$$
$$0.587(G_L) \equiv 1.0(G)$$
$$0.114(B_L) \equiv 1.0(B)$$

Because

$$0.33(R_L) \equiv 0.67(X)+0.33(Y)+0.00(Z)$$

we have

$$1.0(R) \equiv 0.299(R_L) \equiv (0.299/0.33) \, (0.67(X)+0.33(Y)+0.00(Z))$$

Similarly

$$1.0(G) \equiv 0.587(G_L) \equiv (0.587/0.71) \, (0.21(X)+0.71(Y)+0.08(Z))$$
$$1.0(B) \equiv 0.114(B_L) \equiv (0.114/0.08) \, (0.14(X)+0.08(Y)+0.78(Z))$$

Which simplify to:

$$1.0(R) \equiv 0.607(X)+0.299(Y)+0.000(Z)$$
$$1.0(G) \equiv 0.174(X)+0.587(Y)+0.066(Z)$$
$$1.0(B) \equiv 0.200(X)+0.114(Y)+1.111(Z)$$

The reverse transformation equations are given in Section 22.17.

22.5 The effects of variations in chrominance-signal magnitude

The way in which variations in the magnitudes of the two chrominance signals $E_R - E_Y$ and $E_B - E_Y$ differ in their effects on the chromaticities of the reproduced colours can be seen by constructing the grid of lines shown in Fig. 22.1. This construction is facilitated by plotting the positions of the stimuli S_C, C_R, and C_B corresponding to the variables E_Y, E_R-E_Y, and E_B-E_Y of the N.T.S.C. system. To find the positions of S_C, C_R, and C_B we proceed as follows.

The relationships between the signals may be written:

$$E_R = (1.0)\,(E_R-E_Y)+(1.0)E_Y+0(E_B-E_Y)$$

$$E_G = -\frac{l}{m}(E_R-E_Y)+(1.0)E_Y-\frac{n}{m}(E_B-E_Y)$$

$$E_B = 0(E_R-E_Y)+(1.0)E_Y+(1.0)\,(E_B-E_Y)$$

If we assume, for the moment, that the optical signals R, G, B, and L are proportional to the electrical signals E_R, E_G, E_B, and E_Y respectively, we may write $R = kE_R$, $G = kE_G$, $B = kE_B$, and $L = kE_Y$, where k is a constant. Hence:

$$R = (1.0)\,(R-L)+(1.0)L+0(B-L)$$

$$G = -\frac{l}{m}(R-L)+(1.0)L-\frac{n}{m}(B-L)$$

$$B = 0(R-L)+(1.0)L+(1.0)\,(B-L)$$

Hence the corresponding equations connecting the stimuli (see Section 8.4) may be written down by inspection as follows:

$$1.0(C_R) \equiv 1.0(R) -\frac{l}{m}(G)+0(B)$$

$$1.0(S_C) \equiv 1.0(R)+1.0(G)+1.0(B)$$

$$1.0(C_B) \equiv 0(R) -\frac{n}{m}(G)+1.0(B)$$

Therefore, using the (R) (G) (B) to (X) (Y) (Z) transformation equations.

$$1.0(C_R) \equiv 0.607(X)+0.299(Y)+0.000(Z)$$
$$-(0.299/0.587)\,(0.174(X)+0.587(Y)+0.066(Z))$$
$$1.0(S_C) \equiv (0.607+0.174+0.200)\,(X)+(0.299+0.578+0.114)\,(Y)$$
$$+(0.000+0.066+1.111)\,(Z)$$
$$1.0(C_B) \equiv -(0.114/0.587)\,(0.174(X)+0.587(Y)+0.066(Z))$$
$$+0.200(X)+0.114(Y)+1.111(Z)$$

These equations simplify to

$$1.0(C_R) \equiv 0.518(X)+0.000(Y)-0.034(Z)$$
$$1.0(S_C) \equiv 0.981(X)+1.000(Y)+1.177(Z)$$
$$1.0(C_B) \equiv 0.166(X)+0.000(Y)+1.098(Z)$$

The values of x and y are obtained in the usual way by dividing by $X+Y+Z$; the corresponding values of u' and v' can be obtained from x and y by formula or by nomogram, with the following results:

	x	y	u'	v'
(C_R)	1.070	0	4.981	0
(S_C)	0.310	0.316	0.201	0.461
(C_B)	0.131	0	0.191	0

The positions of these stimuli in the u', v' diagram are shown in Fig. 22.1.

These values show that S_C is located at the position of Standard Illuminant C, as was to be expected. For C_R and C_B the values of y (and v') are zero, and this shows that, like the stimuli X, Z, U', and W', the stimuli C_R and C_B affect colour but not luminance. That this must be so can be seen by considering the equation:

$$1.0(C_R) \equiv 1.0(R) - \frac{l}{m}(G)$$

which, when the amounts of R and G are measured in luminance units, becomes:

$$1.0(C_R) \equiv l(R_L) - \frac{l}{m}m(G_L) \equiv l(R_L) - l(G_L)$$

Therefore the luminance of 1.0 (C_R) is equal to $l-l = 0$. Similarly:

$$1.0(C_B) \equiv -\frac{n}{m}m(G_L) + n(B_L) \equiv -n(G_L) + n(B_L)$$

Hence the luminance of 1.0 (C_B) is also zero.

It is clear from Fig. 22.1 that C_B lies on the line GB produced; this feature stems from the fact that the position of C_B corresponds to the case where $E_R - E_Y$ and E_Y are both zero, in which case E_R must be zero, and this means that C_B must therefore by a mixture of G and B only; hence it lies on the line GB or GB produced. Similarly, C_R, which corresponds to $E_B - E_Y$ and E_Y both being zero,

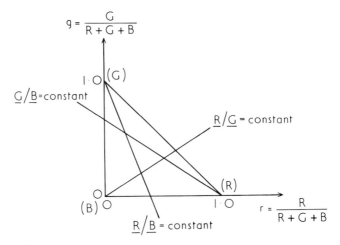

Fig. 22.2. The properties of straight lines through points representing the three matching stimuli.

444

so that E_B is also zero, must be a mixture of R and G only; although C_R is located too far to the right in Fig. 22.1 to be plotted, it does in fact lie on the line GR produced.

By drawing a fan of lines from the point C_B the loci of changes in chromaticity corresponding to variations in $E_B - E_Y$ only are obtained. The fan of lines radiating from C_R has also been drawn and shows the loci of changes in chromaticity corresponding to variations in $E_R - E_Y$ only. The full significance of the fans of lines passing through these points can be seen by referring to any colour triangle such as that shown in Fig. 22.2. It is clear from this figure that for any line passing through B, r is proportional to g, and hence, where s is the slope of the line,

$$g = sr$$

Hence all the points on any line passing through B have the property that r/g is constant. But since $r = R/(R+G+B)$ and $g = G/(R+G+B)$ it is also true that if r/g is constant then R/G is constant. For lines passing through R

$$g = s(1-r)$$
$$= s(r+g+b-r)$$
$$= sg+sb$$

Hence: $\qquad\qquad G/B = g/b = s/(1-s)$

Similarly for lines passing through G

$$B/R = b/r = (1-s)/s$$

Thus lines radiating from R have G/B constant, lines radiating from G have R/B constant, and lines radiating from B have R/G constant. In an exactly similar way, considering the $R-L, L, B-L$ system and a colour triangle in which $(R-L)/(R-L+L+B-L)$ is plotted against $(B-L)/(R-L+L+B-L)$ it can be shown that lines radiating from C_B have $(R-L)/L$ constant, and hence $(E_R-E_Y)/E_Y$ is also constant, lines radiating from C_R have $(B-L)/L$ constant, and hence $(E_B-E_Y)/E_Y$ is also constant, and lines radiating from S_C have $(R-L)/(B-L)$ constant, and hence $(E_R-E_Y)/(E_B-E_Y)$ is also constant.

These general properties of lines passing through mixture stimuli are not changed by linear transformation to a different colour triangle and it is thus possible to assign to each line in each fan in Fig. 22.1 its value of $(E_R-E_Y)/E_Y$ or $(E_B-E_Y)/E_Y$. It is seen that the lines passing through S_C both have a value of zero as required. The values of the other lines can be determined by calculating the positions along the line joining R and G of various values of $(E_R-E_Y)/E_Y$ and the positions along the lines joining G and B of various values of $(E_B-E_Y)/E_Y$. Along the line joining R and G, $E_B = 0$ and hence using $E_Y = 0.299E_R+0.587E_G +0.114E_B$ we have:

$$\frac{E_G}{E_R} = 1.704 \left/ \left(\frac{E_R-E_Y}{E_Y} +1 \right) \right. -0.509.$$

Points on the line joining R and G will divide it, in accordance with the Centre of Gravity Law (see Section 7.6), in the same ratio as the centre of gravity of weights

445

Rl/v_R' and Gm/v_G', where v_R' and v_G' are the v'-co-ordinates of R and G respectively. Since $R = kE_G$ and $G = kE_B$ the same result is obtained with weights $E_R l/v_R'$ and $E_G m/v_G'$. The ratio of the two weights therefore reduces to:

$$\frac{v_R' m E_G}{v_G' l E_R} = \frac{0.528}{0.576} \times \frac{0.587}{0.299} \times \frac{E_G}{E_R} = 1.80 \frac{E_G}{E_R}$$

If l_{RG} is the length of the line joining R and G, the distances from R corresponding to various mixtures of E_R and E_G are then given by:

$$\frac{1.80 E_G/E_R}{1 + 1.80 E_G/E_R} \times l_{RG}$$

The positions corresponding to various values of $(E_R - E_Y)/E_Y$ can therefore be determined. Similarly, when $E_R = 0$

$$\frac{E_G}{E_B} = 1.704 \left/ \left(\frac{E_B - E_Y}{E_Y} + 1 \right) \right. - 0.194$$

and points on the line joining G and B will divide it in the same ratio as the centre of gravity of weights having the ratio:

$$\frac{v_B' m E_G}{v_G' n E_B} = \frac{0.195}{0.576} \times \frac{0.587}{0.114} \times \frac{E_G}{E_B} = 1.75 \frac{E_G}{E_B}$$

If l_{GB} is the length of the line joining G and B, the distances from B corresponding to various mixtures of E_R and E_G are then given by:

$$\frac{1.75 E_G/E_B}{1 + 1.75 E_G/E_B} \times l_{GB}$$

The positions corresponding to various values of $(E_B - E_Y)/E_Y$ can therefore be determined.

22.6 The effect of gamma correction on $E_R - E_Y$ and $E_B - E_Y$

The signals actually transmitted in practical colour television systems have to be gamma corrected as described in Section 19.13. It is therefore of interest to see how gamma correction affects the relation between the chromaticity of the reproduced colour and the values of the chrominance signals.

As was explained in Section 19.13, the dim-surround viewing conditions commonly used for television require an overall gamma of about 1.25, rather than 1.0; since the gamma of colour receivers is usually about 2.8, it is therefore general practice to transmit signals reduced in gamma by about 2.8/1.25, which is equal to 2.2. The effect of this arrangement on colour reproduction is to increase colour purity (and to introduce some hue errors) as shown in Fig. 19.10, but the dim surround may then reduce apparent colour saturation in a way that present knowledge is inadequate to evaluate. However, some idea of the effects that will occur can be obtained by calculating chromaticities for a display having a gamma of 2.2, giving an overall system-gamma of 1.0, such as would be required for viewing with a surround of luminance equal to the average of that of the picture.

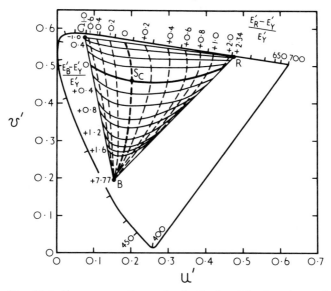

Fig. 22.3 The effect of gamma correction on the distribution of the chrominance signals in the chromaticity diagram. Lines of constant $(E_R'-E_Y')/E_Y'$ and $(E_B'-E_Y')/E_Y'$ are shown where $E_R' = E_R^{1/2.2}$, $E_B' = E_B^{1/2.2}$, $E_G' = E_G^{1/2.2}$, and $E_Y' = 0.299E_R'+0.587E_G'+0.114E_B'$, and a display gamma of 2.2 is assumed.

Therefore in Fig. 22.3 a grid of lines shows the chromaticities produced on a display of gamma 2.2 by various values of $(E_R'-E_Y')/E_Y'$ and $(E_B'-E_Y')/E_Y'$ where

$$E_R' = E_R^{1/2.2}$$
$$E_G' = E_G^{1/2.2}$$
$$E_B' = E_B^{1/2.2}$$
$$E_Y' = 0.299E_R'+0.587E_G'+0.114E_B'$$

It is only legitimate to draw such lines if, for each pair of values of $(E_R'-E_Y')/E_Y'$ and $(E_B'-E_Y')/E_Y'$, there is a unique chromaticity. This is so, because these voltages produce ratios of stimuli amounts $(R-L)/L$ and $(B-L)/L$ and these variables correspond to unique lines on a chromaticity diagram for the variables $R-L$, L, $B-L$. (See Plate 21, page 289.)

The lines that form the grid in Fig. 22.3 are fairly straight around the (S_C) point, but elsewhere they are quite curved, and the two grids of lines of Figs. 22.1 and 22.3 are by no means the same. However, the differences do not represent errors in colour reproduction but only the way in which the effects of gamma correction alter the relationships between the electrical and the optical signals. Large areas of colour will be unaffected by gamma correction (on a receiver operating at a gamma of 2.2).

If a pattern is transmitted for which the variations in $E_R' - E_Y'$ are too fine to be resolved because of their limited band-width, then the result on the display will be an average value of $E_R' - E_Y'$, together with the variations produced by the $E_B' - E_Y'$ signal which will result in chromaticity changes along one of the broken lines of Fig. 22.3 (or a similar parallel line); upon this will be superimposed the usual luminance difference produced by the E_Y' signal (provided that this does not call for a negative amount of green light, see Section 19.7). Similarly, if a pattern is transmitted for which the $E_B' - E_Y'$ variations are too fine to be resolved, the chromaticity displayed will comprise that produced by the average $E_B' - E_Y'$ value together with variations in chromaticity lying along one of the full lines of Fig. 22.3 (or along a similar parallel line), upon which the usual luminance difference will again be produced by the E_Y' signal. The above arguments apply to chromaticity differences in the horizontal direction in the picture (along the lines of the display); the reproduction of chromaticity differences in the vertical direction depends on the line structure of the picture.

22.7 The effect of gamma correction on E_Y

In Fig. 22.4 the effect of gamma correction on the E_Y' signal is shown by plotting $(E_Y')^{2.2}/E_Y$ in the u', v' diagram and drawing contours at the 0.9, 0.8, and 0.7 levels. The three effects noted in Section 19.13 resulting from the use of an E_Y' signal instead of a true E_Y signal were: first, that the constant luminance principle

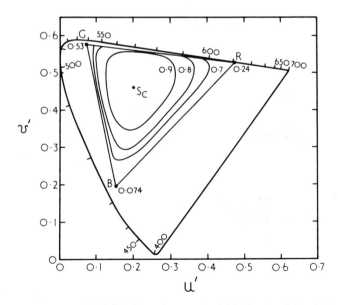

Fig. 22.4. The way in which $(E_Y')^{2.2}/E_Y$ varies over the RGB triangle. Although this ratio falls to less than 0.1 at the point B, over by far the greater part of the triangle it is above 0.8.

would be violated, so that the luminance would be in error (too low) in areas of fine detail, and noise in the chrominance signals would affect luminance; secondly, that coloured areas would be displayed too dark on monochrome receivers; and thirdly, that definition in small coloured areas would suffer, because of the restricted bandwidth of the luminance content carried by the chrominance signals. It is clear from Fig. 22.4, however, that although the values of $(E_Y')^{2.2}/E_Y$ at the corners of the triangle RGB are only 0.24 at R, 0.53 at G, and 0.074 at B, over by far the greater part of the triangle the ratio does not fall below 0.80. Since the extreme corners and edge of the triangle represent colours that are not used very often in typical scenes, the effects of transmitting the E_Y' signal instead of a true luminance signal are likely to be quite small in practice; however, since real receivers have gammas of about 2.8 instead of 2.2 the effects will be somewhat greater in practice than are shown in Fig. 22.4.

22.8 The P.A.L. and S.E.C.A.M. systems

The P.A.L. and S.E.C.A.M. versions of the N.T.S.C. system transmit the signals E_Y', $E_R' - E_Y'$, and $E_B' - E_Y'$ (Roizen and Lipkin, 1965). (We consider these versions before the normal N.T.S.C. version because the latter, although historically earlier, involves additional signal-coding operations.) The E_Y' signal is transmitted at full bandwidth, and the $E_R' - E_Y'$, and $E_B' - E_Y'$ signals at about a quarter of the E_Y' bandwidth. Carrier sharing is used so that some means of identifying the two chrominance signals is required. (See Plate 21, page 289.)

In the S.E.C.A.M. (Sequence and Memory) system, this is achieved by transmitting $E_R' - E_Y'$ and $E_B' - E_Y'$ on alternate lines of each field, together with a suitable identifying signal so that the receiver knows which of the two signals is being transmitted on each line. A circuit that delays a signal for a time exactly equal to that taken to scan one line, known as a *delay line*, is then used to provide the chrominance signal of the previous line of the field. In this way, although at any one time only one of the two chrominance signals is being transmitted, both are available for producing the colour picture. There is, however, a loss of definition of the chrominance information in the vertical direction, because the delay-line technique results, in effect, in an averaging over pairs of successive field lines, which is equivalent to groups of four picture lines. But the resultant vertical chrominance resolution is no worse than the horizontal chrominance resolution, and hence the loss is visually unnoticeable. By alternating the order of the transmission of the $E_R' - E_Y'$ and $E_B' - E_Y'$ signals on successive pictures (pairs of fields) the vertical chrominance resolution for stationary or slowly moving objects is improved.

The S.E.C.A.M. system uses frequency modulation instead of the more usual amplitude modulation for the $E_R' - E_Y'$ and $E_B' - E_Y'$ signals; but the E_Y' signal is amplitude-modulated, and this difference in modulating method further aids the receiver in correctly distinguishing between chrominance and luminance. The $E_R' - E_Y'$ sub-carrier is at 4.40625 MHz above the luminance carrier, but that of $E_B'-E_Y'$ is at 4.250 MHz above the luminance carrier (see Fig. 22.11(c)).

The S.E.C.A.M. system has been adopted by France, where broadcast colour television was begun in 1967, and by Russia and some other countries (see Table 22.5).

In the P.A.L. (Phase Alternation Line) system $E_R'-E_Y'$ and $E_B'-E_Y'$ are transmitted simultaneously on the same carrier, using amplitude modulation, but with a quarter of a cycle difference in phase between the two signals; the two signals are then distinguished at the receiver by virtue of their different phases with the aid of a signal carrying a reference phase, called a *colour-burst signal* (because it consists of a short burst of chrominance sub-carrier frequency). However, the phase of one of the two signals, say $E_R'-E_Y'$, is altered by half a cycle between successive field lines so that the signals transmitted on two successive lines of a field are:

$$\begin{array}{lcc} \text{Line 1} & E_R'-E_Y' & E_B'-E_Y' \\ \text{Line 3} & -(E_R'-E_Y') & E_B'-E_Y' \end{array}$$

The signals from line 1 are then passed through a delay line so that an average for the two lines can be obtained; since the $E_R'-E_Y'$ signals are half a cycle out of phase with one another on the two lines, they cancel one another out and hence $E_B'-E_Y'$ is obtained on its own thus:

$$\tfrac{1}{2}(E_R'-E_Y')+\tfrac{1}{2}(E_B'-E_Y')-\tfrac{1}{2}(E_R'-E_Y')+\tfrac{1}{2}(E_B'-E_Y') = E_B'-E_Y'$$

By also passing the signals from line 3 through a circuit that alters the phase by half a cycle, and then taking another average with the delayed signal from line 1, $E_R'-E_Y'$ is recovered thus:

$$\tfrac{1}{2}(E_R'-E_Y')+\tfrac{1}{2}(E_B'-E_Y')+\tfrac{1}{2}(E_R'-E_Y')-\tfrac{1}{2}(E_B'-E_Y') = E_R'-E_Y'$$

As in the S.E.C.A.M. system it is necessary for the transmission to include a signal that enables the receiver to know which of the two types of signal is being transmitted on each line: in this case the difference consists of the phase of one of the chrominance signals being reversed, and the colour-burst signal is varied to provide an appropriate identification on each line. The lines on which this reversal takes place are alternated on successive pictures (pairs of fields) so that, as in the S.E.C.A.M. system, only every fourth field is the same.

A feature of the P.A.L. system is that, if there is a slight phase error in the relationship between the reference signal and the chrominance signals, the fact that the phase is reversed on alternate lines, results in the error being cancelled out and the system is thus fairly insensitive to slight phase shifts (such as may occur in transmission over long links).

It is possible to make receivers for the P.A.L. system that do not have a delay line. In this case the receiver, using its phase-sensitive detector, demodulates the two chrominance signals separately, at their two phases separated by a quarter of a cycle. Without a delay line, if there is a phase error in the detection of the chrominance signals, they will produce errors in hue in the display; but these errors will be in opposite directions on successive lines of each field, and therefore if they are fairly small the eye is able to average them out and obtain an approximately correct result; if the errors are large enough to be visible they appear as a coarse coloured line structure, referred to as *Hanover bars* because they were first observed at Hanover.

The sub-carrier frequency used in the P.A.L system is equal to $283\frac{3}{4}$ times that of the line frequency (see Fig. 22.11(b)). It is not possible to use an odd multiple of half the line frequency, such as $283\frac{1}{2}$, because the reversal of the phase of the $E_R' - E_Y'$ signal on alternate lines causes its side-bands to be shifted by half a line-frequency multiple: thus if $283\frac{1}{2}$ had been chosen, while the $E_R' - E_Y'$ side-bands would have been interleaved with the luminance signals as required, the $E_R' - E_Y'$ side-bands would have coincided with the luminance signals and caused confusion. By choosing an odd multiple of quarter the line frequency, such as $283\frac{3}{4}$, the $E_B' - E_Y'$ and the $E_R' - E_Y'$ side-bands are both separated from the luminance signals by a quarter of a line frequency multiple. Separation of the luminance and chrominance signals is usually further improved by using a *notch filter* to remove from the luminance signal a band of frequencies corresponding to those at which the chrominance signals have substantial amplitude; the loss of definition that this causes is small compared to the improved isolation of the chrominance signals.

The P.A.L. system was invented by Walter Bruch (Townsend, 1978) and adopted by Great Britain and Germany, where broadcast colour television was begun in 1967, and by many other countries (see Table 22.5).

22.9 The N.T.S.C. system

The N.T.S.C. system, like the P.A.L. system, uses carrier sharing with the two chrominance signals transmitted simultaneously a quarter of a cycle out of phase with one another, but, unlike the S.E.C.A.M. and P.A.L. systems, the signals are treated in the same way on every line. A colour-burst signal provides a reference phase, and the receiver, which is provided with a phase-sensitive detector, recovers the two chrominance signals correctly by virtue of their different phases. A higher degree of accuracy in the phase identification is necessary in the N.T.S.C. system than in the P.A.L. system, however, because phase errors are not now cancelled out: thus phase must be held to about $\pm 5°$ in N.T.S.C. signals but only to about $\pm 25°$ in P.A.L. signals.

For accurate separation of the two chrominance signals by means of their phase difference it is desirable for them both to be transmitted with both side-bands present: but, to accommodate about a quarter of the luminance band-width both above and below the sub-carrier frequency, requires the sub-carrier frequency being nearer that of the main carrier than is desirable. The N.T.S.C. system overcomes this problem by transmitting its two chrominance signals at different bandwidths: one at about a quarter of the luminance bandwidth, and the other at only about a tenth. The higher definition chrominance signal then has *double* side-bands over the bandwidth where *both* signals are operating, but only a *single* side-band over the rest of its frequency range (see Fig. 22.11(a)). In this way the two signals can be distinguished satisfactorily, and the chrominance sub-carrier kept well-spaced from the main carrier. To operate successfully with one chrominance signal having a bandwidth equal to only about one-tenth of that of the luminance signal, the N.T.S.C. system uses not only band sharing and carrier sharing but also blue saving (see Section 19.6).

The N.T.S.C. system has been adopted not only in the U.S.A., where it was introduced in 1953, but also in Japan, and various other countries (see Table 22.5)

22.10 Blue saving in the N.T.S.C. system

As has already been pointed out (see Section 19.6), the eye is much less able to detect fine blue detail than fine red or green detail. Fine patterns of blues and yellows of the same luminances are confused and invisible to the eye, a phenomenon known as *foveal tritanopia* (Willmer, 1944; Willmer and Wright, 1945; Thomson and Wright, 1947; Wright, 1952). It would therefore seem reasonable to transmit the $E_B' - E_Y'$ signal with less bandwidth than that used for the $E_R' - E_Y'$ signal.

In Fig. 22.5 are shown lines connecting the points representing colours that

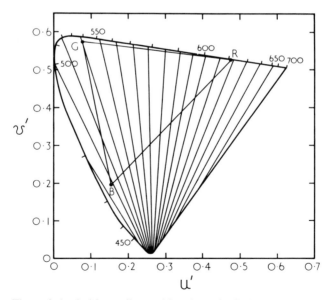

Fig. 22.5. The confusion loci for small-area vision. Any pair of colours whose chromaticities fall on the same line will tend to be confused when seen at small angles of view (Thomson and Wright, 1953).

are most readily confused when seen in small areas (Thomson and Wright, 1953). Comparison with Fig. 22.3 shows that the directions of the lines of Fig. 22.5 are roughly similar to those representing variations in $E_B{}'-E_Y{}'$ only, and this would seem to confirm that the $E_B{}'-E_Y{}'$ signal would be the appropriate one to restrict in bandwidth. However, the restriction in the N.T.S.C. system is not applied in quite this way; instead, it is applied so that, when that fineness of detail is reached such that only one chrominance signal is operating, the colour range remaining is not parallel to the reddish to greenish direction of the $E_R{}'-E_Y{}'$ signal, but approximately parallel to the orange to blue-green direction. This has been found preferable in practical tests. It was also found by Middleton and Holmes that, using pieces of coloured papers subtending very small angles to the eye, there was a tendency for the normal range of colour discrimination to degenerate into orange to blue-green differences only (Middleton and Holmes, 1949; Hacking, 1957). (See Plate 19, page 288.)

An understanding of the way in which this shift in the axis of restriction is achieved in the N.T.S.C. system is facilitated by considering a diagram in which E_R-E_Y is plotted against E_B-E_Y as shown in Fig. 22.6. (The effects of gamma correction will be brought in later.) In this diagram the origin represents all colours having the same chromaticity as standard illuminant C, whatever their luminance, so that blacks, greys, and whites plot in this vicinity. The distance of a

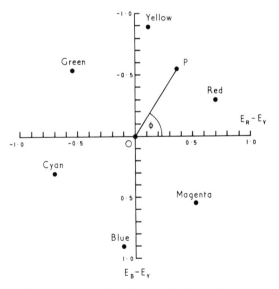

Fig. 22.6. The distribution of colours for which E_R, E_G, E_B equal zero or unity on a plot of E_R-E_Y against E_B-E_Y. The length of a line such as OP depends on the purity and the luminance of the colour plotting at P. The angle ϕ of the line OP is a function of the dominant wavelength of the colour represented by P.

453

point, P, from the origin is a function of the purity multiplied by the luminance of the colour concerned. The way in which colours are distributed in this diagram can be seen by considering the cases where E_R, E_G, and E_B, are allowed to have values of either 0 or 1, as given in Table 22.1 (to be found at the end of this chapter). These colours are plotted in Fig. 22.6. Because, in the N.T.S.C. system, 1 volt is generally the maximum value allowed for E_R, E_G, or E_B at the camera, the colours considered are those of maximum purity and luminance if the axes are considered to represent camera voltages.

In the N.T.S.C. system, the sub-carrier wave is amplitude-modulated by the two chrominance signals a quarter of a cycle out of phase with one another. For any given colour, the combined result of the two separate modulations can be considered as a single amplitude modulation at a particular phase; the amplitude then depends on the luminance and the purity, while the phase indicates the dominant wavelength. Thus in Fig. 22.6 the colour, P, in this carrier sharing system, would be transmitted by an amplitude represented by OP at a phase represented by the angle ϕ. In the N.T.S.C. system this correlation between phase and dominant wavelength is used to alter the directions of the axes representing the colour-difference signals. But, as will be explained more fully later, in order to avoid overloading the transmitter when the chrominance signals are transmitted at maximum amplitude they are first reduced in amplitude by factors of 1.14 for E_R-E_Y and 2.03 for E_B-E_Y. It is therefore necessary to replot the data of Fig.

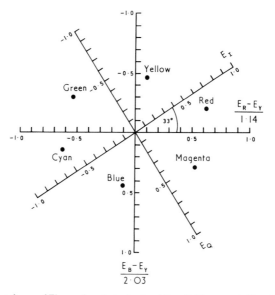

Fig. 22.7. The colours of Fig. 22.6 replotted using $(E_R-E_Y)/1.14$ and $(E_B-E_Y)/2.03$ as co-ordinates. Relative to these axes, the axes of E_I and E_Q are rotated through an angle of 33°.

454

22.6 using $\dfrac{E_R-E_Y}{1.14}$ and $\dfrac{E_B-E_Y}{2.03}$ as variables, as shown in Fig. 22.7. The two chrominance signals used, E_I and E_Q are then shifted in phase by an angle of $33°$ relative to the E_R-E_Y and E_B-E_Y axes as shown. The E_I signal is an in-phase component of the sub-carrier, while the E_Q signal is a quadrature component of the sub-carrier, a quarter of a cycle out of phase with E_I. The two sets of signals are related to one another (in accordance with the usual equations for the rotation of co-ordinate axes) thus:

$$E_I = \frac{E_R-E_Y}{1.14}\cos 33° - \frac{E_B-E_Y}{2.03}\sin 33°$$

$$E_Q = \frac{E_R-E_Y}{1.14}\sin 33° + \frac{E_B-E_Y}{2.03}\cos 33°$$

which (to three significant figures) reduce to:

$$E_I = 0.736(E_R-E_Y)-0.268(E_B-E_Y)$$
$$E_Q = 0.478(E_R-E_Y)+0.413(E_B-E_Y)$$

These two signals E_I and E_Q are used together with the same luminance signal as before, E_Y. (See Plate 19, page 288.)

The values of E_I and E_Q for the colours considered previously are given in Table 22.2 (to be found at the end of this chapter).

The equations relating E_Q and E_I with E_R-E_Y and E_B-E_Y enable the positions of the stimuli C_Q and C_I corresponding to the signals E_Q and E_I to be located in the chromaticity diagram. Thus, solving for E_R-E_Y and E_B-E_Y we obtain

$$E_R-E_Y = 0.956E_I+0.621E_Q$$
$$E_B-E_Y = -1.106E_I+1.703E_Q$$

from which:

$$E_R = 0.956E_I+1.000E_Y+0.621E_Q$$
$$E_B = -1.106E_I+1.000E_Y+1.703E_Q$$

and using $E_Y = 0.299E_R+0.587E_G+0.114E_B$ we obtain:

$$E_G = -0.272E_I+1.000E_Y-0.647E_Q$$

If we put $I = kE_I$, $L = kE_Y$, and $Q = kE_Q$ where k is constant (see Section 22.5) we have

$$R = 0.956I+1.000L+0.621Q$$
$$G = -0.272I+1.000L-0.647Q$$
$$B = -1.106I+1.000L+1.703Q$$

The corresponding equations relating to the stimuli are therefore:

$$1.0(C_I) \equiv 0.956(R)-0.272(G)-1.106(B)$$
$$1.0(S_C) \equiv 1.000(R)+1.000(G)+1.000(B)$$
$$1.0(C_Q) \equiv 0.621(R)-0.647(G)+1.703(B)$$

Using the $(R)(G)(B)$ to $(X)(Y)(Z)$ transformation equations (from Section 22.4) we obtain:

$$1.0(C_I) = 0.956(0.607(X)+0.299(Y)+0.000(Z))$$
$$-0.272(0.174(X)+0.587(Y)+0.066(Z))$$
$$-1.106(0.200(X)+0.114(Y)+1.111(Z))$$

Calculating 1.0 (S_C) and 1.0 (C_Q) similarly, and simplifying, we obtain:

$$1.0(C_I) \equiv 0.312(X)+0.000(Y)-1.247(Z)$$
$$1.0(S_C) \equiv 0.981(X)+1.000(Y)+1.177(Z)$$
$$1.0(C_Q) \equiv 0.605(X)+0.000(Y)+1.849(Z)$$

It is useful for some purposes to re-write these equations in the corresponding form connecting the amounts of the stimuli:

$$X = 0.312I+0.981L+0.605Q$$
$$Y = 1.000L$$
$$Z = -1.247I+1.177L+1.849Q$$

From the equations for (C_I), (S_C), and (C_Q) in terms of (X), (Y), and (Z), their positions in the x, y diagram (and in the u', v' diagram) can be calculated in the usual way with the following results:

	x	y	u'	v'
(C_I)	-0.333	0	-0.365	0
(S_C)	0.310	0.316	0.201	0.461
(C_Q)	0.245	0	0.393	0

The positions of these stimuli in the u', v' diagram are shown in Fig. 22.8. It is clear that C_I and C_Q are also stimuli which affect colour but not luminance, because their values of y (and v') are zero. Lines drawn through C_I have constant Q/L and therefore E_Q/E_Y is also constant; lines drawn through C_Q have constant I/L and therefore E_I/E_Y is also constant. It should be noted that as the value of E_I/E_Y increases the chromaticity moves away from C_I instead of towards it.

In Fig. 22.8 the fans of lines radiating from C_I and C_Q are shown and it is seen that the transformation has rotated the axes as required, variations in E_I now running roughly parallel to the orange-red to cyan direction. The bandwidth restriction is therefore applied to the E_Q signal, which is transmitted at about $\frac{1}{3}$ MHz bandwidth while the E_I signal is transmitted at about 1 MHz, for 525 line systems. Some idea of the restriction in definition can be gauged from the fact that 0.3 MHz corresponds to a blurring of about 6 mms along a line of a 56 cm receiver picture. In some areas where the N.T.S.C. system is used, the E_I signal is also restricted to about $\frac{1}{3}$ MHz; this results in a marked loss of colour definition but allows less sophisticated equipment to be used.

The values of the lines of Fig. 22.8 are indicated and are calculated from the equations for E_I and E_Q in terms of E_R-E_Y and E_B-E_Y, remembering that $E_Y = 0.299E_R+0.587E_G+0.114E_B$.

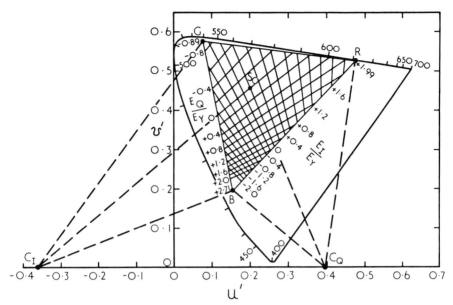

Fig. 22.8. C_I, S_C, and C_Q represent the N.T.S.C. stimuli corresponding to the variables I, L, and Q, respectively, where

$$I = 0.736(R-L) - 0.268(B-L) \text{ and}$$
$$Q = 0.478(R-L) + 0.413(B-L).$$

The lines drawn from C_Q represent I/L constant, those drawn from C_I represent Q/L constant. Since I, L, Q are proportional to E_I, E_Y, E_Q, respectively, these are also lines of constant E_I/E_Y and E_Q/E_Y, the values of which are given in the figure.

When $E_B = 0$

$$\frac{E_G}{E_R} = \frac{1.253}{E_I/E_Y + 0.468} - 0.509$$

The positions on the line joining R to G corresponding to various values of E_I/E_Y can therefore be calculated using as before (in Section 22.5):

$$\frac{1.80 E_G/E_R}{1 + 1.80 E_G/E_R} \times l_{RG}$$

to give the distance from R. (The values of these lines are actually marked on the line joining R to B in Fig. 22.8 for convenience.)

When $E_R = 0$

$$\frac{E_G}{E_B} = \frac{0.704}{E_Q/E_Y + 0.891} - 0.194$$

and the positions on the line joining G to B corresponding to various values of E_Q/E_Y can therefore be calculated using as before:

$$\frac{1.75E_G/E_B}{1+1.75E_G/E_B} \times l_{GB}$$

to give the distance from B.

22.11 Gamma correction in the N.T.S.C. system

We must now consider the effects of gamma correction on the N.T.S.C. signals. Without gamma correction we have

$$E_I = 0.736(E_R - E_Y) - 0.268(E_B - E_Y)$$
$$E_Q = 0.478(E_R - E_Y) + 0.413(E_B - E_Y)$$

But because of gamma correction the signals actually transmitted are

$$E_I' = 0.736(E_R^{1/\gamma} - E_Y') - 0.268(E_B^{1/\gamma} - E_Y')$$
$$E_Q' = 0.478(E_R^{1/\gamma} - E_Y') + 0.413(E_B^{1/\gamma} - E_Y')$$

where
$$E_Y' = 0.299E_R^{1/\gamma} + 0.587_G^{1/\gamma} + 0.114E_B^{1/\gamma}$$

The changes in chromaticity when the fineness of detail is such that only one chrominance signal is transmitted will now correspond to those caused by the E_I' signal instead of by the E_I signal. It is therefore instructive to consider how variations in E_I' and E_Q' affect chromaticity, and this can be done if, as before, E_I' and E_Q' are both divided by the third signal transmitted, which in this case is E_Y'; thus:

$$\frac{E_I'}{E_Y'} = 0.736 \left(\frac{E_R^{1/\gamma}}{E_Y'} - 1 \right) - 0.268 \left(\frac{E_B^{1/\gamma}}{E_Y'} - 1 \right)$$

$$\frac{E_Q'}{E_Y'} = 0.478 \left(\frac{E_R^{1/\gamma}}{E_Y'} - 1 \right) + 0.413 \left(\frac{E_B^{1/\gamma}}{E_Y'} - 1 \right)$$

The positions of lines of constant E_I'/E_Y' and E_Q'/E_Y' can be calculated by inserting various values of E_R, E_G, E_B in the above equations, plotting the corresponding values of E_I'/E_Y' and E_Q'/E_Y', and drawing the appropriate contours. This has been done in Fig. 22.9 for $\gamma = 2.2$, which is the value adopted in the N.T.S.C. system. It is seen that around the S_C point the grid of lines of constant E_I'/E_Y' and E_Q'/E_Y' is similar to that of the lines of constant E_I/E_Y and E_Q/E_Y in Fig. 22.8. As the colour considered departs from S_C it is clear that considerable differences occur between the gamma-distorted and the undistorted signals. It is emphasized, however, that these differences only occur in the transmitted electrical signals; the colours displayed by a receiver operating at a gamma of 2.2 will be the same as those which would be displayed by a (hypothetical) receiver using the undistorted signals and operating at a gamma of unity (however, real receivers have gammas of 2.8±0.3, see Section 19.13 and Fig. 19.10). In detail that is too fine for the E_Q' signal to resolve, the E_Q' signal will assume the average value for the area and the chromaticities displayed in the horizontal direction of the picture will exhibit differences only in the direction of

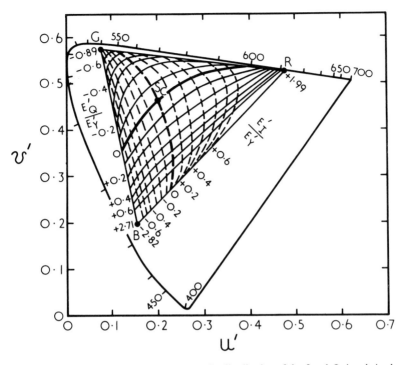

Fig. 22.9. The effect of gamma correction on the distribution of the I and Q signals in the chromaticity diagram. Lines of constant E_I'/E_Y' and E_Q'/E_Y' are shown for a display gamma of 2.2 where

$$E_I' = 0.736(E_R^{1/2.2}-E_Y')-0.268(E_B^{1/2.2}-E_Y')$$
$$E_Q' = 0.478(E_R^{1/2.2}-E_Y')+0.413(E_B^{1/2.2}-E_Y')$$
$$E_Y' = 0.299E_R^{1/2.2}+0.587E_G^{1/2.2}+0.114E_B^{1/2.2}$$

The full lines, representing E_Q'/E_Y' constant, show the directions of chromaticity changes available when the fineness of detail is such that only the E_Y' and E_I' signals are being transmitted.

the full line corresponding to the particular average value of E_Q'/E_Y' concerned. Similarly, when the detail is also too fine for the E_I' signal to resolve, the displayed chromaticity in the horizontal direction of the picture will correspond to the average values of both E_I' and E_Q' for the particular area, upon which will be superimposed fine detail luminance differences provided by the E_Y' signal. In the vertical direction of the picture, of course, the reproduction of chromaticity differences depends on the line structure. It is seen that the directions of the lines corresponding to variations in E_I'/E_Y' are still mainly parallel to the orange to cyan direction, but curve towards red (at the orange end) and towards blue (at the cyan end). (See Plate 19, page 288.)

22.12 Maximum signal amplitudes

It was mentioned earlier that, to avoid overloading the transmitter, it was necessary to reduce the magnitude of the $E_R - E_Y$ signal by a factor of 1.14 and that of the $E_B - E_Y$ signal by a factor of 2.03 before forming the E_I and E_Q signals. In Table 22.3 (to be found at the end of this chapter) the values of the gamma corrected E_Y', E_I', and E_Q' signals that are actually produced are shown for the colours corresponding to E_R, E_G, E_B being equal to 0 or 1.0. Since 1.0 is the maximum value normally permitted for these signals, the values in Table 22.3 show the maximum which will have to be transmitted. The load on the transmitter will be a function of the combined amplitudes of the luminance signal and the chrominance signal and the maximum and minimum values are given by:

$$E_Y' \pm \sqrt{(E_I')^2 + (E_Q')^2}$$

It is seen that the weighting factors of 1.14 and 2.03 result in this combined signal being confined to the range -0.333 to $+1.333$. Strictly speaking, to avoid all overloading, the permitted range should be from 0 to 1.0, but since large values of the chrominance signal occur only for very short times in typical scenes the

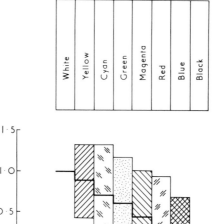

Fig. 22.10. The upper part of the figure depicts a typical 'colour bar' test pattern. The lower part shows the way in which the amplitude of the transmitted signal varies along a single line scan. The heavy line represents the amplitude of the luminance signal; the shaded rectangles represent the amplitude ranges of the combined luminance and chrominance signals. (The ordinate is for the N.T.S.C. system, but it applies equally to the P.A.L. system.)

slightly wider range is permissible. The negative amplitude simply indicates that the signal is using a region of amplitudes normally reserved for the receiver synchronizing signals.

The reduced-amplitude signals $(E_R'-E_Y')/1.14$ and $(E_B'-E_Y')/2.03$, are also used in the P.A.L. system, in order to avoid overloading the transmitter. (They are sometimes referred to as the V and U signals, respectively, and must not be confused with U', V', W' tristimulus values.)

In Fig. 22.10 the pattern of amplitudes of Table 22.3 is shown in diagrammatic form. The top half of the figure shows an array of vertical bars of different colours, known as the *colour bar test signal*. The colours in this test signal correspond to the eight colours for which E_R, E_G, and E_B have values of 0 or 1.0, and hence the corresponding signals have the values of Table 22.3. In the lower half of the figure the range of values from

$$E_Y'+\sqrt{(E_I')^2+(E_Q')^2} \text{ to } E_Y'-\sqrt{(E_I')^2+(E_Q')^2}$$

has been shown for each of the eight colours immediately below them. The result in the lower half of the figure is to depict the changes in amplitude of the transmitted signal during the scanning of a single line, the value at any instant being somewhere between the top and bottom of each of the hatched areas, according to the phase of the chrominance signal; the phases at which the extreme values occur are, of course, different for the different colours. If in fact the amplitude is displayed on an oscilloscope which is synchronized to the line scanning frequency then its appearance is similar to that shown in Fig. 22.10, the hatched area appearing as an area of very high frequency oscillations superimposed on the amplitude representing the luminance. If the signal were replaced by a scale of greys for which $E_R = E_G = E_B$ and E_Y' was the same as for the eight colours, then the display would consist simply of the heavy line in Fig. 22.10 representing the amplitude of the luminance signal, the chrominance signal being zero throughout.

22.13 Cross-talk between E_I' and E_Q'

In Fig. 22.11 and Table 22.4 the exact frequency arrangement is shown for the 525-line N.T.S.C. system and for the 625-line P.A.L and S.E.C.A.M. systems. It is clear that the adoption of bandwidths of 1 MHz for E_I' and $\frac{1}{3}$ MHz for E_Q', at the chrominance carrier frequencies shown, means that the E_I' signal has its upper side-band partially cut off by the upper limit of the total band; hence the E_I' transmission is not fully double side-band. The effect of this on cross-talk (or interference) between the two signals is as follows: E_Q' is double side-band so does not cross-talk to E_I'; also E_I' is transmitted double side-band up to the bandwidth of E_Q', so these lower frequency E_I' components do not cross-talk to E_Q'; but above E_Q' cut-off, E_I' does cross-talk to E_Q', but if these frequencies are removed from the E_Q' channel by a filter that passes only frequencies up to E_Q' limit, the result is effective freedom from cross-talk between E_I' and E_Q'.

461

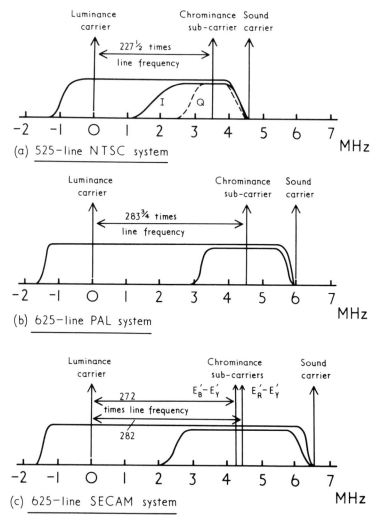

Fig. 22.11. The frequency arrangements adopted (a) in the 525-line N.T.S.C. system, (b) in the 625-line P.A.L. system, and (c) in the 625-line S.E.C.A.M. system.

22.14 The effect of the chrominance sub-carrier on the display

Before transmission, the chrominance sub-carrier frequency is suppressed, leaving only the side-band frequencies. Hence, when a grey area is being transmitted, the chrominance signals are both zero, and therefore the colour signal has no effect on the picture either in colour or in black-and-white. But when a non-grey colour is displayed on a monochrome receiver, the chrominance side-bands appear as modulations of intensity along each line, at spacings equivalent to frequencies near that of the sub-carrier; these modulations produce dot pairs that are displaced along successive lines of each *field* by a distance equal to half (or quarter) a dot pair, and, because the total number of lines in each picture system is odd, the dot pattern is reversed on successive complete pictures. The effect of the dots is therefore reduced by persistence of vision, but their presence can nevertheless be disturbing on monochrome pictures, particularly if they tend to 'crawl' in rows or columns.

When a colour receiver displays a non-grey colour, the intensity of each of the three electron beams does not produce dots along each line because the sub-carrier side-bands are demodulated into low bandwidth colour-difference signals; the result is that the colour sub-carrier has a less disturbing effect on a colour display than it has on a monochrome display.

22.15 Comparison of the N.T.S.C., P.A.L., and S.E.C.A.M. systems

All three systems, N.T.S.C., P.A.L., and S.E.C.A.M., have been shown in practical tests to be capable of producing very similar and satisfactory results, and the differences between them can often be detected only by viewing the reproduced picture at viewing distances appreciably closer than normal. The main differences between the systems, may, however, be summarized as follows.

The P.A.L. and S.E.C.A.M. systems, by delaying chrominance information over a line scan, reduce colour definition in the vertical direction; but since chrominance is severely bandwidth limited anyway, they do not make the vertical colour definition worse than in the horizontal direction. The normal interlacing of the lines on successive fields helps to make the averaging of the colour information over successive lines more uniform, and, by interchanging which signal is transmitted on any given line on successive complete pictures, the definition and smoothness of the final result is further improved except for rapidly moving objects.

The N.T.S.C. and P.A.L. systems, for certain patterns, can result in fine detail luminance signals being misinterpreted as chrominance signals; thus certain black-and-white patterns may be reproduced with spurious coloured fringes. The S.E.C.A.M. system avoids this defect by its use of frequency modulation for the chrominance signals. The P.A.L. system can be considerably improved in this respect by the use of a 'notch filter' that preferentially attenuates luminance signals of frequencies near those of the chrominance sub-carrier; curious edge effects are obtained if the notch filter has too strong an effect but useful advantages can be obtained at a proper compromise level.

Spurious 'crawling dot' or 'herring-bone' patterns can occur in all three systems, typical effects being as follows. In the N.T.S.C. system, crawling dots occur on monochrome displays, but the colour displays are undisturbed. In the P.A.L. system, lines moving at 30° occur on monochrome displays, and less obtrusively on colour displays. In the S.E.C.A.M. system, herring-bone patterns occur on monochrome displays, and less obtrusively on colour displays.

The greater similarity between the N.T.S.C. and P.A.L. systems means that they can more easily be used together than either can with the S.E.C.A.M. system. Thus the P.A.L. system, in which phase changes matter far less than in the N.T.S.C. system, could be used for long distance transmission along co-axial cables; and then the N.T.S.C. system could be used for local broadcasting to avoid the necessity for incorporating delay circuits in the receivers.

Perhaps the most important differences are that the S.E.C.A.M. system is the only one in which luminance and chrominance are never confused, but it is the most complicated system; and the P.A.L. system, because of its automatic phase correction, is easier to receive without hue errors than the N.T.S.C. system, in which the necessity for adjusting a hue control on the receiver is a serious disadvantage for unskilled users.

The P.A.L. and S.E.C.A.M. systems are usually used with 625-line television standards, whereas the N.T.S.C. system is usually used with 525-line standards so that its pictures usually have lower definition; on the other hand, because a field rate of 60 Hz is usually used with 525-lines, the flicker is usually less perceptible than with the 50 Hz field rate usually used with 625-lines.

22.16 Some useful graphical constructions

If reference is made to Fig. 22.12 it is clear that it must be possible to match a colour M, whose chromaticity lies on the line joining G and B, by a suitable mixture of G and B, so that we can write

$$M(M) \equiv G(G) + B(B)$$

Similarly a colour N, whose chromaticity lies on the line joining M and R can be matched by a suitable mixture of M and R:

$$N(N) \equiv M(M) + R(R)$$

The exact position of the point representing N on the line joining R and M will be governed by the centre of gravity law, and will be the same as that of the centre of gravity of weights

$$\frac{mG + nB}{v_M'} \qquad \text{placed at (M), and}$$

$$\frac{lR}{v_R'} \qquad \text{placed at (R).}$$

The same result is obtained using weights

$$\frac{mE_G + nE_B}{v_M'} \qquad \text{and} \qquad \frac{lE_R}{v_R'}$$

But $mE_G + nE_B = E_Y - lE_R$, therefore the weights can become

$$\frac{E_Y - lE_R}{v_M'} \qquad \text{and} \qquad \frac{lE_R}{v_R'}$$

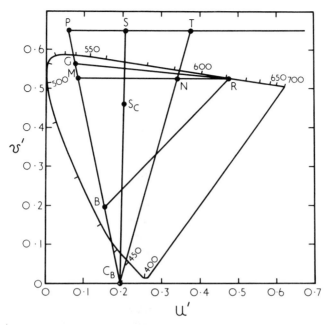

Fig. 22.12. Since the line joining R to M is a line of constant v', mixtures of R and M are represented by points equivalent to the centres of gravities of weights directly proportional to the luminances of R and M, placed at R and M. The proportion of R in the mixtures can therefore be represented by a uniform scale along MR, or along any line parallel to MR such as PT. A uniform scale along PT can therefore be used to find the values of R/L, and therefore of $(R-L)/L$, for lines of constant R/L drawn from the point C_B.

If M and R lie on a line of constant v' so that $v_M' = v_R'$, then the weights can become

$$E_Y - lE_R \qquad \text{and} \qquad lE_R$$

The distance from M of the point representing N is therefore given by

$$\frac{lE_R}{(E_Y - lE_R) + lE_R} l_{RM} = \frac{lE_R}{E_Y} l_{RM}$$

where l_{RM} is the length of the line joining R and M.

Since l and l_{RM} are both constants, the distance of N from M is thus proportional to E_R/E_Y, and hence the line joining R and M can carry a linear scale of E_R/E_Y, which is easily converted to a linear scale of $(E_R - E_Y)/E_Y$ by subtracting 1.0 from each value. The construction can often be made more conveniently by using a line parallel to RM, such as PT in Fig. 22.12 (with P lying on BG or BG produced). If the line from C_B through S_C is produced to meet the line PT at S, and the position of the line PT is chosen so that PS = 1.0 on some convenient scale, then the other values can be very easily marked out directly.

This arrangement gives the correct results because, at the point S_C, $E_R = E_Y$, and therefore the line through S_C is the line for which $E_R/E_Y = 1.0$ (hence $(E_R-E_Y)/E_Y = 0$) and the line from C_B through B and G is the line for which $E_R = 0$, and therefore $E_R/E_Y = 0$ (hence $(E_R-E_Y)/E_Y = -1.0$).

In principle the same type of construction could be used for obtaining the fan of lines from C_R. In this case the distances are given by:

$$n\frac{E_B}{E_Y}l_{BM}$$

where l_{BM} is the length of the line through B, parallel to the u'-axis, which meets GR produced at M. Unfortunately the geometry of the u', v' diagram is such that the method is impracticable for scaling E_B/E_Y but in some other chromaticity diagrams it can be used.

The method can also be used for obtaining the values of lines of constant E_Q/E_Y and E_I/E_Y. For since on the line joining R and M the distances are given by

$$l\frac{E_R}{E_Y}l_{RM},$$

by substituting for E_R/E_Y we obtain:

$$l_{RM}l(0.621\frac{E_Q}{E_Y}+0.956\frac{E_I}{E_Y}+1.0).$$

But for all points on the line joining R and M the value of v' is constant and equal to v_R', the v' co-ordinate of R. Therefore

$$\frac{V'}{U'+V'+W'} = v' = v_R'$$

$$\therefore \qquad V' = v_R'(U'+V'+W')$$

$$\therefore \qquad V'(1-v_R')/v_R' = U'+W'$$

Using equations relating U', V', W', with I, L, Q (Section 22.17), and inserting $v_R' = 0.528$ we obtain

$$L(1-0.528)/0.528 = -0.381I+1.168L+0.683Q$$

Hence $\qquad 0.894L = 0.683Q-0.381I+1.168L$

Therefore $\qquad 0.894E_Y = 0.683E_Q-0.381E_I+1.168E_Y$
which reduces to:

$$\frac{E_Q}{E_Y} = 0.557\frac{E_I}{E_Y} - 0.400$$

But the distances are given by:

$$l_{RM}l(0.621\frac{E_Q}{E_Y}+0.956\frac{E_I}{E_Y}+1.0).$$

Therefore, substituting for E_Q/E_Y, we have the distances given by:

$$l_{RM}l(0.621 \times 0.557 \frac{E_I}{E_Y} - 0.621 \times 0.400 + 0.956 \frac{E_I}{E_Y} + 1.0).$$

Inserting $l = 0.299$ this reduces to

$$(0.389 \frac{E_I}{E_Y} + 0.225)l_{RM}.$$

Thus E_I/E_Y is linear along RM with the zero 22.5 per cent away from M. Similarly, by substituting for E_I/E_Y, we have the distance for E_Q/E_Y given by:

$$(0.699 \frac{E_Q}{E_Y} + 0.504)l_{RM}.$$

Thus E_Q/E_Y is linear along RM with the zero 50.4 per cent away from M.

Once again lines parallel to RM may be chosen instead of RM for more convenient scaling, as shown in Fig. 22.13. The lines through S_C correspond to the zero in each case, and the line from C_Q through R for which $E_I/E_Y=1.99$ is used to fix the scale for E_I/E_Y, and the line from C_I through G for which $E_Q/E_Y=-0.89$ is used to fix the scale of E_Q/E_Y.

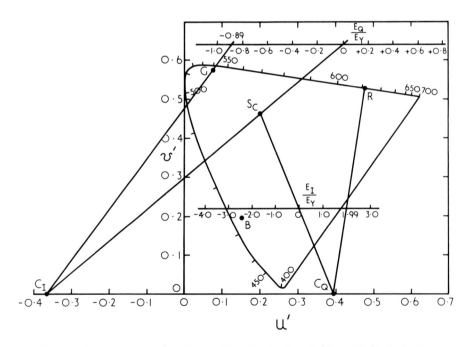

Fig. 22.13. Lines of constant v' used to provide scales of uniform E_I/E_Y and E_Q/E_Y for finding the values of lines of constant E_I/E_Y drawn from C_Q and lines of constant E_Q/E_Y drawn from C_I.

22.17 Some useful equations

In this section are collected together various equations that will be found useful when it is necessary to perform calculations in the N.T.S.C. system.

(a) Connecting luminance and R, G, B.

$$E_Y = 0.299E_R + 0.587E_G + 0.114E_B$$
$$L = 0.299R + 0.587G + 0.114B$$

(b) Connecting R, G, B and X, Y, Z.

$$1.0(R) \equiv 0.607(X) + 0.299(Y) + 0.000(Z)$$
$$1.0(G) \equiv 0.174(X) + 0.587(Y) + 0.066(Z)$$
$$1.0(B) \equiv 0.200(X) + 0.114(Y) + 1.111(Z)$$

$$X = 0.607R + 0.174G + 0.200B$$
$$Y = 0.299R + 0.587G + 0.114B$$
$$Z = 0.000R + 0.066G + 1.111B$$

$$1.0(X) \equiv 1.909(R) - 0.985(G) + 0.058(B)$$
$$1.0(Y) \equiv -0.532(R) + 1.997(G) - 0.119(B)$$
$$1.0(Z) \equiv -0.288(R) - 0.028(G) + 0.902(B)$$

$$R = 1.909X - 0.532Y - 0.288Z$$
$$G = -0.985X + 1.997Y - 0.028Z$$
$$B = 0.058X - 0.119Y + 0.902Z$$

(c) Connecting $R-L$, L, $B-L$ and X, Y, Z.

$$1.0(C_R) \equiv 0.518(X) + 0.000(Y) - 0.034(Z)$$
$$1.0(S_C) \equiv 0.981(X) + 1.000(Y) + 1.177(Z)$$
$$1.0(C_B) \equiv 0.166(X) + 0.000(Y) + 1.098(Z)$$

$$X = 0.518(R-L) + 0.981L + 0.166(B-L)$$
$$Y = \qquad\quad 1.000L$$
$$Z = -0.034(R-L) + 1.177L + 1.098(B-L)$$

$$1.0(X) \equiv 1.909(C_R) + 0.000(S_C) + 0.058(C_B)$$
$$1.0(Y) \equiv -1.532(C_R) + 1.000(S_C) - 1.119(C_B)$$
$$1.0(Z) \equiv -0.288(C_R) + 0.000(S_C) + 0.902(C_B)$$

$$R-L = 1.909X - 1.532Y - 0.288Z$$
$$L = \qquad\quad 1.000Y$$
$$B-L = 0.058X - 1.119Y + 0.902Z$$

(d) Connecting E_I, E_Y, E_Q and E_R, E_G, E_B.

$$E_I = 0.736(E_R - E_Y) - 0.268(E_B - E_Y)$$
$$E_Q = 0.478(E_R - E_Y) + 0.413(E_B - E_Y)$$

$$E_R - E_Y = 0.956 E_I + 0.621 E_Q$$
$$E_B - E_Y = -1.106 E_I + 1.703 E_Q$$

$$E_R = 0.956 E_I + 1.000 E_Y + 0.621 E_Q$$
$$E_G = -0.272 E_I + 1.000 E_Y - 0.647 E_Q$$
$$E_B = -1.106 E_I + 1.000 E_Y + 1.703 E_Q$$

$$1.0(C_I) \equiv 0.956(R) - 0.272(G) - 1.106(B)$$
$$1.0(S_C) \equiv 1.000(R) + 1.000(G) + 1.000(B)$$
$$1.0(C_Q) \equiv 0.621(R) - 0.647(G) + 1.703(B)$$

$$E_I = 0.596 E_R - 0.274 E_G - 0.322 E_B$$
$$E_Y = 0.299 E_R + 0.587 E_G + 0.114 E_B$$
$$E_Q = 0.211 E_R - 0.523 E_G + 0.312 E_B$$

$$1.0(R) = 0.596(C_I) + 0.299(S_C) + 0.211(C_Q)$$
$$1.0(G) = -0.274(C_I) + 0.587(S_C) - 0.523(C_Q)$$
$$1.0(B) = -0.322(C_I) + 0.114(S_C) + 0.312(C_Q)$$

(e) Connecting I, L, Q and X, Y, Z.

$$1.0(C_I) \equiv 0.312(X) + 0.000(Y) - 1.247(Z)$$
$$1.0(S_C) \equiv 0.981(X) + 1.000(Y) + 1.177(Z)$$
$$1.0(C_Q) \equiv 0.605(X) + 0.000(Y) + 1.849(Z)$$

$$X = 0.312 I + 0.981 L + 0.605 Q$$
$$Y = \qquad\qquad 1.000 L$$
$$Z = -1.247 I + 1.177 L + 1.849 Q$$

$$1.0(X) \equiv 1.389(C_I) + 0.000(S_C) + 0.936(C_Q)$$
$$1.0(Y) \equiv -0.826(C_I) + 1.000(S_C) - 1.193(C_Q)$$
$$1.0(Z) \equiv -0.454(C_I) + 0.000(S_C) + 0.235(C_Q)$$

$$I = 1.389 X - 0.826 Y - 0.454 Z$$
$$L = \qquad\qquad 1.000 Y$$
$$Q = 0.936 X - 1.193 Y + 0.235 Z$$

(f) Connecting X, Y, Z and U', V', W'.

$$1.0(X) \equiv \tfrac{4}{9}(U') \qquad\qquad -\tfrac{1}{3}(W')$$
$$1.0(Y) \equiv \qquad\quad 1.0(V') + \tfrac{2}{3}(W')$$
$$1.0(Z) \equiv \qquad\qquad\qquad \tfrac{1}{3}(W')$$

$$U' \equiv \tfrac{4}{9}X$$
$$V' = \qquad Y$$
$$W' = -\tfrac{1}{3}X + \tfrac{2}{3}Y + \tfrac{1}{3}Z$$

$$1.0(U') \equiv \tfrac{9}{4}(X) \qquad\qquad +\tfrac{9}{4}(Z)$$
$$1.0(V') \equiv \qquad 1.0(Y) - 2(Z)$$
$$1.0(W') \equiv \qquad\qquad\qquad 3(Z)$$

$$X = \tfrac{9}{4}U'$$
$$Y = \qquad V'$$
$$Z = \tfrac{9}{4}U' - 2V' + 3W'$$

(g) Connecting I, L, Q and U', V', W'.

$$1.0(C_R) \equiv 0.139(U') + 0.000(V') - 0.520(W')$$
$$1.0(S_C) \equiv 0.436(U') + 1.000(V') + 0.732(W')$$
$$1.0(C_B) \equiv 0.269(U') + 0.000(V') + 0.414(W')$$

$$U' = 0.139I + 0.436L + 0.269Q$$
$$V' = \qquad 1.000L$$
$$W' = -0.520I + 0.732L + 0.414Q$$

$$1.0(U') \equiv 2.096(C_I) + 0.000(S_C) + 2.635(C_Q)$$
$$1.0(V') \equiv 0.083(C_I) + 1.000(S_C) - 1.664(C_Q)$$
$$1.0(W') \equiv -1.362(C_I) + 0.000(S_C) + 0.704(C_Q)$$

$$I = 2.096U' + 0.083V' - 1.362W'$$
$$L = \qquad\qquad V'$$
$$Q = 2.635U' - 1.664V' + 0.704W'$$

TABLE 22.1

Values of camera signals for saturated colours at maximum luminance

	White	Yellow	Cyan	Green	Magenta	Red	Blue	Black
E_R	1	1	0	0	1	1	0	0
E_G	1	1	1	1	0	0	0	0
E_B	1	0	1	0	1	0	1	0
E_Y	1.000	0.886	0.701	0.587	0.413	0.299	0.114	0
$E_R - E_Y$	0	0.114	−0.701	−0.587	0.587	0.701	−0.114	0
$E_B - E_Y$	0	−0.886	0.299	−0.587	0.587	−0.299	0.886	0

TABLE 22.2

Values of E_I and E_Q signals for saturated colours at maximum luminance

	White	Yellow	Cyan	Green	Magenta	Red	Blue	Black
E_R	1	1	0	0	1	1	0	0
E_G	1	1	1	1	0	0	0	0
E_B	1	0	1	0	1	0	1	0
E_Y	1.000	0.886	0.701	0.587	0.413	0.299	0.114	0
E_I	0	0.321	−0.596	−0.275	0.275	0.596	−0.321	0
E_Q	0	−0.312	−0.212	−0.523	0.523	0.212	0.312	0

471

TABLE 22.3

Values of gamma corrected signals for large area saturated colours at full luminance

	White	Yellow	Cyan	Green	Magenta	Red	Blue	Black
E_R	1	1	0	0	1	1	0	0
E_G	1	1	1	1	0	0	0	0
E_B	1	0	1	0	1	0	1	0
$E_R^{1/\gamma}$	1	1	0	0	1	1	0	0
$E_G^{1/\gamma}$	1	1	1	1	0	0	0	0
$E_B^{1/\gamma}$	1	0	1	0	1	0	1	0
E_Y'	1.000	0.886	0.701	0.587	0.413	0.299	0.114	0
$E_R^{1/\gamma}-E_Y'$	0	0.114	-0.701	-0.587	0.587	0.701	-0.114	0
$E_B^{1/\gamma}-E_Y'$	0	-0.886	0.299	-0.587	0.587	-0.299	0.886	0
$V=(E_R^{1/\gamma}-E_Y')/1.14$	0	0.100	-0.615	-0.515	0.515	0.615	-0.100	0
$U=(E_B^{1/\gamma}-E_Y')/2.03$	0	-0.436	0.147	-0.289	0.289	-0.147	0.436	0
$E_Y'+\sqrt{V^2+U^2}$	1.000	1.333	1.333	1.178	1.004	0.931	0.561	0
$E_Y'-\sqrt{V^2+U^2}$	1.000	0.439	0.069	-0.004	-0.178	-0.333	-0.333	0
E_I'	0	0.321	-0.596	-0.275	0.275	0.596	-0.321	0
E_Q'	0	-0.312	-0.212	-0.523	0.523	0.212	0.312	0
$\sqrt{(E_I')^2+(E_Q')^2}$	0	0.447	0.632	0.591	0.591	0.632	0.447	0
$E_Y'+\sqrt{(E_I')^2+(E_Q')^2}$	1.000	1.333	1.333	1.178	1.004	0.931	0.561	0
$E_Y'-\sqrt{(E_I')^2+(E_Q')^2}$	1.000	0.439	0.069	-0.004	-0.178	-0.333	-0.333	0

TABLE 22.4

Frequencies used in various systems

Type of system	N.T.S.C.	P.A.L.	S.E.C.A.M.
Number of lines	525	625	625
Number of complete pictures per second	29.97	25	25
Approximate bandwidth of pictures MHz	4	$5\frac{1}{2}$	6
Frequency of chrominance sub-carrier above luminance carrier MHz	3.579545	4.43361875	4.40625 4.250
Chrominance carrier frequency as a multiple of line frequency	227.5	283.75	282 272
Modulation used for chrominance	Amplitude	Amplitude	Frequency
Frequency of sound carrier above luminance carrier, MHz	4.5	6.0	6.5
E_I' bandwidth (at 3dB below max.)	1.0	—	—
E_Q' bandwidth (at 3dB below max.)	0.3	—	—
E_R'-E_Y' bandwidth (at 3 dB below max.)	—	1.3	1.5
E_R'-E_Y' bandwidth (at 3dB below max.)	—	1.3	1.5
Approximate frequency of luminance carrier above band minimum	1.25	2	2
Approximate total bandwidth used MHz	6	8	8.5

TABLE 22.5

Systems of colour television adopted in different countries

Country	System adopted
Afghanistan	PAL
Albania	SECAM
Algeria (Algerian Democratic and Popular Republic)	PAL
Argentine Republic	PAL
Australia	PAL
Austria	PAL
Bahrein	PAL
Bangladesh (People's Republic of)	PAL
Belgium	PAL
Benin (People's Republic of)	—
Brazil (Federative Republic of)	PAL
Brunei	PAL
Bulgaria (People's Republic of)	SECAM
Burundi (Republic of)	—
Cameroon (United Republic of)	—
Canada	NTSC
Central African Republic	—
Chad (Republic of the)	—
Chile	NTSC

473

TABLE 22.5 (continued)

Country	System adopted
China (People's Republic of)	PAL
Colombia (Republic of)	—
Congo (People's Republic of the)	—
Costa Rica	NTSC
Cuba	—
Cyprus (Republic of)	—
Czechoslovak Socialist Republic	SECAM
Denmark	PAL
Dominican Republic	NTSC
Ecuador	NTSC
Egypt (Arab Republic of)	—
El Salvador	NTSC
Ethiopia	—
Finland	PAL
France	SECAM
French Guyana	NTSC
Gabon Republic	—
German Democratic Republic	SECAM
Germany (Federal Republic of)	PAL
Ghana	—
Greece	SECAM
Group of territories represented by the French Overseas Post and Telecommunications Agency	—
Guatemala	NTSC
Guinea (Revolutionary People's Republic of)	—
Haiti	SECAM
Hong Kong	PAL
Hungarian People's Republic	SECAM
Iceland	—
India (Republic of)	PAL
Indonesia (Republic of)	PAL
Iran (Islamic Republic of)	SECAM
Iraq	SECAM
Ireland	PAL
Israel (State of)	—
Italy	PAL
Ivory Coast (Republic of the)	SECAM
Jamaica	—
Japan	NTSC
Jordan (Hashemite Kingdom of)	—
Kenya (Republic of)	—
Korea (Republic of)	—
Kuwait (State of)	—
Lebanon	SECAM
Liberia (Republic of)	—
Libya (Socialist People's Libyan Arab Jamahiriya)	SECAM
Luxembourg	SECAM/PAL
Madagascar (Democratic Republic of)	—
Malawi	—

TABLE 22.5 (continued)

Country	System adopted
Malaysia	PAL
Mali (Republic of)	—
Malta	—
Mauritania (Islamic Republic of)	—
Mauritius	—
Mexico	—
Monaco	SECAM/PAL
Morocco (Kingdom of)	SECAM
Netherlands Antilles	—
Netherlands (Kingdom of the)	PAL
New Caledonia	NTSC
New Guinea	PAL
New Zealand	PAL
Nicaragua	NTSC
Niger (Republic of the)	—
Nigeria (Federal Republic of)	—
Norway	PAL
Overseas territories for the international relations of which the Government of the United Kingdom of Great Britain and Northern Ireland are responsible	—
Overseas territories of the United Kingdom in the European Broadcasting Area	—
Pakistan (Islamic Republic of)	—
Panama (Republic of)	—
Peru	—
Philippines	NTSC
Poland (People's Republic of)	SECAM
Portugal	PAL
Portuguese Oversea Provinces	—
Puerto Rico	NTSC
Qatar	PAL
Roumania (Socialist Republic of)	—
Rwanda (Republic of)	—
Saudi Arabia (Kingdom of)	SECAM
Senegal (Republic of the)	—
Sierra Leone	—
Singapore (Republic of)	—
Somali Democratic Republic	—
South Africa (Republic of)	—
South Korea	NTSC
South West Africa	PAL
Spain	PAL
Sri Lanka (Democratic Socialist Republic of)	—
Sudan	PAL
Suriname (Republic of)	—
Sweden	PAL
Switzerland (Confederation of)	PAL
Taiwan	NTSC
Tanzania (United Republic of)	—

TABLE 22.5 (continued)

Country	System adopted
Thailand	PAL
Togolese Republic	—
Tunisia	SECAM
Turkey	PAL
Uganda (Republic of)	PAL
Union of Soviet Socialist Republics	SECAM
United Arab Emirates	PAL
United Kingdom of Great Britain and Northern Ireland	PAL
United States of America	NTSC
Upper Volta (Republic of)	—
Uruguay (Oriental Republic of)	—
Venezuela (Republic of)	—
Yugoslavia (Socialist Federal Republic of)	PAL
Zaire (Republic of)	—
Zambia (Republic of)	—
Zimbabwe (Republic of)	—

REFERENCES

Hacking, K., *Acta Electronica*, **2**, 87 (1957).
Loughren, A. V., *J. Soc. Mot. Pic. Tel. Eng.*, **60**, 321 and 596 (1953).
Middleton, W. E. K., and Holmes, M. C., *J. Opt. Soc. Amer.*, **39**, 582 (1949).
Roizen, J., and Lipkin, R., *Electronics*, **38**, 97 (March 22nd, 1965).
Thomson, L. C., and Wright, W. D., *J. Physiol.*, **105**, 316 (1947).
Thomson, L. C., and Wright, W. D., *J. Opt. Soc. Amer.*, **43**, 890 (1953).
Townsend, G. B., *J. Roy. Television Soc.*, **17**, 63 (1978).
Willmer, E. N., *Nature*, **153**, 774 (1944).
Willmer, E. N., and Wright, W. D., *Nature*, **156**, 119 (1945).
Wright, W. D., *J. Opt. Soc. Amer.*, **42**, 509 (1952).

GENERAL REFERENCES

Carnt, P. S., and Townsend, G. B., *Colour Television: N.T.S.C., Principles and Practice*, Iliffe, London (1961), and *Colour Television, Vol. 2, P.A.L., S.E.C.A.M. and other systems*, Iliffe, London (1969).
Fink, D. G., *Color Television Standards*, McGraw Hill, New York (1955).
Kaufman, M., and Thomas, H., *Introduction to Color T.V.*, John F. Rider, New York (1954).
Proc. Inst. Radio Engnrs., **39**, 1124–1331 (1951).
Proc. Inst. Radio Engnrs., **41**, 838–858 (1953).
Proc. Inst. Radio Engnrs., **42**, 5–344 (1954).
Proc. Inst. Radio Engnrs., **43**, 742–748 (1955).
Theile, R., *J. Soc. Mot. Pic. Tel. Eng.*, **72**, 860 (1963).
Townsend, G. B., *P.A.L. Colour Television*, Cambridge University Press (1970).
Wentworth, J. W., *Color Television Engineering*, McGraw Hill, New York (1955).

CHAPTER 23

The Use of Colour Film in Colour Television

23.1 Introduction

A VERY large amount of programme material potentially useful for broadcast television exists in the form of photographic film. This has always been the case, from the earliest days of monochrome television, and equipment capable of producing television signals from film is therefore required in broadcasting facilities; this type of equipment is usually known as *telecine*, and is of three main types, one, using camera tubes, another, the flying-spot technique, and, the third, linear solid-state sensor arrays.

For new programmes that do not have to be broadcast live (and these are often the majority), the availability of telecine equipment gives the television producer a choice between using television cameras and recording on magnetic tape, or using film cameras and recording on photographic film. Magnetic tape has the advantage of providing a record that can be replayed immediately, whereas film involves a processing delay, but the use of film is widespread, for five reasons: first, film cameras require less capital investment, and are usually more convenient to use on location, than television cameras and video tape recorders; secondly, film can be edited more easily; thirdly, film is a cheaper and more permanent means of long-term storage; fourthly, film enables television signals to be readily obtained in the form required for any desired system (e.g. 525-line N.T.S.C., 625-line P.A.L., etc.), a feature termed *free standards conversion*, which is important when the same programme is used in different countries (standards conversion can also be carried out electronically from video tape recordings but only with very costly equipment); and fifthly, duplicate copies can be produced more cheaply from film (Hayer and Verbrugghe, 1972).

For these reasons, even if a programme is made in the first place using a television camera, there is sometimes a requirement to convert it to film: this process is usually known as *telerecording*.

It is thus clear that film has an important role to play in television, and its use is in fact very considerable.

23.2 Filming and televising techniques

When a programme is filmed for television, normal motion picture filming techniques can be used. These techniques have traditionally involved the use of a single camera on a set, the lighting and action being separately arranged for each *take*; this practice takes time, but enables results of very high quality to be achieved. On the other hand, television techniques traditionally involve the use of three cameras on a set, the lighting being a compromise for the three camera positions, and the action being selected from three monitors during the programme; this practice is much quicker but inevitably gives less perfect results. (See Plate 12, page 176.)

Some use has been made of television cameras, and their associated television techniques, even when the final requirement is a film: the programme can be recorded on to magnetic tape and subsequently transferred to film. Advantage can then be taken of the faster, and therefore cheaper, shooting costs of the television cameras, and trick effects peculiar to electronic signal-processing can be incorporated if desired. So far, this approach to film making has only had a limited use (Wayne, 1973), but it may become more attractive if television systems with higher definition are adopted, having, for instance, 1125 or more lines and bandwidths of as much as 20 MHz for luminance and 6.5 MHz for chrominance (see Section 19.15).

23.3 Combined film and television cameras

Another method of using television shooting techniques to produce programmes on film is to use cameras that both expose film and produce television signals at the same time. The television signals are then used for directing and editing the programme, but the film provides the main record. Cameras of this type that have been developed include those used in the *Electronicam* system (Caddigan and Goldsmith, 1956; Spooner, Pryke and Gardiner, 1968), the *Gemini* system, and the *Addavision* system.

In these systems, in order to avoid any parallax errors, the television and the film images are formed by the same objective lens, a beam splitter (which can consist of a rotating sector with reflecting blades) separating the beams of light required for the two different parts of the camera. For economic reasons, the film is only run while each television camera is actually contributing to the programme; but, because the film camera takes about five (photographic) frames to attain its true running speed, and to stop at the end of each take, the cutting from one camera to another occurs about a fifth of a second later in the film records than on the television monitor; however, this delay is usually quite acceptable. Editing of the films is carried out, either with the aid of special marks made to identify the correct sequence of the lengths of film corresponding to the programme, or a film recording from a television monitor is also made and used as a guide for assembling the final film.

In spite of the potential advantages of the above arrangements, however, most film used for television is shot in ordinary film cameras using conventional filming techniques.

23.4 Choice of Film

The film used for the actual broadcasting is preferably a print, and not the original camera film; this is because of the danger of damaging the original film, the undesirability of broadcasting a film with many splices, and the frequent need to supply several identical copies. The negative-positive film system is therefore very suitable, but the reversal-reversal or positive-positive system (see Section 12.11) is also used for some applications, such as newsreel work, where having a positive image on the camera film assists very rapid editing and also facilitates the use of the camera film itself for broadcasting when this is necessary. Negative film itself can also be used for broadcasting (see Section 23.14).

It might be thought that, with the rather limited definition of all broadcast television pictures, it would be unnecessary to use 35 mm motion-picture film for television, and that 16 mm film would provide sharp enough pictures. It must be remembered, however, that in any system some losses of definition occur at frequencies well below the limiting frequency, and that these losses are multiplicative: thus, if, for instance, a 16 mm system, consisting of a photographic lens, a negative film, and a print film, resulted in a 25 per cent loss of contrast of detail spaced at 200 line-pairs per picture width, and the television system also introduced a similar loss, the combined loss of televised 16 mm film would be 44 per cent ($\frac{3}{4} \times \frac{3}{4} = 0.56$). Thus, although the 25 per cent loss might be unnoticeable in either the film or the television system on its own, the effect in the combined system might be quite noticeable, and 35 mm films would then be necessary if the best results were desired.

23.5 Deriving television signals from colour film

Professional motion-picture film is normally shot at 24 pictures per second[1]; but, because a light interrupted at this frequency appears to flicker very noticeably,

---

[1] The rate used in amateur systems is 16 pictures per second, or 18 pictures per second for Super 8 films.

motion-picture projectors usually provide two (or sometimes three) dark periods per picture instead of one, so as to raise the frequency to 48 (or 72) per second when flickering is much less noticeable. One of the dark periods is used to move the film in the gate from one picture to the next, so that the sequence of events is as shown in Fig. 23.1 (a), the picture being projected in flashes of about 1/96 second duration with dark periods of about 1/96 second in between. The sequence of events in a television picture is, however, rather different, as is indicated in Fig. 23.1 (b). The television equivalent to the period when film is moved from one picture to the next (known as the *pull-down* time) is the time taken for the spot to move from the end of the bottom line of the picture to the beginning of the top line of the picture (known as the *fly-back* time). This fly-back time can be very short indeed and the interval between the end of scanning one frame and the beginning of scanning the next is usually less than 10 per cent of one cycle of the frame frequency, that is less than 1/500 second in 50 Hz systems or less than 1/600 second in 60 Hz systems. If, therefore, the light from an ordinary cine projector were shone in to a television camera, the light would only be falling on to the pick-up tubes for roughly half the time occupying each scanned frame. Special considerations therefore apply to televising film, and the telecine equipment used for this purpose has to be designed accordingly.

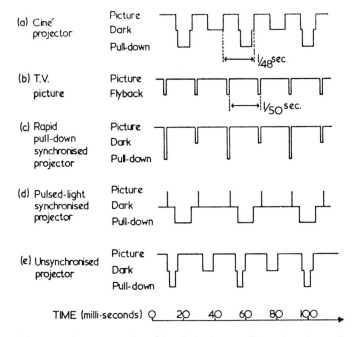

Fig. 23.1. Diagrammatic representation of the relation between film projected at 24 pictures per second and television displayed at 50 fields per second.

480

23.6 Telecines using fast pull-down

It has been found possible in 16 mm film equipment (but not in 35 mm equipment) to design special very fast pull-down mechanisms, that can move the film on in about 1/750 second (Wheeler, 1969) so that the pull-down time can be confined to the fly-back time; alternatively the film may be run continuously and optical devices, such as drums of mirrors or rotating prism blocks, used to present the required series of stationary images with very short intervals between them. If such projectors are run at 25 pictures per second instead of the usual 24, the situation for a 50 frame per second television system is as shown in Fig. 23.1(c). Satisfactory pictures are obtained if the projector pull-down is synchronized with the television fly-back. The slight change in picture-rate, from 24 to 25 per second, is not usually noticeable, but if the film is being shot specifically for television purposes the film cameras can be run at the higher speed.

23.7 Telecines using camera-tubes

In telecines consisting of combinations of cine-projectors and television cameras, simplifications are possible, however, because some television camera tubes, such as the vidicon, are able to store their images of electron charge very well between successive scans of the electron beam, and hence it is not necessary for the optical image to be present throughout the entire electron-beam scanning period. In fact it is possible to confine the optical image to a single flash during each fly-back period, as shown in Fig. 23.1(d), leaving plenty of time between the flashes for a normal pull-down mechanism to operate (the flashes, of about 1/1000 second duration, can be given either by means of a rotating sector or by pulsing the light source electronically). With this type of television pick-up tube (capable of good image storage) it is also possible to work without exact synchronization of the camera pull-down and television fly-back periods, but it is advisable in this case for the pick-up tube to be illuminated for at least 60 per cent of the total time (otherwise those parts of the picture covered by the electron beam during the time when the light was on would be noticeably different from the rest); this type of arrangement is shown in Fig. 23.1(e), a slow drift in phase between the film pull-down and the television fly-back being unimportant. The widespread use of vidicon and Plumbicon tubes in cameras has resulted in this last mode of operation (Fig. 23.1(e)) now being widely used, sometimes with four tubes (see Section 20.8).

23.8 Telecines giving 60 fields per second

When the television system operates at 60 fields per second, the situation is more complicated because the film cannot be speeded up from 24 to 30 pictures per second without obvious distortion of the portrayed motion. The film cannot be projected at 24 pictures per second on to a television camera with tubes capable of

481

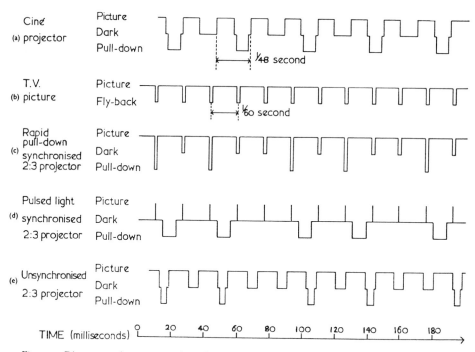

Fig. 23.2. Diagrammatic representation of the relation between film projected at 24 pictures per second and television displayed at 60 fields per second.

storing the electronic image, because the beating between the 24 pictures per second of the film and the 30 pictures per second of the television scan results in a pulsation of the television signal caused by the fact that the interval between successive television frames would alternate between including a whole dark interval, only part of one, and practically no dark interval at all (see Fig. 23.2(a) and (b)). It is therefore necessary to adopt one of the arrangements shown in Fig. 23.2(c), (d), and (e). In these arrangements each pair of pictures on the film provides five television frames: thus the first picture on the film provides two television frames, occupying $2/60 = 1/30$ second, but the second picture provides three television frames occupying $3/60 = 1/20$ second. The two pictures together therefore occupy $5/60 = 1/12$ second, as required to maintain an average of 24 pictures per second, but the television fields are produced at the rate of 60 per second. The unsynchronized system (Fig. 23.2(e)) is now widely used with vidicon or Plumbicon cameras.

23.9 Flying-spot scanners

An alternative method of deriving colour television signals from colour film is to

482

use a *flying-spot scanner* (as shown in Fig. 23.3): in this case the image is broken up into its lines before the light is passed through the film instead of afterwards. An unmodulated (white all over) television raster is displayed on a cathode-ray tube having a very short afterglow (but even the best phosphors available for this purpose require the use of an afterglow correction circuit); this raster of lines is imaged on to the film and the light is then split into red, green, and blue components and made to fall on three photomultiplier tubes. The photomultiplier tubes then generate three simultaneous signals that are functions of the red, green, and blue transmittances of the film at each point. The device has the great advantage that the registration of the three images depends only on the time responses of the three photomultiplier tubes and their associated circuitry which can be made very similar to each other. If the film remained stationary throughout the whole of the scan of each field, the device would have to be used with a rapid pull-down intermittent projector. With continuously moving film, optical devices can be used for immobilizing images of the raster on the film. (If the field frequency is 60 Hz the 3:2 type of scanning arrangements would have to be used.) If no interlacing were required the film could be moved at a uniform speed past an image of a single line produced on a cathode-ray tube operating with no vertical deflection; the movement of the film would provide the vertical scanning effect. The need for interlacing complicates the situation, however, and it is usual practice to form pairs of images of partially collapsed rasters of lines on the continuously moving film; this type of system is widely used for 50 fields per second operation. In the earlier telecines of this type, the two raster images were formed from a single raster display by means of two lenses; but, in more recent equipment, the required rasters have been produced electronically on the same cathode-ray tube (Godden, 1975; Journal of the Royal Television Society, 1976).

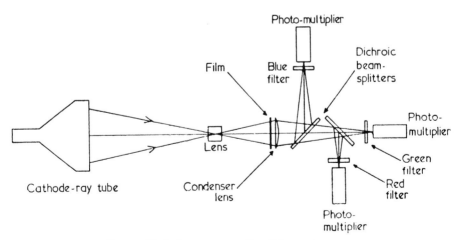

Fig. 23.3. Typical arrangements for a flying-spot scanner.

More complicated arrangements of rasters are required to achieve the same results for 3:2 type of operation (Whitehead, 1965).

Flying-spot scanners are also widely used for deriving colour television signals from colour transparencies: in this case there are no problems of pull-down or image immobilization, of course.

The four-tube camera principle (see Section 20.8) can be used in flying-spot, as well as in projector-type, telecine equipment: but its main advantage is in reducing registration problems in the latter (Taylor, 1965).

23.10 Telecines using solid-state sensors

The compactness, reliability, and simple electrical supply requirements of solid-state sensors make them attractive alternatives to camera-type pick-up tubes or photomultiplier tubes for telecine equipment. Whereas, for use in cameras, solid-state sensor arrays must have a minimum of about 600×575 elements for broadcast television, in telecines it is possible to use linear arrays consisting of only a single line of elements, and these are much easier to manufacture than area arrays, and are available with up to as many as about two thousand elements. The light from a narrow horizontal section of a film frame is focused by means of a suitable lens on to the linear array, which is then used to provide the line scan by clocking out the charges produced by the light, at the appropriate rate, as in other charge-coupled device (CCD) systems; the field scan is then provided by continuous motion of the film. For broadcast television, linear arrays with 1024 elements can provide very adequate definition. It is not possible to make all the sensor elements with sensitivities sufficiently equal to avoid spurious vertical stripes being visible in the picture, and these have to be eliminated by storing a correcting wave form. Interlacing can be provided by storing all the signals from a single film frame in a *frame store*, and then reading out the information for the two interlaced television fields as required. The signals are usually stored in digital form, and the correcting wave form is conveniently incorporated in the digital store. (Childs and Sanders, 1978 and 1980; Poetsch, 1978).

Colour signals are conveniently produced by splitting the light into separate red, green, and blue components by means of a dichroic prism block, such as is used in television cameras (see Section 20.5), and arranging for three dimension-ally identical CCD linear arrays to be in the focal planes of the three light beams; of course, the arrays must be in register with one another relative to the optical image, but their small size is a help in achieving this.

23.11 Telerecording

When it is required to record a programme from a television monitor display on to photographic film, the same problems of differences in frame frequencies and of pull-down are encountered as have just been discussed in connection with deriving television signals from film. It is therefore necessary with 60 frame per second television to use a 2:3 *exposing* sequence: that is, two television frames are used for 1/30 second, and then three television frames for 1/20 second; however, to avoid alternate pictures on the film having different densities because of different exposure times, it is necessary to make adjustments to the exposure levels. The pull-down problems can be solved in one of several ways. A rapid pull-down camera (either mechanical intermittent or continuous optical) can be used. Or, by using phosphors with long afterglows on the monitor, a camera with only a moderately rapid pull-down need be used, the television signals being boosted during the part of the scan during which pull-down is taking place so as to give uniform intensity all over the image recorded by the film. Alternatively, at the sacrifice of definition, one of the two (or three) television frames in each film picture can be obscured and the film moved on in this time: this technique is known as *suppressed-field* telerecording and gives results equivalent to rather better than half the full definition (because interlacing itself degrades definition to some extent, especially in the case of moving subjects). Another alternative, which is used in some devices, is to move the film continuously and to provide interlacing by a twin-raster arrangement similar to those used on telecine equipment. The rapid pull-down technique cannot be used with film of wider gauge than 16 mm, but since this gauge of film is often considered adequate for telerecording, the rapid pull-down technique is widely used.

In the interest of having adequate light for colour telerecording on film, a trinoscope (see Section 21.2) may be used, as in the *Videoprinting* system developed by Colour Video Services (Venis, 1969; Lisk and Evans, 1973). Systems have also been developed in which shadow-mask tubes, modified for high light output, have been used (Lisk and Evans, 1971; Lisk, 1979). Another way of increasing the amount of light available is to abandon the cathode-ray tube altogether and to use suitable lasers to produce red, green, and blue beams of light which are then deflected optically to provide the necessary scanning: the horizontal scanning can be provided by a motor-driven conical spinner having mirror facets on its surface, while the vertical scanning can be provided by a galvanometer mirror; modulation of the light can be effected by electro-optical devices such as ammonium dihydrogen phosphate (ADP) crystals or acousto-optical diffraction cells (Beiser, Lavender, McMann, and Walker, 1971; Swan, 1974).

When, as is usually the case, the television signals to be recorded on film are first recorded on magnetic tape (Anderson and Roizen, 1959), the film can be exposed sequentially to a monochrome display tube through red, green, and blue filters, by running the tape three times, and selecting either the red, or the green, or the blue signals, as required, one at a time. This is the method used in the *Vidtronics* system developed by Technicolor (Mulliner, 1969). In the *Image Transform* system, in order to obtain very high definition, three separation positives are recorded on black-and-white film by means of monochrome electron beam exposure in a vacuum; the positives are then printed in succession on to a colour intermediate negative film, which is then used for the production of positive release prints in the usual way (Comandini and Roth, 1978).

Still pictures can be recorded from conventional displays, such as shadow-mask tubes, using ordinary photographic techniques. The exposure time should be long enough to record at least two complete television fields ($\frac{1}{25}$ or $\frac{1}{30}$ second; but $\frac{1}{8}$ second or more is necessary with focal-plane shutters). With film of 64 A.S.A. speed, $\frac{1}{8}$ second at $f/2.8$ is typical of the required exposure using Daylight type film. Some filtration may be necessary to achieve optimum colour balance (see Table 10.3). Amateur movies can be made from television receivers using 160 ASA speed daylight film at $f/2.8$, but some flicker and banding effects usually occur.

23.12 Electronic adjustment of signals derived from colour film

A motion-picture film that, when projected in a dark auditorium, appears to be of perfectly satisfactory quality, may prove disappointing when transmitted on colour television by means of a telecine. This may be for a number of reasons, but it is possible to apply very useful adjustments to the pictures electronically.

For instance, if a motion-picture print has a slight overall colour cast, this can easily pass unnoticed in a dark auditorium, but, on a television set viewed in a room with considerable ambient light, the colour cast is often easily detected. However, by adjusting the relative amplitudes of the red, green, and blue signals produced by the telecine apparatus, a colour cast of this nature is easily corrected. Mismatch of the gammas of the red, green, and blue pictures can also occur, either because of imperfect control of the motion-picture processes, or because of the spectral sensitivities of the three colour channels of the telecine apparatus being such as to evaluate the gammas differently from the eye: this type of defect can also be corrected electronically, and in fact some control can even be exercised over variations of gamma mismatch at different density levels (Wood, Sanders, and Griffiths, 1965). Equipment for doing this is sometimes referred to as *Tarif* (The Apparatus for the Rectification of Inferior Film).

The insertion of adjustments automatically while a film is being broadcast can conveniently be achieved by previewing the film, adjusting the correction controls during each shot, noting the setting of the controls by recording the values they have when the *end* of each shot is reached, and then applying the same corrections to the *beginning* of each shot when the film is broadcast. This procedure can be effected by running with the film a *cue tape*, that causes the changes in corrections to take place at the correct times, but an electronic *shot-change detector* has also been developed whereby the film itself can cue the changes by detection of the much greater changes in signal that normally occur at scene changes than those that occur during scenes (Kitson, Palmer, Spencer, Sanders, and Weston, 1972). Other attempts to make automatic corrections to the signals derived from film in telecines have included automatic white level and black level devices, and arrangements whereby a tendency for the picture approximately to integrate to grey is achieved (Pay, 1970; Pullinger and Reeves, 1970; Marsden, 1978).

Methods of improving the consistency of film quality for television have included the use of a reference picture printed from a standard negative at the beginning of each film (Knight, 1970; Brown, 1970).

A method of improving blacks analogous to the black printer used in graphic arts (see Plate 32, pages 512, 513) has also been devised, and is particularly useful when working with film exposed at very low light levels (Godden, 1977).

23.13 Electronic masking

Even when the colour balance, gamma match, overall gamma, and curve shape of a colour motion-picture film are all ideally suited to a colour television system, however, dissatisfaction with the final result may still be felt. This can be because both the film system and the television system introduce errors of colour reproduction, particularly losses of colour saturation, and while the errors introduced by either system on its own may be perfectly tolerable, the multiplicative effect of combining them can easily become intolerable. The problem becomes particularly acute if it is desired to include film and live sequences of the same subject matter in the same programme (Corbett, 1969). Here, again, electronic correction of the signals can provide a very useful improvement in the results, by means of a technique known as *electronic masking* (Burr, 1954; Brewer, Ladd, and Pinney, 1954; Wood and Griffiths, 1966).

If it is required that the colour and tone reproduction of the recorded and live parts of the programme should be as nearly alike as possible, it follows that, in the sequences of operations shown in Fig. 23.4, the input to the transmission coding stages should be equal.

If the television camera and the film have different spectral sensitivities, exact equality of the final results is impossible, and a rigorous calculation of the optimum electronic masking requires the arbitrary selection and weighting of test

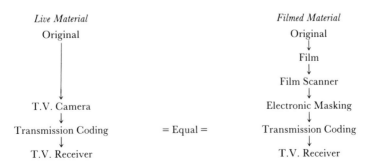
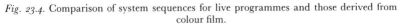

Fig. 23.4. Comparison of system sequences for live programmes and those derived from colour film.

colours, and an evaluation of their reproduction errors that is relevant to overall picture quality. The selection, weighting, and evaluation of test colours can only be done very approximately, and approximate results can in fact be obtained by simpler means. It has, however, been shown (Evans, Hanson and Brewer, 1953) that the effects of altering the spectral sensitivities in colour reproduction systems are often of a fairly minor nature, and it is therefore to be expected that the optimum electronic masking can be calculated to a good approximation by ignoring the differences in spectral sensitivity between the television camera and the film. If this is done, simple calculations lead to results which are otherwise rigorous for colours that are filmed at density levels where the characteristic curves and inter-image effects are linear.

In order to evaluate the degree of the electronic masking required, it is necessary to know the relation between the optical input to the film camera and the electronic output of the telecine apparatus. This can conveniently be determined by varying the exposure of each layer of the film in turn by known amounts above and below a point representing an average medium grey. The rates of change of the logarithms[1] of the red, green, and blue signals from the telecine apparatus, with respect to the logarithms of the film exposures, can then be determined, either directly, or by measuring the film on a densitometer filtered so that its spectral sensitivities match those of the three channels of the telecine apparatus. As a result, nine rates of change, or gammas, are obtained as follows:

	Film layer varied (input)		
Telecine channel (output)	*Red*	*Green*	*Blue*
Red	γ_{rr}	γ_{gr}	γ_{br}
Green	γ_{rg}	γ_{gg}	γ_{bg}
Blue	γ_{rb}	γ_{gb}	γ_{bb}

If this block of gammas is represented by the matrix, M, and the logarithms of the exposures received by the three layers of the film by o_r, o_g, o_b, and the logarithms of the telecine output signals by p_r, p_g, p_b, then

$$\begin{pmatrix} p_r \\ p_g \\ p_b \end{pmatrix} = M \begin{pmatrix} o_r \\ o_g \\ o_b \end{pmatrix}$$

By inverting M (see Appendix 1), we obtain

$$\begin{pmatrix} o_r \\ o_g \\ o_b \end{pmatrix} = M^{-1} \begin{pmatrix} p_r \\ p_g \\ p_b \end{pmatrix}$$

[1] Log signals are used because the major errors to be corrected are usually associated with unwanted dye *densities*, which are approximately proportional to wanted dye *densities*.

488

If, therefore, masking equivalent to the matrix M^{-1} is applied to the signals so that masked signals p_{rm}, p_{gm}, p_{bm} are obtained thus:

$$\begin{pmatrix} p_{rm} \\ p_{gm} \\ p_{bm} \end{pmatrix} = M^{-1} \begin{pmatrix} p_r \\ p_g \\ p_b \end{pmatrix}$$

then the input to the television transmission stage will be the same as that from the original scene viewed by the television camera, apart from the effects of differences in spectral sensitivities between the television camera and the camera film, and any non-linear portions of the characteristic curves used in the photographic steps.

Masking equivalent to the matrix M^{-1} can be set up by practical tests, instead of by calculation, if test film is exposed, as shown in Fig. 23.5(a), to give eight vertical stripes (two grey, and six coloured) in which the exposure of each

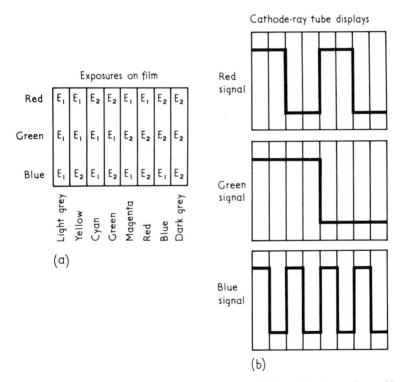

Fig. 23.5. (a) Exposure pattern on test-film to be used for setting electronic masking parameters. (b) Cathode-ray tube displays from the test-film when the masking is correctly adjusted.

489

layer is at one of two values, E_1 and E_2, representing suitable increments above and below a medium grey. The values of the masks are then set so that the magnitudes of the red, green, and blue signals, which can be displayed on cathode-ray tubes, as shown in Fig. 23.5(b), are all at the appropriate one of the two levels produced by the two grey stripes; the nine gammas of the electronic masks are then all proportional to the corresponding values in the matrix M^{-1}. By setting the overall gamma of the system correctly (an adjustment that will alter the two levels on the cathode-ray tubes) the masks can be made equal to the values of M^{-1}. Thus the correct masking can be set up without having to evaluate M^{-1} or calibrate the electronic masking controls (Hunt, 1978). In practice, various non-linearities in the system may make it impossible to get all the levels correct simultaneously, and a compromise may be necessary. Some adjustments may also be desirable to allow for the fact that, in film, reds are normally lightened by enhanced sensitivity in the far red part of the spectrum (see Fig. 4.3), and for this, and other reasons, a reduced level of mask is often used in practice (Staes, 1977).

Electronic masking can also be used to improve the results obtained when broadcasting from a telerecording on colour film. Ideally the picture finally displayed on the receiver should look the same as if the signals had been transmitted directly without going through the camera-film, telecine chain. It therefore follows that in the sequences of operations shown in Fig. 23.6 the input at the transmission coding stages should be the same. Here, the original scene is photographed by a television camera in both cases and therefore the problem is confined to recovering the red, green, and blue signals from the film record; in this case, exact equality of the input to the transmission coding stage is theoretically possible.

Original Programme Display		*Display Derived from Film Recording*
Original		Original
↓		↓
T.V. Camera		T.V. Camera
│		↓
│		Monitor Display
│		↓
│		Film
│		↓
│		Film Scanner
│		↓
│		Electronic Masking
↓		↓
Transmission Coding	= Equal =	Transmission Coding
↓		↓
T.V. Receiver		T.V. Receiver

Fig. 23.6. Comparison of system sequences for live programmes and those derived from telerecording on film.

If the monitor display, and the spectral sensitivities of the layers of the film, are such that there is no *cross-talk* on to the film (in other words if the displayed record corresponding to the red camera signal is recorded only on the red layer of the film, and similarly for the green and blue channels) then the required electronic masking characteristics can be evaluated exactly as described above, and will be inaccurate only in so far as any non-linear portions of the characteristic curves (including inter-image effects) are used in the photographic steps. If appreciable cross-talk does take place, then this can be corrected by using an electronic matrixing circuit (see Section 19.12) on the linear signals before electronic masking is carried out on the log signals. If this is not possible then the effects of the cross-talk can be allowed for approximately by basing the electronic masks on exposures on the camera film made by light having the same spectral power distributions as those of the red, green, and blue displays of the monitor instead of through red, green, and blue filters chosen to isolate the three layers of the film. The effects of cross-talk can only be counteracted approximately by electronic masks, because cross-talk causes additional exposures (log exposure shifts) instead of altering gammas; it is the latter that electronic masking can correct for exactly, but it can be used to provide approximate correction for the former also.

23.14 Overall transfer characteristics

The way in which the magnitude of the recorded signal is related to the magnitude of the optical image in a system is often referred to as the *transfer characteristic*; it is convenient to plot both these magnitudes on logarithmic scales, so that the slope of the transfer characteristic is the gamma of the system. The transfer characteristic of a typical television system is made up of a combination of the individual transfer characteristics of the taking lens, the camera tube, the transmission coding, the display device, and the effect of the ambient lighting on the viewed picture. When film or video tape recordings are included then the transfer characteristics of these media and their associated recording and read-out equipments also contribute.

Because of the effects of flare light in lenses and equipment, and the addition of ambient lighting to the viewed picture, even if the transfer characteristics of all the rest of the systems were linear (on a log-log plot) the overall transfer characteristic would have curvature (Wentworth, 1955). The concept of the slope, or gamma, of the overall transfer characteristic is nevertheless helpful as a broad description of a system.

When transfer characteristics having curvature (on log-log plots) are being discussed, it has been found useful to consider as their gamma their average gradients over the central 1.3 range of log luminances (a ratio of 20:1) involved (DeMarsh, 1972); this will be taken as the basis of the following discussion.

The signals from colour television cameras are nominally gamma-corrected to the power of 1/2.2: this gamma, when combined, with a receiver gamma of 2.8, gives a system of gamma about 1.25, as required for dim-surrounds (see Section 19.13). If pictures derived from film are to look the same as those derived from television cameras operating at this gamma, the signals obtained from the film must also possess a gamma of $1/2.2$ (which is equal to 0.45). Most film is made for projection in darkened rooms, and for this purpose it requires an effective gamma on the screen of 1.5 (see Section 6.5); if the effective gamma of the film in the telecine is also 1.5, then gamma-correction amounting to $(1/1.5) \times 0.45$, which is equal to 0.3, must be introduced in the telecine. If, however, the signals from the television cameras are gamma-corrected, not to 0.45, but to 0.54, so as to give a system gamma of 1.5 (with receivers of 2.8 gamma), as may be necessary to achieve an effective displayed gamma of 1.25 in the presence of ambient viewing light (see Section 19.13), then the equivalent figure for the telecine gamma becomes $(1/1.5) \times 0.54$, which is equal to 0.36: in this case, because the film gamma has the same value as the required system gamma, the telecine gamma is equal to the reciprocal of the gamma of the monitor $(1/2.8 = 0.36)$; in a study in which a monitor having a gamma of 2.5 was used, the optimum telecine gamma was found to be $1/2.5$ or 0.40 (DeMarsh, 1972; Hunt, 1969).

Experience has shown that, with receivers of gamma 2.8, and film of effective gamma 1.5, the optimum telecine gamma is usually about 0.33; but telecines with gammas of about 0.4 are also used, often with their black level adjusted to give a somewhat lower effective gamma.

If colour negative film is used, its gamma is usually about 0.65, and the gamma-correction required in the scanner is then equal to $(1/0.65) \times 0.45$ which is equal to 0.7 (or $(1/0.65) \times 0.54$ which is equal to 0.8) assuming that all the picture is on the straight-line portion of the characteristic curve.

The advantages of using negative film directly in telecines are as follows: the combined negative-film television system can handle the tone scale from highlights to shadows with less distortion, thus resulting in less white and black crushing; the better tone reproduction can also result in better colour reproduction; and the steadiness of the picture, and its resolution, are improved if the use of a negative film eliminates a photographic printing stage. The disadvantages of direct use of the negative film are as follows (Wood, Palmer, and Griffiths, 1972): dirt shows up as white specks, which are more objectionable than the black specks from dirt on positive film; and, if the original camera film is used: splices may be more visible, and may also be more numerous thus bringing greater risk of film breakage; the risk of damage to the original camera film may be increased; and it can only be used when the film producer is also the organization that is broadcasting the film.

If the telecine is of the flying-spot or CCD type (see Sections 23.9 and 23.10), then all the gamma-correction (0.33, 0.4, 0.7 or 0.8, as the case may be) must be provided by suitable circuits; but if a television camera type of telecine is being used (see Section 23.7) some of the gamma correction is sometimes provided by the camera pick-up tubes: thus the vidicon has a gamma of about 0.6 and hence, with this tube, circuits providing gamma correction of about 0.55, 0.7, 1.2, and

1.3, would be required to provide overall values of 0.33, 0.4, 0.7, and 0.8, respectively.

Reduction of gamma because of flare light can cause pictures derived from film to appear 'misty' and lacking in good blacks, particularly in small areas: methods of correcting for these effects have therefore been worked out, such as the addition of a low definition negative picture-signal to the positive picture-signal by electronic mixing (Palmer, 1969).

In any critical judgement of picture gamma, it is important to remember that the setting of the black level ('brightness control') of the monitor or receiver can have a profound effect on the results. The standard setting is normally regarded as placing the display-tube cut-offs at the set-up level of the incoming signal. This setting can be facilitated using special equipment providing signals near the set-up level (*Picture Line-Up Generating Equipment*, or *PLUGE*; Quinn and Siocos, 1967). The amount of ambient light falling on the face of the display tube profoundly affects the displayed gamma, and the luminance of the surround affects the perceived contrast (see Section 6.5). These last two effects are usually approximately self-compensating, since a surround of high luminance is usually accompanied by a high level of ambient light, which reduces the displayed gamma as required from about 1.5 for dark-surround viewing to about 1.25 for the dim-surround viewing typical for television, or even to about 1.0 for a surround of luminance similar to that of the average luminance of the picture (Novick, 1969). But, for assessment-work care should be taken to avoid conditions of viewing that are appreciably different from those of reasonably good home-viewing situations.

23.15 Reviewing colour films for television

As already mentioned (in Section 23.12) films that appear perfectly acceptable when projected in a darkened room may appear unacceptable when displayed on colour television, because of incorrect or variable colour balance. This is because, in typical television viewing situations, the degree to which the observer's colour vision adapts to the colour of the picture is less than in typical projection situations. When reviewing film intended for television display, it has therefore been agreed that it should be projected with an illuminated surround in order to make the observer more critical of the colour balance. The exact projection conditions agreed involve a colour temperature of 5400 K for both the projector and the surround; a luminance of approximately 140 cd/m² for the open-gate screen, giving a peak white of about 70 cd/m²; and a luminance of about one-third of the peak white value for the surround, whose area should be at least nine times that of the picture (Harrop, 1970; Knight, 1972).

It has been found that, for colour films to reproduce well on colour television, they should preferably conform to certain standards (British Standard 4563: 1970). Thus, no areas where detail is to be reproduced should have densities less than 0.3, and any areas intended to be reproduced as fully-lit whites should have

493

densities not exceeding 0.4. Shadow areas in which the reproduction of detail is important should not have densities exceeding 2.0; but blacks of large area should have densities of about 2.5 or higher (Zwick, 1971). The colour balance should be such that neutral greys, when illuminated by a source having a relative spectral power distribution the same as that of a full radiator of colour temperature 5400 K ± 400 K, should be metameric matches to non-selective greys illuminated with the same source, for the C.I.E. 2° Standard Observer. It is found that films that conform to these standards also exhibit good quality when projected, but the converse is not true: for instance, pictures that look good when projected with tungsten light may be too blue for television display (Zwick and Brothers, 1970).

REFERENCES

Anderson, C. E., and Roizen, J., *J. Soc. Mot. Pic. Tel. Eng.*, **68**, 667 (1959).
Beiser, L., Lavender, W., McMann, R. H., and Walker, R., *J. Soc. Mot. Pic. Tel. Eng.*, **80**, 699 (1971).
Brewer, W. L., Ladd, J. H., and Pinney, J. E., *Proc. Inst. Radio Engnrs.*, **42**, 174 (1954).
British Standard 4563: 1970, The Density Range, Contrast and Colour Balance of Films and Slides for Colour Television (1970).
Brown, I., *Brit. Kinematog. Sound Tel.*, **52**, 253 (1970).
Burr, R. P., *Proc. I.R.E.*, **42**, 192 (1954).
Caddigan, J. L., and Goldsmith, T. T., *J. Soc. Mot. Pic. Tel. Eng.*, **65**, 7 (1956).
Childs, I., and Sanders, J. R., *Brit. Kinematog. Sound Tel.*, **60**, 318 (1978).
Childs, I., and Sanders, J. R., *J. Soc. Mot. Pic. Tel. Eng.*, **89**, 100 (1980).
Corbett, J., *Brit. J. Phot.*, **116**, 122 (1969).
Comandini, P., and Roth, T., *J. Soc. Mot. Pic. Tel. Eng.*, **87**, 82 (1978).
De Marsh, L. E., *J. Soc. Mot. Pic. Tel. Eng.*, **81**, 784 (1972).
Evans, R. M., Hanson, W. T., and Brewer, W. L., *Principles of Color Photography*, p. 504, Wiley, New York (1953).
Godden, W. R., *Brit. Kinematog. Sound Tel.*, **57**, 442 (1975).
Godden, W. R., Proceedings of the Montreux Television Conference, paper 5h (1977).
Harrop, L. C., *J. Soc. Mot. Pic. Tel. Eng.*, **79**, 806 (1970).
Hayer, L., and Verbrugghe, R., *Brit. Kinematog. Sound Tel.*, **54**, 324 (1972).
Hunt, R. W. G., *Brit. Kinematog. Sound Tel.*, **51**, 268 (1969).
Hunt, R. W. G., *J. Soc. Mot. Pic. Tel. Eng.*, **87**, 78 (1978).
Journal of the Royal Television Society, **16**, 24 (1976).
Kitson, D. J. M., Palmer, A. B., Spencer, R. H., Sanders, J. R., and Weston, M., *I.E.E. Conference Proceedings No. 88*, p. 237 (1972).
Knight, R. E., *Brit. Kinematog. Sound Tel.*, **52**, 248 (1970).
Knight, R. E., *Brit. Kinematog. Sound Tel.*, **54**, 234 (1972).
Lisk, K. G., and Evans, C. H., *J. Soc. Mot. Pic. Tel. Eng.*, **80**, 801 (1971).
Lisk, K. G., *J. Soc. Mot. Pic. Tel. Eng.*, **88**, 157 (1979).
Lisk, K. G., and Evans, C. H., *J. Soc. Mot. Pic. Tel. Eng.*, **82**, 719 (1973).
Marsden, R. P., *J. Soc. Mot. Pic. Tel. Eng.*, **87**, 73 (1978).
Mulliner, J., *Brit. Kinematog. Sound Tel.*, **51**, 393 (1969).
Novick, S. B., *Brit. Kinematog. Sound Tel.*, **51**, 342 (1969).
Palmer, A. B., *Brit. Kinematog. Sound Tel.*, **51**, 428 (1969).
Pay, D. A., *I.E.E. Conference Proceedings No. 69*, p. 141 (1970).
Poetsch, D., *J. Soc. Mot. Pic. Tel. Eng.*, **87**, 815 (1978).
Pullinger, A. E. H., and Reeves, R. F. J., *I.E.E. Conference Publication No. 69*, p. 272 (1970).
Quinn, S. F., and Siocos, C. A., *J. Soc. Mot. Pic. Tel. Eng.*, **76**, 925 (1967).
Spooner, A. M., Pryke, G. A., and Gardiner, A. V., *J. Roy. Television Soc.*, **12**, 17 (1968).

Staes K., *Brit. Kinematog. Sound Tel.*, **59,** 354 (1977).
Swan, D., *J. Roy. Television Soc.*, **15,** 206 (1974).
Taylor, D. M., *J. Soc. Mot. Pic. Tel. Eng.*, **74,** 930 (1965).
Venis, R. J., *Brit. Kinematog. Sound Tel.*, **51,** 379 (1969).
Wayne, P., *J. Roy. Television Soc.*, **14,** 178 (1973).
Wentworth, J. W., *Color Television Engineering*, p. 156, McGraw Hill, New York (1955).
Wheeler, L. J., *Principles of Cinematography*, 4th Edn., p. 67, Fountain Press, London (1969).
Whitehead, R. C., *Principles of Television Engineering*, Vol. 2, p. 15, Iliffe, London (1965).
Wood, C. B. B., and Griffiths, F. A., *Brit. Kinematog. Sound Tel.*, **48,** 74 (1966).
Wood, C. B. B., Palmer, A. B., and Griffiths, F. A., *J. Soc. Mot. Pic. Tel. Eng.*, **81,** 661 (1972).
Wood, C. B. B., Sanders, J. R., and Griffiths, F. A., *J. Soc. Mot. Pic. Tel. Eng.*, **74,** 755 (1965).
Zwick, D. M., *J. Soc. Mot. Pic. Tel. Eng.*, **80,** 88 (1971).
Zwick, D. M., and Brothers, D. L., *J. Soc. Mot. Pic. Tel. Eng.*, **79,** 31 (1970).

GENERAL REFERENCES

Griffiths, L. H., and Sanders, J. R., *Brit. Kinematog. Sound Tel.*, **56,** 262 (1974).
Gunderson, K. E., *Brit. Kinematog. Sound Tel.*, **53,** 336 (1971).
Palmer, A. B., *J. Soc. Mot. Pic. Tel. Eng.*, **74,** 1069 (1965).
Ross, R. J., *Colour Film for Colour Television*, Focal Press, London (1970).

CHAPTER 24

Video Cassettes and Discs

1. Introduction – *2*. Magnetic tape – *3*. Magnetic tape with helical scanning –
4. E.V.R. (Electronic Video Recording) – *5*. Holographic recording –
6. Recording on discs – *7*. The Teldec system – *8*. Capacitance discs – *9*. Discs
using lasers – *10*. The duplication of programmes on video cassettes and discs

24.1 Introduction

FOR the reproduction of sound, the facilities required in the home for listening
to either live radio broadcasts or prerecorded material (the radio receiver and
the gramophone or phonograph) were developed before simple equipment for
recording. For the domestic reproduction of pictures (with motion), the history
has been somewhat different, in that facilities for recording personal material
were available first, in the form of movie cameras, film, and projectors, with
broadcast television coming later; and the facilities for recording and re-
displaying live television broadcasts, and for displaying prerecorded material on
television, have been more recent developments still. The technology in this area
has developed along two main lines: to be able not only to display prerecorded
material but also to record and replay live broadcasts, the video tape recorder
(VTR), using magnetic tape, has been provided, and with electronic cameras
personal material can also be recorded; it is also used by broadcasting organiza-
tions for recording programmes prior to transmission. To replay prerecorded
material only, in the home, several forms of disc technology have been developed,
which enable multiple copies of a programme to be produced more cheaply than
is possible with magnetic tape.

The basic features of these different forms of *video cassettes* and *video discs* will
now be described, together with some examples of actual systems.

24.2 Magnetic tape

The principles used in recording on magnetic tape are illustrated in Fig. 24.1.
Electric currents corresponding to the information to be recorded are passed

through coils, through which a pair of roughly semicircular metal cores are mounted. These cores are made of easily magnetizable (low coercivity, high permeability) material, generally ferrites or laminated strips of iron alloys. The electric currents cause magnetic fields to be set up in the cores, and, by mounting the cores so that they form a roughly circular path with two small air gaps, the magnetic field spreads into the air in the vicinity of the gaps. The magnetic tape on which the recording is to be made is passed over one of the air gaps, and the magnetic field therefore passes into it.

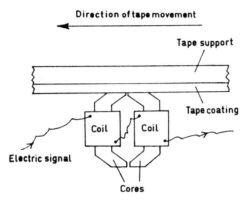

Fig. 24.1. Diagrammatic representation of the principle of magnetic tape recording.

The tape usually consists of a mylar (terylene) support, typically about 0.001 in (0.025 mm) thick, on which has been coated a dispersion of small needle-shaped iron oxide (Fe_2O_3), or chromium dioxide (CrO_2), particles in a suitable binder, the coating being about 0.0002 to 0.0008 in (0.005 to 0.02 mm) thick. The particles can be magnetized, and retain most of their magnetization after the magnetizing field has been removed. Thus, when the tape is passed over the air gap between the cores, a record is obtained of the magnetic field in its vicinity, and hence, as the currents in the coils are varied with time, the magnetization on the tape varies with distance. To reproduce the information, the tape is passed over a similar (or the same) air gap, and now the magnetization in the tape causes changes in the magnetic field across the air gap, and this generates electric currents in the coils; these currents can then be amplified and used to display the information in the required form. The coil and core assembly, with its air gap, is usually referred to as the *head*.

For sound recording, the tape is usually $\frac{1}{4}$ in (6.35 mm) wide, and is used to record two tracks of about 0.09 in (2.3 mm) width with a guard band of about 0.05 in (1.3 mm) between them; alternatively four narrower tracks can be recorded so that two stereo recordings can be made on the same tape (and eight tracks are also sometimes used). The highest frequency that can be recorded is roughly equal to half a cycle per air gap. Typical air gaps are 0.0001 in (0.0025 mm) although special heads may have values less than half this; thus maximum recordable frequencies are about $37\frac{1}{2}$ kHz at a tape speed of $7\frac{1}{2}$ in (19 cm)/sec., 19 kHz at $3\frac{3}{4}$ in (9.5 cm)/sec. and 9 kHz at $1\frac{7}{8}$ in (4.75 cm)/sec. However, progressive loss of signal generally occurs as frequencies approach the maximum, so that for high quality sound recording, in which frequencies up to about 15 kHz should be properly included, a tape speed of $7\frac{1}{2}$ in (19 cm)/sec. is desirable; but reasonably good quality can be recorded at $1\frac{7}{8}$ in (4.74 cm)/sec. The sound has to be recorded, not on its own, but together with a high frequency *bias*; this improves sensitivity, and the maximum attainable undistorted output level is increased. Erasure of recordings can be made by applying a strong high-frequency (ultrasonic) signal to a separate head having a wider gap.

For recording television pictures, maximum frequencies of about 5 MHz must be accommodated instead of about 15 kHz, and hence the head-to-tape speeds necessary are about 300 times higher than those for sound recording. These high speeds (about 1200 in/sec. or 3000 cm/sec.) make simple linear systems, such as those for sound, rather impracticable: the length of tape necessary to record whole programmes becomes unwieldy and costly; and it is difficult to avoid a cushion of air being sucked in between the tape and the capstan roller, thus upsetting the accurate control of the tape speed. For these reasons, video tape recordings for high quality broadcast use are generally made by means of multiple-head transverse systems which produce recorded tracks of the type illustrated in Fig. 24.2. In these video tape recorders (VTRs) a tape of 2 in (51 mm) width is generally used, moved at 15 in (38 cm)/sec. (or sometimes $7\frac{1}{2}$ in (19 cm)/sec.). The tape is forced into a curved shape of radius about 1 in (25 mm) across its width, by means of a suitable guide with vacuum slots. As the tape moves longitudinally, it is scanned transversely by four recording (or replay) heads rotating at 240 revolutions per second for a 60 Hz system, or 250 r.p.s. for a 50 Hz system. Thus, although the tape is only moving at 15 in (38 cm)/sec.

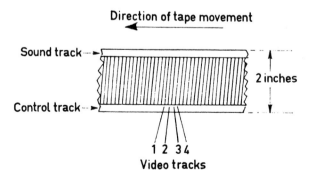

Fig. 24.2. Arrangement of tracks on a 4-head transverse video tape recording of the type used for broadcast television.

longitudinally, the head-to-tape speed can be much higher, in fact usually about 1500 in (3800 cm)/sec. This technique results in each television field being recorded in sixteen adjacent transverse tracks for a 60 Hz system, or twenty for a 50 Hz system, and it is therefore essential that the information in each track be read out with very little error of timing. The signals are often recorded using frequency modulation of an 8 MHz carrier, with a peak-to-peak deviation from about 7 to about 9 MHz, with side bands producing frequencies over a range from about $1\frac{1}{2}$ MHz to about $14\frac{1}{2}$ MHz (Tooms, 1970); the maximum frequency of $14\frac{1}{2}$ MHz corresponds to a spatial wavelength on the tape of 0.0001 in (0.0025 mm).

The technique just described was developed by the Ampex Corporation of the U.S.A. and made available for recording black-and-white programmes in 1956 (Ginsburg, 1957) and colour in 1959 (Anderson and Roizen, 1959). It is now very widely used for broadcast television in systems using 50 or 60 fields/sec. (Machein, 1959), and with N.T.S.C., P.A.L., or S.E.C.A.M. colour coding. A programme recorded on tape in one system can only be used in another system if a *standards converter* is available (Baldwin, Stalley, and Kitchin, 1972).

24.3 Magnetic tape with helical scanning

To record and replay video tapes using four heads moving across the tape, as described in the previous section, requires very accurate control of the movement of the tape and of the heads, and very small differences between the performances of the four heads. These requirements would become less stringent if each transverse track could accommodate a whole television field, instead of only a section of a field. To attempt to do this with a tape of much greater width would be very unwieldy, and the use of only one head would require the tape to be bent into a nearly cylindrical shape so that the single head could pass from the end of one track to the beginning of the next during the flyback time between successive fields. An elegant solution to these difficulties has been found in systems using helical scanning; in one such system the tape is wound round a drum to form a single turn of a helix, as shown in Fig. 24.3(a), the pitch of the helix being such

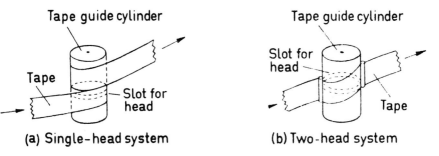

Fig. 24.3. Arrangements of magnetic tape recording using helical scanning: (a) with one head, (b) with two heads.

that the tape is displaced sideways by a distance approximately equal to its width. A single head rotates in a circular slot in the drum, the slot being positioned so that the head starts near one edge of the tape and finishes near the other, but further along the tape. As the head continues along its circular path it starts again at the first edge, but by this time the tape has moved on, so that the next track is further along the tape. In this way the tracks are arranged as a series of diagonal stripes across the tape, as shown in Fig. 24.4. The speed of rotation of the head is arranged to be exactly once per television field, so that each track corresponds to one field. Compared to the four-head system, in which each field is recorded in sixteen tracks (for 60 Hz fields) or twenty tracks (for 50 Hz fields), each single-head helical track must therefore accommodate four or five times, respectively, the amount of information per revolution of the head. If the helix is made to have a large circumference, then the packing density along the track can be reduced, thus alleviating the performance required of the head, but, as the angle of the tracks then becomes more nearly parallel to the edges of the tape, the speed of the tape may have to be increased to avoid adjacent tracks overlapping one another.

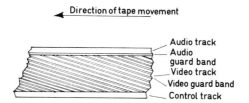

Fig. 24.4. Arrangement of tracks on a typical helical-scan video tape.

An alternative to using the single-head arrangement is to use two heads, and to wrap the tape round only half of the circumference of the drum, as shown in Fig. 24.3(b). Each track again corresponds to one television field, and although it is now necessary to match the performance of two heads, the configuration of the tape round the drum is rather more convenient, and this arrangement is widely used.

One of the advantages of using the helical scan system is that, by keeping the tape stationary, a still picture can be reproduced from two adjacent tracks, but the gradual movement of the head from one track to the next does cause loss of picture quality especially in the middle lines of the picture.

Helical scan systems of various formats are in use: details of some of these using tape-widths of ¾ in or more are given in Table 24.1 (Reynolds, 1970; Sawazaki, Yagi, Iwasaki, Inada, and Tamaoki, 1960; Millward, Guisinger and Roizen, 1972; Ross, 1974).

The formats listed in Table 24.1 are used for high quality broadcasting and,

TABLE 24.1

Formats for video-tape recording

System	Tape width	No. of heads	Tape speed	Track layout	Special features
Ampex	2 in	4	15 in/sec 38 cm/sec 7.5 in/sec. 19 cm/sec.	Transverse	Designed for high quality broadcasting
International Video Corporation	2 in	2	8 in/sec. 20.3 cm/sec.	Helical	Designed for high quality broadcasting
International Video Corporation	1 in	1	6.73 in/sec. 17.1 cm/sec.	Helical	
Victor Company of Japan and Matsushita	$\frac{3}{4}$ in	2	3.75 in/sec. 9.53 cm/sec.	Helical	
Sony U-matic	$\frac{3}{4}$ in	2	3.75 in/sec. 9.53 cm/sec.	Helical	

in the case of those using $\frac{3}{4}$ in tape, for commercial and industrial use. For domestic use, in order to reduce costs, $\frac{1}{2}$ in tape is generally used, and the features of three such systems are given in Table 24.2. All these systems use $\frac{1}{2}$ in tape, a helical scan format, and two heads. In all cases the tape is contained in a cassette, so that loading and unloading it from the players (*Video Cassette Recorders, VCRs*) is easy for the user, and the tape is well-protected by not being handled. All three systems can show still frame, slow-motion, and fast editing pictures. The V2000 system uses a sufficiently narrow video recording track for it to be possible to make two recordings on the tape, side by side, so that a single tape can provide 8 hours of recordings in two 4 hour parts; the narrow tracks require very accurate following by the heads and this is achieved by continuous monitoring and

TABLE 24.2

Formats for domestic video tape recording

System	Manufacturer	Tape speed cm/sec	Writing speed m/sec	Video track μm
Beta	Sony	1.87	5.83	32.8
VHS (Video Home System)	Matsushita and Victor Company of Japan (JVC)	2.34	4.85	49.0
V 2000	Philips	2.40	5.08	22.5

adjustment, a feature that also improves the quality of still frame and slow-motion replay. In the VHS system, the quality of still and slow-motion pictures is improved by having one of the two heads slightly larger than the other. As can be seen from Table 24.2, the VHS system has the slowest writing speed (which implies that the definition might be the lowest), but the widest video track (which implies that the signal-to-noise ratio might be the best). The Beta system has the fastest writing speed (which implies that the definition might be the best), but the lowest tape speed (which implies that the sound quality might be more limited). The three systems are totally incompatible with one another; both the tape format and the cassette design are different in all three systems.

24.4 E.V.R. (Electronic Video Recording)

Before turning to the various disc systems that have been developed, mention should be made of the E.V.R. system, in which television signals are prerecorded on film, and of the holographic method that uses plastic tape. In the E.V.R. system, developed originally by C.B.S. in the U.S.A., advantage was taken of the fact that black-and-white photographic film is cheaper, and can have a higher resolution, than colour film; the luminance and colour parts of the information were therefore recorded side by side on black-and-white film. Advantage was also taken of the higher resolution that can be obtained if films are exposed directly to electron beams, instead of to light produced by phosphors irradiated by electron beams in cathode-ray tubes (Goldmark, 1970). The E.V.R. system met with only limited success.

24.5 Holographic recording

The search for an inexpensive means of making many copies of programmes has led to the experimental use of phase-holograms on transparent vinyl plastic tape. Such holograms could be pressed from a master tape very cheaply.

A system of this type was developed in the U.S.A. by the R.C.A. Corporation. The luminance and chrominance information of each frame of the original programme was transferred by electron beam recording to a 16 mm film frame, from which was then produced a phase-hologram. From this master hologram, the holographic copies were made by a pressing process. The holographic copy was illuminated by means of a laser beam which a vidicon camera then converted into conventional television signals.

An interesting feature of this system was the way in which the luminance and chrominance information was recorded. The luminance signal was restricted in bandwidth to 3 MHz, and the colour-difference signals were then coded on to two sub-carriers of frequencies of about 3.5 MHz and 5.0 MHz above the luminance carrier, each having a bandwidth of 0.5 MHz. The combined signal was then recorded, without any decoding, on to 16 mm black-and-white film of high resolution using an electron beam: the resulting image was sharp enough to record the colour-difference sub-carrier signals as vertical stripes spread about 60 and about 90 μm apart, superimposed on the luminance pictures. This film record was then converted into a hologram in the usual way by dividing a laser beam into two parts, and passing one part, but not the other part, through the film, and then recombining the two parts. The recombined beams were then used to expose a film containing a photo-resist, which, when developed, resulted in a hologram record in terms of depth of plastic, the average depth being about 0.05 μm below the original surface. A nickel-plated master of this relief image was then used to press the vinyl plastic release tapes.

In the playback unit the hologram on the vinyl tape was illuminated with a similar divided laser-beam arrangement to reconstitute the luminance pictures with their vertical stripes. By focusing these images on to a vidicon tube the 3 MHz luminance signal and the two 0.5 MHz colour-difference signals on their two sub-carriers were recovered. These three signals were then re-coded in the form required by the television receiver being used to produce the final picture. The vertical stripes were not visible on the final picture because its bandwidth was limited to 3 MHz, and hence the sub-carriers were removed from the displayed luminance signal.

The restriction of the bandwidths to 3 MHz for luminance and 0.5 MHz for chrominance resulted in pictures that were acceptable, but less sharp than those of standard broadcast quality. Amongst the many technical problems in the system was the difficulty of maintaining to the required accuracy the depth of the holographic indentations in the plastic (Doyle, 1969).

24.6 Recording on discs

The success with which discs have been used for the mass reproduction of sound has prompted attempts to use discs for video recordings also.

The basic technical problem to be solved is how to record on discs the very much higher rates of information required for television as compared to sound. In sound recording, it is necessary to handle about 3×10^5 bits of information per second: a top frequency of 15 000 Hz requires 30 000 samples per second, and to enable 1000 amplitudes to be distinguished requires 10 bits ($2^{10} = 1024$), thus making 30 000 $\times$ 10 or 3×10^5 bits/sec. For video recording a top frequency of about 5 MHz requires about 10^7 bits/sec. and to enable 64 amplitudes to be distinguished requires 6 bits ($2^6 = 64$), thus making 6×10^7 bits/sec., or about 200 times as much as for sound recording. A normal $33\frac{1}{3}$ r.p.m. gramophone record has a data storage capacity of about 5000 bits/mm^2 (as compared to about 1000 bits/mm^2 for magnetic tape) so that for video recording on discs the capacity should be about a million bits/mm^2; however, a figure of about half this value would be adequate if a quality level somewhat below that normally used for broadcast television were regarded as acceptable for video disc applications, or if shorter playing times per disc were acceptable.

24.7 The Teldec system

The Teldec system has been developed by A.E.G.-Telefunken in Germany and Decca in England. It is a video disc recording system which can accommodate television signals of 3 or 4 MHz bandwidth (Gilbert, 1970).

The disc consists of a thin sheet of poly-vinyl-chloride (P.V.C.) foil, about 0.1 mm thick. On the disc is pressed a spiral track, as on an audio disc, but to obtain the higher data-storage capacity necessary, the spiral has a spatial frequency of about 140 grooves/mm (instead of about 18 grooves/mm for audio discs), in which oscillations with a spacing of about $\frac{1}{2}\mu$m can be recorded (instead of about $3\,\mu$m for audio discs); the smoothness of the disc is such that the random unevenness comprises fluctuations of about $1/100\,\mu$m. The shape of the grooves is as shown in Fig. 24.5, the distance between adjacent grooves being $7\,\mu$m. The signal is recorded as changes of about $\pm 0.5\,\mu$m in the depths of the grooves, a form of recording known as *hill and dale* (Dickopp, 1971; Redlich, 1971).

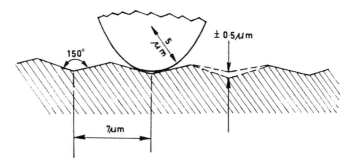

Fig. 24.5. Cross-section of grooves and stylus in the Teldec system.

The disc is mounted on a shaft which rotates it just above a cylindrically shaped plate, at a speed of 1500 r.p.m. for a 50 Hz system, or 1800 r.p.m. for a 60 Hz system. Each revolution corresponds to two television interlaced fields, or one complete television picture. The high speed of revolution sucks air in through an annular space between a circular plate (on which the centre of the disc rests) and a circular hole in the cylindrically-shaped plate, as shown in Fig. 24.6; the air escapes at the outer edge of the disc, but provides a cushion on which the disc rests. The high velocity of the outer edge of the discs accelerates the air, thus creating a region of low pressure, and as a result the disc is pressed by atmospheric pressure towards the cylindrically-shaped plate, and therefore takes up a cylindrical profile. This profile gives the disc stability, especially along the line of the crest of the cylinder, and it is along this line that the stylus travels.

The shape of the stylus can be seen from Fig. 24.5, which shows its appearance from its direction of travel, and from Fig. 24.7, which shows its appearance from the side. The stylus itself is a diamond and it is attached to a

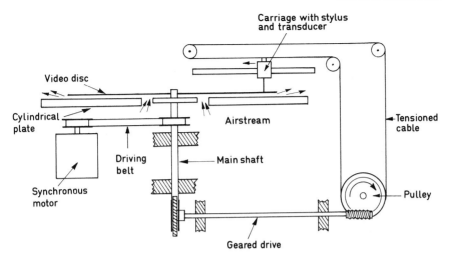

Fig. 24.6. Diagrammatic representation of the Teldec player.

pressure-sensitive piezoelectric ceramic transducer. As a point on the disc travels under the diamond it is pressed downwards until it reaches the trailing edge of the diamond: if this point is at a hill on the hill-and-dale recording, the change in pressure on the trailing edge of the diamond will be greater than if the point is on a dale; it is these changes in pressure that cause the piezoelectric ceramic transducer to produce a voltage from which the television signal can be produced.

With a disc diameter of 210 mm (8.3 in) the recording area is between an outer diameter of 205 mm and an inner diameter of 97.5 mm: at 140 grooves/mm

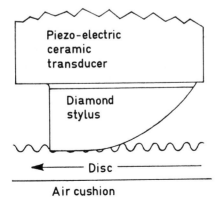

Fig. 24.7. Side-view of the stylus and disc in the Teldec system.

this provides 7820 grooves, or just over 5 minutes playing time. For longer programmes larger diameter discs, or automatic disc changers can be used.

Because it is difficult over the entire length of the groove to maintain disc thickness and groove depth to the very tight tolerances necessary for amplitude modulation, a system of frequency modulation is used, the amplitude being kept constant at about ±0.5 μm. The carrier frequency is such that peak white corresponds to a frequency of 3.75 MHz and sync level to 2.75 MHz. The 2.75 MHz frequency on the outer groove corresponds to a spatial wavelength along the groove of 5.86 μm, and the 3.75 MHz frequency on the inner groove to 2.05 μm along the groove. The video bandwidth of the system is about 3 MHz.

The stylus is assisted in its traverse from the outer turn to the inner turn of the track as it travels down the spiral (which, incidentally, is about 3.5 km or 2 miles long) by a pulley system driven off the main shaft as shown in Fig. 24.6. If the stylus is held stationary, it can jump from one groove to the next quite smoothly and thus a still picture can be displayed. It is also possible to allow the head to follow the spiral for a few seconds and then return rapidly to its initial position, thus enabling a particular sequence to be continually repeated, as may be desirable in an educational or training programme.

The colour coding of the signals is a modification of the *Tripal* system (Bruch, 1967; Bruch, 1972). Instead of using one delay line, as in the P.A.L system, two delay lines are provided. The red, green, and blue signals are then recorded sequentially, and recovered simultaneously by using the delay lines to provide delays corresponding to two line-scans, one line-scan, and zero. In this simple form, the system would provide full bandwidth in each colour, but at the sacrifice of time-resolution and vertical spatial resolution. The system, however, is further elaborated so as to confine these losses to changes in coarse detail.

A signal M is derived, consisting of equal shares of the colour signals, R, G, B. (For simplicity, gamma correction is ignored in this discussion.) Colour difference signals $R-M$, $G-M$, and $B-M$, are then derived and passed through a low-pass filter of band-width 0.5 MHz to obtain signals R_L-M_L, G_L-M_L, and B_L-M_L. These signals are then added to the signal M to obtain:

$$M+R_L-M_L$$
$$M+G_L-M_L$$
$$M+B_L-M_L$$

and these signals are recorded sequentially on successive lines. These signals are equivalent to:

$$M_H+R_L$$
$$M_H+G_L$$
$$M_H+B_L$$

where M_H are the high-frequency components of M, and R_L, G_L, B_L are the low-frequency components of R, G, and B. When the disc is played, a low-pass filter separates R_L, G_L, and B_L, from M_H, and passes them through the delay lines so that all three are available simultaneously; they are then matrixed to obtain the usual chrominance signals R_L-Y_L and B_L-Y_L, and a signal M_L consisting of equal shares of R_L, G_L, and B_L. The signal M_L is then added to M_H to reconstitute M. The receiver then recovers G_L-Y_L, from R_L-Y_L and B_L-Y_L in the usual way (see Section 19.8), and then adds M to all three colour-difference signals to obtain R_L+E+M_H, and G_L+E+M_H, and B_L+E+M_H, where $E=M_L-Y_L$ and represents a residual error: for whites, greys and blacks $R=G=B=M=Y$, and hence M_L-Y_L is zero; M_L-Y_L will also be small for the prevalent desaturated colours.

With this arrangement of recording and replaying, there is no loss of band-width in the M signal at any frequency. For whites, greys and blacks, since $R=G=B=M$, it follows that R_L-M_L, G_L-M_L, and B_L-M_L, are all zero, and hence the signals are the same on every line and there is no loss of time discrimination. For colours, the low frequencies (coarse picture detail) will suffer a progressive loss of time resolution as the saturation increases because of the sequential nature of the R_L-M_L, G_L-M_L, and B_L-M_L, signals; but the high frequencies (fine picture detail) do not lose any time resolution, and the loss of colour resolution is mitigated by the insensitivity of the eye to colour differences at high spatial frequencies at constant luminance (see Section 19.7). The bandwidth of the colour-difference signals is one sixth of that of the luminance signal, and this matches the six to one loss in vertical resolution caused by the field-line sequential nature of the colour-difference signals.

24.8 Capacitance discs

Discs with spiral grooves carrying signals in the form of variations of capacitance have also been developed, notably by R.C.A. (Radio Corporation of America) in the U.S.A. Luminance, chrominance, and audio signals are encoded in carriers and recorded as variations in the width or spacing of pits cut in the bottom of the grooves. The master disc is cut by using an electron beam in a vacuum so that very high resolution can be achieved. The master disc is used to produce release discs in which the bits and grooves are pressed on to a thin insulating dielectric layer situated above a metal conducting layer. A sapphire or diamond pick-up stylus, guided by the grooves, carries a thin metallic electrode that detects the changes in capacitance caused by the pits on the disc. The grooves have a pitch of 218 per radial mm. The discs rotate at 375 r.p.m. for 50 Hz systems and at 450 r.p.m. for 60 Hz systems; each rotation thus corresponds to 8 fields or 4 complete pictures. The discs are 12 in (30 cm) in diameter, and the playing time is 30 minutes, but both sides can be used to give a total of 60 minutes in two parts. This system is known commercially as *Selectavision* or *CED* (*Capacitance Electronic Disc*). A similar system has been developed by the Victor Company of Japan (JVC) and Matsushita, and is known commercially as the *VHD* system; however, in this case, there are no grooves to guide the stylus, and the signals are tracked by special tracking signals which move the stylus in the required manner.

24.9 Discs using lasers

The extremely small diameter that it is possible to achieve with laser beams makes them attractive as a means of recording and reading out closely-packed video information. In the system developed originally by Philips in the Nether-lands, a master disc is recorded by using signals to modulate the intensity of a laser beam that is focused on to a photographic plate, and this is rotated to form a spiral track. The amplitude of the signals is 'clipped' to about a third of the normal maximum, and this results in a signal consisting of pulses, whose variations in frequency carry the luminance information, and whose variations in

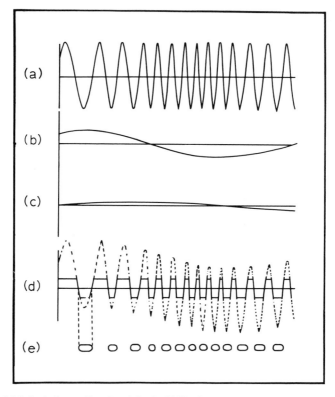

Fig. 24.8. Method of recording signals in the Philips laser disc system.
    (a) Carrier wave with luminance represented by frequency modulation.
    (b) Chrominance represented by amplitude modulation.
    (c) Audio signal represented by amplitude modulation.
    (d) Combined signal with restricted amplitude producing pulses whose frequencies
       represent luminance, and whose lengths represent chrominance and sound.
    (e) Pattern of pits encoding signals.

length carry the chrominance information, as shown in Fig. 24.8. The record of these pulses is then etched, as a series of pits, in the master disc: from this, a nickel pressing-mould is made for the production of the replicas on the same type of plastic as is used for long-playing audio discs. The disc is rotated at 1500 or 1800 revolutions per minute, to give one complete television picture (two fields) per revolution for 50 Hz or 60 Hz operation, respectively, but slower speeds can also be used to extend the playing time. The length of the pits varies from about 0.6 to about 4 μm, their width is about 0.7 μm, and their depth about 1 μm. Frequency modulation is used for the luminance signal, which has a bandwidth of about 4.75 MHz, and amplitude modulation for the chrominance and sound signals, whose bandwidths are about 1 and 0.25 MHz, respectively. The luminance signal occupies the highest band of frequencies used, and the sound signal the lowest, but there is no band sharing amongst the three signals (Kramer and Compaan, 1974).

To reproduce pictures from the disc, another laser is focused on the spiral track in the replica, and the light reflected by the disc is modulated by the pits diffracting the light away from a photodiode detector, which produces the required signals.

In order to obtain playing times of 30 minutes or so on a single disc, the spiral track has a separation of only 2 μm between adjacent turns, and hence very accurate focusing of the laser beam on the track is required. This is achieved in the vertical direction by a servo-operated movement of the microscope objective used to focus the beam; in the radial direction a servo-operated mirror adjusts the position of a beam to equalize the signals from two additional spots of light that are displaced, one very slightly towards, and the other very slightly away from, the centre of the disc. The system is known commercially as *Laservision*, and in its later forms was developed jointly by Philips and M.C.A. of California. Systems using lasers by transmission, instead of by reflection, have also been developed, notably by Thomson-CSF in France.

24.10 The duplication of programmes on video cassettes and discs

When programmes are required in large numbers, for widespread distribution at low cost, it is important to be able to produce the copies at speeds considerably higher than those corresponding to playing the programmes at their normal speed.

Rapid copying of magnetic tape recordings has presented considerable technical problems, because if a tape is run through a tape recorder faster than its proper speed severe loss of performance is caused by the large changes in signal frequencies. Two methods of high-speed contact printing of magnetic tape have therefore been developed. In the first, the two tapes are run together past a suitable magnetic field, and in this way speeds of about 150 in/sec. (3.8 m/sec.) have been achieved. In the second method, the two tapes are heated at the point where they are in contact, and, by using special tapes, printing speeds of about 60 in/sec. (1.5 m/sec.) have been achieved.

Disc systems have been deliberately conceived with very rapid production of copies in mind: they have an important advantage in this respect in that the whole disc is made at the same time, so that the material does not have to be run through a copying station sequentially, as is the case with all the other systems. (Doyle, 1982).

REFERENCES

Anderson, C. E., and Roizen, J., *J. Soc. Mot. Pic. Tel. Eng.*, **68,** 667 (1959).
Baldwin, J. L. E., Stalley, A. D., and Kitchin, H. D., *J. Roy. Television Soc.*, **14,** 3 (1972).
Bruch, W., *Telefunken-Zeitung*, **40,** No. 3 (1967).
Bruch, W., *J. Soc. Mot. Pic. Tel. Eng.*, **81,** 303 (1972).
Dickopp, G., *Brit. J. Phot.*, **118,** 992 (1971).
Doyle, O., *Electronics*, **42,** 109 (November 10, 1969).
Doyle, M. J., *J. Soc. Mot. Pic. Tel. Eng.*, **91,** 180 (1982).
Gilbert, J. C. G., *Wireless World.*, **76,** 377 (1970).
Ginsburg, C. P., *J. Soc. Mot. Pic. Tel. Eng.*, **66,** 177 (1957).
Goldmark, P. C., *Wireless World*, **76,** 366 (1970).
Kramer, P., and Compaan, K., *Audio-Visual*, **3,** 16 (January, 1974).
Machein, K. R., *J. Soc. Mot. Pic. Tel. Eng.*, **68,** 652 (1959).
Millward, J. D., Guisinger, B. E., and Roizen, J., *I.E.E. Conference Publication No.* 88, p. 317 (1972).
Redlich, H., *Brit. J. Phot.*, **118,** 952 (1971).
Reynolds, K. Y., *J. Soc. Mot. Pic. Tel. Eng.*, **79,** 922 (1970).
Ross, P., *J. Roy. Television Soc.*, **15,** 35 (1974).
Sawazaki, N., Yagi, M., Iwasaki, M., Inada, G., and Tamaoki, T., *J. Soc. Mot. Pic. Tel. Eng.*, **69,** 868 (1960).
Tooms, M. S., *J. Roy. Television Soc.*, **13,** 4 (1970).

GENERAL REFERENCES

Abramson, A., *J. Soc. Mot. Pic. Tel. Eng.*, **82,** 188 (1973).
Roizen, J., *J. Roy. Television Soc.*, **16,** 15 (1976).

CHAPTER 25

Pictures from Computers

1. Introduction – *2.* Coloured captions – *3.* Chroma-key – *4.* Teletext – *5.* Colour video display units – *6.* Video graphics – *7.* Computer assisted cartoons – *8.* Colour coding in pictures – *9.* Colour ranges

25.1 Introduction

THE existence of television pictures in the form of electronic signals makes it possible to manipulate them in a very great variety of ways. It also provides opportunities for adding signals that have been derived, not from cameras or telecines, but from computers, to form either parts of pictures or even complete pictures. (Similar techniques can be used with graphic-arts scanners, as will be described in Section 29.9.)

25.2 Coloured captions

Perhaps the simplest form of pictorial content originating from computation is the 'electronic' colouring of a caption. If a caption is produced by a camera viewing white letters on a black background, then it is only necessary to add to the luminance signal, produced by the camera, suitable chrominance signals, in the right time sequence, to produce coloured letters. Thus a chrominance signal giving one colour could be added to the group of television lines corresponding to the first line of the caption, and a signal giving a different colour to the television lines for the second line of the caption. Similarly, subsequent lines of the caption could be coloured differently again. If a change of colour within a single caption line is required, perhaps to emphasize a single word, then the chrominance signal has to be changed appropriately during the line scan in all the corresponding television lines.

It is clear that, when captions are coloured in this way, the colours used do not exist in any original version of the caption, but are created in the form of stored data comprising switching sequences and instructions for signal generation. The colours displayed are therefore limited, not by poster pigments,

511

photographic dyes, or printing inks, but only by the receiver display phosphors; it is therefore possible to use colours corresponding to the maximum chrominance signals listed in Table 22.3, and hence very colourful captions can be produced, which can be prominent and attractive; of course, less colourful results can also be produced if required.

25.3 Chroma-key

The addition of colours by adding chrominance signals is not confined to captions, and simple diagrams can also be coloured by the same method. But, if the switching sequence is very complicated, the method becomes unwieldy. A much more versatile means of switching has therefore been provided by the *chroma-key* method.

The chroma-key method is similar to the *travelling matte* technique used in professional cinematography. If, in a film, it is required to produce a trick shot, such as a girl standing on the hand of a giant, then the following procedure can be used. The person playing the part of the girl is filmed against a black background. From this film, a very high-contrast black-and-white negative is obtained in which the girl is portrayed as a black silhouette against a clear background. When printing the negative of the giant scene, this silhouette film is positioned in contact with it, so that the print film is not exposed in the silhouette area. The print film is then exposed a second time, in this case using the negative of the girl against the black background, thus printing the girl in the unexposed area left by the silhouette. Of course, careful alignment of the positions of the giant's hand and the girl's feet is necessary in making these two shots. It may sometimes be necessary also to use a complementary silhouette, a clear image of the girl against a black background, when making the second exposure on the print film; and the method can be elaborated by using beam splitters and other devices (Wheeler, 1969).

In the chroma-key method used in television, instead of using a black area for the shot to be inserted, a highly coloured area is used. It is then arranged that, whenever the television signal corresponds to the colour of this coloured area, a switch is operated that accepts input from another picture source. Thus, if a commentator is sitting in a studio in front of such a coloured area, pictures of the event being described can be shown as a background, giving the impression that the commentator is present at the event. It is necessary that no part of the scene containing the commentator is of the same colour as the coloured area, otherwise the switch would insert background into the studio scene, producing ludicrous 'holes' in the picture. The colour most often used for the coloured area is saturated blue, because it is usually fairly easy to avoid this colour in studio scenes.

The switch can be made to depend on the values of the two chrominance signals in the television system used; or, to reduce any effects that may be caused by variations in the strengths of these signals for the colour of the area (because of

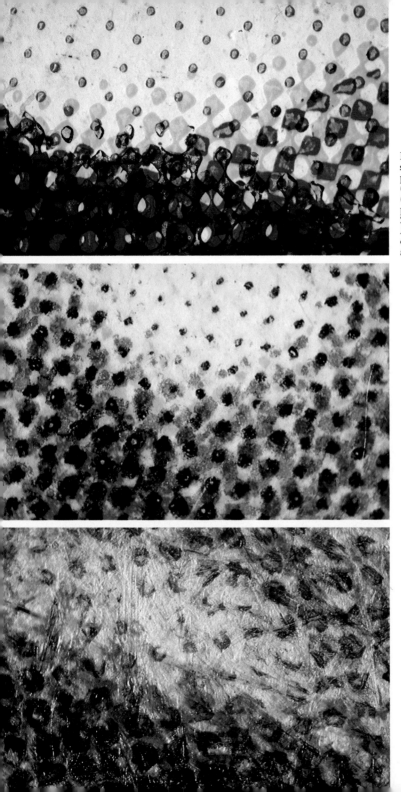

Plate 31
Photomicrographs of small areas of a letterpress (*top*), a lithographic (*centre*), and a gravure (*bottom*), four-colour reproduction (see Chapter 26). Reproduced from *Kodachrome* transparencies on a Crosfield *Magnascan* scanner.

Plate 32

The *top* four illustrations show the cyan, magenta, yellow, and black images of a four-colour lithographic reproduction separately. On the *far left* is shown the result of printing the cyan, magenta, and yellow images without the black; note the poor black obtained from the three coloured inks, as compared to the result *at centre left* where all four images are printed. At the *near left* is shown the type of four-colour reproduction obtained when no masking is used (see Chapter 28) in which the uncorrected unwanted absorptions of the inks result in much duller colours. Reproduced from a 5 × 4 inch *Ektachrome* transparency on a Crosfield *Magnascan* scanner.

Plate 33

The reproduction of faces in sunlight can be improved by providing extra lighting from flash-bulbs, supplementary lights, or reflectors (see Section 13.10). In this case a blue flash bulb was used. Similar effects may be achieved by choosing a location with a light foreground. Reproduced from a 2¼ inch square *Ektachrome* transparency on a Crosfield *Magnascan* scanner.

shadows, for instance), the switch can be made to depend on a measure of the *chromaticity* of the background, derived from both the two chrominance signals and the luminance signal (Davidse and Koppe, 1977). A tendency for spurious background colour fringes to be produced in parts of the picture corresponding to the operation of the switch can be reduced by using gating signals having less steep rise-and-fall times, and also by using a circuit to eliminate from the final composite transmission any signals corresponding to the colour of the area (Nakamura and Kamakura, 1981). Various elaborations of the use of chroma-key are possible that improve the realism of the final composite picture. For instance, by making the coloured area have the same three-dimensional shape as the part of the scene to be replaced, it is possible for shadows from action in a foreground scene to be inserted into a background scene; thus, an actor can walk along a floor near a wall, both the floor and the wall being of the colour operating the switch, and in the final picture appear to be walking along a street by a wall and casting a shadow of the right shape on them.

25.4 Teletext

Insertion of signals from another source is also possible in television by making use of some of the interval between the times when the last line of a field has just been transmitted and the first line of the next field has yet to be started. This interval is called the *field-blanking time*; it is the time, in effect, when the scanning spot flies back from the bottom to the top of the picture; it amounts to a significant proportion of the total field time, and is equivalent to 25 lines in each of the two fields in a 625 line system (leaving only 575 lines actually used in the picture), and to 21 lines in each of the two fields in a 525 line system (leaving only 483 active picture lines). It is necessary, during this time, to transmit signals that enable the receiver to synchronize its field scan phase with that of the transmitted signal, and this *vertical sync signal* is what ensures that each field starts at the top of the receiver display and not part way down the picture. But the vertical sync signal is usually confined to a time equivalent to about 10 lines, and this leaves a period of unused time between each field. It is part of this time, typically equivalent to only two lines, that is used for transmitting *teletext* signals (see Plate 25, page 384.)

Teletext signals comprise information that enable television receivers to display letters and numbers (*alpha-numeric* characters) and simple diagrams (*graphics*), over the whole picture area. It is possible to fill the whole picture area with teletext information derived from a transmission time equivalent to only two lines per field by using appropriate signal storage and read-out devices in the receiver, and by adopting the following special procedures for handling the signals. First, because the display is a still picture, it is not necessary to describe it differently in successive frames; the teletext picture is therefore built up slowly at the rate of two lines of text per field. Secondly, the character and graphic symbols are not transmitted as such, but as a code for interpretation by a *character generator* in the receiver. The character generator forms the symbols by using picture elements (*pixels*) in groups consisting, typically, of seven lines, each having five elements at the same horizontal positions so as to form a seven by five matrix. Usually a choice from 64 different characters or 64 different graphics is provided. The transmitted signal then consists, in effect, of a series of numbers from 1 to 64 that indicate to the receiver the succession of symbols required. The numbers are expressed in binary form, using for each number 6 binary units or *bits* (because $64 = 2^6$). Without the use of the character generator, the 35 pixels in the matrix would require 35 bits. Thirdly, the number of symbols permitted in the display is limited, typically to 24 rows and 40 columns; although a closer packing of symbols would provide more information, legibility would suffer. Fourthly, tone and colour are usually severely restricted to not more than either four choices, needing 2 bits ($4 = 2^2$), or eight choices, needing 3 bits ($8 = 2^3$).

With the above type of arrangements, the time taken to transmit one complete array, or *page*, of information is given by the number of rows in a page divided by the number of rows transmitted per second. Using 2 television lines per field, the number of rows per second is 100 for a 50 Hz system, and 120 for a 60 Hz system; the times to transmit a page of 24 rows are therefore 0.24 and 0.20 seconds, respectively. This assumes that the information for all 40 columns can be packed in to the normal line period. This is about 64 microseconds (in both the 625 line 50 Hz and the 525 line 60 Hz systems); but this is reduced, as in normal television picture signals, by the time required for transmitting a horizontal synchronizing signal (*horizontal sync pulse*) and the colour burst signal (see Section 22.8), leaving a usable time of about 52 micro-seconds. In this time it is necessary to transmit, for each of the 40-columns, the 6 bit number indicating the symbol required, making a total of 240 bits. In addition, it is necessary to include bits for organizing the format of the display: this includes selection of either the alpha-numeric characters or the graphics, and choice of the colours to be used, the arrangements being changed in different parts of the picture to allow for different colours, and for mixtures of simple diagrams (such as crude maps) and text. The total number of bits used per line is typically about 300, and for a 52 microsecond active line period this results in an information transmission rate of about 6 Mega-bits per second.

If a system offers 100 teletext pages (as in the BBC *Ceefax* and the IBA *Oracle* systems used in Britain), then, at 0.24 seconds per page, it takes about 24 seconds to transmit the 100 pages. The user selects the page required by using an index, which indicates a sequence of buttons to be pressed; on pressing these buttons, the receiver selects the required page from the 100 transmitted, and then stores the signals and uses them to generate the corresponding television picture signals required by the receiver display tube. The 100 pages are sequenced through repeatedly in the transmission so that the user has to wait until the selected page is being transmitted before it can be received. In the case of the 24 second cycle time, the maximum delay is 24 seconds (when the required page has just gone), the minimum is almost zero (when the required page is the next one to be transmitted), and the average delay is 12 seconds; however, empty rows and empty pages do not require any transmission time, and if, on the average, a third of the rows and a third of the pages are empty, then the above times are reduced by a factor of $\frac{2}{3} \times \frac{2}{3}$, or roughly halved. The delay times can also be reduced if the receiver is provided with facilities for storing some pages in a memory.

If a television receiver can be connected to an information source in such a manner that two-way interaction is possible, by using, for instance, telephone lines, then selection from a much larger number of pages becomes possible. These systems are called *Viewdata*, an example of which is the *Prestel* system used in Britain.

25.5 Colour video display units

When the information to be displayed is derived from a local source, instead of from signals transmitted by broadcasting or by the use of telephone lines or other long-distance cables, there are usually fewer constraints on the form and information-content of the signal, and more elaborate symbols, and greater ranges of colours and tones, can often be used. Instead of using a television receiver or monitor for the display, it is then possible to use a *video display unit* (*VDU*) specially designed for the purpose. Colour VDUs usually make use of shadow-mask type tubes having a finer pattern of phosphor dots, the triad pitch usually being about 0.3 mm (or sometimes about 0.2 mm) instead of about 0.7 mm (see Section 21.4). This enables the observer to sit nearer to the display without the dots being resolved, and a typical viewing distance for colour VDUs is about a half to one metre. To obtain more legible and more pleasing teletext symbols, the matrix used for character generation may be increased from 5×7 pixels to 7×9 or 9×13 (or be modified to generate by lines instead of by dots, in which case the matrix used for a given level of quality is usually somewhat smaller); better characters can also be obtained by shifting some of the dots (or lines) by half the separation between adjacent elements of the matrix (Brockhurst, Day, Dyer, and Vivien, 1982). The visual comfort of the viewer is usually improved by reducing the specular reflections from the front surface of the VDU to a minimum.

25.6 Video graphics

Diagrams for illustrating television programmes, such as news and current affairs, can be derived from computers, and elaborate systems are often available in broadcasting. One approach used is as follows (Long, 1982). (See Plates 25 and 27, pages 384 and 385.) A computer provides a memory for up to about 1000 different shapes. These shapes can be of any size from a single pixel up to the complete screen. The shapes can be generated from art work, or from computer programmes, or from chroma-key switching outlines, or by moving a 'pen', containing an electrode, on a 'tablet' board (containing wires to detect the position of the pen, which is then displayed on the monitor). Each shape is limited to one colour, but this can be chosen from a pallette of 512 different colours. The display file defining the picture consists of a background colour (or a sequence of them if it is to be striped), and a specification of each shape to be displayed. This specification consists of the identity of the shape, its vertical and horizontal location, its colour, and its *priority*. The priority indicates which shape is to be displayed in those areas where the shapes overlap; in this way symbols and captions can appear to be overlaid on the display. Simple animation sequences can be produced by using shapes as building blocks which can be added or subtracted at successive time intervals; for example, if a rectangle on a histogram is made up of a stack of small rectangles, it can be made to grow or shrink by adding or subtracting rectangles, like building or reducing a pile of identical coins. Realistic depictions of simple three-dimensional objects, such as cubes, pyramids, or rectangular-sided blocks, can be produced by choosing colours of the same chromaticity but lower luminance factors for some of the surfaces. Fully pictorial colour can be added in a shape by using input from colour cameras or telecines with chroma-key type of switching. An example of the use to which this type of system can be put is the coverage of elections. The whole country or region concerned is represented by one shape, and its relevant component parts (states, counties, constituencies, etc., as appropriate) are each represented by their own shapes. At the start of the election, each component area could be coloured grey, for instance; as the results come in, the colour could then be changed, for example, to red for one party or candidate, green for another, blue for a third, and so on. These changes can be made by using the pen on the tablet to identify first a colour and then the area to be coloured. Voting statistics can be superimposed as captions on the appropriate component areas. By leaving a small gap round the shape of each component area, the background colour shows the positions of the boundaries when adjacent parts have the same colour. Another display could show the growth of the support for the different parties as animated histograms.

In avionics, computer generated graphics are often displayed in aircraft to provide important information to the crew, the display device often being a colour VDU using a shadow-mask tube. The information can be presented in conventional *raster* form, as in broadcast television, or in *stroke* form (sometimes also referred to as *calligraphic*, *cursive*, or *vector* form) in which the spot writes the symbols by moving along their lines directly, rather than by building them up line by line with the raster. The advantages of the stroke form are that higher brightnesses are possible (by using slower writing speeds), and jagged edges to lines that should be straight are avoided. The stroke form can also include the writing of closely adjacent parellel lines to fill in a limited area (sometimes referred to as *shading*). The two forms can be used together by switching from one to the other within each frame period, or the stroke information can be inserted during the field-blanking period. In this way pictorial or map-type information can be displayed by raster, with graphic information superimposed by stroke, to give extremely useful aids to avionic operations.

Elaborate video graphics are used in *simulators* for training the crews of aircraft, ships, and space vehicles. From a knowledge of the terrain and environment, including the position of any navigation lights, programs are developed that present the views that the crews would have as their crafts proceed on their courses. By linking the controls of a simulated flight deck or bridge with the computer, the system can be made interactive, so that the visual result of each adjustment to the controls is the same as would have occurred in real life. Corresponding accelerations and decelerations of the training flight deck are also provided in some aircraft simulators to add to the realism. Navigation lights may be displayed in these applications by inserting them between raster scans in the field-blanking interval in the stroke form, thus achieving high brightnesses and freedom from line structure. Very realistic displays are achieved for these applications, using very sophisticated equipment, the high cost of which can be justified by the large reduction possible in real training flights and voyages. Moreover, the computer programs are able to provide an enormous range of different conditions corresponding to different times of day, seasons, and weather, such as could not be covered in any real training programme of practicable length. (See Plate 26, pages 384, 385, and Plate 29, pages 416, 417.)

Elaborate video graphic systems are also used as aids in design and architecture. For instance, in interior design, sketches of rooms can be made, and the areas corresponding to the walls, floor, ceiling, carpets, curtains, furniture, and so on, can be loaded into the computer as identifiable shapes. These shapes can then be displayed in different colours to try out various colour schemes. Similar techniques can be applied to the design of fabrics, carpets, wallpapers, ceramics, and other similar patterned objects in which it is of interest to observe the effects of altering the colours of different elements of the pattern. (See Plate 30, page 417.)

In architecture, computer programs have been developed that produce displays of perspective drawings of buildings, using the usual plan, elevation, and side-view drawings as input. It is necessary in this case for the nearer planes always to have priority over those that are more distant, so that the proper pattern of obscuration is achieved. The perspective presentation also requires the capability of changing rectangles into shapes conforming to the vanishing point for the chosen point of view; and such programs can also include progressive changes in the apparent vanishing point, so that animated presentations are possible in which the building appears to rotate in the display. Added realism can be provided by producing shadows corresponding to those that would be cast by one or more light sources. Different colours can be assigned to different planes in such displays, and these can also be textured to represent different architectural surfaces, such as bricks, concrete, wood, plaster, stones, roof tiles, and so on. The texture patterns are usually most realistic if derived from photographs of real examples of such materials. The patterns are then stored in the computer as digital signals for incorporation into areas as required; their shapes must, of course, be changed so as to include the proper perspective, and they must have any incorporated shadows superimposed upon them. Such systems can be operated in an interactive way by devices for changing the parameters in the program. Finally, it is possible, by mixing the presentation of the building design with signals derived from photographs of the environment in which the building will be situated, to display a view of what the complete scene will look like before any actual building work has started. (See Plate 28, page 416.) Similar techniques can also be used in connection with graphic arts scanners (see Section 29.9 and Plates 37, 38, and 39, pages 576, 577).

The limited resolution of colour television displays and of colour VDUs is often inadequate for the use of computer generated pictures in connection with architecture, design, space exploration, scientific research, and other similar applications. Hence, when pictures having high resolution are required, other means of producing them must be employed: one method often used is to display high definition images on a monochrome cathode-ray tube, and then print them sequentially through red, green, and blue filters on to photographic colour film or paper; arrays of 2000×2000, or 4000×4000, or 5000×5000 pixels are commonly used, but normally only for still pictures.

25.7 Computer assisted cartoons

An animated video graphic can be thought of as a simple cartoon. In normal cartoons, there is usually much more motion and the use of more complicated shapes, but computers can be very useful in helping to 'draw' cartoons by reducing very greatly the number of individual sketches needed to be drawn by hand.

In one type of computer assisted cartoon, the key sketches are line drawings that the artist makes with a pen on a tablet; a colour, chosen from a palette of 250 000, is assigned to each area. The computer is then programmed to draw and colour all the intermediate frames. The computer keeps track of all the changes in geometry and colour of the individual areas as they change size and position, or even disappear. Background and foreground elements are often distinguishable correctly merely on the basis that only the foreground elements change, but, if necessary, priority instructions can be included. Background and foreground elements can be stored in the computer for later use, and, when recalled, can be repositioned, changed in size, recoloured, rotated, and transformed in many other ways. (Halas, 1982.)

25.8 Colour coding in pictures

Colour in pictures can be used for conveying information about scenes that is not related to their normal colour appearance. For example, in astronomy, very small differences in the colour of the light emitted by stars, planets, and other objects in the sky, can be amplified to provide pictures having areas of clearly different colours. Such pictures can provide astronomers with very useful representations of variations of important parameters, such as temperature or reflectivity; but the pictures do not correspond to appearances of the scenes that could ever be seen by a human observer.

In section 12.10, the use of films sensitized so as to reproduce foliage red instead of green were described. When the information in pictures is available in digital form (either by scanning film or by using electronic cameras, and then using an analog to digital converter), computers can be used to provide great flexibility in the choice of colours to be used in the final picture. One application of this technology is in the display of the temperature at each point in a scene. The scene is scanned with a detector sensitive to infra-red radiation, and the data is stored in a computer. In the display, all areas having levels of radiation within one narrow range are assigned to one colour, those within a second range, another colour, and so on. A coloured contour map of the scene is thus produced in which colour and temperature are related (England and Parker, 1972).

Another application of this type of technology is in the exploration of earth resources. Pictures of the earth taken from aircraft, or from satellites, using detectors sensitive not only to the visible, but also to the infra-red and ultra-violet, regions of the spectrum, are stored in computers and then displayed with colour codings designed to emphasize features of special interest. (Goetz, Billingsley, Gillespie, Abrams, Squires, Shoemaker, Lucchitta, and Elston, 1975.)

For space exploration, detectors with an even wider range of spectral sensitivities are used to obtain the maximum possible amount of information from the planets and moons visited by the probes; the resulting pictures are again coded in colour to clarify differences in properties from one area to another. (Huck and Wall, 1976.) Pictures expected to be obtained by space probes can be simulated in advance by computers as an aid to interpreting the real pictures when they become available (see Plate 29, pages 416, 417).

25.9 Colour ranges

It is clear, from the above, that computers are very powerful tools when used in creating pictures, and we must now consider the factors governing the choice of colours in various applications.

For the display of alpha-numeric characters and simple graphic elements, as in captions and teletext, the most important requirement is usually clarity of display. If only 2 colours are permissible, then white symbols on a black background are usually chosen, because monochrome monitors can then be used. The use of black letters on a white background is usually less satisfactory, because the self-luminous nature of the display makes the large background area an unpleasant source of visual glare; another important point is that the large white background area appears to flicker much more than the much smaller areas of the symbols when they are displayed white. If a colour monitor or VDU is used, yellow symbols on a black background may be considered more pleasing, but are not likely to be any more legible than white symbols. Similarly, blue, instead of black, symbols on a white background may be more pleasing, but no more legible than black.

The choice of one from two colours requires one bit of information; with 2 bits of information, the choice can be from four colours ($2^2 = 4$). With a black background, colours of high luminance are required to give the greatest contrast, and green (or cyan), yellow (or white), and red (or orange) are sometimes used on VDUs. If maximum colour difference between the colours is being sought for a VDU display, then it might be thought that red, green, and blue, the colours of the phosphors, would be the best to use; but the colour produced by the blue phosphor has too low a luminance compared to those produced by the red and green phosphors. Blue light is also difficult for the eye to focus and is therefore unsuitable for the display of small symbols. If a bluish colour is required, a desaturated blue, obtained by mixing some light from the red and green phosphors with that from the blue phosphor, can be used. If the red is too dark, an orange can be used instead. With three bits of information, the choice can be from

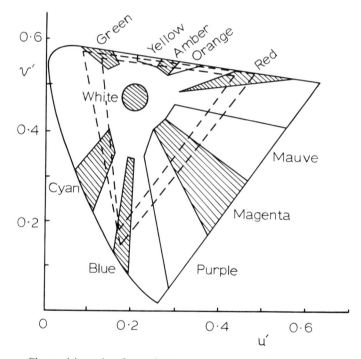

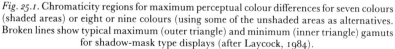

Fig. 25.1. Chromaticity regions for maximum perceptual colour differences for seven colours (shaded areas) or eight or nine colours (using some of the unshaded areas as alternatives. Broken lines show typical maximum (outer triangle) and minimum (inner triangle) gamuts for shadow-mask type displays (after Laycock, 1984).

eight colours ($2^3 = 8$); for a black background, red, green, cyan, white, and yellow (or amber), are good choices, but the addition of two more colours is difficult. There are only four perceptually unique hues, red, yellow, green, and blue (all other hues being describable by combinations of two of these hues), and these unique hues, together with white and black, provide six uniquely identifiable colours; the colours already chosen embrace these six (regarding the cyan as representing blue). Furthermore, it is found that for the display of symbols and simple diagrams, the use of more than five colours and a background can be confusing; hence, the addition of two further colours, such as magenta (sometimes called pink) and blue, is not found to be very advantageous in some applications. (Christ, 1975; Robertson, 1980).

In Fig. 25.1 the shaded areas show chromaticity regions for seven colours chosen to have maximum perceptual colour differences (Laycock, 1984); their colours are white, cyan, green, amber, red, magenta, and blue. The number can be increased to eight or nine by using yellow and orange, instead of amber, and mauve and purple, instead of magenta; the areas of these extra colours are shown by the unshaded areas in the figure. Table 25.1 shows how the colours might be chosen when the number of colours required ranges from two to nine. In Table 25.2 are given the chromaticities of corner points defining the areas shown in Fig. 25.1, except that the area for white is a circle of radius 0.028 (in the u', v' diagram) centred on the chromaticity of Standard Illuminant D_{65}. When colours are displayed at small angular subtense, discrimination in directions towards or away from blue is reduced (see Section 22.10); this reduction can be to as little as one fifth of that in red-green directions for areas subtending a few minutes of arc (Phillips, 1985).

For video graphics, in which more elaborate displays are produced, a wider palette of colours is usually provided. One of the most important additions to provide is a range of colours of the same chromaticity as one of the basic colours, but having different luminance factors; this is because it is these colours that

TABLE 25.1

Choice of colours when various numbers are required

Number	2	3	3	4	5	5	6	7	7	8	8	9
White			x			x	x	x	x	x	x	x
Cyan (turquoise)			x	x	x	x	x	x	x	x	x	x
Green	x	x		x	x	x	x	x	x	x	x	x
Yellow										x	x	x
Amber		x		x	x	x	x	x	x			
Orange			x							x	x	x
Red	x	x		x	x	x	x	x	x	x	x	x
Mauve								x		x		x
Magenta (pink)					x		x		x		x	
Purple								x		x		x
Blue									x		x	x

TABLE 25.2

Chromaticity coordinates of corners defining the boundaries of Fig. 25.1

Corner	1		2		3		4	
	u'	v'	u	$'v'$	u'	v'	u'	v'
Colour								
Cyan	0.0343	0.0000	0.1554	0.3468	0.1455	0.4026	0.0000	0.2196
Green	0.0509	0.6000	0.1352	0.5244	0.1592	0.5429	0.1297	0.6000
Yellow	0.2047	0.6000	0.2022	0.5522	0.2467	0.5366	0.2921	0.6000
Amber	0.2921	0.6000	0.2467	0.5366	0.2719	0.5177	0.3951	0.6000
Orange	0.3951	0.6000	0.2677	0.5149	0.4193	0.5593	0.4574	0.5750
Red	0.4898	0.5500	0.4285	0.5328	0.2884	0.4795	0.6500	0.5243
Mauve	0.7000	0.3747	0.2876	0.4515	0.2662	0.4196	0.7000	0.1104
Magenta	0.7000	0.1104	0.2662	0.4196	0.2386	0.3949	0.4582	0.0000
Purple	0.4582	0.0000	0.2386	0.3949	0.2230	0.3336	0.2854	0.0000
Blue	0.1885	0.0000	0.1951	0.3313	0.1827	0.3405	0.1424	0.0000
White	0.1978	0.4683						
	with circle of radius 0.028							

make it possible to represent solid objects with the effects of directional lighting and shadows included. With 4 bits, the choice can be from 16 colours ($2^4 = 16$), and 5 basic colours could then be used each at 3 different levels of luminance factor, or 3 basic colours each at one level and another 3 basic colours each at 4 levels. One system that has been used provides for the use of 16 colours, which can be selected from a palette of 64 different levels in each of the red, green, and blue display beams, amounting to 262 184 different colours (Brockhurst, Day, Dyer, and Vivian, 1982). Systems using 6 bits (64 colours) and 9 bits (512 colours) have also been described (Long, 1982).

To achieve greater pictorial realism than is usually required in diagrams and cartoons, a larger number of colours is required, and their selection in the most economical way is an interesting problem. The most important requirement is probably a fully graded range of luminance factors. To give complete absence of contouring effects in pictures it is usually agreed that about 256 levels, or 8 bits, are required; 256 levels in all three channels of a display results in about 16.8 million different colours (however, the maximum number of distinguishable surface colours is usually estimated at about 10 million; Judd and Wyszecki, 1975). But quite good results can be obtained with 64 levels, or 6 bits. If these are available at any one of 32 different chromaticities (5 bits), then a total of 11 bits would be needed. The 32 chromaticities could be chosen to be distributed systematically in a uniform chromaticity diagram; thus one chromaticity chosen in each of the areas shown in Fig. 25.2 would give a rough coverage of the whole gamut of chromaticity covered by the phosphors. Where desirable, computer programs can be used that result in the changes in luminance factor and chromaticity that occur, as a colour boundary is crossed, to be gradual rather

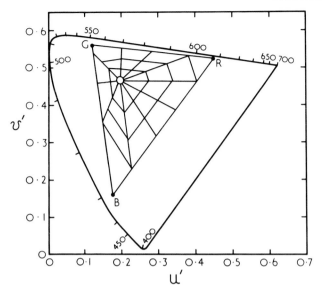

Fig. 25.2. Division of the chromaticity gamut of a typical television display or VDU into 31 discrete areas. Availability of one colour in each area, when combined with a full range of levels of luminance factors, would provide a wide range of display colours.

than abrupt. Of course, subtle fluctuations in chromaticity cannot be reproduced by such a scheme, but, if combined with insertions of texture patterns derived from real objects, then a high degree of realism is attainable. In this way it is possible to derive pictures from computers that are similar in realism to those obtained using cameras, at least for some scenes.

REFERENCES

Brockhurst, D., Day, S., Dyer, M., and Vivien, R. H., *J. Royal Television Soc.*, **19,** 33 (1982).

Christ, R. E., *Human Factors*, **17,** 542 (1975).

Davidse, J., and Koppe, R. P., *J. Soc. Mot. Pic. Tel. Eng.*, **86,** 140 (1977).

England, G., and Parker, A. K., *Photogram. Engng.* **38,** 590 (1972).

Halas, J., *Video*, **8,** 14 (August, 1982).

Goetz, A. F. H., Billingsley, F. C., Gillespie, A. R., Abrams, M. T., Squires, R. L., Shoemaker, E. M., Lucchitta, I., and Elston, D. P., *Tech. Rep. 3L-1597* of the Jet Propulsion Laboratory, California Institute of Technology, Pasadena, California (1975).

Huck, F. O., and Wall, S. D., *Appl. Optics*, **15,** 1748 (1976).

Judd, D. B., and Wyszecki, G., *Color in Business Science and Industry*, 3rd Edn., p. 388, Wiley, New York (1975).

Laycock, J., *Displays*, p. 3 (January, 1984).

Long, C., *J. Soc. Mot. Pic. Tel. Eng.*, **91,** 725 (1982).

Nakamura, J., and Kamakura, K., *J. Soc. Mot. Pic. Tel. Eng.*, **90,** 107 (1981).

Phillips, D. L., *I. E. R. E. Publication No. 61*, p. 85 (1985).

Robertson, P. J., IBM Technical Report G320–6296 (1980).

Wheeler, L. J., *Principles of Cinematography*, 4th Edn., p. 262, Fountain Press, England (1969).

PART FOUR

COLOUR PRINTING

CHAPTER 26

Photomechanical Principles

1. Introduction – *2*. Letterpress – *3*. Lithography – *4*. Gravure –
5. Superimposed dye images – *6*. Superimposed dot images – *7*. Colorimetric
colour reproduction with dot images – *8*. Colour correction by masking –
9. Contact screens – *10*. Autoscreen film – *11*. Colour photocopying

26.1 Introduction

IF a large number of copies of a colour reproduction are required, the cheapest method is usually to transfer colorants from some surface containing the image to a less expensive surface such as mordanted cloth, paper, or gelatin-coated film base, the image-bearing surface then being re-coloured for subsequent transfers. Thus *Technicolor* films (see Sections 12.2 and 12.11), were printed by successively transferring cyan, magenta, and yellow dye images from matrices consisting of gelatin relief-images to suitably prepared gelatin-coated film base. In the textile industries, except when the pattern is woven into the fabric, coloured designs are printed on to the material by rollers embossed with the required design and suitably loaded with dye, or by sublimating dyes from printed paper, or by using stencils as in the silk screen process. And it is the role of the printing industry to provide inexpensive multiple copies of colour reproductions for inclusion in magazines, books, posters, wrappers, and similar items.

The main methods adopted for colour reproduction by the printing trade have been developed from those used for many years in ordinary monochrome printing, known as *letterpress, lithography*, and *photogravure*. It is helpful to consider the characteristics of these three methods in monochrome printing before going on to a consideration of their application to colour reproduction.

26.2 Letterpress

Letterpress, as its name implies, is a method widely adopted for the reproduction of *letters*, and many (though not all) newspapers, and some magazines and books are letterpress productions. The method originated in the hand

527

engraving of wooden blocks so that the areas to be printed light were gouged out of the wood, while those to be printed dark were left untouched. Running an inky roller over such a surface resulted in the untouched parts being inked and the gouged out parts not being inked, as shown in Fig. 26.1(a). By pressing the paper into contact with a block inked in this way, the required pattern of ink was obtained on the paper. Movable type, whereby letters are carved on small blocks that can be arranged and rearranged to form different matter, was invented independently in China in the tenth century and by Gutenberg in Germany in the middle of the fifteenth century, and the letterpress system, in its literal sense, was born.

It is clear that the letterpress system tends to be an all or nothing affair. An area is either inked or not inked, and hence either black (assuming black ink is being used) or white. Thus only two colours can be produced by this system, that of the ink and that of the support. For the printing of letters this is ideal, and results in the well-known clarity of letterpress reading matter. But for printing black-and-white photographs it is necessary to reproduce not only black and white, but also all the tones of grey in between. The way in which this has been done in the letterpress system is very ingenious. The same physiological property of the eye is utilized as in the mosaic processes of additive colour reproduction. As we have seen, in these processes, the fairly sharply defined limit below which the eye ceases to resolve fine detail results in the individual red, green, and blue areas blending into all the intermediate colours. (see Section 3.3.)

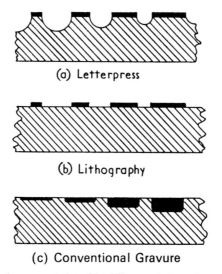

(a) Letterpress

(b) Lithography

(c) Conventional Gravure

Fig. 26.1. Diagrammatic representation of the differences between letterpress, lithographic, and gravure methods of printing. Cross-sections of the three types of printing surface after inking are shown.

Fig. 26.2. Above: Small portion of a half-tone screen, highly magnified. *Below:* Distribution of light (highly magnified) obtained when a half-tone screen is placed a short distance away from the plate. This is also the appearance of a *contact screen.*

The letterpress method of reproducing greys is to print them as mosaics of black and white dots which the eye blends into greys. The formation of the dots is achieved by photographing the original through a *screen* which is placed a short distance in front of the photographic negative plate or film being used. The screen consists of two glass plates, on which have been ruled very fine opaque lines, cemented together with the lines at right-angles to one another, leaving square interstices as shown in the upper part of Fig. 26.2. These interstices can be thought of as being rough pinhole 'lenses' forming out-of-focus 'images' of the camera lens on the plate. An extremely high contrast film is used so that after development the image consists almost entirely of black dots and white spaces. However, the pinhole images of the camera lens in areas light in the original, because they are bright, will produce larger areas of developable latent image than those in areas dark in the original. Thus, in areas light in the original, large black dots will be produced after development, but in areas dark in the original only small black dots will be formed. Hence the appearance of a negative, when highly magnified, is similar to that shown in Fig. 26.4. In areas very light in the original, the black dots are so large as to overlap and leave the 'white-dot' pattern shown. This type of image structure is often referred to as *half-tone*, and the screens from which they are made as *crossline* or *half-tone screens*. Cameras in which screens are used are referred to as *process cameras* and the negatives they produce as *screened*, or *half-tone*, or *dot* negatives.

By various photomechanical processes which need not concern us here, screened negatives are printed on to copper or zinc plates which are then treated so that the black areas of the negatives result in copper being etched away, while the white areas leave it untouched. It is clear that upon inking such a copper plate a positive dot image (as shown in Fig. 26.5) is obtained which can be printed on to paper. If the dot pattern is sufficiently fine it will blend to form grey tones similar to those in the original photograph. From the point of view of successful blending into grey the finer the dot pattern the better. The fineness of the dot pattern is set by that of the screen and can be made as fine as 300 lines per inch (120 lines per cm). Unfortunately, however, dot patterns of this fineness can only be printed successfully on very smooth high quality paper, and 133 lines per inch (53 lines per cm) are generally used for magazine work, while for newspaper work 65 or 85 lines per inch (26 or 34 lines per cm) may be used, which is why the dot structure of newspaper photographs is somewhat obtrusive. Incidentally, the abrupt limit of resolution of the eye is well illustrated by moving gradually further and further away from a newspaper photograph: its dot structure will be found to vanish quite suddenly. If at this distance the photograph is rotated through 45 degrees it will be found that the dot-structure at once reappears; this is because the screen patterns in black-and-white dot reproductions are always arranged at $45°$ (as shown in Fig. 26.2) and not vertically-and-horizontally, and the eye is less able to resolve fine detail at $45°$ (probably because the eye and brain system codes vertical and horizontal information preferentially to diagonal information).

It will be realized that the method of forming grey tones by the letterpress

531

method is sufficiently complicated to induce distortions in the tone reproduction. Thus some greys will be too dark and others too light. Much of the skill in successful letterpress reproduction lies in the careful choice of lens aperture, distance from screen to photographic material, and development, in making the screened negative. In spite of every care at these and subsequent stages, however, hand-correction of important areas of the picture may be required (Bryngdahl, 1978).

26.3 Lithography

In lithography greys are again reproduced by means of physiological blending of dots of different sizes, but the printing plate is not etched as in the letterpress system but is quite flat as shown in Fig. 26.1(b). The early stages of the process are the same as in the letterpress method in that a screened negative is produced. The screened negative is then printed on to a plate in such a way that a greasy ink can be deposited in the interstices but not in the dot areas. (Alternatively, a screened positive can be used with a plate such that a greasy ink is deposited only in the dot areas.) The method of keeping the greasy ink out of the interstices is generally to make them highly water-accepting and then to wet the surface immediately prior to inking. The printing cycle then consists of wetting, inking, printing, wetting, inking, printing, etc. It is customary to print from cylinders rather than flat plates, and the ink image is often transferred from the cylinder to a rubber-covered roller, which then prints it on to the paper, a technique known as *offset printing*.

26.4 Gravure

In contrast to the letterpress and lithographic methods, the gravure method (in its conventional form) does not produce greys by means of varying the relative sizes of black dots and white interstices. The printing plate has hollows or *cells* filled with ink, while unrecessed portions are left clear (Fig. 26.1(c)). Light or dark tones are then printed by transferring small or large quantities of ink from shallow or deep hollows respectively. At first sight, in such a system, there would seem to be no need for any screen pattern, there being no apparent need to break the image into dots. But it is essential in the gravure method that the unetched areas be absolutely free from ink, and this can only be achieved by wiping them clean after the plate as a whole has been inked. But in the wiping operation, large etched areas would tend to be wiped clear of ink also. To avoid this, a very fine screen, usually 175 lines per inch (70 lines per cm), and with interstices three times as wide as the lines, is generally used in making the plate in order to provide a fine honeycomb of unetched walls in the large etched areas. The purpose of this screen is thus quite different from those used in the letterpress and lithographic methods where the object is to form dots of different sizes; in conventional gravure the 'dots' are all the same size, and remain square at all densities. In colour gravure there are practical difficulties in controlling the colour in the shallow hollows, and processes have been devised to try to overcome this by making the shallower hollows smaller, somewhat like the letterpress half-tone, but of course with the dots sunk into the surface instead of raised above it.

The gravure method is well suited to paper of only medium smoothness, because there is no need for the 175-line screen to be sharply reproduced, because it has no (or little) image-producing function. For this, and other reasons, gravure is widely used for printing fairly inexpensive weekly magazines, where good quality paper is precluded by its expense, but where the more costly gravure etched printing cylinders are justified by the large circulation of the periodical. The use of gravure for newspaper work is precluded by the difficulty of making changes to the printing cylinders once they have been etched.

26.5 Superimposed dye images

We have seen that, in colour photography, the subtractive methods employ three superimposed dye images, of cyan, magenta, and yellow colours. In Chapter 18 we considered the microscopic nature of these images. The black-and-white photographic images from which all colour photographs are ultimately derived are made up of minute particles, or *grains* as they are usually called, of finely divided metallic silver. In some transfer processes, such as *Technicolor*, the Kodak *Dye Transfer* process, and the in-camera systems (see Section 17.10), the dye diffuses somewhat and the granular structure of the original silver image (from which the dye image has been obtained) may be almost entirely blurred over. In such cases the process is a truly subtractive one. From the point of view of picture sharpness, however, such dye diffusion is undesirable, and most subtractive processes result in the three dyes being deposited in granular form, either closely following the structure of the parent silver image, as in the case of *Kodachrome*, for instance, or modifying it slightly by means of a superimposed coupler structure as in the case of *Ektachrome* and *Kodacolor*, for instance, as shown in Figs. 12.4, 12.5, and 12.6. (See also Plate 10, pages 112, 113, and Plate 16, pages 240, 241.)

These discontinuities in photographic dye images are generally of so fine a pattern that to the naked eye they are quite invisible, and indeed for this reason some subtractive processes have carried a reputation of being 'grainless'. A powerful microscope, however, soon reveals that the image has a random structure, as was also the case with the random mosaic additive processes (although their structure was about ten times as coarse).

26.6 Superimposed dot images

It has been seen in the previous section that most subtractive photographic images consist of superimposed granular dye patterns. Their success encourages the hope that, in printing, successful colour reproduction can be achieved simply by printing three dot images one on top of the other. In fact, trichromatic colour reproductions depending on this principle were made by Jakob Christoffel LeBlon as long ago as the early 1700s (see Birren, 1981). The three dot images then have to be made from red, green, and blue separation negatives and printed in cyan, magenta, and yellow inks. This, of course, is widely practised, although, as mentioned in Section 4.2, the colours are often termed blue, red, and yellow in the printing trade. Very frequently a black dot image is also printed, partly to make up for certain deficiencies in the colours of some printing inks which make a good black difficult to attain, and partly for various other reasons which include: facilitating registration, reducing variations of colour balance in blacks, and using the technique of *under-colour removal* (to be described in section 29.14) to reduce ink costs (see Plate 35, pages 544, 545). In order to avoid undesirable patterns (*moiré patterns*) caused by superimposed parallel lines of dots of different colours, the dot images are always printed with the lines of dots running at different angles, and a common arrangement is for the black ink to be printed at the least obtrusive angle, that is 45°, for the cyan to be printed at 30° to one side of the black, the magenta at 30° to the other side, and the yellow, being the colour for which the dots are least noticeable, midway between the cyan and magenta, that is, vertically and horizontally, as in Fig. 26.3.

In conventional gravure work the image is not broken into dots of different sizes, and since the original silver grain structure is generally far too fine to be obtrusive in the final reproduction, the process tends to be truly subtractive rather like the photographic transfer systems. In letterpress and lithographic

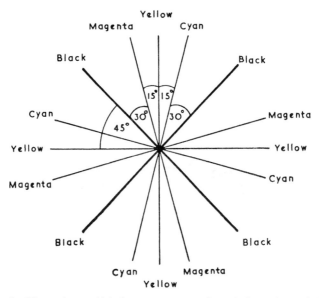

Fig. 26.3. The angles at which the screens are usually set in four-colour printing.

colour reproduction, however, the result is a cross between a subtractive and an additive mosaic process. For, consider three dot images, cyan, magenta, and yellow, superimposed on one another. A mosaic of eight different colours will be formed: white, where three gaps in the dot images coincide; black, where three dots are superimposed; cyan, magenta, and yellow, where the ink dots are seen against gaps in the other images; red, where magenta and yellow dots are superimposed; green, where cyan and yellow dots are superimposed; and blue where magenta and cyan dots are superimposed. (See Plate 31, page 512.)

In practice, the distortions in tone rendering and the deficiencies of the inks used in all three processes generally wreak such havoc with the colour reproduction that a great deal of colour correction is usually desirable. Various methods of correcting such errors automatically have been devised and several of these are described in Chapters 28 and 29.

26.7 Colorimetric colour reproduction with dot images

In 1937 Neugebauer (Neugebauer, 1937) treated the problem of eight-colour mosaics quite generally in the following way. Let c be the area of paper covered by the cyan dots, m the area covered by the magenta dots, and y the area covered by the yellow dots, per unit area of paper. It follows that the area not covered by cyan is $1-c$, that not covered by magenta $1-m$, and that not covered by yellow $1-y$, per unit area of paper. The probability of any particular point on the paper being covered by a cyan dot is obviously equal to c, by a magenta dot, equal to m, and by a yellow dot, equal to y. Hence the probability of any particular point on the paper being covered by all three colours is equal to the product cmy; hence the area covered by all three inks, which of course is the area that is black, is equal to cmy per unit area of paper. Similarly the probability of any particular point on the paper being not covered by any of the three colours is equal to $(1-c)(1-m)(1-y)$; hence the area that is white is equal to $(1-c)(1-m)(1-y)$ per unit area of paper. By similar reasoning the areas of each of the eight colours can be evaluated, and the results are as follows, the symbols in brackets preceding each colour name representing the CIE tristimulus values for that colour:

$$(X_1, Y_1, Z_1)\text{White} \quad (1-c)(1-m)(1-y) = f_1$$
$$(X_2, Y_2, Z_2)\text{Cyan} \quad c(1-m)(1-y) = f_2$$
$$(X_3, Y_3, Z_3)\text{Magenta} \quad m(1-c)(1-y) = f_3$$
$$(X_4, Y_4, Z_4)\text{Yellow} \quad y(1-c)(1-m) = f_4$$
$$(X_5, Y_5, Z_5)\text{Red} \quad my(1-c) = f_5$$
$$(X_6, Y_6, Z_6)\text{Green} \quad cy(1-m) = f_6$$
$$(X_7, Y_7, Z_7)\text{Blue} \quad cm(1-y) = f_7$$
$$(X_8, Y_8, Z_8)\text{Black} \quad cmy = f_8$$

Now suppose we have some patch, P, of colour in our original, the tristimulus values of which were X_P, Y_P, Z_P. It is clear that colorimetric colour reproduction would result if the reproduction had the same tristimulus values X_P, Y_P, Z_P. The conditions for this to be so can be set out quite simply as follows:

535

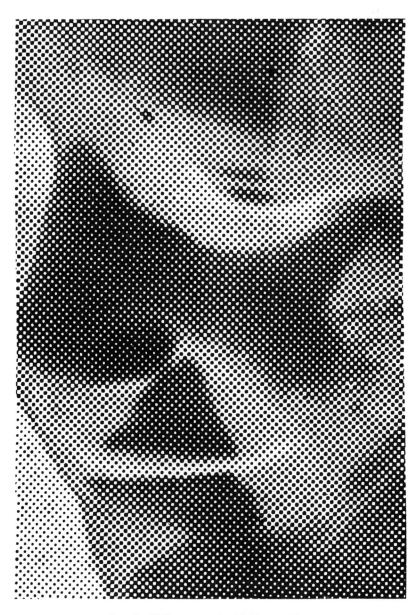

Fig. 26.4. Half-tone negative, highly magnified.

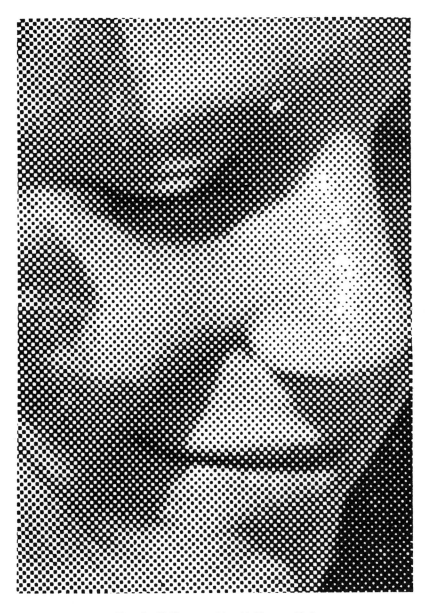

Fig. 26.5. Half-tone positive, highly magnified.

$$f_1X_1+f_2X_2+f_3X_3+f_4X_4+f_5X_5+f_6X_6+f_7X_7+f_8X_8 = X_P$$
$$f_1Y_1+f_2Y_2+f_3Y_3+f_4Y_4+f_5Y_5+f_6Y_6+f_7Y_7+f_8Y_8 = Y_P$$
$$f_1Z_1+f_2Z_2+f_3Z_3+f_4Z_4+f_5Z_5+f_6Z_6+f_7Z_7+f_8Z_8 = Z_P$$

That these must be the conditions follows from the additivity of colour equations (see Section 7.4). For in our patch of superimposed dots we have f_1 units of white, and f_2 units of cyan, which add together thus:

$$f_1 \text{units of white} \equiv f_1X_1(X)+f_1Y_1(Y)+f_1Z_1(Z)$$
$$f_2 \text{ units of cyan} \equiv f_2X_2(X)+f_2Y_2(Y)+f_2Z_2(Z)$$

Hence f_1 units of white additively mixed with f_2 units of cyan $\equiv$

$$(f_1X_1+f_2X_2)\,(X)+(f_1Y_1+f_2Y_2)\,(Y)+(f_1Z_1+f_2Z_2)\,(Z)$$

It is clear, therefore, that the additive mixture of all eight colours will have as coefficients of (X), (Y), and (Z) the expressions given in the left hand sides of the set of three equations shown above. If, then, these coefficients are identical to X_P, Y_P, and Z_P, colorimetric colour reproduction will have resulted.

But the three equations given above are really equations for c, m, and y, and, if X_1, Y_1, Z_1, X_2, Y_2, Z_2, etc. and X_P, Y_P, Z_P are known, they can be solved for c, m, and y. These values of c, m, and y are, then, the fractional areas of inks that must be printed in order to produce the colour of the original in the patch P. If therefore the original colours could all be analysed in terms of their tristimulus valves X, Y, and Z, all the corresponding values of c m, and y calculated, and the printing plates etched so that these amounts of ink were printed at each point of the picture, colorimetric colour reproduction would be achieved. In spite of the obvious complexity of such a procedure, Hardy and Wurzburg (Hardy and Wurzburg, 1948) invented a method of achieving this, which will be described in Section 29.2.

Of course, some colours in the original may be too saturated to be matched by any mixture of the three inks being used, and in this case one or more of the values of c, m, and y will become negative. As it is impossible to print a negative amount of ink on the paper the best that can be done is to print no ink at all at these points, and this can result in errors in hue and saturation. Hardy and Wurzburg also considered the case of four-colour printing, involving a black ink as well as the cyan, magenta, and yellow inks (Hardy and Wurzburg, 1948).

26.8 Colour correction by masking

The Hardy and Wurzburg method is of great interest for both its elegance and its theoretical possibilities, but its practical realization is obviously complicated. Simpler methods of obtaining some improvements in colour reproduction have therefore been sought, and, apart from tedious hand retouching of individual areas on the printing plates, the use of masking, either manually or in scanners, has been the method most widely used. For instance, Pollak, by making a few assumptions concerning the nature of the inks, has solved the Neugebauer

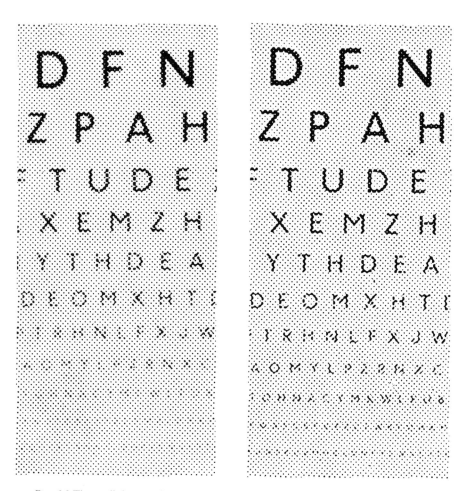

Fig. 26.6. The small charts at the top are reproduced by means of a conventional ruled screen (on the left), and by a contact screen (on the right). The lower charts are photo-micrographs of parts of the upper charts and show that the contact screen reproduces more fine detail because fine lines are less broken into dots.

540

equations and derived a system of masking based upon them (Pollak, 1955). So far, however, the methods most widely used have been worked out empirically, and some of them are described in Chapter 28, while Chapter 29 deals with masking in scanners. (See Plate 32, pages 512, 513, for an example of masking.)

Much thought has been given to the validity of applying the continuous tone type of masking theory, dealt with in Chapter 15, to half-tone images. The topics discussed have included the additivity of half-tone densities, and the necessity of using non-linear mask characteristics (Pollak, 1955 and 1956; Yule and Clapper, 1955; Preucil, 1953; Pollak and Hepher, 1956).

26.9 Contact screens

An important device is the *contact screen* (Yule, Johnston, and Murray, 1942). In normal practice in the gravure method, the screen is printed in contact with the photographic material, but its only function is to prevent large inked areas being wiped clean. But in the letterpress and lithographic methods the half-tone screen is deliberately printed out of contact with the photographic material so that dots of different sizes can be produced.

At a given lens aperture, a screen at a given distance from the photographic material results in a certain distribution of light. If a film exposed under these conditions were developed in an ordinary developer, instead of in a very high contrast developer, and a positive made from the negative thus obtained, an approximate record of the original light distribution would result. If now this 'photographic screen' were placed in contact with a suitable film in a process camera, the ordinary screen could be removed, for the photographic screen gives approximately the same light distribution on the film. The appearance of such a *contact screen* is as shown in the lower half of Fig. 26.2. Unlike ordinary half-tone screens, or the screens used in gravure, the contact screen is 'vignetted' with a range of intermediate densities.

At first sight there may not seem to be any advantages in such a system, but there are in fact several. First, the production of the conventional type of screen, by the traditional ruling methods, is a very costly process. But large numbers of contact screens can be produced from one screen used as a master.

Secondly, the fact that the contact screen is used *in contact* with the photographic film means that fine detail is reproduced more clearly. Thus with the conventional type of screen, which has to be used *out of contact*, a long fine line, for instance, can only be reproduced as a line of dots, all of which are approximately circular or square in shape. With a contact screen, however, a fine line will be reproduced as a line of dots, each of which is elongated in the direction of the line. The reproduction of the line will therefore be finer and less broken up than with the conventional screen method (Hepher, 1953). This is illustrated in Fig. 26.6.

Thirdly, by making the contact screen a magenta dye-image instead of a black image, a very simple method of controlling contrast is obtained. If such a magenta screen is viewed through a red filter, since magenta dyes absorb little or no red light, the screen pattern becomes virtually invisible. If it is viewed through a blue filter, the blue absorption that all magenta dyes exhibit enables the screen pattern to be seen at a low contrast. If it is viewed through a green filter, the magenta dye being a heavy absorber of green light, the screen pattern is seen at its maximum contrast. If, therefore, the magenta screen is used in conjunction with an orthochromatic film, which is sensitive to both blue and green light, the contrast can be varied by altering the colour of the exposing light from blue to green; and intermediate contrasts can be obtained by using blue-green filters or by giving part of the exposure through a blue filter and part through a green filter. (In practice, since orthochromatic films are insensitive to red light a magenta filter can be used instead of a blue filter, and a yellow instead of a green; magenta and yellow filters are in fact preferable because they are generally more efficient transmitters of the required light.)

Paradoxical as it seems, when this system is adopted, it is the exposure to green light which gives low contrast, and that to blue light which gives high contrast. The reason for this can best be understood by referring to Fig. 26.7. In this figure the density of the screen along part of a line of dots is plotted for green light in the upper diagram and for blue light in the lower diagram. Two exposure levels E_1 and E_2 are indicated on both diagrams, together with the dot sizes d_1 and d_2 that result from them in both cases. It is clear from the figure that it is the lower

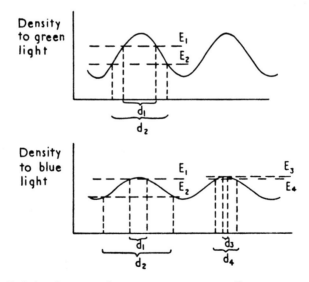

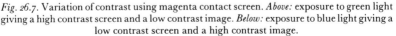

Fig. 26.7. Variation of contrast using magenta contact screen. *Above:* exposure to green light giving a high contrast screen and a low contrast image. *Below:* exposure to blue light giving a low contrast screen and a high contrast image.

contrast screen (obtained by exposure to blue light) that results in the larger differences $d_2 - d_1$ between the dot sizes produced by the given exposure difference $E_1 - E_2$. Hence the blue light exposure results in a higher half-tone contrast than the green light exposure. This very simple means of controlling the contrast of half-tone images is obviously a very valuable tool in the hands of the printer, and properly handled can result in considerable improvement in quality.

In spite of the advantage of easy control of contrast offered by magenta contact screens, grey contact screens are widely used, the contrast then being controlled by *flashing* (giving part of the exposure without the original, thus effectively reducing the contrast range of originals). When it is required to carry out colour separation by *direct screening* of coloured originals (see Section 28.6) magenta screens cannot be used, and grey screens are used with high contrast panchromatic films.

The fourth advantage of the contact screen is that, when it is used, highlights tend to be reproduced at a somewhat higher gamma and therefore gain in clarity and brilliance. This is because, in the portions of the screen that are responsible for highlight rendering, the modulating density pattern passes through a maximum and is therefore of very low contrast, resulting as before in contrasty reproduction. This is illustrated in Fig. 26.7 by the two highlight exposure levels E_3 and E_4, which, although very similar, result in a large difference $d_4 - d_3$ in dot-size.

The fifth advantage of contact screens is that, when no change in image-size is required, they can be used in vacuum contact-printing frames. They can also be used on the easels of enlargers.

26.10 Autoscreen film

Another interesting device is the autoscreen film (Yule and Maurer, 1954). In this material the photographic sensitivity is not constant over its area, but varies in the same pattern as that of the light distribution from a half-tone screen. It is therefore possible with this material, to use it in the camera or in a contact-printing frame, without either a conventional or a contact screen, and to obtain a half-tone, instead of a continuous-tone, image. The *Kodalith Autoscreen Ortho Film*, for example, thus enables half-tone images to be obtained without the need for the skill necessary for the successful manipulation of half-tone or contact screens (Maurer, 1956). However, for colour work, the necessity for using different screen angles makes its use rather inconvenient.

26.11 Colour photocopying

In many monochrome photocopiers, a black pigment is transferred to plain paper from a surface carrying an image in the form of a pattern of electrostatic charges. By having three such images, representing the red, green, and blue information of the original document, and transferring from them, in succession, the appropriate patterns of cyan, magenta, and yellow toners, respectively, copies in colour can be reproduced. However, just as in monochrome copiers the image characteristics are not such as to produce good copies from pictorial originals, so, in colour copiers, it is not possible to obtain good quality copies of colour pictures unless special techniques are used.

REFERENCES

Birren, F., *Color Res. Appl.*, **6,** 85 (1981).
Bryngdahl, O., *J. Opt. Soc. Amer.*, **68,** 416 (1978).
Hardy, A. C., and Wurzburg, F. L., *J. Opt. Soc. Amer.*, **38,** 300 (1948).
Hepher, M., *Penrose Annual*, **47,** 116 (1953).
Maurer, R. E., *Penrose Annual*, **50,** 97 (1956).
Neugebauer, H. E. J., *Z. tech. Phys.*, **36,** 22 (1937).
Pollak, F., *J. Phot. Sci.*, **3,** 180 (1955).
Pollak, F., *J. Phot. Sci.*, **4,** 65 (1956).
Pollak, F., and Hepher, M., *Penrose Annual*, **50,** 106 (1956).
Preucil, F., *Tech. Assoc. Graphic Arts*, **5,** 102 (1953).
Yule, J. A. C., and Clapper, F. R., *Tech. Assoc. Graphic Arts*, **7,** 1 (1955).
Yule, J. A. C., Johnston, F. B., and Murray, A., *J. Franklin Inst.*, **234,** 567 (1942).
Yule, J. A. C., and Maurer, R. E., *Penrose Annual*, **48,** 93 (1954).

GENERAL REFERENCES

Cartwright, H. M., *Ilford Graphic Arts Manual*, Ilford, London (Vol. 1, 1961; Vol. 2, 1966).
Gamble, C. W., *Modern Illustration Processes*, Pitman, London (1953).
Smith, W. J., Turner, E. L., and Hallam, C. D., *Photo Engraving in Relief*, Pitman, London (1932).
Yule, J. A. C., *Principles of Color Reproduction*, Wiley, New York (1967).

Plate 34

Binocular vision is particularly valuable in providing three-dimensional perception of objects nearer than about a metre: at greater distances, perception of solidity and depth is achieved mainly by clues provided by size, perspective, and lighting effects, as in the apparent shape of the boat in this example. Because most pictures are of objects more than a metre from the camera, stereoscopic reproductions using two cameras, and the viewing of two pictures simultaneously by the left and right eyes, has been little used (see Section 5.9). Reproduced from a 2¼ inch square *Ektachrome* transparency on a Crosfield *Magnascan* scanner.

Plate 35

Each black image has been used with the combination of cyan, magenta, and yellow images shown on its *left*, to provide the final result shown on its *right*. Under-colour removal (see Section 29.14) is absent in the *top* row, partially present in the *middle* row, and present to its maximum extent in the *bottom* row. The advantages of under-colour removal can include sharper rendering of fine detail, better consistency in the grey scale, quicker ink drying, and savings in ink costs. Maximum under-colour removal can increase registration problems, so that a partial level is usually preferred. Reproduced from a $2\frac{1}{4}$ inch square *Ektachrome* transparency on a Crosfield *Magnascan* scanner.

Plate 36
The enhancement of the contrast of edges is a useful means of increasing the apparent sharpness of pictures, and is widely practised by various means in photography, television, and printing. An example of the sort of difference in sharpness caused by enhancing edge-contrast on a scanner is shown by comparing (*top*) the result obtained without, and (*below*) with, such enhancement (see Sections 15.3 and 29.12). Reproduced from a 36 × 24 mm *Ektachrome* transparency on a Crosfield *Magnascan* scanner.

CHAPTER 27

Preparing the Copy and Checking the Results

Introduction

IN the publishing industry, it is common for editorial departments to select for reproduction 'originals' from many sources. Sometimes art work, which really is 'original', will be used, but very often colour transparencies, negatives, or reflection prints, will be used; and, although these are themselves reproductions, the term 'original' is generally also used to cover all such pictorial matter that the printer has to copy. These originals, in addition to varying in form, usually vary in several other respects. Thus there is often no uniformity of size, either of the originals, or of their reproductions, so that many different degrees of magnification have to be used. Some originals may require to be reproduced only in part, and therefore require to be cropped along one or more edges or sometimes to a complicated pattern if a composite of several original pictures is required (as is often the case for advertising work). If the originals are colour transparencies or negatives they often require correction to bring their density level, density range, and colour balance, into line with one another, or with some standard required by the reproduction system; reflection copy may also require some adjustments in these respects but is usually less variable.

    Uniform results from this wide range of original input can be achieved by making the necessary adjustments at the stages where the separation negatives are made; as will be described in the next two chapters, masking and adjustments on scanners can provide plenty of scope for making the necessary changes. This

545

procedure is widely used, especially where only a single original is printed on a page. But, if several originals are to be printed on the same page, different manipulation of the separations for each original may be a costly and time-consuming procedure.

27.2 Duplicating and converting originals

A technique widely used when several different originals have all to be printed on the same page is to duplicate or convert each original so as to produce corrected transparencies or reflection prints of the required size. These second originals are then mounted in their correct positions for page layout, and then separation negatives made for the complete page, either by a suitable masking procedure (see Chapter 28) or by means of a scanner (see Chapter 29).

The main advantage of the above procedure is that the page layout of the second originals provides an opportunity to check the quality of the page as a whole at an early stage. The method also usually reduces costs.

When composite layouts are required involving adjoining transparencies, the individual pieces of film can be cemented along their edges to form a butt-joint, or the film base can be dissolved and removed from the images and the parts of the images required cemented on to a fresh piece of film base. A smooth composite without any ridges is particularly necessary for use with scanners. When reflection prints are used, the task of producing the composite layout is easier, but it must be flexible if it is to be used with drum types of scanner.

Other advantages of making the separations from second originals are as follows. Extra sets of identical originals can be made and sent to other printers; this may be desirable in advertising work, for instance, where it may be necessary to reproduce from the same originals in various forms, such as gravure colour magazines, litho showcards, and letterpress cartons. By making the separations from complete page layouts, instead of making them from each picture individually, the number of pieces of film needing handling and registering in position (*planning*) is much reduced. The use of standardized reproduction procedures is facilitated by working from the same film or paper type for all the pictures, thus eliminating complications caused by the presence of various dye sets and differences in ultra-violet transmissions of films. Finally, scanner time can be used more economically.

27.3 Duplicating transparencies

If an original transparency is of good photographic quality, then the aim in duplicating it will usually be to match it as closely as possible, apart from any necessary change in size. To do this it is possible either to use a special duplicating film having a gamma of about 1.0, or to use a camera type film of gamma about 1.25 and to use a contrast-reducing mask, together with a highlight mask if necessary (see Sections 15.2 and 27.8). If the transparency is of too high a contrast, this must be corrected, and this can be done by using a contrast-reducing mask with the duplicating film, or by using more masking with the camera film. If the transparency is of too low a contrast, the masking can be reduced, or even omitted, when using the camera film.

Original transparencies that are too light or too dark are corrected for density level by adjusting the exposure when making the duplicate and adjustments of colour balance are made by inserting colour correcting filters in the enlarger, preferably in the lamp-house rather than over the lens so as not to impair definition.

For some colours, losses of colour saturation, and modifications of hue and lightness, may occur when transparencies are duplicated. Although compensation for these effects can, in principle, be incorporated in the masking or scanning procedures adopted for making the separations, it is desirable for the duplicating step to be capable of reproducing the colour and tone qualities of a good original transparency with as few differences as possible. Modern duplicating and camera films generally have extensive inter-image effects which usefully counteract the effects of the unwanted absorptions of their image dyes; and, in duplicating films, the spectral sensitivities of the layers can be separated more widely along the wavelength axis than is the case for camera films, and this can provide some further compensation for losses in saturation; camera films can provide similar effects by virtue of their higher gamma. In these ways the inherent deficiencies of subtractive colour reproductions can be largely overcome and duplicates that match originals remarkably closely can sometimes be achieved. There are, however, nearly always some residual differences.

A practical method of transparency duplication is outlined in Section 27.8.

27.4 Converting reflection prints to transparencies

If the original is a reflection print, it is usually found that a satisfactory transparency can be made from it using a camera type of film: in this case the original will have a gamma of about 1.0, so that its use with a camera film will give a result having a gamma similar to that obtained when the same film is used with an actual scene. Corrections for density and colour balance will usually be fairly small, but can be made by adjusting the exposure and using colour correcting filters when exposing the film to the print.

27.5 Producing second originals on paper

Although it is more difficult to maintain excellence of sharpness, and tone and colour reproduction quality, in reflection prints than in transparencies, second originals are often produced on paper because paper is easier to assemble into page layouts; and the layouts resemble the final result more closely, and this facilitates their critical assessment. The most convenient photographic material to use is a reversal colour paper, such as *Ektachrome* paper (see Section 13.3), or

547

the reversal version of the *Ektaflex* transfer system (see Section 17.10). The difference in gamma between transparencies and reflection prints (see Section 6.5) may necessitate the use of reversal materials of different contrasts in the two cases, but the effective gamma of reflection print originals can be raised to be more like that of transparencies by using very specular illuminating conditions when copying them, so as greatly to reduce their 'viewing flare'.

27.6 Working from colour negatives

If the original is a colour negative, there are four different ways in which separations can be made from it. First, they can be made direct from the negative. Secondly, a reflection print made from the negative can be used. Thirdly, a transparency made from the reflection print (as described in Section 27.4) can be used. Fourthly, a transparency made directly from the negative, using a suitable print film, can be used. Generally speaking, the fewer the photographic stages used, the lower the costs and the better the quality: on this basis the first method would be the best, and the third the worst. A disadvantage of the first method, however, is that it does not provide a positive colour image which can be checked for quality before the separations are made; a positive image could be provided by a television type viewer (see Section 16.2), but this does not enable any necessary retouching to be done.

The main area in which negatives are used for the direct production of separations is in the newspaper industry. In newspapers, black-and-white pictures are usually reproduced by making half-tone printing surfaces from black-and-white photographic reflection prints: it therefore fits the method of working to make the separations for colour reproductions also in the form of black-and-white reflection prints, and considerable use is made of this method, particularly in the U.S.A. (Austin, 1968). In a plant in which this type of system is operated, any transparencies received for reproduction may be converted to negatives by copying them on to suitable internegative film: in this way all originals can be handled through the same system.

27.7 Facsimile transmission

The rapid transmission of photographs from one part of the world to another has long been a practice in the newspaper industry. For black-and-white pictures, the usual practice has been to mount a reflection print on a drum, which is rotated on its axis while slowly advancing along its length past a point of light imaged on its surface: in this way the whole of the area of the print is scanned in a series of parallel lines. Signals proportional to the reflectance of the print at each point are then transmitted, usually over telephone links, to receiving stations: each station then reproduces the pictures by exposing a suitable sheet of black-and-white photographic paper, wrapped round a similar drum, the exposure being made by controlling the intensity of a suitable lamp forming a point of light on the paper.

The speed with which the scanning operations can be performed is limited by the bandwidth (see Section 19.2) of the telephone links: the lines used in these links may be limited to a bandwidth of 3 kHz. With this bandwidth, the rates of scanning used are in the region of one line per second, and the spacing of the lines is usually in the range of 100 to 135 lines per inch (40 to 54 lines per cm).

Colour photographs can be transmitted using exactly the same technique, if they are in the form of sets of three black-and-white separation prints: the receiving station then generates a duplicate set of separations from which the printing surfaces can be prepared. If masking is to be carried out (see Chapter 28) it may be advantageous to generate the separations at the receiving station on film instead of on paper. These arrangements have the advantages that they use existing equipment and techniques, and are well suited to the newspaper industry. However, the equipment can be modified to suit other needs: for instance, if the original to be transmitted is in the form of a colour transparency or a colour reflection-print it can be scanned first by a spot of red light, then by a green spot, and finally by a blue spot; similarly, the receiving apparatus can be arranged to scan a sheet of colour film or paper with first a spot of red light, then green, and then blue, to produce, instead of three separations, a colour picture, which is more useful for origination purposes in television news programmes, for instance.

The time taken to transmit three colour separations is typically about 60 minutes, or about 80 minutes if a black separation is also transmitted. At one line per second this provides about 1200 lines, or a picture height of about 10 inches (25 cm) at 120 lines per inch (48 lines per cm).

If it is required to transmit signals that incorporate masking (see Chapter 28) and under-colour removal (see Section 29.14), then it is necessary for the red, green, and blue signals to be available simultaneously. This can be achieved by scanning the original (usually a transparency) with white light, and then dividing the light, by means of dichroic beam-splitting mirrors, into red, green, and blue components, as in flying-spot scanners (see Section 23.9). In one such arrangement the scanning is carried out by imaging, on to the original, a raster of about a thousand lines formed on a cathode-ray tube, using special slow-scan circuits operating at a rate of one line per second (Smith,1973). With this type of equipment, fully-corrected separations can be transmitted, from which printing surfaces can be made directly without any further correction.

27.8 A practical system of transparency duplication

As mentioned in Section 27.2, transparency duplication is widely used in preparing copy prior to making separations. In this section we consider practical means of obtaining duplicates of satisfactory quality.

It was pointed out in Chapter 6 that transparencies intended for projection in dark surrounds must result in pictures having gammas of about 1.5 and those intended for viewing in dim surrounds gammas of about 1.25. Camera films are designed to give these results from original scenes, and, if used for duplicating transparencies, they therefore result in excessive gamma, because a true duplicating film should obviously have a gamma of about 1.0. Special duplicating films having gammas of about 1.0 are available (such as Kodak *Ektachrome Duplicating Film*) and are widely used, but, by using a camera film with a contrast-reducing mask, a system of somewhat greater flexibility is attained. A particular way in which such a system can be operated in practice is now given by way of example (Bethell, 1968).

The sequence of operations can be as follows:

1. A highlight mask is made from the original.
2. The highlight mask is combined with the original and this combination is contact-printed on to a low-contrast masking film.
3. The highlight mask is removed and the original is bound up in register with the contrast-reducing mask made in step 2.
4. The film used for duplicating is put on the baseboard of an enlarger and is exposed to the original-plus-mask combination, using colour filters (in the lamp-house) to give the required colour balance.
5. The film is given a post-exposure uniform flash through a suitable colour filter.
6. The film is processed.
7. The duplicate is retouched, as necessary, usually on the film-base side.

The purpose of the highlight mask (as explained in Section 15.2) is to prevent the contrast-reducing mask from reducing the contrast of highlights in the transparency, which, being usually recorded on the toe of the transparency film, are generally of rather low contrast.

It is in fact possible to use a highlight mask in such a way as to result in a duplicate transparency having highlights of higher contrast than those in the original transparency. Thus, if the gamma of the film used for making the highlight mask is greater than 1.0, the original-transparency and highlight-mask combination will render the highlights as a negative image instead of as a positive image. This negative image will then print on to the contrast-reducing mask as a positive, and thus the original-transparency and contrast-reducing mask combination will render the highlights with greater contrast than is the case for the original transparency on its own; this enhancement of contrast will also occur in the duplicate transparency (unless the highlights are exposed on a low-contrast toe region of the colour film used for duplicating). Enhancement of highlight contrast in duplicates in this way may sometimes be desirable for obtaining the best quality in the final printed result.

It is sometimes advantageous to expose the highlight mask through a medium orange filter (such as a Wratten 85B). This results in most highlight masking in reds (and least in blues) and this helps to avoid too much cyan ink being added to light warm colours such as skin tones.

In under-exposed, dense, transparencies, highlights are reproduced at densities above those of the low-contrast toe of the camera film, and therefore require little or no highlight masking; but in over-exposed, thin, transparencies, highlights are reproduced very much on the toe and therefore require a lot of highlight masking. A convenient way of allowing for this is to give a fixed exposure when making the highlight mask, no matter what the density of the transparency. Thus, dense transparencies will record little or nothing on the highlight mask, whereas thin transparencies will record quite heavily upon it, as required: the type of relationship required between the density in the original transparency and that in the highlight mask is as shown in Fig. 27.1. It is found in practice that some transparencies do not require a highlight mask at all.

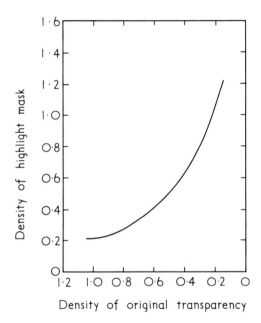

Fig. *27.1*. Relationship between density in an original transparency and density in a typical highlight mask used for transparency duplication.

The gamma required for the contrast-reducing mask depends on several factors, including the gamma of the film on which the duplicate is being made, and the amount of reduction in contrast required for the particular original transparency in question. A convenient way of operating is to obtain a series of characteristic curves by developing a suitable masking film for a series of times, as shown in Fig. 27.2. For each type or batch of film used for duplicating, one of these curves is selected as optimum (for instance, the curve with the heavy line in Fig. 27.2), and all the masks are then developed for the corresponding time. The type of contrast reduction required for different original transparencies will be largely a function of their levels of exposure: dense, under-exposed, originals will require most contrast reduction for the lighter parts of the picture; thin, over-exposed, originals will require most contrast reduction for the darker parts of the picture. Once again, this can be allowed for by giving a fixed exposure when making the mask, no matter what the density of the transparency. Thus, if, as shown in Fig. 27.2, that fixed exposure is chosen so that densities on the original above about 2.0 fall towards the toe of the mask-film characteristic then the mask will tend to give less reduction in contrast for dark parts of dense originals: this is desirable because these parts tend to be lacking in contrast because of shouldering of the film used for making the original. On the other hand, the dark parts of thin transparencies will fall away from the toe of the mask film and hence receive full contrast reduction as required.

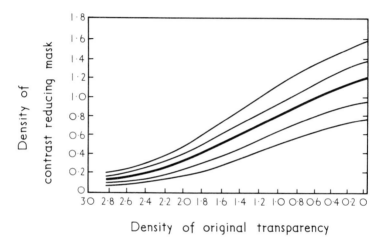

Fig. 27.2. Relationship between density in an original transparency and density in a typical contrast-reducing mask used for transparency duplication. The different curves are for different development times, one of which (such as that shown by the heavy line) is usually chosen as standard for a particular batch of colour film used for making the duplicates.

If the contrast-reducing mask is also made through a medium orange filter, such as a Wratten 85B, then warm colours will expose the mask more than cold colours and hence be reproduced relatively darker on the duplicate: this is found to be a useful way of preventing the cyan printer from losing modulation in reds, an eventuality that results in objectionable flat red areas in certain types of subject matter. If the subject matter is predominantly blue or green, with no important red areas, then the contrast-reducing mask can be made with white light.

The contrast-reducing mask is normally made slightly unsharp, so as not to reduce the contrast of fine detail and thus improve apparent sharpness. (See Section 15.3.)

The contrast-reducing mask is bound up in register with the original transparency (the highlight mask having been removed) and inserted in the enlarger to expose the colour film used for making the duplicates.

The post-exposure uniform flash is given to the colour film to reduce shadow gradation in the duplicates, and, although this may adversely affect the appearance of the duplicates, it generally makes it easier to print them. By using colour correcting filters in the light when giving the uniform flash, it is possible to correct any tendency of the shadows to exhibit a colour bias.

Original transparencies on different types or brands of film may need different treatment in order to obtain a standard type of duplicate. This presents a practical problem, because it is often impossible to tell by visual inspection of the original transparency what type of film has been used. One source of difficulty is that the cyan dyes used in different types of film often have different absorption in the far red part of the spectrum where the eye has low sensitivity but where colour films tend to have high sensitivity (see Fig. 4.3); if this far red light is absorbed by means of a suitable filter (which may have to be of the dichroic interference type in order to get a sharp enough absorbing band) the problem can be somewhat alleviated. Density measurements on the films, made through red filters having different far-red transmissions, can be used to identify films having different types of cyan dye (Graebe, 1976).

27.9 Comparing transparencies

When transparencies have been duplicated, it is, of course, necessary to check them. If the original transparency and its duplicate are composed of the same cyan, magenta, and yellow dyes, and if the aim is for the duplicate to match the original transparency as closely as possible, then a visual check can be made by side-by-side comparison using a uniform area illuminated by any reasonably normal source of white light.

If, on the other hand, the duplicate has to deviate from the original, so as to match some standard of density, colour balance, and contrast, represented by transparencies of other subject matter, then the light source for making the comparison must be chosen with more care; and if transparencies with different cyan, magenta, or yellow dyes are involved, then it is essential to choose the light source for making the comparison with considerable care, because transparencies that match under one light source might look significantly different under another.

There is now general agreement that the light source to be used for the appraisal of colour quality in the printing industry should have the same chromaticity as the CIE daylight illuminant D_{50} (see Section 8.2), which has a correlated colour temperature of about 5000 K (200 mireds); the source may deviate from this chromaticity only within tolerances corresponding to changes in colour temperature of about ± 5 mireds, and within changes of similar size in the u, v diagram in other directions (American National Standards Institute, PH 2.32, 1972; International Standards Organization, 3664, 1975).

It is, however, also necessary to define the spectral power distribution of the source in some way, and this has been done by adopting that of the CIE daylight illuminant D_{50} (as given in the table of Spectral Power Distributions in Appendix 2). The tolerances allowable for the spectral power distribution for the source (see Section 10.12) are defined as being such that the CIE General Colour Rendering Index (CIE, 1965) lies between values of 90 and 100, and the special indices for samples 1 to 8 each have a value of over 80; alternatively the tolerances can be defined in the spectral band system as ± 15 per cent deviation for the light in single bands and $\pm 7\frac{1}{2}$ per cent deviation for the light in contiguous pairs of bands (with ± 30 per cent for each of the ultra-violet bands) (Crawford, 1963; British Standard 950, Part II, 1967).

27.10 Comparing reflection prints and transparencies

If it is necessary to compare a reflection print and a transparency, considerable care must be taken with the viewing conditions. The necessity for this type of comparison arises when the original is in the form of a transparency and the final printed result is on a reflection material; but it also arises when the original is a reflection print or artwork, and it is required to check, against it, the quality of a transparency made from it as an intermediate for the production of separations.

There is now a widely agreed method of making such comparisons, which can be summarized as follows. First, the correlated colour temperature and spectral power distribution of the illumination falling upon both the transparency and the reflection print must be the same. Secondly, the correlated colour temperature of this illumination must be 5000 K, and the tolerances for chromaticity and spectral power distribution must be as defined in the previous section. Thirdly, the transparency must be illuminated from behind by diffuse light, and this diffuse light must be present beyond the edge of the transparency for at least two inches (51 mm) on at least three sides (except that its area must not exceed four times that of the transparency); or, alternatively, the transparency should be surrounded by an opaque sheet of the print substrate, illuminated at the same level as the print. Fourthly, the surface luminance of the transparency illuminator must be 1270 ± 320 cd/m^2. Fifthly, the level of illumination of the print viewing plane must be 2000 ±500 lux. (International Standards Organization, ISO 3644, 1975). However, because these high levels can result in print densities being chosen that look too dark at typical domestic illumination levels (50 to 100 lux), it has been suggested that 635 cd/m^2 for the transparency and 1000 lux for the print may be preferable.

These conditions are designed so that transparencies and reflection prints can appear closely similar to one another. Thus, using the same spectral power distributions makes similar colour rendering possible, and the light levels chosen means that a perfectly reflecting white corresponds to a transparency density of 0.3. The light present round three sides of the transparency provides enough flare light to reduce the effective gamma of the displayed transparency from 1.25 to about 1.0 as is required to match the gamma of the reflection print. The reduction of effective gamma by flare light in this way can only be approximate, but if a more elaborate viewing situation were provided in which a uniform veiling flare corresponding to 2 per cent of the luminance of the illuminator were provided, then, as can be seen from Fig. 27.3, the tone reproduction of the displayed transparency could be made to match that of typical reflection prints very closely (Hunt, 1968); however, this more elaborate approach is not usually followed in practice.

This agreed method of comparing transparencies and reflection prints is widely used, and provides a consistent method of appraising colour work in the printing industry, which is essential for obtaining high quality results at reasonable cost. It also enables different groups of people, such as customers,

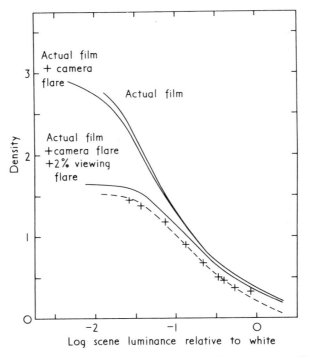

Fig. 27.3. The way in which the characteristic curve of a cut-sheet transparency film ('Actual film') is modified by the addition of camera flare (at 0.4 per cent of the luminance of white) and a degree of viewing flare sufficiently great (actually 2 per cent of the zero-density luminance) to reduce the maximum gamma to about 1.0. The dotted line shows the effect of increasing the luminance slightly; the crosses are reproduced from Fig. 6.5 and show the densities actually seen by an observer in typical room-viewing conditions when looking at a reflection print. The near coincidence of the crosses and the dotted line shows that under suitable conditions a transparency can be made to resemble a reflection print quite closely.

advertisers, and printers, to use a common framework within which jobs can be appraised, and this helps retouchers in their task of modifying the printing surfaces to achieve the required results (Harris, 1973).

When the final copies are being printed, it is necessary for the pressman to check the press sheets against the proofs, and for this purpose an illuminant of higher colour temperature than 5000 K is preferred because variations in yellow printing are then more visible: an illuminant of chromaticity similar to D_{65}, having a correlated colour temperature of about 6500 K, is widely used for this purpose (British Standard 950, Part I, 1967), and, as the inks used in the proof and in the press sheets are normally identical, the change of illuminant does not involve any problems of metamerism in this case.

27.11 Pre-press colour proofing

When screened separations have been made, it is useful to be able to produce a full colour reproduction from them, in order to check them before printing surfaces are made and the first results obtained from the press. This can be done by using special proofing materials.

In one widely used system, known as *Cromalin* (by DuPont), the light-sensitive side of a special film is laminated to the substrate selected for proofing, and then exposed using ultra-violet radiation by contact-printing with a screened positive separation. The film is then peeled off the substrate; in the unexposed areas a tacky image is left on the substrate, and this is toned with a dry powder whose colour is chosen to match the printing ink that will be used for that separation. The laminating, exposing, peeling, and toning sequence is then repeated for each of the other separations using appropriately coloured toners. Excess toner is wiped into a trough around the work and drawn away by suction. A final protective laminate is added to complete the proof. A problem with this type of system is to match the size of the dots on the proof with those that will occur on the press (*dot-gain*, see Section 28.15); but the optical spread occurring in the proofing system can sometimes be made to be similar to the amount of ink-spreading on the press.

In some other pre-press proofing systems, pigments imitating the inks are coated on to transfer sheets by the manufacturer so that no toning stage is necessary; such systems include *Transfer-Key* (by the 3M Company) and *Gevaproof* (by Agfa-Gevaert). The absence of the toning stage makes for easier handling, but the method is less flexible in that different ink colours cannot be simulated by changing the colours of the toners.

The above systems are directly applicable to letterpress and litho printing, but not to gravure, because the amount of ink printed is not then simply related to dot size on the separations. But the systems can be used for gravure by converting continuous tone gravure separations to screened separations prepared specially for proofing. Alternatively, the continuous tone separations can be printed sequentially through red, green, and blue filters on to a reversal colour material, such as *Ektachrome* paper, or the reversal version of the *Ektaflex* system (see Section 17.10); this method of proofing has the advantage of being simple, but the cyan, magenta, and yellow dyes of the photographic materials do not usually match the colours of the printing inks very well.

Another method of pre-press proofing is to scan the separations so as to obtain their images in the form of electronic signals, and then to display a colour image derived from them on a television monitor or colour VDU (see Section 25.5); examples of this type of method are the Hazeltine *Separation Viewer* and the Chromos *Toppan* equipments. This method has the advantage that the image can be adjusted while being viewed to correct for any deficiencies and a computer can be programmed to translate the adjustments into the corresponding changes required on the separations; but the method has the disadvantage that correlation between the self-luminous display and the final printed page introduces some uncertainties. For this reason, electronic proofing systems of this type may also provide the option of *hard* copy (on paper) as well as *soft* copy (on the monitor); the hard copy can be obtained by using displays on cathode-ray tubes, or displays produced by lasers, to expose photographic colour papers.

Methods of pre-press proofing are sometimes referred to as *surprint* (when the colorants are transferred to a substrate), *overlay* (when the colorants are on separate supports and are superimposed), and *electronic* (when a picture is viewed on a television monitor or VDU).

REFERENCES

American National Standards Institute, PH2.32, Viewing conditions for the appraisal of color quality and color uniformity in the graphic arts industry (1972).

Austin, M., *Conference Proceedings 1968*, p. 31, Institute of Printing, London (1968).

Bethell, P., *Conference Proceedings 1968*, p. 19, Institute of Printing, London (1968).

British Standard 950: 1967, Artificial daylight for the assessment of colour, Parts I and II (1967).

C.I.E., *Method of measuring and specifying colour rendering properties of light sources*, CIE Publication No. 13 (1965).

Crawford, B. H., *Brit. J. Appl. Phys.*, **14**, 319 (1963), and *Trans. Illum. Eng. Soc. (London)*, **28**, 50 (1963).

Graebe, C. M., *Professional Printer*, **20**, 2 (1976).

Hunt, R. W. G., *Conference Proceedings* 1968, p. 5, Institute of Printing, London (1968).

Harris, I., *Printing Technology*, **17**, 17 (July, 1973).

International Standards Organization, ISO 3664, Photography, illumination conditions for viewing colour transparencies and their reproductions (1975).

Smith, J. H., *Wireless World*, **79**, 215 (1973).

CHAPTER 28

Practical Masking in Making Separations

1. Introduction – *2.* A two-mask system – *3.* A four-mask system *4.* Masking procedures – *5.* Special colour films for masking – *6.* A direct screening system – *7.* Two-stage masking – *8.* Highlight masking in making separations – *9.* Camera-back masking – *10.* Choice of filters for making masks and separations – *11.* Patches for controlling masking procedures – *12.* Inks used in practice – *13.* The subtractive colour triangle – *14.* Standard inks – *15.* Effects of printing procedures

28.1 Introduction

THE use of standardized procedures to produce colour printing surfaces that require little or no hand correction of local areas has for long been a goal in photomechanical processes. The use of scanners for this purpose will be described in Chapter 29; in this chapter we consider the use of masking. Amongst earlier attempts to solve the problems involved mention may be made of the Gresham-McCorquodale system (Gresham, 1952a, 1952b and 1956) and the Kodak *Short-Run* system (Clark, 1952; Staehle, 1952; Yule, 1953). Some systems of more recent origin will now be described; they all assume that the 'original' to be reproduced is a reversal colour transparency (which may be a duplicate transparency), since the majority of photomechanical colour reproductions are in fact made from colour transparencies; but similar methods can be used with reflection originals. (The general principles of masking were discussed in Chapter 15. See Plate 32, pages 512, 513, for an example of masking.)

In the majority of cases the aim is to produce sets of separation negatives or positives that are fully corrected, so that the printing surfaces can then be made from them without any individual adjustments being necessary. The corrected separations may be unscreened (continuous-tone) or screened (consisting of dot images), but, if the former, then screening has to be carried out subsequently.

28.2 A two-mask system

When masking is carried out manually the procedures are usually simplified as much as possible and the details established empirically. One procedure, illustrated in Fig. 28.1, is to make only two masks (in addition to a highlight mask, see Section 28.8, if this is necessary), one through a green filter which is used when making the blue separation, and the other through a magenta filter which is used when making the green and red separations.

Mask exposed to	Mask used when making	Mask gamma	Provides correction for
Green light	Blue separation	0.5	Blue absorptions of magenta and cyan inks
Red and blue light (Magenta filter)	Green separation	0.5	Average green absorption of cyan and yellow inks
	Red separation		Gamma of cyan image

Fig. 28.1. Simplified masking procedure in which masks are made through green and magenta filters.

The mask made through the green filter and used when making the blue separation provides correction for the unwanted blue absorption of the magenta ink; but if the inks are such that the ratio of the green density to the blue density is the same for the cyan ink as for the magenta, then this mask will also automatically correct for the unwanted blue density of the cyan. Inks that have this common ratio are sometimes referred to as inks of *balanced hue* (Hartsuch, 1958; Yule, 1967, page 53); all magenta inks have higher green than blue densities and as this is also true of most cyan inks, an approximate correction for their unwanted blue absorption usually takes place. The reason why a green-filter mask is able to correct for a cyan ink deficiency is as follows: the green-filter mask makes no allowance for the fact that the other mask will be reducing the amount of magenta required in areas where the cyan will be present; hence, in these areas, more masking will be provided than is required to correct for the magenta deficiencies, and it is this excess masking in cyan-abundant areas that provides correction for the unwanted blue absorption of the cyan (Pollak, 1956).

The mask made through the magenta filter and used when making the green separation provides correction for the average green absorption of the cyan and yellow inks. The unwanted green absorption of the cyan ink is usually much greater than that of the yellow ink, so that what is really required is that the red light transmitted by the magenta filter should result in a higher contrast mask than that produced by the blue light; two different contrasts cannot be properly achieved in only one mask, of course, but, if the magenta filter is chosen so that

559

the blue light exposes the mask material less than the red light, some rough allowance for the difference in the green absorptions of the two inks can result.

Although the unwanted red absorptions of the yellow and magenta inks are usually small, it is convenient to use a mask when making the red separation, so that similar gammas can be provided for all three separations, which then need only a single film-process combination. Strictly speaking, a mask whose function is only to control cyan gamma should be made through a red filter but in practice it is found that the mask made through the magenta filter can be used with fairly good results. The gammas of the film-process combinations used for making the masks in this procedure are usually both about 0.5.

A modified version of this two-mask system is sometimes used when making separation positives in the form of black-and-white reflection prints for colour reproductions in newspapers. In this case, originals in the form of colour negatives may be used. The procedure is to place the colour negative in the gate of an enlarger, and to compose the picture for size and cropping on the baseboard. A mask is then made by exposing through a red filter a piece of suitable masking film placed on the baseboard; another mask is then made on a second piece of masking film, but using a green filter. The blue separation is then made by placing a piece of suitable panchromatic photographic paper underneath the green-filter mask on the baseboard and exposing it from the colour negative using a blue filter; the green separation is made similarly, using the red-filter mask and a green filter; finally, the red separation is exposed using a red filter, but without any mask.

A separation for a black printer can also be made, if required, by giving partial exposures through all three filters. It is usually necessary to use a black ink, because of the difficulty of producing good blacks with cyan, magenta, and yellow inks alone, and for other reasons (see Section 26.6).

28.3 A four-mask system

The two-mask procedure outlined above has been elaborated for use in the preparation of half-tone printing surfaces by using four masks, as illustrated in Fig. 28.2, although one of the masks is used only to facilitate the preparation of the printing surface used for the black ink. Of the other three masks, one is made through a green filter and used as before when making the blue separation; another is made through a magenta filter but is now used only when making the green separation; the third is used mainly for reducing cyan gamma, but by making it through an orange filter instead of through a red filter, some correction for the unwanted red absorption of the magenta ink is also achieved. Once again, this single mask cannot properly achieve the two different functions required of it, but if the orange filter is chosen so that the exposure levels produced on the mask material by its red and green transmissions bear some relation to the relative importance of cyan gamma-control and correction for magenta unwanted red-absorption, then both functions will be served approximately. The gammas of all the film-process combinations used for making the masks in this procedure are also usually about 0.5.

Mask exposed to	Mask used when making	Mask gamma	Provides correction for
Green light	Blue separation	0.5	Blue absorptions of magenta and cyan inks
Red and blue light (Magenta filter)	Green separation	0.5	Average green absorption of cyan and yellow inks
Red and green light (Orange filter)	Red separation	0.5	Gamma of cyan image and red absorption of magenta ink
Narrow-band yellow-green light	Black separation	0.5	Black printing surface

Fig. 28.2. Simplified masking procedure in which masks are made through green, magenta, and orange filters (and through a narrow-band yellow-green filter in connection with the black printer).

28.4 Masking procedures

The sort of masking procedures often used in practice will now be described, using the four-mask system as an example.

If a highlight mask is needed (see Section 28.8), this is made first, and bound up in register with the transparency. The transparency is then placed in contact with a sheet of unexposed, low contrast black-and-white masking film and exposed through an orange filter, such as Wratten 85B together with Wratten 81EF. The transparency is placed with its *base side* (not *image side*) in contact with the emulsion side of the mask film so as to introduce a small degree of unsharpness into the mask; this eases registration problems and increases the apparent sharpness of the final reproduction (see Section 15.3). If necessary, additional unsharpness can be obtained by inserting a thin spacer between the two films. After exposure, the mask material is removed for subsequent processing, and the second mask is exposed. The second mask is made using the same techniques and type of film as for the first mask, but with a magenta filter, such as Wratten 33 together with Wratten 81EF; the third and fourth masks are made similarly using, respectively, a green filter, such as Wratten 58, and a narrow-cut yellow-green filter, such as two thicknesses of Wratten 90. The four masks are then processed together (to ensure uniform treatment).

The separation negatives and the black-printer negative are then made as follows. The transparency (without any highlight mask) is bound up in register with the mask made through the orange filter and printed by contact or by enlargement on to a sheet of unexposed black-and-white negative film of gamma suitable for making the separations. The exposure is made with red light, using a red filter, such as Wratten 25. The various films are arranged so that the light passes through in the following sequence: mask base, mask, transparency base, transparency (then the lens if enlargement is being used), negative emulsion, negative base; this ensures that the mask is slightly unsharp but that the transparency prints (or enlarges) sharply on to the separation emulsion. After exposure, the sheet of separation film is removed for subsequent processing, and the mask made through the orange filter replaced by the one made through the magenta filter. Using the same techniques and type of film, the green separation is then exposed through a green filter, such as Wratten 58. The mask made through the magenta filter is then replaced by that made through the green filter, and the blue separation exposed through a blue filter, such as Wratten 47B. Finally, the green-filter mask is replaced by the yellow-green filter mask, and the black-printer separation exposed to white light (a pale green filter, such as Kodak *Colour Compensating* filter CC50G, is sometimes used for this exposure). The four exposed negatives are then processed: it may be necessary to develop them for slightly different times in order to obtain those gammas that will result in a correctly matched grey scale in the final reproduction. (The choice of filters for making the masks and separations is discussed in Section 28.10.)

28.5 Special colour films for masking

When black-and-white masks are used in contact with colour transparencies for the production of separation negatives, it is necessary (as described in the previous Section) to change the masks appropriately for exposing each separation; as the masks have to be accurately registered with the transparency this is a tedious business. Multi-layer colour films have therefore been made available that produce mask images consisting of cyan, magenta, and yellow dyes; these special films are designed so that when a transparency is printed on to them the cyan, magenta, and yellow images result in the red-light transmission constituting the required mask for making the red separation, and the green and blue transmissions the required masks for making the green and blue separations, respectively. A contact print of the transparency is therefore made on such a film and bound up in register with it: all three separation negatives can then be made by printing from the same transparency-mask combination, using red, green, and blue filters in the ordinary way. Two masking colour films of this type have been produced: Agfa-Gevaert *Multimask* film, and Kodak *Tri-Mask* film; their functions are shown diagrammatically in Figs. 28.3 and 28.4 respectively.

In *Multimask* film, Fig. 28.3, layers 1 and 3 perform exactly the same functions as the masks made through the magenta and green filters, respectively, in the procedure illustrated in Fig. 28.2; layer 2 performs a similar function to that of the mask made through the orange filter, but its blue sensitivity is greater than the small residual blue sensitivity usually provided by the orange filter, and hence a correction for unwanted red absorption of the yellow ink takes place. The gammas of the cyan, magenta, and yellow mask images are all about 0.5, so that a

Multimask mask, when bound up with a transparency does not upset the grey-scale contrast match.

Layer	Layer sensitive to	Mask colour	Separation affected	Mask gamma	Provides correction for
1	Red and blue light	Magenta	Green	0.5	Average green absorption of cyan and yellow inks
2	Red, green and blue light	Cyan	Red	0.5	Gamma of cyan image, and red absorptions of magenta and yellow inks
	Yellow filter layer				
3	Green light	Yellow	Blue	0.5	Blue absorptions of magenta and cyan inks
	Film base				

Fig. 28.3. Masking procedure provided by Agfa-Gevaert *Multimask* film.

In *Tri-Mask* film, as shown in Fig. 28.4, the masking is slightly more elaborate. Layer 3 performs the same functions as those of the green-filter mask of Fig. 28.2. Layers 1 and 5 perform the same two functions as those of the magenta-filter mask; but the different unwanted green absorptions of the cyan and yellow inks are corrected by mask images of different gammas, instead of by critical

Layer	Layer sensitive to	Mask colour	Separation affected	Mask gamma	Provides correction for
1	Blue light	Magenta	Green	0.1	Green absorption of yellow ink
	Yellow filter layer				
2	Green light	Cyan	Red	0.25	Red absorption of magenta ink
3	Green light	Yellow	Blue	0.5	Blue absorptions of magenta and cyan inks
4	Red light	Cyan	Red	0.25	Gamma of cyan image
5	Red light	Magenta	Green	0.4	Green absorption of cyan ink
	Film Base				

Fig. 28.4. Masking procedure provided by Kodak *Tri-Mask* film.

choice of the magenta filter to 'read' the cyan and yellow transparency images at the correct ratio: the *Tri-Mask* system is thus less sensitive to changes in the nature of the cyan and yellow transparency dyes. Layers 2 and 4 perform the same two functions as those of the orange-filter mask; but, once again, because the two mask images are formed in different layers, their gammas are adjusted independently, instead of depending on a critical choice of orange filter to 'read' the cyan and magenta transparency dye-images at the correct ratio; the *Tri-Mask* system is thus also less sensitive to changes in the nature of the cyan and magenta transparency dyes. The *Tri-Mask* film, therefore, not only achieves the same functions as the masks made through the orange, magenta, and green filters, but actually provides more elaborate masking in that the correction is based on five mask images instead of on only three. The gammas of the various mask images in *Tri-Mask* film are shown in Fig. 28.4: the total image gamma in each dye-colour adds up to 0.50, so that a *Tri-Mask*, when bound up with a transparency does not upset the grey scale contrast-match.

The procedure when using *Tri-Mask* film is as follows. The transparency (with a highlight mask if required) is placed in contact with a sheet of unexposed *Tri-Mask* film and exposed with white light; as before, the transparency is placed with its base side in contact with the *Tri-Mask* film so as to make the mask slightly unsharp (with additional unsharpness from a thin spacer inserted between the two films, if required). The *Tri-Mask* film is processed and then re-registered with the transparency (without any highlight mask). Separation negatives are then made by printing (by contact or by enlargement) the masked transparency on to black-and-white negative film of suitable contrast using red, green, and blue filters, such as Wratten 25 (plus a neutral filter of density 1.0 for exposure equalization) for the red, Wratten 58 (plus a neutral filter of density 0.5) for the green, and Wratten 47B for the blue; the black separation is exposed by giving a fourth piece of film a suitable composite exposure through each of the three filters successively. As before, the separations are then developed (for different times if necessary) to obtain the desired gammas, for subsequent screening. A similar procedure is adopted when using *Multimask* film.

The advantage of using multi-layer colour masking films over the four-mask system are as follows. First, a more elaborate degree of colour correction can be provided; secondly, only one exposure is necessary to produce all the masks; thirdly, registration problems are reduced because only one mask requires registering with the transparency; fourthly, all four separations are made from the same transparency-mask combination, thus eliminating tiresome manipulations between the exposure of each separation; fifthly, because of the smaller number of separate films involved, colour balance and tone reproduction are more easily controlled. A disadvantage is that the colour masking films cannot be processed in the machines used for processing the black-and-white films.

28.6 A direct screening system

Further simplifications to the process of obtaining corrected printing surfaces result if the corrected separation negatives are already screened, so that a separate screening does not have to be introduced subsequently. A major reason why the introduction of this apparently obvious simplification was delayed was that, even with white light, the screen exposure tended to be quite lengthy, so that screen exposures made from masked transparencies using red, green, and blue exposures were very inconveniently long; this difficulty is aggravated whenever a size-change is required between the transparency and the screened negatives, because the exposure then has to be made in an enlarger and not by contact. However, the high contrast black-and-white negative films necessary for producing satisfactory half-tone images became available with higher photographic speed, and condenser-type enlargers fitted with high-intensity pulsed-xenon lamps provided more light. Hence the exposure of screened negatives direct from masked colour transparencies, *direct screening*, became feasible (Clapper, 1964). The availability of multi-layer colour masking films also facilitated direct screening, by easing registration problems in the enlarger, and by providing non-scattering masks whose contrasts, unlike those of silver masks, are not dependent on the degree of specularity of the light in the enlarger.

The sequence of operations in one direct-screening system (Clapper, 1964) is as follows. First, a mask is exposed (with slight unsharpness) on Tri-Mask film by contact (using a highlight mask if necessary) and, after processing, the mask is bound up in register with the colour transparency (without any highlight mask). The masked transparency is then placed in a suitable enlarger and a grey contact-screen placed upon a suitable high-contrast panchromatic black-and-white film on the enlarger easel (in order to obtain good contact between the contact screen and the film being exposed, a vacuum printing frame is generally used). Exposures are then made through red, green, and blue filters on to three separate sheets of film; typical filters are Wratten 23A for the red, Wratten 58 for the green, and Wratten 47B for the blue. Contrast control is carried out by using an additional uniform exposure (*flashing*) made through the contact screen (in the case of the red separation a small additional exposure is also made from the transparency without the contact screen in position; this *no-screen* exposure increases the contrast of the red-screened-negative and this is necessary to obtain a balanced grey scale in the final result (Pollak, 1955)). The black-printer separation is made using a single exposure with a Wratten 85B filter, the exposure level being such that in the final reproduction black ink is only printed at reflection densities above about 0.8. (See Plate 32, pages 512, 513.)

28.7 Two-stage masking

In most of the discussion in this chapter, it has been tacitly assumed that the colour reproduction systems with their masks can be represented by relationships that are always proportional to density; in other words that the equations relating densities are always linear. It was pointed out in Sections 15.2 and 27.8, however, that the tone reproduction was sometimes sufficiently non-linear to require a special non-linear highlight mask (performing a tone-correcting function similar to that of the upswept shoulder of the internegative film described in Section 14.16); the possibility of the failure of densities to obey the additivity and proportionality rules was mentioned in Section 15.7; and the non-linear relationships between transmission and reflection densities in reflection print materials was discussed in Section 14.22. In half-tone colour printing, other sources of non-linear density distortion can be important, such as the tone-reproduction characteristics of the steps involved in preparing and using the printing surfaces, differences in gloss between one ink and another, and the fact that the way an ink-image prints often depends on the amount of ink, if any, already printed on each area (a form of inter-image effect).

There are, therefore, a number of reasons why non-linear masking may be desirable. For this reason, it is sometimes advantageous to use fairly broad red, green, and blue filters (instead of the usual narrow ones) when exposing the separation negatives, because the increased non-additivity and non-proportionality thus introduced is sometimes useful in correcting other non-linearities in the system. One difficulty of using non-linear *masks*, however, is that unless the non-linearity introduced is exactly the same in all three separations, unpleasant distortions will be produced in the grey scale. In the case of the highlight mask, the neutrality of the grey scale may be ensured by using the same highlight mask when making all three separations; but colour-correcting masks must, of course, be different for the three separations, and making sets of non-linear masks matched closely enough to avoid grey-scale distortion is difficult. This difficulty can, however, be largely overcome by using a technique known as *two-stage masking*.

In two-stage masking a set of separation negatives is first made without masking. These separations are then contact printed on to a black-and-white film that can be processed to give a gamma of 1.0, to yield three separation positives. If one of these separation positives is then bound up with the negative from which it was made the two images will cancel one another and only a uniform grey will result: but if a positive is bound up with one of the *other* negatives the two images will only cancel for colours that exposed equally the two separation negatives concerned; such equal exposure will occur for all grey colours (as well as for some others) and hence such negative-positive combinations will reproduce the grey scale as a uniform grey of a single density. If, therefore, this type of negative-positive combination is used for making colour-correcting masks, even if they are non-linear they cannot affect the grey scale reproduction.

One way in which it has been found useful to apply the two-stage masking technique is as follows. The ink images used in half-tone reflection printing often show marked non-additivity of their densities. For example, the density to blue light of a patch of yellow ink printed over a patch of magenta ink is frequently less than the sum of the blue densities of the two patches printed side by side. This means that if the gamma of the green-light mask (to be used in making the blue printer) is adjusted so as to give the right degree of correction for the unwanted blue absorption of the magenta ink on its own, it will give too much correction when the magenta ink has yellow printed over it (the unwanted absorption of an ink is sometimes called an *unwanted colour* unless it is printed with the ink which is meant to absorb in the region of unwanted absorption in which case the total absorption is called a *wanted colour*). Hence the green-filter colour-correcting mask is required to have a gamma that *decreases* as the amount of yellow ink present *increases*. This can be achieved by making this mask from a combination of the green separation negative and the blue separation positive and under-exposing when making the mask: this is so because in areas where there will be no yellow ink (unwanted colours in the blue separation) the separation positive will be light, and the mask will be well-exposed and therefore reproduced at normal gamma; but in areas where there *will* be yellow ink (wanted colours in the blue separation) the separation positive will be dense and the mask will be under-exposed and therefore reproduced at low gamma (on the toe of the characteristic curve of the masking film). The mask thus provides non-linear colour-correction, but has no effect on the grey scale. The mask can then be bound up with the uncorrected separation negative to give a corrected combination: however, the grey scale contrast will be higher than with the normal one-stage methods of masking, because in this case the colour-correcting mask does not reduce gamma; any gamma reduction required, therefore, has to be provided by other means.

It is possible for a single mask to correct for the unwanted blue absorptions of both the magenta and the cyan inks in two-stage masking, if, in addition to the reproduction inks being of balanced hues (see Section 28.2), the effect of the presence of yellow ink on the non-additivity of the blue absorption of the cyan ink is similar to that with the magenta ink; this similarity does sometimes occur, enabling one mask to perform both functions.

The principles of two-stage masking can be incorporated into the operations performed by scanners (see Chapter 29).

28.8 Highlight masking in making separations

In Sections 15.2 and 27.8, it was shown that, when using a contrast-reducing mask in duplicating transparencies, reduction in the contrast of the highlights could be avoided by using a highlight mask: this is sometimes important because highlights in transparencies tend to be reproduced on the low-contrast toe of the characteristic curve of the film. Since the masks used for colour correction described in this chapter usually also have the effect of lowering contrast, it may again be desirable to use highlight masking to obtain adequate contrast in the highlights. A single black-and-white mask is normally used for this purpose.

Although useful improvements in the rendering of highlights are obtained by using a single highlight mask, such a mask can only increase contrast equally in all colours (and hence cannot restore saturation lost because of low toe contrast). Yule has shown that better results can be obtained by using three highlight masks: a highlight mask exposed through a red filter when making the mask for the red separation, one exposed through a green filter when making the mask for the green separation, and one exposed through a blue filter when making the mask for the blue separation (Yule, 1967, page 70), but this complication is not usually included in practice.

28.9 Camera-back masking

When the original to be reproduced consists of reflection copy, or is a very small transparency, registration of masks in contact with it may not be convenient. Masking can still be carried out, however, by registering masks with an image of the original in the back of a camera, or on the easel of an enlarger, a technique known as *camera-back masking*.

The basic principles in camera-back masking are the same as those described in the previous sections; but the image will differ from the original because of the effects of flare from the lens, and it may, therefore, be necessary to allow for this in choosing the gammas to be used when making the masks or the separation negatives.

A problem with camera-back masking is that the light forming the image has to pass through the mask before exposing the film, and this may cause some loss of sharpness unless precautions are taken; thus, when silver masks are made from reflection originals, they should be exposed through the film base so that their emulsion sides can be in contact (preferably by vacuum) with the film when masking it in the camera back. When multilayer colour masking films are used, exposure through the base is not possible (because the yellow filter layer would not then perform its proper function) but they can be used in the camera back with their base sides in contact without affecting the sharpness of the main image too much because, unlike silver images, they do not scatter light appreciably; in this case, the mask itself will not print sharply, of course, but this does not usually matter since unsharp masking is normally used.

28.10 Choice of filters for making masks and separations

Sets of cyan, magenta, and yellow dyes have spectral absorption curves that differ according to the brand of film or reflection print material used. This metamerism amongst originals means that only if the spectral sensitivities of the filtered mask and separation films are sets of colour-matching functions (see Section 9.5) will transparencies that look alike be reproduced alike. From this point of view, the filters mentioned in Section 28.4 are rather narrow in their spectral transmission bands, and this can cause problems, especially when materials having cyan dyes with markedly different far-red absorptions (as mentioned in Section 27.8) are encountered. The suitability of various filters can be studied by assessing the

568

degree to which they result in approximations to colour-matching functions by evaluating their *Colorimetric Quality Factors*, or *q-factors* (see Section 9.5). As a result of such studies, different filters can be recommended for different applications, such as Wratten 23A instead of a Wratten 25 for making the red separations.

28.11 Patches for controlling masking procedures

Control of the processing, in masking, is very important. In the Kodak *Three-Aim-Point* method of control, standardized procedures are drawn up so that separation negatives can be produced that will yield results having consistent tone reproduction and colour balance (Clapper, 1962). The aim is for the three colour separation negatives to be made always with the same tone reproduction relative to a standard original; the black printer is then used to accommodate originals of varying density ranges, short-range originals using little black printer, long-range originals using more. A standard 'original' is provided in the form of three neutral density patches: patch A represents a minimum reproducible density in an average transparency or reflection print; patch B a similar maximum density; and patch M a similar medium density. These patches are then mounted alongside the original before the masks are made. When the separations are made, the masked patches are printed along with the picture. As the result of experience, standard values and tolerances have been arrived at for the densities of the A, M, and B patches on typical masks and separation negatives. These are given in Table 28.1. The values for the cyan separation negative are different from those for the magenta and yellow, because, in typical printing systems, it is often necessary for the cyan printer to be made from a slightly higher-contrast separation negative in order to achieve a good grey scale (this results, at least in

TABLE 28.1

Typical densities in the Kodak *Three-Aim-Point* method of controlling processing when making colour-correcting masks and separation negatives

Patch	On transparencies	On reflection prints	On masks from transparencies	On masks from reflection prints	On separation negatives		
					C	M and Y	Black
A	0.4	0	1.15	0.80	1.70	1.55	
B	2.4	1.6	0.25	0.20	0.30	0.30	0.50 to 0.90
A-B			0.90±.05	0.60±0.05	1.40	1.25	
M	1.3	0.7	0.80	0.50	0.90	0.90	
A-M			0.35	0.30	0.80±.05	0.65±.05	
M-B			0.55	0.30	0.60±.05	0.60±.05	0.60 to 0.90
(M-B)− (A-M)			0.20±.05	0.00±0.05			

All the masks in a single set should have values of A-B, the *range*, and (M-B)−(A-M), the *mask number*, within 0.05.

part, from the fact that most inks have very low unwanted red-absorptions; hence, in a grey, nearly all the red-aborption has to be provided by the cyan ink).

When masking procedures are being determined empirically, it is often very helpful to position a few special colour patches (in addition to neutral density patches) so that they are reproduced on the edges of the masks and separations. If the inks used in the reproduction obey the proportionality and additivity rules, then patches of cyan, magenta, and yellow ink suffice: the masking is then usually adjusted so that the cyan patch only reproduces on the corrected red separation, the magenta on the green, and the yellow on the blue; full correction will then have been made for the unwanted absorptions of the inks used in the patches, and a correct copy could be made of an original consisting of mixtures of the same inks. When the original is a transparency it is not convenient to use inks for the patches, but dyes having similar unwanted absorptions can be used instead. If the proportionality and additivity rules are not obeyed, then patches of the dyes or inks both singly and in pairs can be used; the masking can then be adjusted so that single patches reproduce on their one appropriate separation only, and combination patches on their two appropriate separations only, a result that may require non-linear masks produced, for instance, by the two-stage masking technique. It must be remembered, however, that even when patches of both single and pairs of dyes or inks are reproduced correctly, the intermediate colours can still show errors, although these are often fairly small.

28.12 Inks used in practice

In Fig. 28.5, the spectral density curves of a set of typical cyan, magenta, and yellow inks are given, together with the spectral density curves of Kodak Wratten filters 25 (red), 58 (green), and 47 (blue). It is clear that these inks exhibit unwanted absorptions: thus the magenta ink, in addition to absorbing green light as it should, also absorbs blue light to a considerable degree, and the cyan ink absorbs not only red light as it should but also a considerable amount of green light. The yellow ink is the best of the three, having a negligible unwanted red absorption, and only a small unwanted green absorption.

Although these colour properties of inks are obviously very important there are other factors that affect the choice of inks to be used for a particular job, such as viscosity, drying rate, light fastness, and cost. Consequently, quite a wide variety of cyan, magenta, and yellow inks are used in practice.

Since the masking used in making separations is usually intended mainly to correct for the unwanted absorptions of inks, it has been found helpful to classify the colour properties of inks in terms of their densities to red, green, and blue filters (such as Kodak Wratten 25 for red, 58 for green, and 47 for blue). A typical set of such densities is given in Table 28.2 (Yule, 1967, page 161).

It is helpful to work out from these densities, two quantities known as *per cent hue error* and *per cent greyness* (Preucil, 1957), as follows:

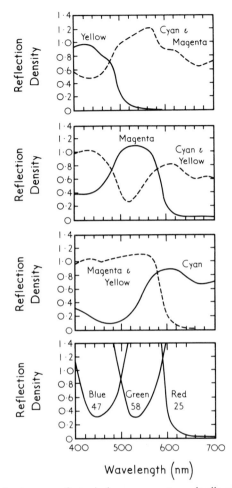

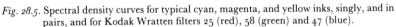

Fig. 28.5. Spectral density curves for typical cyan, magenta, and yellow inks, singly, and in pairs, and for Kodak Wratten filters 25 (red), 58 (green) and 47 (blue).

$$\text{Per cent hue error} = 100\,\frac{M-L}{H-L}$$

$$\text{Per cent greyness} = 100\,\frac{L}{H}$$

where, for any one ink, H is the highest, M the middle, and L the lowest, of its densities to the red, green, and blue filters.

If an ink had no unwanted absorptions, M and L would both be zero, and the per cent hue error and per cent greyness would therefore both also be zero. If an

571

TABLE 28.2

Densities of typical inks measured through Kodak Wratten filters 25 (red), 58 (green) and 47 (blue)

Colour	Density		
	Red (25)	Green (58)	Blue (47)
Cyan	1.20	0.37	0.17
Magenta	0.11	1.09	0.56
Yellow	0.01	0.06	0.96
Cyan+Magenta	1.33	1.44	0.67
Magenta+Yellow	0.11	1.22	1.61
Cyan+Yellow	1.29	0.43	1.19
Cyan+Magenta+Yellow	1.31	1.53	1.66
Cyan+Magenta+Yellow+Black	1.60	1.83	1.90
	Cyan	Magenta	Yellow
Per cent greyness	14.2	10.1	1.0
Per cent hue error	19.4	45.9	5.5

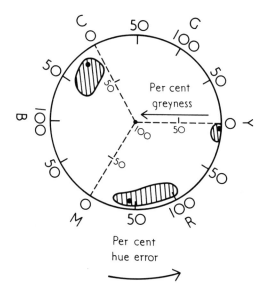

Fig. 28.6. Per cent hue error and per cent greyness for the inks of Table 28.2 (dots) and for the range of inks used in practice (hatched areas).

ink had unwanted absorptions, but they were equally large, then M and L would be equal (but not zero), and then the per cent hue error would be zero but there would be some per cent greyness: in this case the ink can be thought of as a perfect ink (with no unwanted absorptions) combined with a grey ink, so that it is greyed but not altered in hue. If an ink has unequal unwanted absorptions, then M will be greater than L and the per cent hue error will not be zero: the ink can now be thought of as perfect ink, combined with a grey ink and with a subsidiary amount of a perfect ink of one of the other colours. Thus, in the case of the magenta ink in Table 28.2, it can be thought of as being equivalent to a combination of a grey ink of density 0.11, a perfect yellow ink of density 0.45 (0.56–0.11), and a perfect magenta ink of density 0.98 (1.09–0.11): it therefore differs from a perfect magenta ink by being greyed (to the extent of $0.11/1.09 = 10.1$ per cent) and being yellower (to the extent of $0.45/0.98 = 45.9$ per cent). The values for per cent greyness and per cent hue error are given in Table 28.2 for each of the three inks, and it is seen that the yellow has the lowest greyness and hue error, while the highest error is the hue error of the magenta. The hue errors of inks are virtually always in the same directions, cyans towards blue (equivalent to a magenta addition), magentas towards red (equivalent to a yellow addition), and yellows towards red (equivalent to a magenta addition). In Fig. 28.6 the per cent greyness and per cent hue error of the inks of Table 28.2 are plotted together with areas indicating the range of values of these parameters that practical inks cover; in this figure, hue error is plotted around a circle, and greyness as distance in from its circumference (zero greyness) towards the centre (100 per cent greyness).

28.13 The subtractive colour triangle

If the densities of inks to red, green, and blue light, D_R, D_G, D_B, are expressed as the proportions

$$r = D_R/(D_R+D_G+D_B)$$
$$g = D_G/(D_R+D_G+D_B)$$

it is possible to construct a useful diagram by plotting g against r, as shown in Fig. 28.7(a) which uses axes inclined at 60° to one another; this is often referred to as the *subtractive colour triangle* (Preucil, 1960). If the densities obey the Additivity and Proportionality Rules (see Section 15.7), then any ink will be represented by a single point on the diagram no matter what its thickness on the substrate; and combinations of pairs of inks will be represented by points on the line joining the points representing the two individual inks. The apices of the triangle, C_o, M_o, Y_o, represent inks having no unwanted absorptions.

On this diagram loci of constant per cent hue error and per cent greyness are all straight lines, as shown in Fig. 28.7(b), and hence a useful way of plotting these parameters is provided as an alternative to that shown in Fig. 28.6.

The subtractive colour triangle also provides a simple graphical method of determining the contrasts of the masks required for correcting for the unwanted

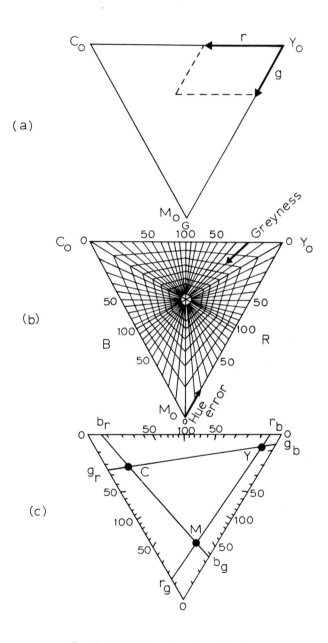

Fig. 28.7. The Subtractive colour triangle.

absorptions of inks, as shown in Fig. 28.7(c). The positions of the inks to be corrected are shown by the points C, M, and Y; b_g is the value of the green filter mask to be used when making the blue separation negative, and has a value of 37 per cent; g_r is the value of the red filter mask to be used when making the green separation negative, and has a value of 24 per cent; r_g, b_r, r_b, and g_b, give the values for the other masks similarly. These percentages are those of the mask gammas relative to the separation gammas.

28.14 Standard inks

The use of inks of different colour properties affects not only the range of colours that can be produced, but also the gammas of the masks that will give the best results. Because corrected colour separations may be made in one location, and used for the preparation of printing surfaces in a different location, sometimes even in a different country, some standardization of the colours of inks used is clearly desirable.

The Comité Européen d'Imprimerie (C.E.I.) has standardized the colours produced, under Illuminant D_{65} by cyan, magenta, and yellow inks used singly and also in pairs (to produce red, green, and blue); tolerances, ΔD, have also been standardized in terms of the $U^*V^*W^*$ colour difference formula (see Section 8.8); standard values and tolerances for these colours are given in Table 28.3. The values refer to films of each ink of 1 micrometre thickness, printed on to white two-sided coated paper, light-fast, with no optical bleaching agent, free from mechanical wood pulp, and of basic weight not less than 150 gm/m² (European Standard C.E.I. 12 : 66, 1966; British Standard 4160: 1967; European Standard C.E.I. 13 : 67, 1967; British Standard 4666 : 1971; International Standard Organization, ISO 2845 and 2846, 1975).

TABLE 28.3

Standard colours produced by C.E.I. inks

Inks	x	y	Y	U*	V*	W*	ΔD
Yellow	0.437	0.494	77.8	19.1	70.9	89.7	2.3
Magenta	0.464	0.232	17.1	112.0	−12.7	47.4	5.0
Cyan	0.153	0.196	21.9	−54.5	−50.9	52.9	3.0
Magenta over Yellow	0.613	0.324	16.3	139.8	21.8	46.4	7.3
Cyan over Yellow	0.194	0.526	16.5	−69.1	28.1	46.7	5.3
Cyan over Magenta	0.179	0.101	2.8	−3.6	−35.4	18.1	8.0

Sets of inks meeting these standards could have a variety of spectral absorption curves: this could result in such inks producing different colour reproduction, from identical printing surfaces, for colours other than the six specified; if the ink colours were specified only singly, this could be a considerable problem, but, by specifying, in addition, their colours when used in pairs, this effect is reduced. Similarly, the amount of metamerism when viewed in illuminants of different spectral compositions, between different sets of inks meeting the standard, is also reduced.

The colours produced by inks depend on other factors in addition to their colours in the standard conditions given above: in particular, the type of paper (or other substrate) on which they are printed will affect the colours by reason of the colour of the paper, its absorbency, and its surface texture; the thickness of the film of ink used in the printing will also affect the colour reproduction, and the thicknesses achieved in practice are usually less than 1 micrometre.

28.15 Effects of printing procedures

Masking will only give the expected result in the printed material if various parameters of the actual printing process are correctly anticipated. Mention has already been made of the importance of knowing which inks are to be used, but it is also necessary to know the weight at which the inks will be printed, and the effects of phenomena such as *dot-gain, trapping,* and *slur.* Dot-gain is related to spreading of the ink when a dot is printed on to the substrate so that the density produced is greater than would be expected from the size of the dot on the printing surface; the major factors affecting dot-gain are the thickness of the ink film, the physical properties of the ink (such as viscosity), and the nature of the surface of the substrate (such as whether it is glossy or matt). Trapping is the effect whereby the density produced by a dot pattern of ink depends on differences in ink acceptability according to whether the ink is being printed directly on to the substrate or on top of a previously printed colour. Slur is directional distortion of dot-shape caused by the action of the press.

Because of these factors, practical printing procedures may have to involve the following steps. First, the optimum ink weight has to be decided. This may be a compromise between high quality obtained with high ink weights, and good printability on the press obtained with lower weights. Secondly, by printing a series of near neutral patches of different densities, the dot sizes required on the separation negatives to give a truly neutral grey scale are determined, including the effects of dot-gain and trapping. Thirdly, the grey scale tone reproduction is set up using a suitable method, such as the *Three-Aim-Point* method described in Section 28.11. This usually involves some compression of the tone range of the original in order for it to be accommodated in the more restricted tone range of the reproduction system. Fourthly, the colour correcting masking is chosen; for the masking methods discussed in this Chapter, the choice is usually confined to one of a few standardized procedures, but, when scanners are used (to be described in the next Chapter), a wider choice is available and tests with colour patches can be made to optimize masking parameters.

Plate 37

In the *top* picture, an operator is programming an input scanner to scan original pictures. The signals produced can be stored in digital electronic form for later use. In the *bottom* picture, an operator is sitting at a planning table, and is comparing original reflection copy with a colour cathode-ray tube display derived from stored signals obtained from a previous scan. The *centre* picture shows the control cursor employed on the planning table. This cursor can be used to identify particular areas of pictures by positioning it on the planning table, which contains a mesh of wires to detect its location; the position of the cursor is indicated on the CRT. By moving the cursor on the planning table, it is possible to enclose any chosen area of the picture, and then, by placing the cursor on a 'menu' of features and commands, as shown in the bottom picture, to manipulate that area as desired (see Section 29.9). The final manipulated array of images is stored for future use in deriving separation negatives on an output scanner. Pictures by courtesy of Crosfield Electronics Limited. Separations made on a Crosfield *Magnascan* scanner system.

Plate 38
This composite planned design provides an example of the type of manipulation that can be carried out on electronic page make-up systems for catalogue or advertising work (see Section 29.9). The picture of the girl holding the drum has been reproduced with various changes in the colours of her face, her clothing, and the background. All the other 'art work' was

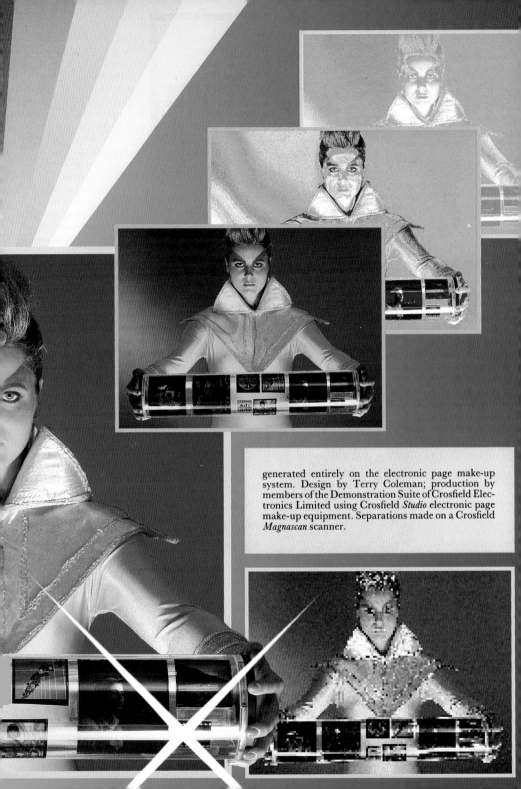

generated entirely on the electronic page make-up system. Design by Terry Coleman; production by members of the Demonstration Suite of Crosfield Electronics Limited using Crosfield *Studio* electronic page make-up equipment. Separations made on a Crosfield *Magnascan* scanner.

Plate 39
Several pictures of the same scene assembled directly on a scanner. Produced by members of the Demonstration Suite of Crosfield Electronics Limited using Crosfield *Magnascan Tints and Borders*. Separations made on a Crosfield *Magnascan* scanner.

REFERENCES

British Standard 4160: 1967, Inks for letterpress three- or four-colour printing (1967).
British Standard 4666: 1971, Inks for offset three- or four-colour printing (1971).
Clapper, F. R., *Tech. Assoc. Graphic Arts*, **14,** 107 (1962).
Clapper, F. R., *J. Phot. Sci.*, **12,** 28 (1964).
Clark, W., *Penrose Annual*, **46,** 125 (1952).
European Standard C.E.I. 12 : 66, European range of primary colours for letterpress printing (1966).
European Standard C.E.I. 13 : 67, Gamme européen d'encres d'imprimerie pour l'offset (1967).
Gresham, D. C., *Phot. J.*, **92B,** 91 (1952a).
Gresham, D. C., *Penrose Annual*, **46,** 77 (1952b).
Gresham, D. C., *Penrose Annual*, **50,** 102 (1956).
International Standards Organization, Sets of printing inks for letterpress and offset printing colorimetric characteristics, ISO 2845 and 2846 (1975).
Hartsuch, P. J., *Tech. Assoc. Graphic Arts*, **10,** 29 (1958).
Pollak, F., *J. Phot. Sci.*, **3,** 112 (1955).
Pollak, F., *J. Phot. Sci.*, **4,** 65 (1956).
Preucil, F., *Research Progress*, No. 38, Lithographic Technical Foundation, Chicago (1957).
Preucil, F., *Tech. Assoc. Graphic Arts*, **12,** 151 (1960).
Staehle, H. C., *Tech. Assoc. Graphic Arts*, **4,** 143 (1952).
Yule, J. A. C., *Tech. Assoc. Graphic Arts*, **5,** 94 (1953).
Yule, J. A. C., *Principles of Color Reproduction*, Wiley, New York (1967).

CHAPTER 29

Colour Scanners

29.1 Introduction

THE facility with which electrical signals can be manipulated to correspond to a wide variety of algebraic equations has led to the use, in graphic arts processes, of a number of devices known as *scanners*; in these, either all or part of the picture information is converted point by point into electrical signals at some intermediate stage, and the picture then subsequently reconstituted in a more conventional form. During the electrical stage the equivalent of tone-correction and masking procedures are carried out, with almost limitless flexibility. Scanners can be used to produce fully-corrected separations that are either continuous-tone or screened: screened separations can be made by exposure through a contact screen, or by a digital electronic system capable of generating the required dot shape at high resolution. In order to obtain the picture in the form of convenient electrical signals, it is necessary, as in television, to convert the picture from a two-dimensional array, to a one-dimensional array, and this, as in television, is most conveniently done by scanning it in successive lines.

In graphic arts it is not necessary to scan pictures with the same rapidity as is required in television, but it is necessary to scan in such a way as to provide much better definition. Scanning times of a few minutes for each picture are therefore customary, and the number of lines in the scanned picture varies from about 200 to about 1000 per inch (8 to 40 per mm). The first two scanners to be constructed were those invented by Hardy and Wurzburg (Hardy and Wurzburg, 1948) and by Murray and Morse (Murray and Morse, 1941). The Hardy and Wurzburg scanner was developed initially by the Interchemical Corporation and subsequently by the Radio Corporation of America; the Murray and Morse scanner was developed in its early stages by the Eastman Kodak Company and subsequently by Time Incorporated and its subsidiary, Printing Developments Incorporated (P.D.I.). These two scanners will now be described to provide a historical introduction to the subject, after which various modified methods derived from them will be outlined.

29.2 The Hardy and Wurzburg scanners

As originally conceived, the Hardy and Wurzburg scanner resulted in the direct production of screened photographic plates, from which the printing surfaces were obtained. Later, however, the emphasis swung to the production of continuous-tone photographic plates from which the printing surfaces were obtained in the conventional way; but the photographic plates made full correction for all distortions of tone and colour introduced by the characteristics of the printing surfaces and of the inks, so that no handwork or individual treatment of the printing surfaces was intended.

The Interchemical implementation of the Hardy and Wurzburg method is illustrated in Fig. 29.1. On a single carriage were mounted four separate photographic plates. One of these was an unexposed plate on which a fully corrected image was exposed; the other three plates were separation positives (or

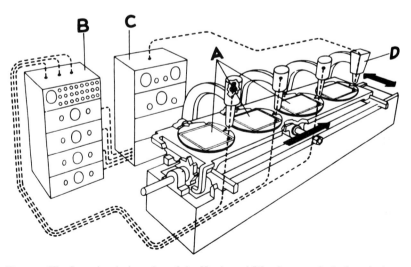

Fig. 29.1. The Interchemical version of the Hardy and Wurzburg method of producing colour-corrected, screened or continuous-tone, separations on a flat-bed mechanical type of scanner.

579

negatives), A, of the original scene which it was desired to reproduce. The carriage oscillated from side to side with an amplitude equal to the width of the separations and slowly progressed a distance equal to the length of the separations, thus enabling the whole area of the separations to be covered by an array of parallel lines.

Above the three separations were rigidly fixed three projectors which focused sharp points of light on to them. The projectors and separations were so located, of course, that corresponding parts of the picture were illuminated on each of the three separations. If the three separations were obtained from the original using plates having effective spectral sensitivity curves equal to the colour-matching functions, $\bar{x}(\lambda), \bar{y}(\lambda), \bar{z}(\lambda)$, the transmission of the separations at each point would be records of the tristimulus values X_P, Y_P, Z_P of the original at each point. Hence, by allowing the light transmitted by the separations to fall on three photocells, three signals were obtained, representing X_P, Y_P, and Z_P. The three signals were fed into an electronic circuit network, B, in which were stored the tristimulus values X_1, Y_1, Z_1; X_2, Y_2, Z_2, etc., of the eight printing colours (produced by the eight different ways in which the dot images can overlap); the amounts of ink necessary to produce colours having the tristimulus values X_P, Y_P, Z_P, were continuously evaluated by electronic computing circuits, in terms of the corresponding values of c, m, and y in the Neugebauer equations described in Section 26.7.

The three images were exposed one at a time, and when it was required to expose the cyan image, for instance, the continuously evaluated value of c was fed into another electronic circuit network, C, which resulted in the exposing light being modulated in such a way as to produce the required image on the unexposed plate, above which the exposing light, D, was rigidly fixed.

The actual operations involved in making a colour reproduction by this means were as follows:

(1) The original was photographed on plates (or films) having effective spectral sensitivities equal to the $\bar{x}(\lambda), \bar{y}(\lambda), \bar{z}(\lambda)$ curves (or any linear combination of them, since the electronic networks can solve the extra equations that result).
(2) The plates were developed, and from the three separation negatives thus obtained, three separation positives were made.
(3) The three separation positives were mounted in register on the scanning machine and the continuously evaluated value of c fed into the electronic network so that the required image was exposed.
(4) Similarly the image corresponding to m was exposed.
(5) Similarly the image corresponding to y was exposed.
(6) The three negatives were developed.
(7) Three screened printing surfaces were made.
(8) The three printing surfaces were inked and finally printed.

The conditions that have to be fulfilled, in order that colorimetric colour reproduction is achieved with this system, are as follows:

(1) The three separation positives must at all points have transmissions related to the tristimulus values X_P, Y_P, Z_P, of the original (or to linear combinations of them). This is generally practicable to within the required accuracy.

(2) The final coloured dot mosaic must contain the eight colours in the required amounts. Owing to the non-linearities of the etching processes this is not easily achieved, but it can be fairly well approximated to, if special compensations are introduced in the electronic stages.

(3) Only the eight expected colours must be present in the final dot mosaic. Clapper and Yule (Clapper and Yule, 1953) have pointed out that inter-reflections of light, within the layers of ink and the paper fibres, introduce other colours which upset the simple eight-colour theory.

(4) The paper and inks used must be capable of reproducing all the tristimulus values for which the electronic networks call. Of course, some colours may be too saturated to be reproduced, but in addition there is the limitation common to all reflection print systems, as mentioned in Section 13.10, that the range of tones ordinarily visible is limited to about 35 to 1 in intensity, that is, a density range of about 1.55 (with some inks and papers the density range is only just over 1.0).

In the Radio Corporation of America (R.C.A.) version of the Hardy and Wurzburg method, the separations and the plate being exposed were stationary, and the scanning was achieved by focusing on to them images of spots on cathode-ray tubes which were scanned in a suitable raster pattern (Rydz and Marquart, 1954).

Consideration was also given, in the various forms of the Hardy and Wurzburg method, to the need for producing four corrected separations, for printing with a black ink as well as with cyan, magenta, and yellow inks (Rydz and Marquart, 1955).

These scanners, although not now in commercial use, are important for their historical and theoretical interest. In spite of their limitations, as enumerated under the four headings given above, colour reproduction by means of their scanning method is capable of producing results of very high quality (Haynes, 1952; Ohler, 1955).

29.3 The P.D.I. scanner

The P.D.I. (Printing Developments Incorporated) scanner (known also at one time as the Time-Life Springdale scanner) is similar to the Hardy and Wurzburg scanner in that it breaks the image into a series of lines, converts it into electrical signals, carries out correction operations with them, and then exposes fully-corrected separations; but in almost all other respects there are fundamental and important differences. Thus the P.D.I. scanner scans a colour transparency instead of three black-and-white separations, it scans cylindrically instead of on a flat-bed, its correcting functions depend on masking theory and not on the Neugebauer equations, and it exposes all three corrected separations

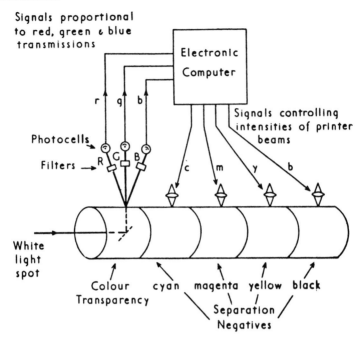

Fig. 29.2. Diagrammatic representation of the P.D.I. scanner. Fully corrected separations are made from colour transparencies wrapped round a rotating drum.

simultaneously (Bishop, 1951). The main features of the P.D.I. scanner are shown diagrammatically in Fig. 29.2.

A long cylinder, which has a transparent section at one end, is slowly rotated on its axis. The colour transparency to be reproduced is wrapped round the transparent section of the cylinder, while round the rest of the cylinder four unexposed sheets of film are wrapped. A small spot of light is focused on to the colour transparency from the inside of the cylinder and, after passing through the transparency, the light is split into three beams and falls on to three separate photocells, after passing through red, green, and blue filters. As the cylinder rotates it also travels longitudinally by means of a fine screw-thread and by this means the small spot of light eventually scans the entire area of the colour transparency, at either 250, 500, or 1000 lines per inch (10, 20, or 40 mm), taking proportionately longer for the finer scans.

The three photocells give rise to three electrical signals that, for each point of the transparency, are proportional to its red, green, and blue transmittances. By means of an electronic computer these signals are transformed into four related signals which, by modulating the intensities of four spots of light focused on the four unexposed films, result in fully corrected separation negatives being

582

exposed. From these separation negatives, cyan, magenta, yellow, and black printing surfaces are made by orthodox methods.

It is clear that the P.D.I. scanner must work from a colour transparency, and, if the original consists of reflection copy, a colour transparency of it has to be made (see Section 27.4). The colour transparency can be of any size up to 11 in X 14 in. The final reproduction can be the same size as the separation negatives, or it can be enlarged about $1\frac{1}{2}$ times if 250-line scanning per inch is used, about 3 times if 500-line scanning is used, or about 6 times if 1000-line scanning is used (or 10, 20, and 40 lines/mm, respectively).

29.4 Other drum scanners

Working on principles similar to those used in the P.D.I. scanner, the Fairchild *Scan-a-color* (Sigler, 1964) also had the facility of handling flexible reflection, as well as transmission, originals.

Less complicated, and smaller, drum scanners can be made if the separations are made one at a time instead of all four together; in this way it has been possible for scanners to be manufactured at a price low enough for it to be economic for many printing works to have their own equipment on the premises. Scanners of this general type included the Crosfield *Diascan*, the original models of the Hell *Chromagraph*, and the Linotype-Paul *Linoscan* (Nash, 1965), all of which produced fully-corrected separations, one at a time, from originals consisting of transparencies, which were wrapped round the drum.

A disadvantage of drum scanners is that they can only handle flexible originals, but most subjects can be copied on to colour transparency film so that the limitation is not too severe; moreover, such a copying step provides an opportunity for adjusting the sizes of the originals and this is useful when, as is common practice, a number of individual pictures are mounted together for common scanning so as to produce several scenes together on the same separations (see Section 27.2). Any changes in tone and colour reproduction introduced by the copying step may have to be allowed for in the correcting circuits, however.

29.5 Other flat-bed mechanical scanners

Flat-bed scanners, such as the Hardy and Wurzburg machine described in Section 29.2, are not restricted to the use of flexible originals, and a number have been developed. The Hell *Colorgraph* (Allen, 1958a), produced fully-corrected separations, all four at the same time, from either separations, colour transparencies, or flat copy, according to the particular model. The models using separations as originals carried all seven films (three originals and four being exposed) on the same reciprocating bed, which moved under three light-beams illuminating photocells, and four light-beams exposing the plates.

Some flat-bed mechanical scanners such as the Hell *Vario-Klischograph* (Hell, 1954 and 1957) were designed to produce letterpress plates by arranging for the output signal from the correction circuits to cause a tool to engrave a printing plate physically with a dot structure (the size of the dots varying as a suitable function of the signal strength), instead of varying the intensity of a beam of light falling on to a photographic material; the dots can be formed by the tool at rates of up to about 1000 dots per second. In this way, such scanners can produce the actual letterpress printing plates directly. By using the tool to remove appropriate proportions of an opaque layer coated on a transparent support, half-tone images suitable for making conventional printing surfaces by screened photographic intermediates can be obtained, so that this type of scanner can be used for making litho plates, for instance, as well as letterpress plates. Changes in size between the original and the corrected printing surfaces are possible on the Hell Vario-Klischograph by means of a pantographic linkage, and either transparent or reflection originals can be used.

29.6 Optical feed-back scanners

One of the disadvantages of scanners that convert the whole of the picture into electrical information is that to provide adequate definition a very fine line-structure has to be used, and this calls for very high precision in the scanning mechanism. Some scanners have therefore been developed in which separations are made optically, and only the correction from the normal optical result are passed through the electronic stages. One such device was the Hunter-Penrose *Autoscan* (Kilminster, 1956), in which correcting signals were applied to the intensity of the spot scanning the original; another was the Crosfield *Scanatron* (Allen, 1958a and b), in which corrected separations were produced by scan-printing through uncorrected unscanned separations; and the Log-Etronic *Color-Separator* was similar to the Scanatron but printed corrected separations one at a time through a colour transparency instead of through separations (Craig and Street, 1960). But most scanners now in use convert the whole of the picture into electronic signals.

29.7 Scanners with variable magnification

The early drum scanners provided means for making separations of the same size as the original; facilities were introduced subsequently to allow enlarged separations to be made from small transparencies, such as those of 6×6 cm and 24×36 mm image size.

A later development in scanners has been the provision of facilities for obtaining a very wide range of different magnifications and reductions between the original and the separations. Examples of this type of equipment are the Crosfield *Magnascan* series (Wilby and Pugsley, 1970), the P.D.I. scanners, the Hell *Chromagraphs*, the Linotype-Paul *Linoscans*, and the Dainippon *Direct Scanagraph* models. (See Plate 13, pages 176, 177.)

In the Magnascan equipment, because original transparencies are usually smaller than the separations made from them, the drum diameter for making the latter has been chosen to be twice that of the drum for the former. If the larger drum lead-screw had a pitch twice that of the smaller, then all separations would be magnified by a factor of 2. By driving the two lead screws at different speeds, degrees of enlargement in the axial direction either greater or less than 2 can be achieved. To produce the same degree of enlargement in the transverse direction, the signals obtained for a complete line in the picture are stored and then read out at such speed as will stretch or compress the picture as required. The storage is digital in nature, the magnitude of the signal being assigned to one of 256, or 2^8, discrete levels (128 levels were used in the original Magnascan); the position of the signal along the picture line is also stored digitally, by sampling the analogue signal from the original at discrete points. Storage and read-out operations run concurrently so that scanning proceeds continuously. The speed of the machine is at least as fast as that of non-enlarging scanners. The range of magnification obtainable is from 0.3 times to about 20 times.

The inclusion of an enlarging buffer, and in some cases multiple output heads, enables some scanners to produce two or four separations simultaneously when the required format is sufficiently small compared with the output drum.

In one of the Linoscan scanners, the transparency drum rotates three times as fast as the exposing drum, and red, green, and blue signals are read from the transparency on successive rotations and stored, so that only one photomultiplier tube has to be used.

Some scanners (such as the Magnascan 530 and 540, and the Direct Scanagraph 701 and 708) are made with electronic synchronization between the input and output drums, enabling the input and output signals to be separated so that the input can be in an open working area while the output is in a darkroom. In some cases (for example, the Magnascan 690) one input module can drive multiple modules enabling large separations to be made simultaneously.

29.8 Scanner outputs

Scanner outputs can be such as to produce continuous-tone separations; or screened separations can be made, by using a contact screen or by generating the dots electronically. By carrying out the screening on the scanner, the subsequent steps in making the printing surface are simplified. But continuous separations have to be made when it is required to originate from them a series of printing surfaces having different magnifications. They are also necessary for the production of gravure printing cylinders. Light sources used for exposing the separations include high pressure Xenon lamps, tungsten-halogen lamps, and lasers. Modulation of the light intensity may be by voltage change, or by electro-optic or acousto-optic devices, according to the requirements and the light source used.

When screened negatives are made by wrapping a contact screen round the exposing drum, higher levels of light are required for exposing, and Xenon arcs and lasers are often used. Electronic dot generation is carried out by suitable control of the intensity of exposing light spots as the drum rotates. In one system, six fibre-optic cables, aligned in a row, carry the light to the film to produce micro-dots which make up the half-tone dot; there may be as many as twelve rows of micro-dots, both vertically and horizontally, to form one half-tone dot, so that various dot shapes are possible, such as square, rectangular, round, or elliptical.

29.9 Electronic retouching

In many scanners, the signals are at one stage transformed entirely into digital form, and hence they can be readily stored and interfaced with computers between the scanning and exposing ends of the equipment. This opens up the possibility of manipulating the signals in ways similar to those described in Chapter 25 in connection with television.

In one system, in which a colour television monitor is used to display a picture derived from the original, a position indicator is displayed on the monitor and can be moved about by the operator. By tracing round an outline of part of the picture, a closed area can be produced which can be treated differently from the rest of the picture, as in video graphics (see Section 25.6) or Chroma-key (see Section 25.3). Thus, in this area, the colour reproduction can be changed completely, to produce, for instance, pictures of a range of cars of the same model but having quite different colours; in this way printed catalogues can show a much wider range of colours of goods than are available in photographs of them or even in the goods themselves.

Another facility is to define a rectangular area in the picture and then to enlarge it greatly. One use of this facility is the removal of scratches or other blemishes: this can be done by instructing the computer to regard all picture elements (*pixels*) in the area as identical, and hence blemishes are filled in with the surrounding colour. In an improved technique, which is also available, the area of the blemish is filled by copies of a variety of surrounding pixels. In this way texture is preserved and the production of an unnatural-looking 'flat' spot avoided.

By blending together in different ways the original information in an area and new information derived from computer instructions or from other picture sources or from computer generated typography, a whole variety of effects can be produced. For instance, it is possible to change the eye make-up and hair colour of a subject without any visible discontinuities at the edges of the altered areas, and to add appropriate captions where required. (See Plate 23, pages 352, 353.)

Of course, local adjustments in colour and tone reproduction of areas of the original picture are also easily possible, thus making it feasible to make corrections to the colours of individual objects and to the tone reproduction of local areas. (See Plates 37, 38, and 39, pages 576, 577.)

29.10 Electronic page make-up

When, as is often the case, it is required to position reproductions from several different originals on the same page, the *page make-up* or *planning* can be carried out electronically (Pugsley, 1981). One way of doing this is to use a back-lit planning table on which the required positions of the pictures are drawn in outline. These outlines are then traced by means of a cursor so as to record in the computer a series of areas. These areas are then allocated to the various pictures and to any computer-generated borders or other material. When all the information has been supplied to the computer for the whole page, it then assembles the digital signals in the correct sequence for feeding to the exposing end of the scanner to produce the separations for the entire page. (See Plates 37, 38, and 39, pages 576, 577.)

29.11 Logic circuits in scanners

In photographic masking procedures, the degree of masking provided is usually represented by a single set of equations, which are regarded as applying to all colours. In practice, the actual effects usually vary somewhat from one colour to another, but such variations occur gradually throughout the distribution of colours. In scanners, however, it is possible to use logic types of circuit that will apply one set of masks to colours in one domain, and a different set to colours in a different domain, with a discontinuity in the masking equations at the boundary between the two domains. Thus yellowish colours might be treated in one way and bluish colours in a different way, with the transition taking place across the grey scale. The ability to incorporate such effects can provide useful degrees of freedom in adjusting the colour reproduction characteristics.

29.12 Unsharp masking in scanners

The enhancement of fine detail by the use of masks that are unsharp (see Section 15.3) was achieved in the optical feed-back type of scanners simply by making the size of the scanning spot larger than that representative of the finest detail in the image obtained optically. But in the fully electronic scanners it is not quite so easy: if scanning is carried out with a large spot the whole image simply becomes unsharp. The effects of unsharp masking *along* each line can be simulated by suitably designing the frequency response of the electronic circuits, but the effect *across* the lines can only be simulated electronically if the scanner has a memory from one line to the next at each point along each line. Another way of simulating unsharp masking is to scan the image both with a small spot from which the image is derived, and with a large spot (about three times the diameter), from which the unsharp mask is derived (Hall and Yule, 1956; Nash, 1965). (See Plate 36, page 545.) Both these techniques are used, but the decreased cost of providing the scanner with a memory of several lines renders this a more important method. In some cases the digital electronic circuits employed in

587

electronic unsharp masking can be specially programmed: for example the unwanted accentuation of grain in shadow areas can be reduced.

29.13 Differential masking in scanners

If the image is also scanned with a spot of very much larger size, something can be done to lighten areas of the pictures that are generally rather dark, and darken those that are rather light (Hall and Yule, 1964). This can be very useful when transparencies contain important parts of a scene illuminated at different levels, and corresponds to the individual shading, or *dodging*, of different parts of an image during photographic enlarging, a procedure which can be carried out by hand, or electronically as in the Log-Etronic type of equipment (Cox, 1959).

29.14 Under-colour removal

An interesting facility that can be provided on scanners is the means of making separations such that, of the three coloured inks, only two are printed heavily at the same point in the reproduction, any required darkening of the colour being achieved mainly by means of the black image and not by means of the third colour, a technique known as *under-colour removal*. This means that parts of the picture that are neutral in colour are rendered mainly by the black printer, so that the correct rendering of a grey scale is greatly facilitated. Moreover, the variations in luminance in the reproduction are then controlled in large measure by the black printer, and since impressions of sharpness and resolving power are dependent almost entirely on differences in luminance, rather than colour, some improvement in these respects arises from the fact that most of the luminance differences stem from a single image, rather than from four superimposed images. It is interesting to note that, in colour television, luminance is transmitted as a separate high-definition signal for much the same reasons (see Section 19.7). Under-colour removal also facilitates rapid drying of the inks during printing, and can save ink costs. Full under-colour removal can increase registration problems, so that a partial level is usually preferred. (See Plate 35, pages 544, 545.)

29.15 Under-colour correction

The electronic computing stage in scanners makes possible refinements in masking that in the ordinary way are often omitted. One such refinement is known as *under-colour correction* (Smith, 1954). In wet letterpress printing, succeeding images are applied before the previous images are dry, and the wetness of the ink already printed prevents the next ink from transferring to the paper in the proper amount. The effect of this *trapping* (see Section 28.15) obviously depends on the order in which the inks are printed, but once this order has been established allowance can be made for it, and this can be done on scanners.

29.16 Typical scanner signal sequences

The electrical input to the computer in a scanner generally consists of three d.c. currents proportional to the red, green, and blue transmittances (or reflectances) of the original material. It is usual, before scanning commences to set into the computer the levels of the signals that correspond to the white level on each original, and sometimes the black level also. The signals may then be passed through a circuit that converts them from a linear to an approximately logarithmic form so that they become approximately proportional to density instead of to transmittance; and it may be necessary to compress the range of the signals so as to make them easier to handle in the circuits and more suitable for the final result which will usually be on a reflection support with its limited luminance range.

Masking is carried out by circuits, corresponding to three simultaneous equations, which add and subtract different proportions of the logarithmic signals; the circuits may also perform operations equivalent to two-stage masking and the use of non-linear masks (see Section 28.7); and logic circuits may be used to switch the computer from one mode to another as the colour being scanned varies. Partial or nearly complete under-colour removal and any under-colour correction (see Sections 29.14 and 29.15), and the computation of the signals needed for the black printer, may be carried out next; then each of the four signals may be passed through 'curve-shaping' circuits that ensure that the final logarithmic signals have the required gradation of tones: it is sometimes necessary to reproduce both highlights and shadows at fairly high gamma, in order to retain good visibility of detail in these areas, and middle-tones then have to be reduced in gamma; scanner circuits can usually produce the high-low-high gamma characteristic necessary to achieve this. Finally, the signal may go through an anti-logging circuit prior to being used to control the intensity of the exposing light (or lights) or the depth of a cutting tool. In some scanners some or all of these processes are carried out under the control of a digital computer. This enables a greater variety of characteristics, such as curve shapes, to be achieved, and it also enables adjustment settings to be stored in the computer memory and recalled when desired. For instance, colour correction adjustments to suit more than one set of printing inks may be made immediately available in this way.

29.17 Television image display

In those scanner systems that include colour monitors for viewing pictures, visual checks on many aspects of quality are possible as the work proceeds. This visual 'proofing' is a valuable control facility, and, as mentioned in Section 27.11, can also include pre-press proofing of the final separations. It can also be used to assess adjustments necessary to bring originals to a standard level for ease of correcting whole pages of work. It is also possible to connect a special-purpose camera to some systems, to obtain, as a photographic colour print, a permanent record of the image currently displayed on the colour monitor. This is especially useful when several departments must approve the work before final separation films are made.

This combination of photographic and television technologies is an example of the way colour reproduction is increasingly being developed.

REFERENCES

Allen, G. S., *J. Phot. Sci.*, **6,** 125 (1958a).
Allen, G. S., *Penrose Annual,* **52,** 123 (1985b).
Bishop, N., *Penrose Annual,* **45,** 92 (1951).
Clapper, F. R., and Yule, J. A. C., *J. Opt. Soc. Amer.,* **43,** 600 (1953).
Cox, H. W., *Penrose Annual,* **53,** 88 (1959).
Craig, D. R., and Street, J. N., *Tech. Assoc. Graphic. Arts,* **12,** 175 (1960).
Hall, V. C., and Yule, J. A. C., *U.S. Patent* 2,744,950 (1956).
Hall, V. C., and Yule, J. A. C., *U.S. Patent* 3,153,698 (1964).
Hardy, A. C., and Wurzburg, F. L., *J. Opt. Soc. Amer.,* **38,** 300 (1948).
Haynes, H. E., *Penrose Annual,* **46,** 83 (1952).
Hell, R., *Penrose Annual,* **48,** 101 (1954).
Hell, R., *Penrose Annual,* **51,** 117 (1957).
Kilminster, R., *Penrose Annual,* **50,** 111 (1956).
Murray, A., and Morse, R. S., *U.S. Patent* 2,253,086 (1941).
Nash, C. F., *Litho Printer,* **8,** 33 (August, 1965).
Ohler, A. E., *Penrose Annual,* **49,** 80 (1955).
Pugsley, P. C., *I.E.E.E. Transactions and Communications, Com.* **29,** 1891 (1981).
Rydz, J. S., and Marquart, V. L., *Tech. Assoc. Graphic Arts,* **6,** 139 (1954).
Rydz, J. S., and Marquart, V. L., *Tech. Assoc. Graphic Arts,* **7,** 15 (1955).
Sigler, H., *Tech. Assoc. Graphic Arts,* **16,** 192 (1964).
Smith, F. H., *Penrose Annual,* **48,** 131 (1954).
Wilby, W. P. L., and Pugsley, P. C., *Printing Technology,* **14,** 142 (1970).

GENERAL REFERENCES

Allen, G. S., *J. Phot. Sci.,* **6,** 125 (1958).
Nash, C. F., *Litho Printer,* **8,** 33 (August, 1965).
Nash, C. F., *Litho Printer,* **8,** 43 (September, 1965).
Yule, J. A. C., *Principles of Color Reproduction,* Chapter 12, Wiley, New York (1967).

APPENDICES

APPENDIX I

Matrix algebra

A1.1 General principles

$\mathbf{M}$ATRIX algebra is useful in colorimetric calculations, in the evaluation of colour correcting masks, and in the formulation of colour reproduction theory. In this Appendix, therefore, a short explanation of matrix algebra is given, together with an example of its application to a common colorimetric problem. A matrix is an array of numbers or symbols; thus

$$\begin{matrix} 271 & 18 \\ 671 & 12 \end{matrix} \quad \text{and} \quad \begin{matrix} x_1 & x_2 & x_3 \\ x_4 & x_5 & x_6 \end{matrix}$$

are both matrices. If two matrices are equal, each term of the first matrix is equal to the corresponding term of the second. Thus the single matrix equation:

$$\begin{pmatrix} x_1 & x_2+b \\ y_1+c & y_2 \end{pmatrix} = \begin{pmatrix} 61 & d+3e \\ 12 & 6f \end{pmatrix}$$

represents the four equations:

$$x_1 = 61 \qquad x_2+b = d+3e$$
$$y_1+c = 12 \qquad y_2 = 6f$$

By adopting a set of rules for multiplying matrices, sets of simultaneous equations, when written as single matrix equations, can be simplified by factorizing. For example, the equations:

$$a_1x+a_2y = a_5$$
$$a_3x+a_4y = a_6$$

when written in matrix algebra take the form:

$$\begin{pmatrix} a_1x+a_2y \\ a_3x+a_4y \end{pmatrix} = \begin{pmatrix} a_5 \\ a_6 \end{pmatrix}$$

or after factorizing:

$$\begin{pmatrix} a_1 & a_2 \\ a_3 & a_4 \end{pmatrix} \begin{pmatrix} x \\ y \end{pmatrix} = \begin{pmatrix} a_5 \\ a_6 \end{pmatrix}$$

The multiplication rule in this case is therefore that the terms of the first row of the first matrix are multiplied successively by the terms of the column of the second matrix and summed, to give the term for the first row of the product matrix; the term for the second row of the product matrix is similarly derived from the second row of the first matrix. The rule can be stated quite generally as follows: the term in the pth row and qth column of the product matrix, is given by the sum of the successive products of the terms of the pth row of the first matrix and the qth column of the second. Thus:

$$\begin{pmatrix} a_1 & a_2 \\ a_3 & a_4 \end{pmatrix} \begin{pmatrix} b_1 & b_3 \\ b_2 & b_4 \end{pmatrix} = \begin{pmatrix} a_1b_1+a_2b_2 & a_1b_3+a_2b_4 \\ a_3b_1+a_4b_2 & a_3b_3+a_4b_4 \end{pmatrix}$$

Two of the most important uses of matrix algebra occur when variables have to be changed in equations and when equations have to be solved. Thus if

$$a_1x+a_2y = a_5$$
$$a_3x+a_4y = a_6$$
$$b_1x'+b_3y' = x$$
$$b_2x'+b_4y' = y$$

Then in matrix algebra we have:

$$\begin{pmatrix} a_1 & a_2 \\ a_3 & a_4 \end{pmatrix} \begin{pmatrix} x \\ y \end{pmatrix} = \begin{pmatrix} a_5 \\ a_6 \end{pmatrix}$$

$$\begin{pmatrix} b_1 & b_3 \\ b_2 & b_4 \end{pmatrix} \begin{pmatrix} x' \\ y' \end{pmatrix} = \begin{pmatrix} x \\ y \end{pmatrix}$$

Therefore
$$\begin{pmatrix} a_1 & a_2 \\ a_3 & a_4 \end{pmatrix} \begin{pmatrix} b_1 & b_3 \\ b_2 & b_4 \end{pmatrix} \begin{pmatrix} x' \\ y' \end{pmatrix} = \begin{pmatrix} a_5 \\ a_6 \end{pmatrix}$$

That this substitution is valid is easily checked by multiplying out this triple matrix product and comparing the two equations obtained with the results of ordinary algebra. It should be noted, however, that the order of the matrices is important, and must not be changed. Thus if two matrices are represented by A and B, then

$A.B$ is not equal to $B.A$

Matrix algebra is often very useful when sets of simultaneous equations have to be solved, as is sometimes the case in colorimetric calculations. In order to simplify the solution of equations, two derived matrices are used, as follows:

$A' = $ *the transpose* of A, obtained by writing the rows as columns and the columns as rows.

adj. $A = $ the *adjugate* of A, obtained by replacing each term of the matrix by the determinant formed by all the rows and columns of the matrix not containing that term, and transposing the result, a negative sign being prefixed to all terms situated an odd number of non-diagonal moves from the first term.

Thus if $\quad A = \begin{pmatrix} a_1 & a_2 & a_3 \\ a_4 & a_5 & a_6 \\ a_7 & a_8 & a_9 \end{pmatrix}$

$$A' = \begin{pmatrix} a_1 & a_4 & a_7 \\ a_2 & a_5 & a_8 \\ a_3 & a_6 & a_9 \end{pmatrix}$$

$$\text{adj. } A = \begin{pmatrix} |A_1| & -|A_4| & |A_7| \\ -|A_2| & |A_5| & -|A_8| \\ |A_3| & -|A_6| & |A_9| \end{pmatrix} \qquad \begin{array}{l} \text{where } |A_1| = a_5a_9 - a_6a_8, \\ \qquad\ |A_2| = a_4a_9 - a_6a_7, \\ \qquad\qquad \text{etc.} \end{array}$$

The *inverse* or *reciprocal* matrix A^{-1} is the matrix that expresses solution equations. Thus if

$$\begin{pmatrix} x \\ y \\ z \end{pmatrix} = A \begin{pmatrix} x' \\ y' \\ z' \end{pmatrix} \text{ then } \begin{pmatrix} x' \\ y' \\ z' \end{pmatrix} = A^{-1} \begin{pmatrix} x \\ y \\ z \end{pmatrix}$$

It may be shown by simple algebra that the reciprocal matrix is given by:

$$A^{-1} = \frac{\text{adj. } A}{|A|}$$

where $|A|$ is the determinant corresponding to the matrix A. Hence if

$$A = \begin{pmatrix} a_1 & a_2 & a_3 \\ a_4 & a_5 & a_6 \\ a_7 & a_8 & a_9 \end{pmatrix} \qquad |A| = \begin{vmatrix} a_1 & a_2 & a_3 \\ a_4 & a_5 & a_6 \\ a_7 & a_8 & a_9 \end{vmatrix}$$

thus $\qquad |A| = a_1(a_5a_9 - a_6a_8) - a_2(a_4a_9 - a_6a_7) + a_3(a_4a_8 - a_5a_7).$

A1.2 Application to colorimetry

A common problem in colorimetry is as follows: given the position of three stimuli R, G, and B in some colour triangle, for instance the XYZ triangle, it is required to find the transformation equations necessary to transfer to that triangle results obtained using R, G, and B as matching stimuli, using units such that equal quantities are required to match some white stimulus W. The basic data therefore consist of equations of the type:

$$
\begin{aligned}
(R) &\propto a_1(X) + a_2(Y) + a_3(Z) \\
(G) &\propto a_4(X) + a_5(Y) + a_6(Z) \\
(B) &\propto a_7(X) + a_8(Y) + a_9(Z) \\
(W) &\propto h_1(R) + h_2(G) + h_3(B) \\
(W) &\propto j_1(X) + j_2(Y) + j_3(Z)
\end{aligned}
$$

where in each equation the coefficients sum to unity $(a_1 + a_2 + a_3 = 1$ etc.$)$.

595

It is convenient to insert constants k_1, k_2, and k_3 so as to avoid the proportional signs thus:

$$k_1(R) \equiv a_1(X) + a_2(Y) + a_3(Z)$$
$$k_2(G) \equiv a_4(X) + a_5(Y) + a_6(Z)$$
$$k_3(B) \equiv a_7(X) + a_8(Y) + a_9(Z)$$

and to rewrite the equations for (W) in the form

$$k_4(W) \equiv H_1(R) + H_2(G) + H_3(B)$$
$$k_4(W) \equiv J_1(X) + J_2(Y) + J_3(Z)$$

where H_1, H_2, H_3 are proportional to h_1, h_2, h_3 respectively, but represent the actual amounts of (R), (G), (B) required to match the white stimulus, W; and J_1, J_2, J_3 are proportional to j_1, j_2, j_3 respectively, but J_2 is the luminance factor of the white, W. It is now required to evaluate k_1, k_2, and k_3, and in order to do this it is necessary to solve the above equations for (X), (Y), and (Z).

A convenient systematic way of doing this is by means of matrix algebra. It is required to find the reciprocal of the matrix,

$$\begin{pmatrix} a_1 & a_2 & a_3 \\ a_4 & a_5 & a_6 \\ a_7 & a_8 & a_9 \end{pmatrix}$$

If this matrix is represented by A, then:

$$A^{-1} = \frac{1}{|A|} \begin{pmatrix} a_5a_9 - a_6a_8 & -(a_4a_9 - a_6a_7) & a_4a_8 - a_5a_7 \\ -(a_2a_9 - a_3a_8) & a_1a_9 - a_3a_7 & -(a_1a_8 - a_2a_7) \\ a_2a_6 - a_3a_5 & -(a_1a_6 - a_3a_4) & a_1a_5 - a_2a_4 \end{pmatrix}'$$

$$= \frac{1}{|A|} \begin{pmatrix} b_1 & b_2 & b_3 \\ b_4 & b_5 & b_6 \\ b_7 & b_8 & b_9 \end{pmatrix}' = \frac{1}{|A|} \begin{pmatrix} b_1 & b_4 & b_7 \\ b_2 & b_5 & b_8 \\ b_3 & b_6 & b_9 \end{pmatrix}$$

When, as is usually the case, $a_1 + a_2 + a_3 = a_4 + a_5 + a_6 = a_7 + a_8 + a_9 = 1$, then:

$$|A| = b_1 + b_4 + b_7 = b_2 + b_5 + b_8 = b_3 + b_6 + b_9$$

which, as well as evaluating $|A|$ very simply, provides a check on every term of the matrix.

Dividing each term of the matrix by $|A|$, we obtain:

$$A^{-1} = \begin{pmatrix} c_1 & c_2 & c_3 \\ c_4 & c_5 & c_6 \\ c_7 & c_8 & c_9 \end{pmatrix}$$

and as a final check: $c_1 + c_2 + c_3 = c_4 + c_5 + c_6 = c_7 + c_8 + c_9 = 1$ should be true.

We can now write:

$$1.0(X) \equiv c_1 k_1(R) + c_2 k_2(G) + c_3 k_3(B)$$
$$1.0(Y) \equiv c_4 k_1(R) + c_5 k_2(G) + c_6 k_3(B)$$
$$1.0(Z) \equiv c_7 k_1(R) + c_8 k_2(G) + c_9 k_3(B)$$

and substituting (X), (Y), and (Z) in the equation:

$$k_4(W) = J_1(X) + J_2(Y) + J_3(Z)$$

and comparing the result with the equation

$$k_4(W) = H_1(R) + H_2(G) + H_3(B)$$

we obtain:

$$k_1 = H_1/(J_1c_1 + J_2c_4 + J_3c_7)$$
$$k_2 = H_2/(J_1c_2 + J_2c_5 + J_3c_8)$$
$$k_3 = H_3/(J_1c_3 + J_2c_6 + J_3c_9)$$

Hence k_1, k_2, k_3 are evaluated and the transformation equations are given by:

$$1.0(R) \equiv (a_1/k_1)\,(X) + (a_2/k_1)\,(Y) + (a_3/k_1)\,(Z)$$
$$1.0(G) \equiv (a_4/k_2)\,(X) + (a_5/k_2)\,(Y) + (a_6/k_2)\,(Z)$$
$$1.0(B) \equiv (a_7/k_3)\,(X) + (a_8/k_3)\,(Y) + (a_9/k_3)\,(Z)$$

and the reciprocal transformation equations by:

$$1.0(X) \equiv c_1k_1(R) + c_2k_2(G) + c_3k_3(B)$$
$$1.0(Y) \equiv c_4k_1(R) + c_5k_2(G) + c_6k_3(B)$$
$$1.0(Z) \equiv c_7k_1(R) + c_8k_2(G) + c_9k_3(B)$$

In general, the coefficients of these equations will not sum to unity.

It is often more convenient to re-write these equations as relationships between tristimulus values (see Section 8.4) as follows:

$$X = (a_1/k_1)R + (a_4/k_2)G + (a_7/k_3)B$$
$$Y = (a_2/k_1)R + (a_5/k_2)G + (a_8/k_3)B$$
$$Z = (a_3/k_1)R + (a_6/k_2)G + (a_9/k_3)B$$
$$R = c_1k_1X + c_4k_1Y + c_7k_1Z$$
$$G = c_2k_2X + c_5k_2Y + c_8k_2Z$$
$$B = c_3k_3X + c_6k_3Y + c_9k_3Z$$

In general, the coefficients of these equations will not sum to unity.

If $H_1 = H_2 = H_3 = H$ and $J_1 = J_2 = J_3 = J$, then: on the right hand sides of the above sets of equations, the sums of the coefficients of (R), (G), and (B) are all equal to H/J, and this is also true for the coefficients of X, Y, and Z; and the sums of the coefficients of (X), (Y), and (Z) are all equal to J/H, and this is also true for the coefficients of R, G, and B. If, in addition, $J = H$, then these coefficient-sums are all equal to unity.

APPENDIX 2

Colorimetric Tables

A2.1 Calculating colorimetric measures

In this Appendix sufficient information is given to enable colorimetric specifications to be evaluated from spectrophotometric data. The data may be in one of two forms: either the amount of light (in photometric units) at each wavelength may be known; or the amount of energy or power (in radiometric units) at each wavelength may be known.

In the first case the calculation proceeds by applying the Centre of Gravity Law of colour mixture as described in Sections 7.6, 8.5, and 8.6. If the amounts of light at successive wavelengths, λ_1, λ_2, λ_3, etc. are L_1, L_2, L_3, etc., then the chromaticity of the resultant mixture is given by calculating the centre of gravity weights:

$$L_1/y_1 \text{ at } x_1, y_1$$
$$L_2/y_2 \text{ at } x_2, y_2$$
$$L_3/y_3 \text{ at } x_3, y_3 \text{ etc.}$$

where x_1, y_1, etc., are the chromaticity co-ordinates in the XYZ system of the wavelengths $\lambda_1, \lambda_2, \lambda_3$, etc. The co-ordinates x_m, y_m, of the centre of gravity of such a system of weights is given by:

$$x_m = \frac{x_1 L_1/y_1 + x_2 L_2/y_2 + x_3 L_3/y_3 + \dots}{L_1/y_1 + L_2/y_2 + L_3/y_3 + \dots}$$

$$y_m = \frac{y_1 L_1/y_1 + y_2 L_2/y_2 + y_3 L_3/y_3 + \dots}{L_1/y_1 + L_2/y_2 + L_3/y_3 + \dots}$$

$$= \frac{L_1 + L_2 + L_3 + \dots}{L_1/y_1 + L_2/y_2 + L_3/y_3 + \dots}$$

In the tables, values of x and y (for the 2° Standard Observer) are given at 10 nm intervals from 380 to 780 nm so that the above type of calculation can be made.

In the second case, where the amount of power or energy, $e(\lambda)$, at each wavelength is known (in radiometric units), we could convert this to the amount of light at each wavelength by multiplying each value of $e(\lambda)$ by the appropriate value of the spectral luminous efficiency function $\bar{y}(\lambda)$, which is the same as $V(\lambda)$, see Section 8.5; and the calculation would then proceed as above:

598

$$x_m = \frac{x_1 e_1 \bar{y}_1/y_1 + x_2 e_2 \bar{y}_2/y_2 + x_3 e_3 \bar{y}_3/y_3 + \ldots \ldots}{e_1 \bar{y}_1/y_1 + e_2 \bar{y}_2/y_2 + e_3 \bar{y}_3/y_3 + \ldots \ldots}$$

$$y_m = \frac{e_1 \bar{y}_1 + e_2 \bar{y}_2 + e_3 \bar{y}_3 + \ldots \ldots \ldots}{e_1 \bar{y}_1/y_1 + e_2 \bar{y}_2/y_2 + e_3 \bar{y}_3/y_3 + \ldots \ldots \ldots}$$

But, because the chromaticity co-ordinates, x, y, z, of spectral colours are related to the colour-matching functions $\bar{x}(\lambda), \bar{y}(\lambda), \bar{z}(\lambda)$, by expressions of the type $x_1 = \bar{x}_1/(\bar{x}_1 + \bar{y}_1 + \bar{z}_1), y_1 = \bar{y}_1/(\bar{x}_1 + \bar{y}_1 + \bar{z}_1)$, and $z_1 = \bar{z}_1/(\bar{x}_1 + \bar{y}_1 + \bar{z}_1)$, it follows that $x_1/y_1 = \bar{x}_1/\bar{y}_1$; similarly $x_2/y_2 = \bar{x}_2/\bar{y}_2$, etc; and $z_1/y_1 = \bar{z}_1/\bar{y}_1$, etc. Hence the summations simplify to:

$$S_m x_m = e_1 \bar{x}_1 + e_2 \bar{x}_2 + e_3 \bar{x}_3 + \ldots \ldots \ldots \ldots = X_m$$
$$S_m y_m = e_1 \bar{y}_1 + e_2 \bar{y}_2 + e_3 \bar{y}_3 + \ldots \ldots \ldots \ldots = Y_m$$
$$S_m z_m = e_1 \bar{z}_1 + e_2 \bar{z}_2 + e_3 \bar{z}_3 + \ldots \ldots \ldots \ldots = Z_m$$
$$S_m = e_1 \bar{y}_1/y_1 + e_2 \bar{y}_2/y_2 + e_3 \bar{y}_3/y_3 + \ldots \ldots \ldots \ldots$$

It is therefore more convenient when the data is in radiometric units, to use the tabulated values of $\bar{x}(\lambda), \bar{y}(\lambda), \bar{z}(\lambda)$; these are therefore also given in the tables (for the 2° Standard Observer), at every 10 nm from 380 to 780 nm. If X_m, Y_m, Z_m and S_m are all evaluated, x_m, y_m, z_m can be obtained and the computation can be checked by making sure that $x_m + y_m + z_m = 1$. (Individual entries in the computation can also be checked by seeing whether at each wavelength $X_1 + Y_1 + Z_1 = S_1$ etc.) Alternatively, the more usual procedure is to ignore S_m altogether and to obtain x_m, y_m, and z_m from:

$$x_m = X_m/(X_m + Y_m + Z_m)$$
$$y_m = Y_m/(X_m + Y_m + Z_m)$$
$$z_m = Z_m/(X_m + Y_m + Z_m)$$

When the spectrophotometric data is in radiometric units, it often takes the form of percentage spectral reflectance or transmittance readings, $t(\lambda)$, and the spectral power or energy distribution, $E(\lambda)$, of an illuminant; in this case $e(\lambda) = E(\lambda) t(\lambda)$. The calculation then proceeds as follows:

$$X_m = E_1 t_1 \bar{x}_1 + E_2 t_2 \bar{x}_2 + E_3 t_3 \bar{x}_3 + \ldots \ldots \ldots$$
$$Y_m = E_1 t_1 \bar{y}_1 + E_2 t_2 \bar{y}_2 + E_3 t_3 \bar{y}_3 + \ldots \ldots \ldots$$
$$Z_m = E_1 t_1 \bar{z}_1 + E_2 t_2 \bar{z}_2 + E_3 t_3 \bar{z}_3 + \ldots \ldots \ldots$$

The corresponding values of x_m, y_m, z_m are then evaluated as before. The total spectral reflectance (or transmittance), Y, relative to the perfect diffuser (or transmitter), is given by

$$Y = \frac{E_1 t_1 \bar{y}_1 + E_2 t_2 \bar{y}_2 + E_3 t_3 \bar{y}_3 + \ldots}{E_1 \bar{y}_1 + E_2 \bar{y}_2 + E_3 \bar{y}_3 + \ldots}$$

so that to obtain this result $E_1 \bar{y}_1 + E_2 \bar{y}_2 + E_3 \bar{y}_3 + \ldots = F$ must also be evaluated. If the values of t are in the form of percentages, the above formula gives the value of Y as a percentage as is customary. The corresponding set of the three tristimulus values are then given by:

599

$$X = X_m/F$$
$$Y = Y_m/F$$
$$Z = Z_m/F.$$

To facilitate the above type of calculation, values of the spectral power distributions, $E(\lambda)$, are given in the tables for the illuminants S_A, S_B, S_C, and D_{65}, together with those of a full radiator of colour temperature 3250 K (which is representative of the light emitted by tungsten-filament projector lamps), and of D_{55} (which is representative of sunlight and skylight as often used for outdoor pictures), and of D_{50} (which is representative of a slightly yellower daylight), and of D_{75} (which is representative of north skylight).

For plotting chromaticity it is often also required to evaluate

$$u' = 4x/(-2x + 12y + 3)$$
$$v' = 9y/(-2x + 12y + 3)$$

and the CIELUV colour space can be used by evaluating.

$$L^* = 116(Y/Y_n)^{1/3} - 16$$
$$u^* = 13L^*(u' - u'_n)$$
$$v^* = 13L^*(v' - v'_n)$$

where Y_n, u'_n, v'_n are the values of Y, u', v' respectively for a specified reference white. To facilitate this evaluation, the chromaticity co-ordinates u'_n, v'_n, of various illuminants are given in the tables, and also values of L^* corresponding to various values of Y/Y_n. The difference between two reflecting samples having values L_1^*, u_1^*, v_1^* and L_2^*, u_2^*, v_2^* can then be evaluated as

$$[(L_1^* - L_2^*)^2 + (u_1^* - u_2^*)^2 + (v_1^* - v_2^*)^2]^{\frac{1}{2}}$$

A worked example is included to clarify the actual procedures involved: the values of L^*, u^*, v^* are found corresponding to a reflecting sample whose percentage spectral reflectance, $t(\lambda)$, is known, when it is illuminated by standard iluminant A. (A digital calculator greatly facilitates this type of work, or of course a computer can be used.) From the table of results for the worked example (page 602) we have:

$$X_m = 29540$$
$$Y_m = 38933$$
$$Z_m = 27930$$
$$100F = 107896$$

Hence:

$$x_m = X_m/(X_m + Y_m + Z_m) = 29540/96403 = 0.3064$$
$$y_m = Y_m/(X_m + Y_m + Z_m) = 38933/96403 = 0.4039$$
$$z_m = Z_m/(X_m + Y_m + Z_m) = 27930/96403 = 0.2897$$

The values of x_m, y_m, z_m sum to unity. The tristimulus values, X, Y, Z, are obtained thus:

$$X = 29540/1078.96 = 27.38$$
$$Y = 38933/1078.96 = 36.08$$
$$Z = 27930/1078.96 = 25.89$$

Hence, if the reference white is the perfect diffuser (for which $Y = 100$), then $Y/Y_n = 36.08/100 = 0.3608$, and, using the Table of values of L^*, we obtain $L^* = 66.58$. The values of u' and v' are calculated as:

$$u' = 4(0.3064)/[-2(0.3064)+12(0.4039)+3] = 0.1694$$
$$v' = 9(0.4039)/[-2(0.3064)+12(0.4039)+3] = 0.5025$$

The values of u' and v' for Standard Illuminant A are $u'_n = 0.2560$ and $v'_n = 0.5243$, and hence:

$$u^* = (13)(66.58)(0.1694-0.2560) = -74.96$$
$$v^* = (13)(66.58)(0.5025-0.5243) = -18.87$$

If another sample, having a slightly different spectral reflectance curve, resulted in values for illuminant A as follows:

$$u^* = -75.13$$
$$v^* = -16.54$$
$$L^* = 62.31$$

then the difference between the two samples would be given by:

$$[(-74.96+75.13)^2+(-18.87+16.54)^2+(66.58-62.31)^2]^{\frac{1}{2}}$$
$$= [(0.17)^2+(-2.33)^2+(4.27)^2]^{\frac{1}{2}}$$
$$= (0.03+5.43+18.23)^{\frac{1}{2}}$$
$$= (23.69)^{\frac{1}{2}} = 4.87$$

It will be seen that in the above example the difference in lightness, L^*, contributes most to the total colour difference; however, if the two samples are not seen side by side across a narrow dividing line, less weight should be given to the lightness difference. In this case a difference formula of the type

$$[(u_1{}^*-u_2{}^*)^2+(v_1{}^*-v_2{}^*)^2+(kL_1{}^*-kL_2{}^*)^2]^{\frac{1}{2}}$$

should be used, where k is chosen appropriately for the particular case. Thus if k is put equal to $\frac{1}{2}$, then in the above example the difference becomes:

$$[(0.17)^2+(-2.33)^2+\tfrac{1}{4}(4.27)^2]^{\frac{1}{2}}$$
$$= (0.03+5.43+4.56)^{\frac{1}{2}}$$
$$= (10.02)^{\frac{1}{2}} = 3.17$$

The L^* a^* b^* colour difference formula is used in a similar way (see Section 8.8).

Worked Example

λ	$t(\lambda)$	$E(\lambda)$	$t(\lambda)E(\lambda)\bar{x}(\lambda)$	$t(\lambda)E(\lambda)\bar{y}(\lambda)$	$t(\lambda)E(\lambda)\bar{z}(\lambda)$	$100E(\lambda)\bar{y}(\lambda)$
380	51.3	9.80	1	0	3	0
390	56.2	12.09	3	0	14	0
400	60.5	14.71	13	0	60	1
410	66.5	17.68	51	1	244	2
420	72.5	20.99	205	6	982	8
430	75.3	24.67	527	22	2547	29
440	76.2	28.70	762	50	3821	66
450	75.9	33.09	844	95	4451	126
460	74.8	37.81	822	170	4721	227
470	73.4	42.87	615	286	4052	390
480	71.6	48.24	330	480	2808	671
490	69.5	53.91	120	779	1743	1121
500	66.7	59.86	20	1290	1086	1933
510	63.9	66.06	39	2123	668	3323
520	60.8	72.50	279	3130	345	5148
530	57.0	79.13	746	3888	190	6821
540	52.6	85.95	1313	4313	92	8200
550	48.0	92.91	1933	4437	39	9245
560	42.8	100.00	2544	4259	17	9950
570	37.0	107.18	3022	3775	8	10204
580	30.6	114.44	3209	3047	6	9956
590	25.5	121.73	3186	2350	3	9215
600	20.9	129.04	2865	1702	2	8142
610	16.8	136.35	2297	1152	1	6858
620	12.9	143.62	1583	706	0	5472
630	10.0	150.84	969	400	0	3997
640	7.8	157.98	552	216	0	2765
650	6.7	165.03	313	118	0	1766
660	6.2	171.96	176	65	0	1049
670	5.9	178.77	92	34	0	572
680	5.4	185.43	47	17	0	315
690	4.9	191.93	21	8	0	157
700	5.0	198.26	11	4	0	81
710	6.2	204.41	7	3	0	43
720	9.3	210.36	6	2	0	21
730	17.4	216.12	5	2	0	11
740	27.5	221.67	4	1	0	7
750	42.7	227.00	3	1	0	2
760	56.2	232.12	3	1	0	2
770	66.1	237.01	2	0	0	0
780	76.0	241.68	0	0	0	0
Totals			29540	38933	27930	107896

A2.2 Predicting colour appearance

The scaled appearance of surface colours, when seen against medium grey backgrounds at normal levels of illumination that are not highly coloured, can be predicted by the following method. (See end of Section 8.11.)

From the tristimulus values, X, Y, Z, of the sample, estimates of the retinal cone responses, R, G, B, are obtained by a set of transformation equations. These responses are then normalized to allow for adaptation to the colour of the illuminant. From the normalized cone responses, an achromatic signal, A, and three colour-difference signals, C_1, C_2, C_3, are obtained. From these signals, predictions of appearance are obtained in terms of hue H, lightness L (which includes a contribution from the colour-difference signals), and chroma C.

The method involves ten steps; these steps will now be described, and illustrated by using, as an example, a sample whose tristimulus values in Standard Illuminant A are $X = 19.31$, $Y = 23.93$, $Z = 10.14$.

Step 1. From X_n, Y_n, Z_n, the tristimulus values for the reference white in the illuminant being used, derive:

$$R_n = 0.4002X_n + 0.7076Y_n - 0.0808Z_n$$
$$G_n = -0.2263X_n + 1.1653Y_n + 0.0457Z_n$$
$$B_n = 0.9182Z_n$$

For the example, the reference white is taken as the perfect diffuser (for which $Y_n = 100$) in Standard Illuminant A (for which $x = 0.4476$, $y = 0.4074$, $z = 0.1450$). Hence its tristimulus values are:

$$X_n = 100(0.4476/0.4074) = 109.87$$
$$Y_n = 100.00$$
$$Z_n = 100(0.1450/0.4074) = 35.59$$

Then using the transformation equations, we obtain:

$$R_n = 111.85 \qquad\qquad G_n = 93.30 \qquad\qquad B_n = 32.68$$

Step 2. From X, Y, Z, the tristimulus values of the colour considered, derive:

$$R = (0.4002X + 0.7076Y - 0.0808Z)(100/R_n)$$
$$G = (-0.2263X + 1.1653Y + 0.0457Z)(100/G_n)$$
$$B = (0.9182Z)(100/B_n)$$

The division by R_n, G_n, B_n, provides the chromatic adaptation adjustment. Thus, for the example, we obtain:

$$R = 23.84(100/111.85) = 21.31$$
$$G = 23.98(100/93.30) = 25.70$$
$$B = 9.31(100/32.68) = 28.49$$

Step 3. Derive the signals:

$$A = 2R^{\frac{1}{2}} + G^{\frac{1}{2}} + (1/20)(B^{\frac{1}{2}})$$
$$C_1 = R^{\frac{1}{2}} - G^{\frac{1}{2}}$$
$$C_2 = G^{\frac{1}{2}} - B^{\frac{1}{2}}$$
$$C_3 = B^{\frac{1}{2}} - R^{\frac{1}{2}}$$

The values for the sample are: $A = 14.569$, $C_1 = -0.453$, $C_2 = -0.268$, $C_3 = 0.721$. From the values of R_n, G_n, B_n, the value of $A_n = 31.097$.

Step 4. Derive a correlate of hue,

$$h_s = \arctan \{[-\tfrac{1}{2}(C_3 - C_2)/4.5]/[C_1 - C_2/11]\}$$

For the sample, the value of h_s is found to be arctan $[(-0.110)/(-0.429)]$.

When both the quantities in this expression are positive, h_s lies between $0°$ and $90°$; when the upper is negative and the lower is positive, between $90°$ and $180°$; when both are negative, between $180°$ and $270°$; when the upper is positive and the lower negative, between $270°$ and $360°$. Thus, in this case, h_s is equal to $14.38° + 180°$, that is $194.38°$.

Step 5. Derive the value of an eccentricity function, e_s, as follows. Identify between which pair of the four unique hues, R, Y, G, and B, the value of h_s occurs, and for the unique hue having the lower value of h_s note its h_s value h_1 and its e_s value e_1; similarly, for the unique hue having the higher value of h_s, note its h_s value h_2 and its e_s value e_2. Then:

$$e_s = e_1 + (e_2 - e_1)(h_s - h_1)/(h_2 - h_1).$$

The values of h_s and e_s for the four unique hues are:

	R	Y	G	B
h_s	20.14	90.00	164.25	237.53
e_s	0.8	0.7	1.0	1.2

For the sample, $h_1 = 164.25$; $h_2 = 237.53$; $e_1 = 1.0$; $e_2 = 1.2$; and the value of e_s is 1.082.

Step 6. Derive M_{BY}, a blueness-yellowness response, and M_{RG}, a redness-greenness response:

$$M_{BY} = \tfrac{1}{2}(10/13)e_s(C_3 - C_2)/4.5$$
$$M_{RG} = (10/13)e_s(C_1 - C_2/11)$$

and a response

$$M = (M_{BY}^2 + M_{RG}^2)^{\frac{1}{2}}$$

For the sample, $M_{BY} = 0.091$; $M_{RG} = -0.357$; and $M = 0.368$.

Step 7. Derive the correlate of saturation:

$$s = M/(R^{\frac{1}{2}} + G^{\frac{1}{2}} + B^{\frac{1}{2}})$$

For the sample, $s = 0.0245$.

Step 8. Evaluate the correlate of lightness:

$$L = 116(A + M)^{2/3}/A_n^{2/3} - 16$$

For the sample, $L = 55$.

Step 9. Evaluate the correlate of chroma:

$$C = 50Ls$$

The constant 50 is included to give numbers of convenient size. For the sample, $C = 68$.

Step 10. Evaluate the correlate of hue:

$$H = H_1 + 100[(h_s - h_1)/e_1]/[(h_s - h_1)/e_1 + (h_2 - h_s)/e_2]$$

where H_1 is either 0, 100, 200, or 300, according to whether R, Y, G, or B, respectively, was the hue found in Step 5 as having the nearest lower value of h_s. For the sample, $H_1 = 200$, and $H = 246$; thus the sample is between unique green ($H = 200$) and unique blue ($H = 300$), and its hue can be expressed as 54% green and 46% blue.

It is thus clear, from the above ten steps, that a colour appearance specification for the sample has been obtained, showing that it is a slightly greenish blue-green (246 being on the green side of the blue-green at 250 that is equally green and blue), of medium lightness (55), and of medium chroma (68). If this same sample is now illuminated with Standard Illuminant D_{65}, and its tristimulus values are recalculated, and found to be:

$$X = 19.13 \qquad Y = 27.04 \qquad Z = 29.56$$

then its new colour appearance can be calculated as follows. Taking the reference white now as the perfect diffuser (for which $Y_n = 100$) in Standard Illuminant D_{65} (for which $x = 0.3127$, $y = 0.3290$, $z = 0.3583$) its tristimulus values are:

$$X_n = 100(0.3127)/(0.3290) = 95.05$$
$$Y_n = 100.00$$
$$Z_n = 100(0.3583)/(0.3290) = 108.91$$

Using the same ten steps again, we now find that $H = 228$, $L = 59$, and $C = 64$. It is thus clear that, by changing the Standard Illuminant from A to D_{65}, the appearance of the sample has changed: it has become greener (72% green and 28% blue, instead of 54% green and 46% blue), and slightly lighter (lightness 59, instead of 55), and slightly weaker (chroma 64, instead of 68).

No method of predicting colour appearance has yet been standardized, but the method described in this section has been found to give useful results in some applications.

A2.3 Formulae and tables

RELATIONSHIPS BETWEEN THE XYZ, UVW, AND U'V'W' SYSTEMS

$x = 1.5u/(u-4v+2)$
$y = v/(u-4v+2)$
$z = \dfrac{-0.5u - 5v + 2}{u - 4v + 2}$

$X = (3/2)U$
$Y = V$
$Z = (3/2)U - 3V + 2W$

$u = u'$
$v = \frac{2}{3}v'$
$w = w' + \frac{1}{3}v'$

$U = (3/2)U'$
$V = V'$
$W = (3/2)W' + \frac{1}{2}V'$

$x = 9u'/(6u' - 16v' + 12)$
$y = 4v'/(6u' - 16v' + 12)$
$z = \dfrac{-3u' - 20v' + 12}{6u' - 16v' + 12}$

$X = (9/4)U'$
$Y = V'$
$Z = (9/4)U' - 2V' + 3W'$

$u = 2x/(-x+6y+1.5)$
$v = 3y/(-x+6y+1.5)$
$w = \dfrac{-3x+3y+1.5}{-x+6y+1.5}$

$U = \frac{2}{3}X$
$V = Y$
$W = -\frac{1}{2}X + (3/2)Y + \frac{1}{2}Z$

$u' = u$
$v' = (3/2)v$
$w' = w - \frac{1}{2}v$

$U' = \frac{2}{3}U$
$V' = V$
$W' = \frac{2}{3}W - \frac{1}{3}V$

$u' = 4x/(-2x+12y+3)$
$v' = 9y/(-2x+12y+3)$
$w' = \dfrac{-6x+3y+3}{-2x+12y+3}$

$U' = (4/9)X$
$V' = Y$
$W' = -\frac{1}{3}X + \frac{2}{3}Y + \frac{1}{3}Z$

$u = 4X/(X + 15Y + 3Z)$
$v = 6Y/(X + 15Y + 3Z)$
$u' = 4X/(X + 15Y + 3Z)$
$v' = 9Y/(X + 15Y + 3Z)$

THE U*V*W*, L*u*v*, AND L*a*b* SYSTEMS

$$U^* = 13W^*(u-u_n)$$
$$V^* = 13W^*(v-v_n)$$
$$W^* = 25Y^{1/3}-17$$

$$u^* = 13L^*(u'-u'_n)$$
$$v^* = 13L^*(v'-v'_n)$$
$$L^* = 116(Y/Y_n)^{1/3}-16$$

$$a^* = 500[(X/X_n)^{1/3}-(Y/Y_n)^{1/3}]$$
$$b^* = 200[Y/Y_n)^{1/3}-(Z/Z_n)^{1/3}]$$
$$L^* = 116(Y/Y_n)^{1/3}-16$$

X_n, Y_n, Z_n, u_n, v_n, u'_n, v'_n, are the values of X, Y, Z, u, v, u', v', for a specified reference white. In the formula for W^* it is necessary to express Y as a percentage. Colour differences are expressed as:

$$\Delta E = [(\Delta U^*)^2+(\Delta V^*)^2+(k\Delta W^*)^2]^{\frac{1}{2}}$$
$$\Delta E^*_{uv} = [(\Delta u^*)^2+(\Delta v^*)^2+(k\Delta L^*)^2]^{\frac{1}{2}}$$
$$\Delta E^*_{ab} = [(\Delta a^*)^2+(\Delta b^*)^2+(k\Delta L^*)^2]^{\frac{1}{2}}$$

where $k = 1$ for samples in close proximity, but may have a lower value for other situations. When Y/Y_n is less than 0.008856, L^* is evaluated as 903.3 (Y/Y_n).

CHROMATICITY CO-ORDINATES OF VARIOUS ILLUMINANTS

Illuminant	x	y	u	v	u'	v'
S_A	0.4476	0.4074	0.2560	0.3495	0.2560	0.5243
3250 K	0.4201	0.3976	0.2424	0.3442	0.2424	0.5163
S_B	0.3484	0.3516	0.2137	0.3234	0.2137	0.4851
S_C	0.3101	0.3162	0.2009	0.3073	0.2009	0.4609
D_{50}	0.3457	0.3585	0.2092	0.3254	0.2092	0.4881
D_{55}	0.3324	0.3474	0.2044	0.3205	0.2044	0.4807
D_{65}	0.3127	0.3290	0.1978	0.3122	0.1978	0.4683
D_{75}	0.2990	0.3149	0.1935	0.3057	0.1935	0.4585
S_E	0.3333	0.3333	0.2105	0.3158	0.2105	0.4737
9300 K	0.2848	0.2932	0.1915	0.2957	0.1915	0.4436

COLOUR MATCHING FUNCTIONS AND CHROMATICITY CO-ORDINATES

λ(nm)	$\bar{x}(\lambda)$	$\bar{y}(\lambda)$	$\bar{z}(\lambda)$	x	y	u'	v'
380	0.0014	0.0000	0.0065	0.1741	0.0050	0.2569	0.0165
390	0.0042	0.0001	0.0201	0.1738	0.0049	0.2564	0.0163
400	0.0143	0.0004	0.0679	0.1733	0.0048	0.2558	0.0159
410	0.0435	0.0012	0.2074	0.1726	0.0048	0.2545	0.0159
420	0.1344	0.0040	0.6456	0.1714	0.0051	0.2522	0.0169
430	0.2839	0.0116	1.3856	0.1689	0.0069	0.2461	0.0226
440	0.3483	0.0230	1.7471	0.1644	0.0109	0.2347	0.0349
450	0.3362	0.0380	1.7721	0.1566	0.0177	0.2161	0.0550
460	0.2908	0.0600	1.6692	0.1440	0.0297	0.1877	0.0871
470	0.1954	0.0910	1.2876	0.1241	0.0578	0.1441	0.1510
480	0.0956	0.1390	0.8130	0.0913	0.1327	0.0828	0.2708
490	0.0320	0.2080	0.4652	0.0454	0.2950	0.0282	0.4117
500	0.0049	0.3230	0.2720	0.0082	0.5384	0.0035	0.5131
510	0.0093	0.5030	0.1582	0.0139	0.7502	0.0046	0.5638
520	0.0633	0.7100	0.0782	0.0743	0.8338	0.0231	0.5837
530	0.1655	0.8620	0.0422	0.1547	0.8059	0.0501	0.5868
540	0.2904	0.9540	0.0203	0.2296	0.7543	0.0792	0.5856
550	0.4334	0.9950	0.0087	0.3016	0.6923	0.1127	0.5821
560	0.5945	0.9950	0.0039	0.3731	0.6245	0.1531	0.5766
570	0.7621	0.9520	0.0021	0.4441	0.5547	0.2026	0.5694
580	0.9163	0.8700	0.0017	0.5125	0.4866	0.2623	0.5604
590	1.0263	0.7570	0.0011	0.5752	0.4242	0.3315	0.5501
600	1.0622	0.6310	0.0008	0.6270	0.3725	0.4035	0.5393
610	1.0026	0.5030	0.0003	0.6658	0.3340	0.4691	0.5296
620	0.8544	0.3810	0.0002	0.6915	0.3083	0.5202	0.5219
630	0.6424	0.2650	0.0000	0.7079	0.2920	0.5565	0.5165
640	0.4479	0.1750	0.0000	0.7190	0.2809	0.5830	0.5125
650	0.2835	0.1070	0.0000	0.7260	0.2740	0.6005	0.5099
660	0.1649	0.0610	0.0000	0.7300	0.2700	0.6108	0.5084
670	0.0874	0.0320	0.0000	0.7320	0.2680	0.6161	0.5076
680	0.0468	0.0170	0.0000	0.7334	0.2666	0.6200	0.5070
690	0.0227	0.0082	0.0000	0.7344	0.2656	0.6226	0.5066
700	0.0114	0.0041	0.0000	0.7347	0.2653	0.6234	0.5065
710	0.0058	0.0021	0.0000	0.7347	0.2653	0.6234	0.5065
720	0.0029	0.0010	0.0000	0.7347	0.2653	0.6234	0.5065
730	0.0014	0.0005	0.0000	0.7347	0.2653	0.6234	0.5065
740	0.0007	0.0002	0.0000	0.7347	0.2653	0.6234	0.5065
750	0.0003	0.0001	0.0000	0.7347	0.2653	0.6234	0.5065
760	0.0002	0.0001	0.0000	0.7347	0.2653	0.6234	0.5065
770	0.0001	0.0000	0.0000	0.7347	0.2653	0.6234	0.5065
780	0.0000	0.0000	0.0000	0.7347	0.2653	0.6234	0.5065

L* FOR VARIOUS VALUES OF Y/Y_n

Y/Y_n	.000	.001	.002	.003	.004	.005	.006	.007	.008	.009
1.00	100.00	100.04	100.08	100.12	100.15	100.19	100.23	100.27	100.31	100.35
.99	99.61	99.65	99.69	99.73	99.77	99.81	99.85	99.88	99.92	99.96
.98	99.22	99.26	99.30	99.34	99.38	99.42	99.46	99.50	99.53	99.57
.97	98.83	98.87	98.91	98.95	98.99	99.03	99.06	99.10	99.14	99.18
.96	98.43	98.47	98.51	98.55	98.59	98.63	98.67	98.71	98.75	98.79
.95	98.03	98.07	98.11	98.15	98.19	98.23	98.27	98.31	98.35	98.39
.94	97.63	97.67	97.71	97.75	97.79	97.83	97.87	97.91	97.95	97.99
.93	97.23	97.27	97.31	97.35	97.39	97.43	97.47	97.51	97.55	97.59
.92	96.82	96.86	96.90	96.94	96.98	97.02	97.07	97.11	97.15	97.19
.91	96.41	96.45	96.49	96.53	96.57	96.62	96.66	96.70	96.74	96.78
.90	96.00	96.04	96.08	96.12	96.16	96.20	96.25	96.29	96.33	96.37
.89	95.58	95.62	95.66	95.71	95.75	95.79	95.83	95.87	95.91	95.96
.88	95.16	95.20	95.25	95.29	95.33	95.37	95.41	95.45	95.50	95.54
.87	94.74	94.78	94.82	94.87	94.91	94.95	94.99	95.03	95.08	95.12
.86	94.31	94.36	94.40	94.44	94.48	94.53	94.57	94.61	94.65	94.70
.85	93.88	93.93	93.97	94.01	94.06	94.10	94.14	94.18	94.23	94.27
.84	93.45	93.49	93.54	93.58	93.62	93.67	93.71	93.75	93.80	93.84
.83	93.01	93.06	93.10	93.15	93.19	93.23	93.28	93.32	93.36	93.41
.82	92.57	92.62	92.66	92.71	92.75	92.80	92.84	92.88	92.93	92.97
.81	92.13	92.18	92.22	92.27	92.31	92.35	92.40	92.44	92.49	92.53
.80	91.68	91.73	91.77	91.82	91.86	91.91	91.95	92.00	92.04	92.09
.79	91.23	91.28	91.32	91.37	91.42	91.46	91.51	91.55	91.60	91.64
.78	90.78	90.83	90.87	90.92	90.96	91.01	91.05	91.10	91.14	91.19
.77	90.32	90.37	90.41	90.46	90.51	90.55	90.60	90.64	90.69	90.73
.76	89.86	89.91	89.95	90.00	90.04	90.09	90.14	90.18	90.23	90.28
.75	89.39	89.44	89.49	89.53	89.58	89.63	89.67	89.72	89.77	89.81
.74	88.92	88.97	89.02	89.06	89.11	89.16	89.21	89.25	89.30	89.35
.73	88.45	88.50	88.54	88.59	88.64	88.69	88.73	88.78	88.83	88.88
.72	87.97	88.02	88.06	88.11	88.16	88.21	88.26	88.30	88.35	88.40
.71	87.49	87.53	87.58	87.63	87.68	87.73	87.78	87.82	87.87	87.92
.70	87.00	87.05	87.09	87.14	87.19	87.24	87.29	87.34	87.39	87.44
.69	86.50	86.55	86.60	86.65	86.70	86.75	86.80	86.85	86.90	86.95
.68	86.01	86.06	86.11	86.16	86.21	86.26	86.31	86.36	86.40	86.45
.67	85.50	85.55	85.60	85.66	85.71	85.76	85.81	85.86	85.91	85.96
.66	85.00	85.05	85.10	85.15	85.20	85.25	85.30	85.35	85.40	85.45
.65	84.48	84.54	84.59	84.64	84.69	84.74	84.79	84.84	84.89	84.95
.64	83.97	84.02	84.07	84.12	84.17	84.23	84.28	84.33	84.38	84.43
.63	83.44	83.49	83.55	83.60	83.65	83.70	83.76	83.81	83.86	83.91
.62	82.91	82.97	83.02	83.07	83.13	83.18	83.23	83.28	83.34	83.39
.61	82.38	82.43	82.49	82.54	82.59	82.65	82.70	82.75	82.81	82.86
.60	81.84	81.89	81.95	82.00	82.06	82.11	82.16	82.22	82.27	82.32

L* FOR VARIOUS VALUES OF Y/Y_n

Y/Y_n	.000	.001	.002	.003	.004	.005	.006	.007	.008	.009
.59	81.29	81.35	81.40	81.46	81.51	81.57	81.62	81.67	81.73	81.78
.58	80.74	80.79	80.85	80.91	80.96	81.02	81.07	81.13	81.18	81.24
.57	80.18	80.24	80.29	80.35	80.40	80.46	80.52	80.57	80.63	80.68
.56	79.61	79.67	79.73	79.78	79.84	79.90	79.95	80.01	80.07	80.12
.55	79.04	79.10	79.16	79.21	79.27	79.33	79.39	79.44	79.50	79.56
.54	78.46	78.52	78.58	78.64	78.69	78.75	78.81	78.87	78.93	78.98
.53	77.87	77.93	77.99	78.05	78.11	78.17	78.23	78.29	78.34	78.40
.52	77.28	77.34	77.40	77.46	77.52	77.58	77.64	77.70	77.76	77.82
.51	76.68	76.74	76.80	76.86	76.92	76.98	77.04	77.10	77.16	77.22
.50	76.07	76.13	76.19	76.25	76.31	76.38	76.44	76.50	76.56	76.62
.49	75.45	75.51	75.58	75.64	75.70	75.76	75.82	75.88	75.95	76.01
.48	74.82	74.89	74.95	75.01	75.08	75.14	75.20	75.26	75.33	75.39
.47	74.19	74.25	74.32	74.38	74.44	74.51	74.57	74.64	74.70	74.76
.46	73.55	73.61	73.68	73.74	73.80	73.87	73.93	74.00	74.06	74.13
.45	72.89	72.96	73.02	73.09	73.15	73.22	73.29	73.35	73.42	73.48
.44	72.23	72.30	72.36	72.43	72.50	72.56	72.63	72.69	72.76	72.83
.43	71.55	71.62	71.69	71.76	71.83	71.89	71.96	72.03	72.09	72.16
.42	70.87	70.94	71.01	71.08	71.15	71.21	71.28	71.35	71.42	71.49
.41	70.18	70.25	70.32	70.39	70.46	70.52	70.59	70.66	70.73	70.80
.40	69.47	69.54	69.61	69.68	69.75	69.82	69.89	69.97	70.04	70.11
.39	68.75	68.82	68.90	68.97	69.04	69.11	69.18	69.26	69.33	69.40
.38	68.02	68.09	68.17	68.24	68.31	68.39	68.46	68.53	68.61	68.68
.37	67.28	67.35	67.43	67.50	67.58	67.65	67.72	67.80	67.87	67.95
.36	66.52	66.60	66.67	66.75	66.82	66.90	66.98	67.05	67.13	67.20
.35	65.75	65.83	65.90	65.98	66.06	66.14	66.21	66.29	66.37	66.44
.34	64.96	65.04	65.12	65.20	65.28	65.36	65.44	65.51	65.59	65.67
.33	64.16	64.24	64.32	64.40	64.48	64.56	64.64	64.72	64.80	64.88
.32	63.34	63.43	63.51	63.59	63.67	63.75	63.84	63.92	64.00	64.08
.31	62.51	62.59	62.68	62.76	62.84	62.93	63.01	63.09	63.18	63.26
.30	61.65	61.74	61.83	61.91	62.00	62.08	62.17	62.25	62.34	62.42
.29	60.78	60.87	60.96	61.05	61.13	61.22	61.31	61.39	61.48	61.57
.28	59.89	59.98	60.07	60.16	60.25	60.34	60.43	60.52	60.60	60.69
.27	58.97	59.07	59.16	59.25	59.34	59.43	59.53	59.62	59.71	59.80
.26	58.04	58.13	58.23	58.32	58.41	58.51	58.60	58.70	58.79	58.88
.25	57.08	57.17	57.27	57.37	57.46	57.56	57.66	57.75	57.85	57.94
.24	56.09	56.19	56.29	56.39	56.49	56.58	56.68	56.78	56.88	56.98
.23	55.07	55.18	55.28	55.38	55.48	55.58	55.69	55.79	55.89	55.99
.22	54.03	54.13	54.24	54.34	54.45	54.55	54.66	54.76	54.87	54.97
.21	52.95	53.06	53.17	53.28	53.38	53.49	53.60	53.71	53.81	53.92
.20	51.84	51.95	52.06	52.17	52.29	52.40	52.51	52.62	52.73	52.84
.19	50.69	50.80	50.92	51.04	51.15	51.27	51.38	51.50	51.61	51.72
.18	49.50	49.62	49.74	49.86	49.98	50.10	50.22	50.33	50.45	50.57

L* FOR VARIOUS VALUES OF Y/Y$_n$

Y/Y$_n$	.000	.001	.002	.003	.004	.005	.006	.007	.008	.009
.17	48.26	48.39	48.51	48.64	48.76	48.88	49.01	49.13	49.25	49.37
.16	46.97	47.11	47.24	47.37	47.49	47.62	47.75	47.88	48.01	48.13
.15	45.63	45.77	45.91	46.04	46.18	46.31	46.45	46.58	46.71	46.84
.14	44.23	44.38	44.52	44.66	44.80	44.94	45.08	45.22	45.36	45.50
.13	42.76	42.91	43.06	43.21	43.36	43.51	43.65	43.80	43.94	44.09
.12	41.22	41.37	41.53	41.69	41.84	42.00	42.15	42.31	42.46	42.61
.11	39.58	39.75	39.92	40.08	40.25	40.41	40.57	40.74	40.90	41.06
.10	37.84	38.02	38.20	38.38	38.55	38.73	38.90	39.07	39.24	39.41
.09	35.98	36.18	36.37	36.56	36.74	36.93	37.11	37.30	37.48	37.66
.08	33.98	34.19	34.40	34.60	34.80	35.00	35.20	35.40	35.60	35.79
.07	31.81	32.03	32.26	32.48	32.70	32.92	33.14	33.35	33.56	33.77
.06	29.41	29.66	29.91	30.16	30.40	30.64	30.88	31.11	31.35	31.58
.05	26.73	27.02	27.30	27.57	27.85	28.11	28.38	28.64	28.90	29.16
.04	23.67	24.00	24.32	24.64	24.95	25.26	25.56	25.86	26.16	26.45
.03	20.04	20.44	20.83	21.12	21.58	21.94	22.30	22.65	23.00	23.34
.02	15.49	16.00	16.50	16.99	17.46	17.92	18.36	18.80	19.22	19.64
.01	8.99	9.80	10.56	11.28	11.96	12.61	13.23	13.83	14.40	14.95

RELATIVE SPECTRAL POWER DISTRIBUTIONS

λ (nm)	S_A	3250 K	S_B	S_C	D_{50}	D_{55}	D_{65}	D_{75}
300	0.93				0.02	0.02	0.03	0.04
310	1.36				2.05	2.07	3.29	5.13
320	1.93		0.02	0.01	7.78	11.22	20.24	29.81
330	2.66		0.50	0.40	14.75	20.65	37.05	54.93
340	3.59		2.40	2.70	17.95	23.88	39.95	57.26
350	4.74		5.60	7.00	21.01	27.82	44.91	62.74
360	6.14		9.60	12.90	23.94	30.62	46.64	62.98
370	7.82		15.20	21.40	26.96	34.31	52.09	70.31
380	9.80	16.59	22.40	33.00	24.49	32.58	49.98	66.70
390	12.09	19.63	31.30	47.40	29.87	38.09	54.65	69.96
400	14.71	22.95	41.30	63.30	49.31	60.95	82.75	101.93
410	17.68	26.55	52.10	80.60	56.51	68.55	91.49	111.89
420	20.99	30.42	63.20	98.10	60.03	71.58	93.43	112.80
430	24.67	34.53	73.10	112.40	57.82	67.91	86.68	103.09
440	28.70	38.87	80.80	121.50	74.82	85.61	104.86	121.20
450	33.09	43.42	85.40	124.00	87.25	97.99	117.01	133.01
460	37.81	48.15	88.30	123.10	90.61	100.46	117.81	132.36
470	42.87	53.04	92.00	123.80	91.37	99.91	114.86	127.32
480	48.24	58.06	95.20	123.90	95.11	102.74	115.92	126.80
490	53.91	63.19	96.50	120.70	91.96	98.08	108.81	117.78
500	59.86	68.40	94.20	112.10	95.72	100.68	109.35	116.59
510	66.06	73.67	90.70	102.30	96.61	100.70	107.80	113.70
520	72.50	78.97	89.50	96.90	97.13	99.99	104.79	108.66
530	79.13	84.27	92.20	98.00	102.10	104.21	107.69	110.44
540	85.95	89.56	96.90	102.10	100.75	102.10	104.41	106.29

550	92.91	94.81	101.00	105.20	102.32	102.97	104.05	104.90
560	100.00	100.00	102.80	105.30	100.00	100.00	100.00	100.00
570	107.18	105.12	102.60	102.30	97.74	97.22	96.33	95.62
580	114.44	110.14	101.00	97.80	98.92	97.75	95.79	94.21
590	121.73	115.05	99.20	93.20	93.50	91.43	88.69	87.00
600	129.04	119.83	98.00	89.70	97.69	94.42	90.01	87.23
610	136.35	124.48	98.50	88.40	99.27	95.14	89.60	86.14
620	143.62	128.99	99.70	88.10	99.04	94.22	87.70	83.58
630	150.84	133.33	101.00	88.00	95.72	90.45	83.29	78.75
640	157.98	137.51	102.20	87.80	98.86	92.33	83.70	78.43
650	165.03	141.52	103.90	88.20	95.67	88.85	80.03	74.80
660	171.96	145.35	105.00	87.90	98.19	90.32	80.21	74.32
670	178.77	149.00	104.90	86.30	103.00	93.95	82.28	75.42
680	185.43	152.46	103.90	84.00	99.13	89.96	78.28	71.58
690	191.93	155.74	101.60	80.20	87.38	79.68	69.72	63.85
700	198.26	158.83	99.10	76.30	91.60	82.84	71.61	65.08
710	204.41	161.73	96.20	72.40	92.89	84.84	74.35	68.07
720	210.36	164.44	92.90	68.30	76.85	70.24	61.60	56.44
730	216.12	166.96	89.40	64.40	86.51	79.30	69.89	64.24
740	221.67	169.30	86.90	61.50	92.58	84.99	75.09	69.15
750	227.00	171.46	85.20	59.20	78.23	71.88	63.59	58.63
760	232.12	173.43	84.70	58.10	57.69	52.79	46.42	42.62
770	237.01	175.23	85.40	58.20	82.92	75.93	66.81	61.35
780	241.68			59.10	78.27	71.82	63.38	58.32
790	246.12				79.55	72.94	64.30	59.14
800	250.33				73.40	67.35	59.45	54.73
810	254.31				63.92	58.73	51.96	47.92
820	258.07				70.78	64.99	57.44	52.92
830	261.60				74.44	68.31	60.31	55.54

RELATIVE SPECTRAL POWER DISTRIBUTIONS OF FLUORESCENT LAMPS

λ (nm)	F_2 (low CRI)	F_7 (high CRI)	F_{11} (three-band)
380	1.92	4.15	1.23
390	3.66	7.69	0.96
400	12.36	18.09	7.81
410	13.56	20.57	8.92
420	8.37	15.43	5.02
430	24.86	35.42	23.12
440	32.44	45.27	32.58
450	13.24	23.97	14.23
460	14.36	26.09	13.34
470	15.05	27.37	10.92
480	15.27	27.85	15.21
490	15.15	27.60	26.59
500	14.58	26.87	10.36
510	14.15	26.16	3.18
520	14.42	25.54	1.86
530	16.22	24.90	4.05
540	26.89	33.18	78.46
550	36.38	38.03	72.79
560	32.24	25.16	7.57
570	38.14	26.92	4.87
580	43.17	30.90	20.88
590	37.16	25.31	24.98

λ			
600	33.00	24.35	17.06
610	27.58	23.22	81.42
620	21.95	22.28	41.05
630	16.88	21.45	21.40
640	12.69	20.34	5.80
650	9.41	20.10	11.00
660	6.94	19.25	5.05
670	5.12	14.81	3.27
680	3.80	11.75	3.09
690	2.98	10.02	3.57
700	2.27	8.28	4.00
710	1.76	6.88	8.89
720	1.36	5.51	1.97
730	1.13	4.52	0.47
740	1.02	3.81	0.46
750	0.93	3.27	0.44
760	0.92	2.88	0.57
770	0.80	2.33	0.35
780	0.43	1.31	0.15
x	0.3721	0.3129	0.3805
y	0.3751	0.3292	0.3769
u'	0.2203	0.1979	0.2251
v'	0.4996	0.4685	0.5017
CCT	4230 K	6500 K	4000 K
CRI(R_a)	64	90	83

These spectral power distributions are examples for fluorescent tubes of the types indicated (CRI = colour rendering index); each type occurs in a variety of forms having somewhat different spectral power distributions and various correlated colour temperatures (CCT).

Photometric Units

A3.1 Relations between units of luminance

	candelas per sq. foot	candelas per sq. inch	candelas per sq. metre (or nit)	candelas per sq. cm (stilbs)	foot-lamberts (equivalent foot-candles, or e.f.c.)	lamberts	millilamberts
1 candela per sq. foot =	1	$1/144$	3.281^2	$10.76/100^2$	π	$\pi/929$	$\pi/0.929$
		0.00694	10.76	0.001076	3.142	0.003382	3.382
1 candela per sq. inch =	12^2	1	144×10.76	$1550/100^2$	144π	$\dfrac{144\pi}{929}$	$\dfrac{144\pi}{0.929}$
	144		1550	0.1550	452.5	0.4871	487.1
1 candela per sq. metre = (or nit)	0.3048^2	$\dfrac{0.3048^2}{144}$	1	$1/10\,000$	0.0929π	$\dfrac{\pi}{10\,000}$	$\dfrac{\pi}{10}$
	0.0929	0.0006451		0.0001	0.2919	0.0003142	0.3142
1 candela per sq. cm = (stilb)	929	$929/144$	$10\,000$	1	929π	π	1000π
		6.451			2919	3.142	3142
1 foot-lambert (equivalent foot-candle, or e.f.c.) =	$1/\pi$	$1/144\pi$	$10.76/\pi$	$\dfrac{0.001076}{\pi}$	1	$1/929$	$1/0.929$
	0.3183	0.002210	3.426	0.0003426		0.001076	1.076
1 lambert =	$929/\pi$	$929/144\pi$	$\dfrac{10\,000}{\pi}$	$1/\pi$	929	1	1000
	295.7	2.053	3183	0.3183			
1 millilambert = 10 apostilbs =	$0.929/\pi$	$0.929/144\pi$	$10/\pi$	0.0003183	0.929	$1/1000$	1
	0.2957	0.002053	3.183			0.001	

A3.2. Relations between units of luminance and illumination

A surface of luminance factor β under an illumination E has a luminance:

$L = E.\beta/\pi$ candelas per sq. metre when E is measured in lux (lumens per sq. metre).
$L = E.\beta/\pi$ candelas per sq. foot when E is measured in lumens per sq. foot (foot-candles).
$L = E.\beta$ foot-lamberts when E is measured in lumens per sq. foot (foot-candles).
$L = E.\beta/10$ millilamberts when E is measured in lux (lumens per sq. metre).
$L = E.\beta$ apostilbs when E is measured in lux (lumens per sq. metre).

A3.3 Some useful conversion factors

Illuminance

To change	into	multiply the value by
foot-candles (lumens/ft²)	lux (lumens/m²)	10.76
lux (lumens/m²)	foot-candles (lumens/ft²)	0.0929

Luminance

To change	into	multiply the value by
foot-lamberts	candelas/m²	3.426
candelas/m²	foot-lamberts	0.2919
foot-lamberts	milli-lamberts	1.076
milli-lamberts	foot-lamberts	0.929
foot-lamberts	apostilbs	10.76
apostilbs	foot-lamberts	0.0929
candelas/m²	apostilbs	3.142
apostilbs	candelas/m²	0.3183

Relationships

luminance in cd/m² = illuminance in lux x luminance factor/π
luminance in apostilbs = illuminance in lux x luminance factor

A3.4. Typical levels of luminance and illumination

	Illuminance (lux)			*Luminance** (cd/m²)		
Bright sun	50 000	to	100 000	3 000	to	6 000
Hazy Sun	25 000	,,	50 000	1 500	,,	3 000
Cloudy Bright	10 000	,,	25 000	600	,,	1 500
Cloudy Dull	2 000	,,	10 000	120	,,	600
Very Dull	100	,,	2 000	6	,,	120
Sunset	1	,,	100	0.06	,,	6
Full Moon	0.01	,,	0.1	0.0006	,,	0.006
Star Light	0.0001	,,	0.001	0.000006	,,	0.00006
Operating Theatre	5 000	,,	10 000	300	,,	600
Shop Windows	1 000	,,	5 000	60	,,	300
Drawing Offices	300	,,	500	18	,,	30
Offices	200	,,	300	12	,,	18
Living Rooms	50	,,	200	3	,,	12
Corridors	50	,,	100	3	,,	6
Good Street Lighting	20			1.2		
Poor Street lighting	0.1			0.006		

* For a luminance factor of 20 per cent (average reflectance of a typical scene).

Photographic parameters

A4.1 Film speeds

ASA/ISO	DIN	BS log	Weston I and II
10	11	21	8
12	12	22	10
16	13	23	12
20	14	24	16
25	15	25	20
32	16	26	24
40	17	27	32
50	18	28	40
64	19	29	50
80	20	30	64
100	21	31	80
125	22	32	100
160	23	33	125
200	24	34	160
250	25	35	200
320	26	36	240
400	27	37	320
500	28	38	400
640	29	39	500
800	30	40	640
1000	31	41	800
1250	32	42	1000
1600	33	43	1250
2000	34	44	1600
2500	35	45	2000
3200	36	46	2400
4000	37	47	3200
5000	38	48	4000
6400	39	49	5000
8000	40	50	6400
10000	41	51	8000

The correct exposure for a typical scene in average bright sunlight is 1/ASA seconds at f/16, where ASA is the speed of the film on the ASA scale. (The corresponding aperture for 'hazy sun' is f/11, for 'cloudy bright' is f/8, and for 'cloudy dull' is f/5.6.) For correct exposure at f/2.8 and 1/50 second, the illumination level in lux multiplied by the ASA film speed is equal to 100 000; thus, for example, if the film speed is 400 ASA, the illumination level should be 250 lux.

A4.2 Film dimensions

Sizes of film formats (mm)

Motion Picture	Camera aperture	Projection aperture	Reproduced in television
8 mm	3.68× 4.88	3.28× 4.37	
Super 8	4.22× 6.22	3.99× 5.32	
9.5 mm	—	6.15× 8.00	
16 mm	7.49×10.26	7.21× 9.65	7.01× 9.35
35 mm silent (obsolete)	—	17.25×23.00	
35 mm academy	16.03×22.05	15.24×20.96	15.10×20.12
35 mm wide-screen (1.85 to 1)	—	11.33×20.96	
35 mm Cinemascope (2.35 to 1)	—	18.16×21.31	
70 mm	—	22.10×48.56	

Still	Nominal	Typical area used
Disc film	8.3× 10.6	7.8× 10.0
110 (Pocket Instamatic)	12.9× 17.0	12.0× 15.8
Half-frame 35 mm (still)	18 × 24	17 × 23
35 mm (still)	24 × 36	23 × 34.5
828 (Bantam)	28 × 40	27 × 39
126 (Instamatic)	28 × 28	28 × 28
Half 127	30 × 40	29 × 38
127 square	40 × 40	39 × 39
127	40 × 65	39 × 60
Half 120 and half 620	45 × 60	40 × 56
120 and 620 square	60 × 60	56 × 56
70 mm (still)	60 × 90	56 × 72
120 and 620 ($2\frac{1}{4} \times 3\frac{1}{4}$ in)	60 × 90	56 × 81
Quarter plate	82 ×108	80 ×105
9×12 cm	90 ×120	87 ×117
4×5 in	102 ×127	99 ×124
Half plate	120 ×165	117 ×162
Whole plate	165 ×216	162 ×213
8×10 in	203 ×254	200 ×251

A4.3 Motion picture parameters

Format	Pictures per second	Pictures per foot	Feet per second	Metres per minute	Sound relative to picture (frames)	
					Optical	Magnetic
8 mm (silent)	16	80	0.2	3.66	—	—
8 mm (sound)	18	80	0.225	4.11	—	56 ahead
8 mm (sound)	24	80	0.3	5.48	—	56 ahead
Super 8 (silent)	18	72	0.25	4.56	—	—
Super 8 (sound)	24	72	0.33	6.08	22 ahead	18 ahead
16 mm (silent)	16	40	0.4	7.31	—	—
16 mm (sound)	24	40	0.6	10.97	26 ahead	28 ahead
35 mm (silent)	16	16*	1.0*	18.24	—	—
35 mm (sound)	24	16*	1.5*	27.36	21 ahead	28 behind
65 mm and 70 mm	24	12.8*	1.875*	34.20	—	24 behind

* For 35 mm and 70 mm film, lengths are customarily measured in film-feet: 1 film-foot = 11.968 in.

A4.4 Lens apertures

Nominal value	Approximate relative exposure
$f/32$	1
$f/26$	1.5
$f/22$	2
$f/18$	3
$f/16$	4
$f/13$	6
$f/11$	8
$f/9$	12
$f/8$	16
$f/6.3$	25
$f/5.6$	32
$f/4.5$	50
$f/4$	64
$f/3.2$	100
$f/2.8$	128
$f/2.2$	200
$f/2$	256
$f/1.6$	400
$f/1.4$	512
$f/1.1$	800
$f/1$	1024
$f/0.8$	1600
$f/0.7$	2048

A4.5 Flash guide numbers

The guide number, G, for a flash unit used with a film of ASA speed, S, is given by

$$G = \sqrt{k.M.S.J.(L/W)}$$

where M is the reflector factor, J is the energy supplied for the flash in joules or watt-seconds, L/W is the efficiency in lumen-seconds per watt-seconds, and k is a constant.

The correct f-number, f/n, to be used for an object at a distance, d, from the flash unit, is given by

$$n = G/d$$

If d is expressed in metres, then $k = 0.0005$ (approximately); if d is expressed in feet, then $k = 0.005$ (approximately).

The guide number is proportional to the square-root of the film speed; thus, for example, an increase in film speed of four times results in a doubling of the guide number and a doubling of the distance for a given f-number.

INDEX

Index

Bensen, K. B., 403, 406, 419
Bent, R. L., 335
Berger, H. J., 213, 217, 218
Bermaine, D., 266, 268
Bertero, E. P., 408, 419
Beta video cassette recorder, 501, 502
Betacam television recording camera, 418
Bethell, P., 550, 557
Beyer, B. E., 307, 308
Billingsley, F. C., 519, 523
Billmeyer, F. W., 97, 133
binocular matching, see haploscopic matching
Birkenshaw, D. C., 375, 398
Birren, F., 10, 19, 534, 544
Bishop, N., 582, 590
black-body locus, 168
        sources, 152, 168, 169
black image in printing, 534, 538, 560, 561, 564,
    565, 569, 581, 583, 588
        level, in television, 396, 422, 492, 493
black-out bias, in television, 422
blacks, 226, 228, 229, 287, 384, 492, 493, 494
Blaxland, J., 300, 307
bleaching, of silver, 207, 324
blixes, 207, 324
block dyes, 136–144
Bloom, S. M., 266, 268
blue, 5, 32, 130, 535
        saving, 380, 452
        sky, 43, 44, 175, 191–195
        water, 43
Bocock, W. A., 166, 175
boiling, of graininess, in motion pictures, 339
Bomback, E. S., 218
borders, effects of, see surrounds
Borgan, H., 417, 419
Bouwis, G., 409, 419
Bowmaker, J. K., 12, 19
Boyce, P. R., 174, 175
Boynton, R. M., 19
Bray, C. P., 44, 48, 191, 195, 196
B.R.E.M.A. phosphors, 92, 94, 95, 137, 139–143,
    150, 389, 398, 419, 434, 437
Breneman, E. J., 49, 51, 57, 62, 68, 184, 185, 188,
    196, 231, 232, 256, 268, 395, 398
Breton, H. E., 368, 370
Brewer, W. L., 19, 89, 95, 145, 150, 213, 218, 224,
    232, 253, 268, 269, 287, 290, 317, 335, 488, 494
Brewster's law, 174
Bridgewater, T. H., 374, 398
bright, of brightness, 71
brightness, 39, 43, 69–71, 127, 229, 422
        television, 422
British Standards, 172, 173, 175, 238, 268, 493,
    494, 554, 557, 575, 577
Brockhurst, D., 515, 522, 523
Brocklebank, R. W., 95
Brothers, D. L., 351, 370, 494, 495
Brown, G., 395, 398
Brown, G. H., 335

Brown, I., 487, 494
Brown, P. K., 12, 19
Brown, W. R. J., 25, 27
Brownstein, S. A., 143, 150, 286, 290
Bruch, W., 451, 506, 510
Bryngdahl, O., 532, 544
buffers, in photographic solutions, 323
Bullock, L. M., 130, 131, 134
Burch, J. M., 130, 134
Burgin, R., 176
Burnham, R. W., 124, 125, 133, 184, 190, 196
Burr, R. P., 487, 494

Caddigan, J. L., 478, 494
Callier co-efficient, 342
calligraphic form, in television displays, 516
cameras, process, 531
        colour television, 24, 379, 400–419
        combined film and television, 478
camouflage detection, 214
Campbell-Swinton, A. A., 373, 398
Capacitance Electronic Disc (CED), 516
Capstaff, J. G., 211, 217
captions, coloured, in television, 511
carbon arcs, 166, 169, 429
Carnahan, W. H., 53, 68, 226, 232
Carnt, P. S., 398, 476
carrier frequencies, 377
        sharing, 388, 439, 449, 451
cartoons, computer assisted, 518
Cartwright, H. M., 544
Castellain, A. P., 166, 175
cathode-ray tubes, 22–27, 420–423
cathodes, of television tubes, 384, 420
Cayless, M. A., 176
CCD (change coupled device), 417, 484
CED (Capacitance Electronic Disc), 507
Ceefax, 514
centre of gravity law, 82–83, 109–110, 445, 598
character generator, 513
characteristic curve, in photography, 50, 242,
    247–259, 271, 280, 302–304
charge coupled device (CCD), 417, 484
checking the results, in printing, 545–557
chemistry of colour photography, 309–336
chequer-board pattern, in television, 385, 386
Childs, I., 484, 494
chlorophyll, 4, 214
Christ, R. E., 521, 523
chroma, 70, 71, 129
chroma-key, 512
Chroma, Munsell, 72, 121, 141
chromaticity, 71, 72
    co-ordinates, 82
        of illuminants, 171, 607, 615
        of the spectrum, 608
    diagram, 82
        uniform, 110–113
    limits of colours, 92, 93
chromaticness, 71

626

633

LOCATION OF PLATES

Plate numbers	Between pages
1, 2, 3, 4	48 and 49
5, 6, 7, 8	80 and 81
9, 10, 11	112 and 113
12, 13, 14	176 and 177
15, 16, 17	240 and 241
18, 19, 20, 21	288 and 289
22, 23, 24	352 and 353
25, 26, 27	384 and 385
28, 29, 30	416 and 417
31, 32, 33	512 and 513
34, 35, 36	544 and 545
37, 38, 39	576 and 577